Walter Bradford Cannon seated before a kymograph in the physiology laboratory, Harvard Medical School, 1928.

Walter B. Cannon, Science And Society

ELIN L. WOLFE
A. CLIFFORD BARGER
SAUL BENISON

PUBLISHED BY THE BOSTON MEDICAL LIBRARY IN THE
FRANCIS A. COUNTWAY LIBRARY OF MEDICINE,
AND DISTRIBUTED BY THE HARVARD UNIVERSITY PRESS
CAMBRIDGE, MASSACHUSETTS,
AND LONDON, ENGLAND,
2000

Printed in the United States of America
10 9 8 7 6 5 4 3 2 1

This book is printed on acid-free paper, and its binding materials have been chosen for strength and durability.

ISBN 0-674-00251-2

Contents

Part IV World War II:

Frontispiece and photographs between pages 224–225 and pages 422–423 are from the Cannon Papers and the Harvard Medical Archives in the Countway Library of Medicine, except as noted.

Author's Note

After publication of this biography's first volume, *Walter B. Cannon: The Life and Times of a Young Scientist* (The Belknap Press of Harvard University Press, 1987), Cliff Barger and Saul Benison worked steadily, but separately, in order to digest the staggering amount of material available for the period covered by the projected second volume. Although I kept a hand in the Cannon Research Project, I was working full-time as associate curator of manuscripts and archives in the Countway Library's Rare Books Department. After Cliff's death in 1996, the project still had quite a way to go. I agreed to work with Saul Benison and try to produce a final draft of a manuscript; it was our wish that the second volume bear the same authorship as the earlier one.

When it became evident that Saul's failing health would not allow him to continue, I carried on to the end. Since I bear full responsibility for the finished work, as well as for all of the final decisions that it entailed, my name comes first in the trio of authors. The important point is that here we are, the three of us together again in a fruitful collaboration, for this, the second volume to complete Cannon's life.

Of far more interest than the authorship is the subject of this biography. Here we have never faltered in our belief that Walter Bradford Cannon, the teacher and scientist, the humanitarian, the unique individual, deserves a full-length treatment that is worthy of his many achievements. The fondest hope of each of us when we began the project was that we would be able to do justice to him and to his works. A good many of those works on which Cannon's fame rested took place after he had reached the age of forty-five, and a full description of his scientific and social activities has been eagerly—need I say patiently?—awaited. As with the previous volume, we consciously chose not to interpret but rather to let the life speak for itself through letters and diaries. Now that the biography is finally finished with publication of *Walter B. Can-*

non, Science and Society, it is up to you, our readers, to judge if we have succeeded in reaching our goals.

Several reviewers of our first volume noted that we had consulted, in addition to Cannon's papers, many other collections—no less than fifty—in our search for comprehensive coverage of the early part of the biography. That has not been the case for *Walter B. Cannon, Science and Society.* We have had a superabundance of materials for both world wars and the two decades in between. Indeed, one would be hard pressed in find a scholarly resource that is richer than the Cannon Papers in the Countway Library of Medicine in Boston. As ever, we owe a large debt to the staff of the Countway, and especially to those who worked over a period of time in the department of Rare Books and Manuscripts. We are obliged, too, for the courtesy extended to us by staff members in other libraries. At Harvard, we consulted holdings at the Widener, Pusey, and Schlesinger Libraries, as well as in those at the law and business schools. Of the other libraries that provided information and materials, we acknowledge assistance from staff members at Yale, Columbia, Johns Hopkins, Washington University in St. Louis, the University of Cincinnati, the American Philosophical Society, and the Rockefeller Archives Center. In England, we were aided in our research at the Wellcome Library, the Royal Society, and the archives of the (former) Medical Research Council. We also wish to thank the Trustees of the Wellcome Trust, the Johns Hopkins University Press, and the American Physiology Society for granting permission to reproduce parts of our articles previously published in *Medical History, Bulletin of the History of Medicine,* and *NIPS (News in Physiological Sciences),* which are fully cited in the notes.

There were many individuals who contributed greatly to the completion of the Cannon Research Project. It is not possible to recapture all of their names, but some are the same ones who gave support and encouragement to our earlier effort. First and foremost are members of Dr. Cannon's family—his daughters Marian, Wilma, and Linda (Helen died in 1992), and especially their brother, Bradford and his wife Ellen, who not only gave their memories but also provided us with additional materials. A resource of immeasurable value, for which we owe them a heavy debt, has been Mrs. Cannon's letters and manuscript papers, which now reside in the Cannon Family Papers in the Schlesinger Library. Bradford Cannon remained a steadfast friend throughout the years, a veritable fountain of encouragement and inspiration, and provided financial assistance to Cliff Barger's "Cannon Fund." Others who donated to the fund were Cliff's friends and former students, Victor Dzau and Edward Miller.

We wish we could give credit to everyone whom Cliff consulted when working on the scientific chapters, but a few names I have gleaned from his files must suffice: they are Arthur Loewy, Sam Hellman, and Laurie Green. No doubt, he would wish to acknowledge the help of Cannon's former students who shared their memories when this endeavor was first begun. Among friends and colleagues whom Saul Benison wishes to thank for their interest and assistance are Maurice Fox, Edward Atwater, Daniel Todes, Leonard Wilson, Sherwin Nuland, Bernard Sears, Stanley Troop, and especially Kenneth Berger.

All of us benefited from discussions with Horace Davenport, and with our neighbors on the fourth floor of the Countway: Oglesby Paul, Herman Lisco, Francis Moore and Fred Stare. Later conversations with Clark Sawin and George Thorn enlightened and helped me to understand some of the salient issues involved in the early days of endocrine research. We owe a debt to another endocrinologist, Jim Pittman, who maintained his enthusiastic interest throughout the project and turned up at my office door at a most propitious time.

We have acknowledged in the notes the scholarly efforts that shed light on events and persons in this volume, but we particularly want to single out the valuable attempts of various students to capture the spirit of Cannon's life and times; they often succeeded to an extraordinary extent. We are grateful to Marian Cannon Schlesinger for her Cambridge memoir, *Snatched from Oblivion,* which was even more of a guide in this volume than in the first one; to Maria Dietrich, who generously shared parts of her manuscript for a forthcoming book on Cornelia James Cannon; and to Henry Dale's daughter, Lady Allison Todd, who provided a copy of her father's unpublished autobiography for our use.

For editorial assistance, thanks are due to Jodi Simpson for leading me to Laura Ricard, whose effort with the early chapters was instrumental in helping to get this finished book off the ground; I thank her as well. My gratitude goes to Martha Baldwin, whose final reading of the manuscript before it went to press was most beneficial. Steve Borack deserves praise for his amiable production of photographs of high quality for the book. Readers of earlier drafts, or parts of drafts, included Horace Davenport, Oglesby Paul, Maurice Goodman, Jerome Lettvin, and Gordon Wolfe; their comments helped enormously to catch mistakes and get rid of rough spots. I can never thank Horace Davenport enough for his expertise (and fresh approach) in giving a careful critique of the scientific chapters—and for urging me to go back and rewrite certain parts of the text.

During the course of this work, the Countway Library underwent a number of significant changes, among them a large-scale renovation that threatened to disrupt, or even bring to a halt, the preparation of this volume. It was through the influence of friends and scholars like Amalie Kass, Oglesby Paul, Alan Brandt, and, of course, Brad Cannon that the Cannon Project remained in operation until the biography was complete. We are certain there were others who acted in our behalf whom we would thank if we knew their names. It gives us great pleasure, however, to note the unanimous support of the trustees of the Boston Medical Library, under the presidency of Leonard Morse, which assured publication of this book and its distribution by the Harvard University Press. The world of scholarship is fortunate to have the support of individuals and institutions like these.

Many years ago, we expressed deepest gratitude to three of our nearest and dearest supporters: Mark D. Altschule, Claire Barger, and Richard J. Wolfe. Mark died in 1988, a great loss that we, and many of his other admirers, have felt keenly. Claire continued to maintain an interest in the project and always had a word or two of encouragement. And without Dick, I would not be writing these words. Among his many helpful contributions was his organization and reworking of several of the scientific chapters. He is responsible, too, for shepherding this book through the publication process.

Elin L. Wolfe, 1999

WALTER B. CANNON, SCIENCE AND SOCIETY

PART I

World War I

1 The Mystery of Shock

I: Acidosis

IN AUGUST OF 1914, when the German, French and English sent armies into Flanders and northern France, their governments believed that the war into which they had plunged would be a short one, crowned by a quick victory. That illusion was quickly shattered. Within months the fighting became mired in mile after mile of trenches protected by barbed wire and the massed firepower of rifles, mortars, machine guns, and artillery. Such warfare not only prolonged the fighting, it also exacted a staggering toll of casualties. No one was prepared for this situation, least of all the physicians and surgeons who often had to deal with problems far beyond their understanding and skill. In 1915, day after day British surgeons in Flanders saw appalling numbers of grievously wounded men. To their dismay, more and more of the soldiers in their care were dying of the shock that accompanied severely lacerated and contaminated wounds.

Shock was not an unknown condition. More than 300 years before, Ambrose Paré, France's pioneer military surgeon, had described some of its major symptoms. "The signs of sowning [shock]," he wrote, "are a paleness, a dewy and sudden sweat arising, the failing of the pulse, a sudden falling of the body on the ground without sense and motion, a coldness possessing the whole body so that the patient may seem rather dead than alive. For many who fall into a sowne dye, unless they have present help." Yet despite the advances of medical science and surgery by the early twentieth century, the cause and onset of shock remained as much of a mystery as it had been for Paré.[1]

Late in 1916, increasingly alarmed British medical officers in the field joined forces with clinical researchers in England to undertake an investigation of the deadly condition. The British were not alone in their concern. Several months before, the Rockefeller Foundation had invited Walter B. Cannon at the Harvard Medical School to undertake an experimental study of shock in France in conjunction with the distinguished surgeon Alexis Carrel. Cannon turned down the offer. The problem, he explained, was so complex that he felt he could do little that would be of immediate use. The Foundation thereupon approached his colleague William T. Porter who promptly agreed to serve with Carrel.

But some months later, when the National Research Council asked Cannon if he would establish a physiological committee to develop research programs to maintain the health and safety of the armed forces, Cannon, in a turn-around, made the study of shock one of its priorities. While most members of the committee devoted themselves to seeking the cause of shock, Cannon began a search for a therapy that might reduce its mortality.[2]

Unlike some of those in the United States who had preceded him in this research, chief among them the Cleveland surgeon George Crile, Cannon had never examined or treated a human patient in shock. Crile had extensive clinical experience with shock cases before the war, many of them victims of industrial or transport accidents. Cannon, in contrast, was dependent for his understanding of the onset and development of shock on the clinical descriptions left by surgeons and physicians who had previously treated patients in shock.

One description that caught his attention was that made by Hermann Fischer in 1870 of a man who was injured in an accident with a carriage drawn by runaway horses. Fischer reported that his patient lay perfectly quiet, oblivious to events about him; his pupils were dilated and reacted slowly to light; he stared purposely and apathetically straight before him. He was conscious, but replied slowly and only when repeatedly questioned. His skin was as pale as marble, his hands and lips had a bluish tinge. Large drops of sweat hung on his forehead and eyebrows, his whole body felt cold to the touch. Sensibility was blunted over the entire body; if the limbs were lifted and then let go, they immediately fell as if dead. His pulse was almost imperceptible and very rapid, his respiration "characterized by long, deep, sighing inspirations alternating with very superficial ones," barely visible or audible.[3]

Fischer's description set the agenda for Cannon's research. Cannon knew that if he were to make any headway, he would have to determine the primary and secondary features of shock from the multiplicity of signs and symptoms detailed by Fischer. But first, he would have to salvage from extant theories of the cause of shock an approach that might serve him as a guide in his investigations.

During the American Civil War, S. Weir Mitchell, with the assistance of George Morehouse and William W. Keen, had studied wounded Union soldiers in shock and concluded that their condition was due to a paralysis or exhaustion of nerve force. Such an exhaustion, they asserted, could affect one or more nerve centers and, if severe enough, might under certain conditions produce a paralysis of sensation or motion. More than a generation later, George Crile also suggested that the cause of shock might be due to an exhaustion of nerve force; but in his animal experiments he focused on the circulatory system and in particular, on directly measured arterial blood pressure. The decrease or paralysis of vasoconstrictor tone in shock, he maintained, allows the blood to accumulate and stagnate in the veins of the splanchnic bed (i.e., abdominal viscera), and it was this stagnation that brought on the various effects seen in shock.[4]

Crile's notion that exhaustion of vasomotor centers is a primary factor in shock was promptly criticized by physiologists, who were especially vigorous in advancing their own theories. W.T. Porter at Harvard, following his work with Carrel, stated that fractures of the femur or wounds through subcutaneous fat lead to fat embolism and shock. Yandell Henderson at Yale suggested that shock is due to a decrease of venous return to the heart, a condition he attributed to a loss of carbon dioxide brought on by an excessive overventilation of the lungs. Samuel J. Meltzer at the Rockefeller Institute, taking another tack, argued that the functions of the body are supported on a dual principle of well-balanced antagonism between excitation and inhibition and that injuries which produced shock disturb equilibrium, "favoring the development of the inhibitory side of all of the functions of the body."[5]

Cannon, faced with the necessity of developing a program that would further the quest for an effective therapy against shock, came to believe that a failure in the circulation of blood was central to the development of shock. Like Crile before him, he accepted the notion that the blood pooled in the splanchnic bed, where capacity was so ample that it was possible for a person to bleed to death into his own splanchnic vessels. He could not, however, accept Crile's notion of vasomotor exhaustion as the cause of circulatory failure, for there was clear experimental evidence that the vasomotor reflexes were still responsive in shock. The evidence left Cannon, as it had others, with a puzzle. What prevented blood in the splanchnic area from returning to the general circulation?

Investigators before Cannon had devised a variety of strategies to improve the circulation in shock. Some, in an effort to increase blood pressure, injected warm saline solutions into the veins with only transient effect. Others tried adrenalin and found it counterproductive; although it contracted splanchnic arterioles, it dilated arterioles elsewhere. Cannon decided to focus his attention on the splanchnic portal area. Here, he reasoned, blood flows between two capillary beds: one, the capillaries of the stomach, spleen, pancreas, and intestines from which blood enters the mesenteric branches of the portal vein, and the other, the capillaries of the liver through which blood must flow before reaching the inferior vena cava. The application of a constrictor agent to the walls of the veins might well supply the force to move the stagnant blood forward. But instead of injecting such an agent into the veins, he believed that it might be better to inject it into the abdomen to increase vascular smooth muscle tone. And after some animal experiments, he determined to try pituitrin (pituitary extract) as his constricting agent.

Soon after the United States entered the war in April of 1917, Cannon joined the Harvard Base Hospital No. 5 unit with an understanding that he would be sent to a forward area where he could continue his research on soldiers in shock. "I do not wish to elaborate at the present time on the possibility of using pituitrin . . . in shock cases," he wrote in his Shattuck Lecture before leaving for Europe. "But before anything definite and practical can be made of this suggestion, it must be tested out carefully on a considerable scale. This I hope to do in the course of future work, and if favorable results arise from these tests, I shall report on them."[6]

Cannon's departure from Boston to New York—the first leg of his trip to France—proved less of a wrench than he had anticipated.[7] Save for Harvey Cushing's chauffeur who cried inconsolably, no tears were shed at the railroad station. In an effort to spare her husband the sadness of leaving family and friends, Cornelia

Cannon contrived to create an atmosphere of gaiety and even laughter to lighten the final goodbyes.

Moments after he checked into his New York hotel, Cannon, as a first lieutenant in the A.E.F., received $200 in gold to cover the cost of his uniform. Since he had already bought a uniform at a considerably lower price, he promptly sent the windfall off to Cornelia to bolster the family finances.

When Cannon had first tried on his uniform at home, one of his little daughters was enchanted. "Mother, look," she called out, "here comes Father dressed up like a Boy Scout." The uniform in New York drew a far different response. While being shown around the city by his friend Frederick Lee, he noted, "I was saluted at intervals by lieutenants and enlisted men . . . by police and others . . . much to the diversion of the Lees who wanted to know how it was properly done. And when they asked how long I had been receiving such homage, I had to answer 'one half day,' to their great amusement."

Military etiquette was new to Cannon and at times its requirements bothered him. A meeting with his recent assistant McKeen Cattell, then a sergeant standing guard duty, made him uneasy and left him with ambivalent feelings. "It grates on my sensibilities not to have the free and easy ways with Cattell," he wrote home. Dealing with longtime friends who outranked him proved even more troublesome. "I made a horrible mistake this noon," he admitted in one letter, "in the presence of others I called Major Cushing, Harvey. I will never do that again." Still, these incidents took little away from his euphoria at being part of a unique company on the edge of a grand adventure, and he told Cornelia to think of him "as having the time of my old age."[8]

On May 11, 1917, two days after his arrival in New York, Cannon boarded the troopship *Saxonia.* To set his wife's mind at ease about the dangers involved in the impending voyage, he related a conversation he had with General Goethals, the father of one of his students, who told him "that any boat going at 13 knots had little to fear from submarines and that the speed of the *Saxonia* was 15 knots heavily loaded. You can bet he'd move heaven and earth to make things safe for that son of his." He gave no intimation, however, that the *Saxonia,* in addition to carrying Base Hospital No. 5 personnel, also carried a cargo of ammunition.

The moment the *Saxonia* left the New York Narrows, Captain H.M. Benison, the ship's master and a veteran of war zone crossings, announced that there would be daily lifeboat and life preserver drills. Since the members of Base Hospital No. 5 had not received any basic military training, there were also daily programs that included calisthenics for the officers, talks on first aid for members of the hospital staff, and drills for both officers and enlisted men so they would respond correctly to orders like "parade dress" and "about face."

When Major R.U. Patterson called a conference to discuss the organization of the hospital, he discovered that his staff differed sharply with the plans he had outlined. Cannon was particularly annoyed to learn that he was put in charge of organizing the hospital laboratory, for it appeared to be the first step in saddling him with routine diagnostic work. Several days later he received a pleasant surprise. "I've just had a very satisfactory talk with Major Patterson," he wrote home. "I told him the reason for my being present and expressed the hope that I might be free to go where I could get the best chance to serve in my own way. He said that there would be no trouble at all and that he would at once give me a month's leave of absence from the hospital and would write to Washington for a roving commission which could be given to me by cable. It is only necessary now to get the English and French surgeons to cooperate."

As the *Saxonia* plied its way across the Atlantic, Cannon snatched time to work on the Shattuck Lecture he had promised to deliver to the Massachusetts Medical Society and talked to Harvey Cushing about his plans. On May 23, after eleven days at sea, Base Hospital No. 5 landed safely at Falmouth, England, and he noted, "When we touched terra firma, it was with a distinct sense of relief . . . Elliott Cutler threw himself on the grass, rolled over on it and chewed it out of pure delight." Cannon quickly posted his lecture to David Cheever, who had promised to read it in his behalf at the annual meeting of the Massachusetts Medical Society in June.[9]

Upon arrival in London, Cannon found himself in a hectic round of activities arranged by the British medical establishment. He met political and scientific notables, attended meetings of the Royal Society, and visited the headquarters of the Medical Research Committee (M.R.C.) where he heard first-hand reports of the progress of the Committee's various investigations into shock and its allied conditions. When he learned in conversation with Walter

Fletcher, the secretary of the M.R.C., that his old friend T. R. Elliott was serving as a consultant on medical research with the British Army in France, his hopes for a collaboration soared. He could hardly wait to get to France.[10]

For all of his anticipation, Cannon found his early experiences in France depressing. The Base Hospital No. 5 unit, sent to relieve British Base Hospital No. 11, was well behind the front lines at Camiers, a northern village not far from Boulogne. The encampment—actually a tent city spread over acres of sandy dunes and cesspools—was, in Harvey Cushing's words, "a down-at-the-heels hospital in a shockingly dirty, unkempt camp."[11]

For Cannon, any attempt to do experimental work appeared hopeless; there were no instruments, few animals, and the laboratory was far too small. Worse, unable to make contact with Elliott, he had to content himself with a pledge from Colonel Cuthbert Wallace, surgical consultant to the British First Army, that he would try to arrange a transfer for Cannon to a unit at the front. Frustrated that he could not begin his research at once, and oppressed by a feeling that he was of little value to the hospital, Cannon asked Robert Osgood, his friend, Harvard colleague, and chief of surgery, if he could assist in the operating room.

Cannon's offer could not have come at a better time, for the very next day the first wounded from the British drive at Messines came pouring into the hospital. In the days that followed, he assisted in operations, helped on the wards and in the discharge tent, and in one instance, hypnotized a patient dumb from shell shock and enabled him to speak. "It has been an immense reassurance of self-respect to be in service and actually desired, as I now am, for all sorts of uses," he revealed to Cornelia. "But it is not, as you will probably say at once, getting on with the work I came over to do."

Several days later, his luck changed when an unexpected visit from T. R. Elliott brought support for his plan of going to a British unit at the front. And then to his delight, Elliott agreed that if circumstances dictated a need for further experiments, Cannon should go to London to continue his research in William Bayliss's laboratory at University College—an idea that almost immediately was endorsed by a friendly letter from Bayliss.[12]

Expecting to soon have an opportunity for testing the effect of pituitrin in preventing stagnation of blood in actual shock victims,

Cannon visited a mortuary in a nearby village to see if he could use a blunt trocar for intra-abdominal injection of pituitrin without puncturing the intestines. He noted in his diary on June 15 that nineteen punctures through the anterior abdominal wall had no effect on the intestines; of six injections on the lateral aspect, only one entered the caecum and one the descending colon.

At about the same time, in cooperation with a former student, Reginald Fitz, he began a study on trench nephritis. Fitz, who worked for several years at the Rockefeller Institute with Donald Van Slyke applying techniques of physical chemistry to the study of diseases of the kidney and pancreas, had become interested in examining such matters as acid-base balance in the blood, respiratory gas exchange, and the transport of oxygen and carbon dioxide in the blood. Equally important, he had become adept in the use of the Van Slyke apparatus for measuring oxygen and carbon dioxide content in blood. Fitz welcomed Cannon's help, but by the same token Cannon was equally well served in the collaboration.

To his family, Cannon explained that his purpose in working with Fitz was to learn how to use the Van Slyke apparatus. Then he added, "There is another matter of interest in this connection, and that is the suggestion on the part of certain English investigators, who have recently reported that they find in experimental shock conditions which might be caused by blood changes—such blood changes as we have been studying. It may be necessary, therefore, or possible to continue these studies if I ever get into relation with shock cases in human beings."

On June 24, Cannon received word that Elliott and Wallace had arranged for him to work for three weeks at British Casualty Clearing Station (C.C.S.) No. 33 at Béthune, about 20 miles from Lille. The selection of this particular clearing station as headquarters for Cannon's investigation of shock was hardly haphazard. Its senior surgeon, Captain John Fraser of Edinburgh, was deeply interested in the problem of shock. Moreover, C.C.S. No. 33 generally received the most serious cases of shock in that area of the front, thus affording Cannon a unique environment in which to carry on his research. Realizing the need he would have for a supply of physiological instruments, before going he went to one of his friends, Captain C.G.L. Wolf, at the British Hygiene Laboratories at Boulogne and borrowed a viscometer, a colorimeter, and a Van Slyke apparatus.[13]

When Cannon arrived at Béthune, Wallace promptly took him on a tour of various aid and casualty clearing stations and introduced him to some of the surgeons in the area—including Captain Ernest Cowell, chief surgeon of C.C.S. No. 23. A graduate of University College Hospital in London, Cowell had started an investigation of the onset of shock several months before. Armed with clinical experience, a sphygmomanometer and stethoscope, he recorded the systolic and diastolic pressures of soldiers in rear areas who lived under relatively normal conditions, and compared them with those of soldiers in front-line trenches who were deprived of food and water and lived with constant fear and anxiety. Then he compared both with the pressures of wounded soldiers in shock.

Cowell's study produced a number of important findings, not the least of which was the establishment of parameters for the normal values of blood pressure found in fighting men, as well as indices for measuring the seriousness of the fall in blood pressure often found in shock victims. Perhaps more important, it allowed him to illustrate for surgeons in the field the different ways that shock might appear. It might appear as primary shock, where soldiers after being wounded showed immediate symptoms of shock with a low blood pressure. Or it might appear as secondary shock, where, on first examination, soldiers would have a normal blood pressure and wounds that appeared trivial and yet, several hours after being sent to the rear (with little attention paid to their needs for food, water, or warmth), they suffered from rapidly falling blood pressure and shock that was pernicious and uncontrollable.[14]

Taking blood pressure, which today seems natural and easy, was, save for some well-trained physicians and surgeons like Harvey Cushing and George Crile, still an uncommon procedure in the United States up to this time. Although Cannon's laboratory syllabus in physiology before World War I introduced students to the method that Carl Ludwig had used for directly measuring blood pressure in anesthetized animals and to the oscillatory sphygmomanometer devised by Joseph Erlanger to estimate blood pressure in man, he emphasized the importance of feeling the radial pulse "for finding the rate of heart beat and for judging roughly the state of circulation in the arterial system."[15] Furthermore, prior to coming to Casualty Clearing Station No. 33, Cannon had no experience with the palpatory or auscultatory methods of measuring blood pressure then being used by British surgeons in their study of hypotension and shock.

His first opportunity to deal with patients in either primary or secondary shock occurred during his visit with Cowell when soldiers from the battle of Lens came streaming into C.C.S. No. 23. Cannon, observing a gravely injured German soldier with a big hole in his chest who was suffering from shock with low blood pressure, asked Cowell for permission to inject the patient intraperitoneally with pituitrin. Despite the injection, the soldier died several hours later. After autopsy, Cannon learned two things from Cowell that changed the direction of his research: first, that his injection had not harmed the intestines; second, and more important, that there was no visible pooling of blood in the splanchnic bed of a patient in shock.[16]

British surgeons in the field who had operated on wounded soldiers had long known that there was no apparent excess of blood in the splanchnic veins of their shocked patients. Cannon and other physiologists had been led astray by the engorged veins seen in the abdomens of their shocked laboratory animals and assumed that the blood lost from the general circulation had stagnated in the splanchnics. Cowell's findings raised a new question for Cannon. If the blood that had disappeared from the general circulation was not in the arteries or veins of the abdomen, where had it gone?[17]

Using the Van Slyke apparatus, Cannon began an examination of the blood of patients in shock as well as those suffering from gas gangrene infection, and discovered that the blood in both cases was acidic. Indeed, it appeared to him that the more marked the acidosis, the lower the blood pressure and the graver the shock. Stimulated by these observations as well as by what he had learned from his collaboration with Cowell, Cannon went back to Béthune and asked Captain Fraser and another young surgeon, Captain A.N. Hooper, to join him in a study of the altered distribution of blood in cases of shock complicated by hemorrhage and gas gangrene.[18]

The implication of the acidosis in cases of gas gangrene and shock particularly gripped Cannon, and he wrote home to share some of the excitement of the new turn in his research: "Fraser and I had a long talk about the cases of infection with the gas bacillus which I had become interested in at Cowell's place, and which were turning up here and going down to death with nothing done for them. I have been examining the blood of these men and finding that it is loaded with fixed acid. I have so free a hand, and the cases here are so desperate, that anything hopeful can properly be tried. So we are going ahead quite as we should go in an experimental study."[19]

Others took notice of the turn in Cannon's research as well. Colonel Wallace, who had previously undertaken an investigation of the relation of the toxemia of gas gangrene and shock, immediately asked Cannon's commanding officer to extend his tour of duty at Béthune. Walter Fletcher, at the Medical Research Committee, wrote Cannon of the great satisfaction in England with the cooperative spirit he had shown at C.C.S. No. 23 and No. 33 and intimated that Cannon, working together with Henry Dale and William Bayliss, might well solve the mystery of shock.[20]

Such interest and encouragement gave Cannon momentary pause, and he hinted to his wife of his need for a biochemist to help with his research. "I wish that Lawrence Henderson were here," he wrote. "The condition of acidosis in the cases of gas infection is especially important, and he is a prime authority on the subject. If you are still in Cambridge, ask him to send me a reprint of his general discussion of the acid-base equilibrium in the blood."[21]

Cannon did not permit his lack of expertise in biochemistry to weigh on him, for he felt keenly that his research had to lead to findings that would have practical applications for the human beings who were in such desperate need. Determined to go forward, he threw himself into the problem of sequestration of the blood in cases of shock by making careful measurements of the degree of acidosis, concentration of hemoglobin, and number of red blood cells in the veins and capillaries of the wounded. In addition, he undertook a daily regimen of examining all incoming patients in shock, reading whatever papers on shock and gas gangrene he could lay his hands on, and discussing his findings with his colleagues and visitors to the casualty clearing station.

Some days into his new research, Cannon noted in his diary on July 11, "Read Bayliss and get ideas regarding effect of acid and cold on increasing blood viscosity so that it might stagnate in capillaries." Almost immediately, he became convinced that he was on the track of locating the blood lost in the circulation. "Up at one o'clock in morning," he wrote the next day, ". . . to examine a new case. Hooper and I study case together and find marked discrepancy between venous and capillary blood." In cases of low blood pressure, blood from the capillaries contained 2,000,000 more corpuscles per cubic millimeter than that taken from the arm veins at the same time—evidence of the leakage of plasma. The finding, Cannon believed, was a breakthrough that would lead to a practical benefit. In mid-July he exultantly related to Cornelia that acidosis was a major factor in shock:

> All the evidence accumulated during the past week indicates that in shock and allied conditions a factor of first significance is the development of an acidosis. It would account for the discrepancy in the blood distribution mentioned earlier . . . A patient in shock, with 41 volumes percent CO_2 capacity in his blood, was taken to the operating room at 9:40 and subjected to a severe operation. The blood after the operation at 10:50, just 70 minutes after the first sample was taken, had a CO_2 capacity of only 28 volumes percent—an astounding drop. Surgeons have reported that operating on shock cases is almost invariably fatal and have refused to do it. This looks like the secret of the trouble—and one that can be dealt with by very simple means. Only cooking soda is needed! Well, this is first enthusiasm, but certainly we have got new lines out that seem sure to bring important results.[22]

In the weeks that followed, Cannon reviewed the data he had accumulated, and discussed with Wallace, Fraser and others his idea of using sodium bicarbonate to reverse the acidosis during operations on patients in shock. When he later prepared his reports, he gave credit to Sir Almroth Wright, who had used sodium bicarbonate to treat patients with acidosis accompanying gas gangrene infections. Cannon expanded on Wright's concept, however, noting that acidosis was not peculiar to the toxemia produced by gas gangrene but rather was general for various states of low blood pressure, whether caused by hemorrhage or wounds without any apparent infection. In essence, the lesson he drew from Wright's work was that it was reasonable to treat shock with alkali, regardless of its pathogenesis.[23]

A British drive north of Ypres that began on July 22 gave Cannon new shock cases. After a week of intense work tending the wounded brought into the operating room and on the wards, he described for his wife the circumstances that led to the application of this therapy and its initial results:

> As I think I told you in the last letter, we decided to put our patients to practical test. Well, on Monday there was a patient with a blood pressure of 64 (the normal is about 120) millimeters of mercury and in a bad state. We gave him soda, a teaspoonful every two hours and the next morning the pressure was 130. And on Wednesday a fellow came in with his whole upper arm in a pulp. Fraser said such cases usually die. At the end of the operation he had the incredibly low pressure of 50; soda was started at once and the next morning the pressure was 112. Both these cases turned out very well. But they were not sufficiently decisive for me. I wanted a sure sign of the efficacy of the method. It came Friday night.
>
> A poor fellow was brought in with terrible wounds—his left knee shattered, his right calf torn open, half of his right hand mashed and a ghastly rip in his right forearm. After the operation his blood pressure was 68, his pulse was 148, and his respiration 34 and over. About ten o'clock he was much worse. His respirations, which tell the tale of accumulating acid in the blood, had risen to more than 40 per minute. I urged that this was an emergency case and that radical treatment was necessary. We decided to inject the soda intravenously. When we went to him at 10:25 he was gasping for breath at the rate of 48 per minute, lifting himself on his elbow in spite of his wounds and crying in his gasps, "I can't breathe. Give me air. I can stick the pain but I must have air." We ran 35 ounces of the warm sodium bicarbonate solution into an arm vein. I never saw anything in my life so dramatic as the change that occurred. His respirations promptly fell from 48 to 26 per minute, and his pulse from 148 to 126. And in ten minutes he was quietly sleeping. He had another similar crisis in the morning about 3:30 which was met similarly, and the next day his blood pressure rose 86, 102, 114, and he sat slanting in his bed, smoking with much satisfaction a cigarette.
>
> Two other cases, quite as decisive, though not so dramatic, testified to the efficacy of the treatment Friday night and Saturday morning. The unhappy thought is that I have seen men die, and Fraser says he has seen hundreds of them die within a few hours or less after they have reached the stage from which we recovered these men. Fraser's great experience is a strong support, for he says that these men were literally "snatched from death". . . when it was clear that the men were really secure, he declared, "This is a red letter day in my surgical experience."

When she received Cannon's news, Cornelia was ecstatic. "The wonder and joy of it doesn't grow less in reading the letter over and over," she replied. "Papa says he can't believe it, cooking soda is too simple! And I must say we all feel its incredible quality . . . to think that you were not in the least affected by your prepossessions for pituitrin, but began over again quietly and modestly at the beginning, and in four weeks accomplished what the rest of them have been working on for years."[24]

Cannon's experimental use of sodium bicarbonate appeared to be so promising that Colonel Wallace reported it to British General Headquarters at once. But while Cannon was gratified by the reception given his research, he recognized that some of his insights needed further corroboration. Whereas he fully believed that the blood which had "disappeared" from the circulation in shock stagnated in the capillaries, he did not know whether capillary capacity was sufficient to contain so much blood. There was still much to learn not only about the relationship between low blood pressure and acidosis but also about the extent to which the oxygen-carrying capacity of the blood was reduced with increasing acidosis. There were other puzzles, too, including the practical matter of what factors in a surgical operation might lead to increased acidosis.

At the end of July, the Medical Research Committee dispatched Bayliss to Béthune so that, in company with Cuthbert Wallace, T. R. Elliott, and C.G.L. Wolf, he might review the methods and data that led to the use of sodium bicarbonate in treatment of shock. At first Cannon had some trepidation about the outcome of such a review, but after it was over he wrote home: "They have seen my method and found nothing at fault, and they have discussed the suggestions which I have made and come to agree with them. And in consequence, though isolated from all access to literature on the subject, I feel quite secure in my stand . . . The methods which we have tried here have been described and are being tested in neighboring clearing stations. I suppose that in a month or two we shall know more about their workings."[25]

The meeting had an equally positive effect on Bayliss. Soon after his return to London, he wrote Cannon that the M.R.C. had just formed a special committee on shock and extended a formal invitation to him on behalf of the committee to become one of its members.[26]

Emboldened by the outcome of the conference, Cannon began a new series of experiments to find answers to some of the questions that still troubled him. He quickly discovered that blood pressure and acidosis were closely, and inversely, related—that is, the greater the degree of acidosis, the lower the blood pressure. The practicality of finding a solution to the problem of the effect of anesthesia on acidosis during operations on patients in shock particularly appealed to him. In fact, the answer appeared to be close at hand when Geoffrey Marshall, an English surgeon, reported that he had found much less deleterious effects in shock cases if the operation was done under an anaesthetic mixture of nitrous oxide and oxygen than if done under ether or chloroform. By mid-August after studying the effect of different anaesthetics used by Fraser in operations on patients in shock, Cannon was able to establish that the use of ether not only increases the acidosis but lowers the blood pressure as well, and that nitrous oxide and oxygen cause little if any increase in acidosis and scarcely influence blood pressure.[27]

Cannon's note to Bayliss about these results evoked an intriguing rejoinder. "It is very puzzling to know what is the real cause of 'shock,' other than that of haemorrhage," Bayliss mused in reply. "One wonders whether the low blood pressure produces the acidosis, or vice versa. And whichever it is that is primary, what is it that brings this about? . . . I am somewhat inclined to think, after all, that perhaps afferent stimuli may play a part, but the difficulty in experimental work will be to produce a state of shock similar to the clinical one."[28]

Colonel Wallace remained enthusiastic about Cannon's work and requested his commanding officer at Camiers to extend his leave. Cannon, seeking more substantial support, decided to go to authorities at American Medical Headquarters in Paris to explain the substance of his research and to request official sanction of his transfer to Bayliss's laboratory in London.

The visit to Paris late in August produced the results he had hoped for and more. Colonel Alfred E. Bradley, chief of surgery for the American Army, listened intently to Cannon's account of his work at Béthune, and then urged him to continue his research in London and prepare suggestions for him on how to deal with shock before American troops got into combat. For Cannon, Bradley's keen interest not only served to authorize his plan to extend his investigation of shock with Bayliss and others, it also held out the hope that his findings would be applied by the American Army when it went into the line in France.

A chance meeting with Alexander Lambert, head of the Division of Medical and Surgical Services of the American Red Cross in France, brought more good news. Lambert, who had known Cannon for some years, said he had taken the liberty of appointing him to an Advisory Committee of the Red Cross—an important post since the Red Cross was in charge of medical research for the American Army. Before leaving Paris, Cannon paid a visit to the Pasteur Institute. It brought on a wave of nostalgia for the life he had left behind him in Boston and he wrote Cornelia, "We laboratory hermits are not intended to be men of the world. I found it a most gratifying pleasure to return for a moment out of the hurly burly into the calm and serenity of a scientific workshop."[29]

After he returned to Béthune, Cannon decided that he needed to address three further problems before departing for London. First, he hoped that determining the blood sugar level of patients in shock might reveal whether the inception of acidosis in the wounded was a "starvation acidosis" brought on by a carbohydrate-deficient diet or a lack of food and water before and after action on the battlefield. Second, he examined the possibility that providing the badly-wounded with a reserve of material in the blood would fortify them during operations against dangerous drops in blood pressure and increase in acidosis. And third, he tried to develop efficient methods to prevent the onset of secondary shock in wounded patients during their transit from the trenches to the rear.

The first problem evaporated after several days of measuring the sugar content of the blood of the badly-wounded men brought into C.C.S. No. 33, when Cannon found that their blood sugar was normal. As his investigation of the second problem entered its final stage, he took pride in detailing for his wife some of his new findings. "As soon as the operation starts," he wrote on September 28, "I have been running a pint of a 4 percent solution of sodium bicarbonate into a vein. The results have been extraordinarily good. The patients survive the ordeal of operation with a higher blood pressure than they had before going under the anaesthetic and their acidosis instead of being increased may be changed back to normal with good margin to spare. Fraser is now so pleased with the clinical results that he insists on this prophylaxis now in all critical cases."

The solution to the third problem was in large part due to the splendid cooperation of Fraser and Cowell. Indeed, Cowell, at Cannon's request, returned to the trenches again and again to develop effective methods to prevent the onset of secondary shock in seriously injured patients during transit to the rear. It was Cowell who later emphasized that one of the chief factors in the initiation of secondary shock was the loss of body heat. The recommendation that Cannon, Cowell, and Fraser later jointly made that the wounded be protected from exposure to cold for prolonged periods of time, provided with warm liquids, and transported to the rear in specially prepared dry stretchers clearly emerged from the observations of Cowell.[30]

In the midst of these final investigations, Cannon, to his utter amazement, received orders from American Headquarters to report for duty with the Gas Service. Several weeks before, he had received a letter from one of his former students, Walter Boothby, asking if he would be willing to direct the laboratories then being organized in Paris to study the physiological effects of war gases and to develop treatment for the conditions they caused. Cannon quickly replied that while he was flattered by the offer, he had little expertise in respiratory physiology and, more important, was deeply engaged in and committed to research on shock. He had assumed that his reply to Boothby's query would put an end to the matter.

Upon receipt of these new orders, Cannon gave vent to his frustration and anger. His research at Béthune had gone well, and he was expecting to collaborate with Bayliss and Henry Dale in London in order to corroborate the results of his findings. Indeed, Dale had informed him that experiments were already underway. "What a disappointment it will be to leave all this," Cannon wrote home. "If I am compelled to report at American Headquarters willy-nilly I shall leave my apparatus here, with the firm belief that I can make them see that my best service lies in an elucidation of shock and its treatment which has been going on so auspiciously and which is so important for war surgery."

At first he tried appealing to the commanding officers at Béthune to intercede in his behalf. When that failed, he considered another tactic. "If the Army insists on gas work," he continued the next day, "I shall be tempted to throw up my commission, go over to the Rockefeller arrangement and thus be free to continue the shock work." Several days later when he was ordered to report immediately to the Gas Service, Cannon took matters into his own hands.[31]

Once again he went to Headquarters in Paris. He reminded Col. Bradley of their previous conversation when he had told Cannon to go ahead with the work and send him whatever practical suggestions were justified for use in the medical service. On the basis of that authorization, Cannon explained, "I had asked the British to supply me with the necessary apparatus, I had purchased instruments, I had got a man sent to the trenches to cooperate with me in tracing the history of shock cases, I had planned to get in touch with the Shock Committee in London—in short, that he had given me a warrant which involved me in obligations which I could give up only with awkwardness and a taint of bad faith." Cannon's impassioned argument had its effect. Although Bradley was annoyed that a mere captain (and newly promoted at that) was upbraiding a colonel, he relented and told Cannon to speak to Col. A.R. Fries, the head of the Gas Service, since the final decision would be his.

To his relief, Cannon found that Fries grasped the importance of his research on shock and had no objection to his continuing the work in London. Indeed, as his new commanding officer, he promptly put Cannon on detached duty with the British Army again. Cannon, having got what he wanted, promised that once he completed his research with Bayliss, he would help organize the laboratories of the Gas Service. If the delay created difficulties, he suggested that Fries might ask A.N. Richards, an excellent pharmacologist who had recently come to London from the United States to work with H.H. Dale, to serve temporarily in his place.[32]

When Richards subsequently was called up by the Gas Service, Dale quickly wrote Cannon in his behalf. Richards, he explained, had been specifically recruited by the M.R.C. to help him with his research chores until January 1918, at which time he was to return to his teaching duties at the University of Pennsylvania. "I would not ask him to stay a day longer with me if he felt that inclination, duty, or both, would take him to the post in France. If his going were the condition of you being left to finish your work, I would urge him to go. His own desire, however, is to stay." After Cannon

sis of the shock-like condition produced by histamine; W.M. Bayliss favored the addition of sodium bicarbonate to the infusion of gum acacia he was preparing for widespread distribution as a replacement for blood.

At the next meeting, however, Cannon virtually held the floor. He outlined the results of his investigations at the front, reviewed the theories of the nature of shock advanced by George Crile, Yandell Henderson, and W.T. Porter, and reported cases that had benefited from his alkali treatment. His presentation provoked considerable debate. The discussion apparently was not heated, but the questions raised were so important that the committee recommended further investigation of the different lines of inquiry already underway.[36] For Cannon, the response meant that his efforts in France were proving not only useful but also stimulating to future work. At the same time, it began to dawn on him that his British colleagues had deep-rooted convictions of their own about both the nature and the treatment of shock.

For instance, most of the members of the M.R.C. believed that oligaemia (a deficit of the volume of blood in circulation) was a central feature of the circulatory collapse in shock. None, however, felt that any of the theories put forward thus far adequately explained the pathogenesis of the syndrome. Bayliss, like Cannon, anxious to improve treatment of shock, had developed an infusion containing a colloid, gum acacia, that could be injected intravenously to replace the plasma that had escaped from the circulation. Although he kept an open mind on acidosis as a cause of shock, at no time during his weeks of collaboration with Cannon did Bayliss put his research on blood substitutes aside.[37]

Dale also focused on the problem of loss of blood volume, but he was primarily concerned with cause rather than treatment. A brilliant pharmacologist, from the beginning of the M.R.C.'s shock program Dale had searched for a chemical trigger that might be responsible for the onset of shock. That search was sparked by a study of the pharmacological actions of histamine—an amine derived from histadine and isolated from ergot—that he and an associate, Patrick Laidlaw, had begun some years before.

Dale, first with Laidlaw and then with Richards, found that histamine caused profound hypotension in cats, dogs, and monkeys by two major actions: one, by vasodilatation of small blood vessels (arterioles, capillaries, and venules), the second, by increasing capillary permeability with resultant loss of plasma into the

received a diplomatic letter from Richards stressing the importan
of his research with Dale compared to the unknown prospects
working in the Gas Service, he once more informed Fries that h
would try to take up his post as head of the laboratories of the Ga
Service in December—a concession, he later wrote Cornelia, whic
he thought wise to make.[33]

On the eve of leaving Béthune late in October and ending
shared experience with people he had come to like and depend o
Cannon expressed a twinge of regret: "Fraser learned today that h
is to be relieved, and so we dismantled the laboratory togethe
packing our instruments, gathering up our books and papers, an
in general making the little room look like chaos. We have ha
nearly four very pleasant months together in the room and we bot
felt a dim shadow of sadness in pulling the place to pieces. No tru
friendship could grow between two men than has sprung up be
tween us."[34]

With the results of his latest experiments in hand, Cannon se
out for London certain that acidosis was a primary cause of shoc
and that an intravenous infusion of sodium bicarbonate was
therapy that could help severely wounded soldiers through neede
operations.

London was a crossroads that provided Cannon with welcom
change. Soon after his arrival in October, he had opportunities t
dine at the Royal Society and to exchange ideas about medica
research and education with Lord Haldane, Walter Fletcher, an
Joseph Barcroft. He also enjoyed chance meetings with Harvar
colleagues and former students. Best of all, he was able to talk wit
Bayliss and Starling about the current status of work on shock. I
preparation for his research on acidosis with Bayliss, Cannon bega
to write up the results of his investigations at Béthune, and to stud
and review the literature on shock that had not been available t
him in the field. He looked forward eagerly to participating in th
discussions of the M.R.C. Shock Committee.[35]

When the committee met, Cannon said little but listened prudently to the views and observations of others. Ernest Cowell and Keith Walker reported their experiences at the front; F.A. Bainbridge and E.H. Starling described their efforts to produce shock experimentally; H.H. Dale and A.N. Richards explained their analy-

tissues, thus reducing venous return and cardiac output. Shock produced by histamine, he concluded, appeared "to be a condition in which the circulation failed, and the arterial pressure fell to a low level . . . because the blood drained away into the capillary network and tended to stagnate there, instead of returning to the heart through the veins."[38]

Convinced that he was on the right track with his research on histamine, Dale from the outset was skeptical of Cannon's belief that acidosis was the cause of shock. Earlier that fall, some weeks before Cannon arrived in London, Dale informed the Shock Committee that experiments he had conducted with Richards indicated that acidosis did not lead to a serious fall of blood pressure, as Cannon had claimed.[39] At this time, however, Dale, a consummate politician, did not press an attack against Cannon's theory. Instead, he decided to wait for the results of the latter's corroborative experiments with Bayliss.

Cannon and Bayliss began their work together in late December of 1917 and as the experiments continued into the new year, they appeared to confirm Cannon's original observations that as the alkali reserve diminished, blood pressure fell to shock levels. In one series of experiments in mid-January, they injected hydrochloric acid of varying concentrations into cats that were decerebrated to obviate the need for continuing anaesthesia, and Cannon recorded in his diary, "Bayliss and I again show acid lowers blood pressure to state of shock." A week later, in a new experiment that simulated wounds occurring in combat, Cannon tried to produce shock by traumatizing the muscles of an anaesthetized cat. "A great day," he observed afterwards. "Acidosis developed and as it reached about 38 percent CO_2 capacity—shock blood pressure."[40]

Success followed success, and by the end of January Cannon wrote home: "Confirmation of the crucial experiment on the effects of injured muscles in producing shock and in reducing the alkali reserve has continued. I wrote to Surgeon General Wallace about the observations, and was pleased to see in his answer an agreement with my own opinion that we have probably got the key to the mystery of shock. These are exciting days, almost every one of them brings forth something interesting and significant."[41]

Brimming with confidence, Cannon wrote up the results of the experiments he had conducted with Bayliss, thinking to press the Shock Committee to publish a report of their work and circulate it without delay. What happened next made such an impression on Dale that many years later he was able to recall the entire incident in vivid detail:[42]

> In the evidence, as put before the Committee, I could find no adequate ground for believing that the condition produced by infusing acid into the cats, under urethane, had any necessary relation to the genuine wound shock with its circulatory failure admittedly due to oligaemia; or, on the other hand, for believing that the latter condition was due to acidosis, instead of the acidosis being the result, or the concomitant of the oligaemic shock . . .
>
> I was deeply concerned, however, to avoid a hasty distribution of a Report, leading to conclusions which, put forward by physiologists of such standing, would be immediately accepted and applied, with results which I should have expected to be, at best, tragically disappointing, to involve us in endless arguments and excuses, and then to produce indefinite delay and difficulty in persuading the army authorities to accept other and more surely based methods for trial.

The committee subsequently agreed to postpone any action for several days. Then, certain that only a practical demonstration would change Cannon's mind, Dale suggested a precise repetition of the experiments with acid infusions. He continued his memoir:

> I proposed that Bayliss and Cannon should come round to my laboratory at the Lister Institute on the following morning (a Saturday); Richards and I would have a cat available and Cannon, who was justly proud of his own expert technique for the purpose, should enable us to get the cat tied down peacefully on its back without anaesthesia . . . I assured them that Richards and I would accept the result of such a test as decisive, and hope that they would feel able to do the same. They accepted, as, I think, they were bound to do.
>
> Everything went according to plan; the unanaesthetized cat received its standard infusion of acid with no obvious effect, except the dyspnoea, which was to be expected. I offered to continue the infusion further, but accepted Cannon's decision that it should be stopped at the completion of the standard dose; and the cat, being then released, jumped down from the table and ran into a corner. As we had a boy available, who could be left to look after the cat and ensure that it did not stray, I suggested that we might all go out to luncheon, and come back to see whether symptoms had by then developed.

When they returned to the laboratory, the cat seemed to be perfectly fit, and they decided to examine it under ether anaesthesia. It was found to have, indeed, an unusually high arterial pressure with a pulse of full volume. In order to determine the alkali reserve, they took a sample of blood and used the Van Slyke outfit Richards had brought with him—"and this was found to be still far under the 'critical level' below which, according to Cannon, the animal should by now have been in a state of profound shock."

Cannon's contemporaneous view of the event, not surprisingly, was more restrained than Dale's memory. "The week has been a strenuous one, and rather exciting," he wrote home, giving a brief account of the experiments. "It looks as if there were two factors at work, the acidosis and something else, possibly a substance given off by the injured tissues, which cooperate, and that under urethane the cooperating substance is provided."[43]

If Cannon appeared to hold onto acidosis as a factor in producing shock, Dale was more definite about his interpretation of the experiments conducted in his laboratory. "By this time," he later recalled, "our friends were ready to 'throw up the sponge' and to agree with the view which we had been urging them to consider, that the low alkali-reserve, which had been observed in men suffering from wound shock, must be a consequential symptom of the circulatory defect, and could not be the cause of it."

Within days of Dale's demonstration, Cannon, in response to orders to report to the American Gas Service, left for France. Although he had been preparing to go for some weeks, he departed with mixed emotions. Just before leaving, he wrote Cornelia, "This is the first time that I have had to break away from a deep interest and do something I don't like to do. I appreciate anew the value of university freedom of research."[44]

2 The Mystery of Shock

II: Resuscitation

WHEN IN THE FALL of 1917 Cannon gave his word that he would join the Gas Service after the completion of his experiments on shock, he began to study the rapidly growing literature on gas warfare and to talk with English physiologists who were engaged in research on its problems. Although he had doubts that his abilities and previous experience would be useful, some of his British friends thought otherwise. Lord Haldane, for one, assured him that by means of an unconventional approach he might see things that others had overlooked—and even urged that it was his duty to go. Talks with Joseph Barcroft, who was deeply involved with research on war gases, had an even more telling effect, and at one point Cannon grew almost enthusiastic about his prospects. He explained to his wife that at least one aspect of the subject was related "with uncanny nicety" to his work on shock, and that some ideas were "almost directly transferable to the new field of interest."[1]

But in February 1918, soon after he arrived in Paris, any hope that Cannon had of doing useful work in the Gas Service went up in smoke. When he asked to inspect his new laboratories, he was told that the instruments and materials on order for physiological research were not likely to arrive before April; worse, there was no place yet for him to work. Never one to idle in any circumstances, Cannon sought to continue his investigations on shock and, through the kindness of André Mayer at the Collège de France, was given laboratory space "only a few hundred yards from the little rooms where Claude Bernard carried on his researches for more than thirty years."[2]

By the end of the month, however, worn down by a cold and unable to work, he became depressed about his prospects. His frame of mind grew even gloomier when he discovered through conversations with officials in the Gas Service that he was expected to test the new gases devised by chemists and to report on their effects. In his view, such research was at best routine, the kind that anyone could do.

When he assumed the duties previously bestowed on him by Alexander Lambert in behalf of the American Red Cross Society, however, Cannon's gloom lifted. As chairman of the Society's advisory committee, he presided over the monthly scientific meetings and conferences held in Paris, taking part in the discussions on all of the surgical, physiological, and bacteriological research conducted by the allied armies. The meetings that took place in March presented him with an opportunity to speak publicly about the research he had conducted at Béthune and London on shock. While he candidly reported that Dale and Richards had not been able to corroborate the results of his experiments with Bayliss, he did not indicate that he was willing to discard his theory of acidosis as a cause of shock. Instead, he offered an emendation based on the results of his experiments on traumatizing the muscles of anesthetized cats; namely, that in addition to acidosis there were toxic substances released from injured muscles that could lead to shock.[3]

As he was preparing for these meetings, Cannon received news that presaged a turn in his fortunes. The War Department informed him that he had been promoted to Major—a rank that he had long believed was more commensurate with his duties and responsibilities. Of greater consequence, J.M.T. Finney of the U.S. Army Surgical Service asked if he would direct a center for surgical research then being planned in association with the Central Medical Department Laboratory of the A.E.F. at Dijon. The new post would allow him not only to continue his experimental research on shock but also to transmit first-hand to American medical personnel the lessons he had learned at Béthune in caring for shocked cases at the front. Cannon immediately accepted. The only problem that remained was his transfer from the Gas Service to the Surgical Service, a matter that Finney arranged with little difficulty.[4]

When Cannon learned of his appointment to the surgical laboratories at Dijon, he notified A.N. Richards that he had again recommended him for his job at the Gas Service. Cannon's letter, however, did not reach Richards for two weeks. In the interim, a Colonel Delancey, on behalf of the Gas Service, personally informed Richards of Cannon's recommendation, adding that he would have to accept or reject the offer within twenty-four hours.

Richards was furious. He had heard nothing from Cannon; he knew nothing of the job requirements or if he even had the expertise to carry out his future duties. Indeed, he felt that Cannon had, for his own purposes, saddled him with a job he did not want, and worse, could not decline. Walter Fletcher and Henry Dale at the M.R.C. were aghast when they heard of Richards' situation, and with good reason. From his arrival in London during the summer of 1917 Richards had become Dale's valued collaborator, and it was plain that his departure for the Gas Service would seriously affect the progress of Dale's research on shock.[5]

Fletcher, sensing the prospect of a personnel crisis, immediately took up the cudgels. If he thought he was writing in Richards' behalf, he had a very strange way of expressing himself. "The ideal man for your job," he informed Cannon in a letter of April 2, 1918,

> is a man with good general chemical and physiological knowledge, practical training in pathological and clinical work, having great vigour and push, combined with a thick skin and the ability to suffer fools gladly and answer letters promptly. Now Richards has very few of these qualities. He is a highly specialized laboratory pharmacologist. He is not a good chemist, nor a pathologist, nor a clinician. He could not attempt to organize or criticise other men's work in these directions and could probably not make a creditable report upon an autopsy himself, up to any good hospital standard. He is a man of native modesty with a good deal of diffidence, he has no particular vigour and decision in speech. He is a highly strung and sensitive man and he has no training in the push and go side of administration.

"On the grounds both of personality and expert knowledge," he concluded, "you are trying to put a round man in a square hole if you ask for Richards."

It must be said that these comments were uncharacteristic of Fletcher, a man of sterling character who was normally known for his tact and common sense. Cannon was well aware that Fletcher's letter had less to do with Richards than with protecting the productivity of Dale's laboratory, and his rejoinder was written in a tone to disarm rather than encourage controversy. He took special pains to reassure Fletcher that Richards, even if called to report to the Gas Service, would have ample time to complete his work with Dale because the physiological laboratories were still unfinished, and apt to remain so for some time.[6]

Dale, too, wrote Cannon, in effect charging him with arranging matters for his own benefit and to Richards' great disadvantage. "I know that he is feeling that . . . you have lost interest in him and the post into which you have landed him. Probably postal delay accounts for his not hearing from you; but, in case you haven't written, I thought I ought to write frankly as to the impression your apparent silence is giving. Do what you can to get Richards free to go home."

When Cannon's letter to Fletcher arrived at the Medical Research Council on April 20, Dale quickly wrote Cannon again to explain as best he could the circumstances that had led him to write his accusatory note. He pointed out the difficulty that Richards had of getting any clear indication of the kind of work he would be doing in the Gas Service, as well as the doubts instilled in him by Fletcher's evaluation of his qualities as an investigator and administrator. "I shall do everything possible," Dale promised Cannon, "to confirm his confidence in his qualifications for the work and its importance, for the interests which after all are the only ones which count with any of us . . . I hope you will forgive me. I am very anxious to see matters straightened out in all directions and [Richards] in the frame of mind to take up the new work keenly and eagerly."[7]

Notwithstanding the reasonable tone of Dale's new letter, the damage had been done. In his reply, Cannon confessed to feeling hurt, and he took pains to counter the suggestion that he had somehow contrived Richards' appointment to the Gas Service to get out of an unpleasant job. "You state," he concluded,

> that Richards came over with quite other views as to the length of his stay than those that are opened by the prospects of this new job and that I should do what I can to get Richards free to go home. Do you really think that this is the proper attitude for any one to take at the present moment when there is every need for such service as can be performed by disciplined men? I confess that it does not seem so to me . . . It will be a year next week since I left home and I see no prospect of returning before the end of the war. I should enjoy as

> much as anybody else the pleasures of living in my own home and the delights of watching the growth of my own children, delights which are being forever lost, but this is not a time when any one who can take his part in the great struggle should consider for a moment his own personal wishes.

The depth of Cannon's anger can best be gauged by a comment he made the very same day in a letter home: "It is all very well for Dale to write as he does, he hasn't been away at all since the war began . . . At least he will get a letter that he will have to answer, and I shall be much interested in seeing what he writes."[8]

Fletcher, for his part, passed off the entire affair with administrative aplomb. "All of the particulars of the work which you were good enough to sketch for me," he wrote Cannon on May 7, 1918, after receiving his letter, "make the greater part of my previous letter to you irrelevant. I am very glad that all this appears now to have been happily settled." Dale, however, had a more difficult time. The accusatory note he had written Cannon had bitten too deeply to be passed over. More would have to be done if the ongoing collaboration between the M.R.C. and Cannon was to continue.

Dale met the problem with a carefully crafted letter of apology. He told Cannon that Richards was not responsible for what he (Dale) had written, that the "consideration that he was unexpectedly caught for this work while over here for my benefit was one that was worrying me rather than him. Believe, if you can, that I was actuated solely by the desire to see the position made definite and his mind at rest, and to avert a possibility of misunderstanding between you and him." Admitting that he had behaved like "a foolish busy body," Dale ended, "If your generosity can go so far as to forget the whole affair I shall be more than grateful."[9]

Despite all the fuss, the problem of Richards' prospective transfer to the American Gas Service soon melted away. Richards wrote Cannon several letters, explaining from his perspective how the tangle had arisen. "It seems to me," he observed in one letter, "that if your first note had reached me a day earlier so that I might have known that an instant decision was not so vitally necessary as I was led to believe, or if your note had contained any information which would have given me some ground for decision as to my ability to fill the post, and which thereby might have headed off Fletcher's criticisms, the misunderstanding need not have arisen."[10]

Whatever private feelings may have remained unvented by each of the principals in this brief but sorry episode, all were united in their desire to let bygones be bygones and get on with the work of the war effort. In a matter of weeks the furor ended as abruptly as it had started.

Cannon's call to Dijon was a great stroke of luck. For the first time since he had returned to France, he did not have to persuade his superiors of the importance of the research he wished to do. Colonel Merritte W. Ireland, the chief medical adviser to General Pershing and soon-to-be chief surgeon of the A.E.F., and Colonel J. F. Siler, who was to be the military commander of the Dijon laboratories, knew well the importance of meeting the needs of those engaged in investigation. Cannon was clearly elated when he wrote home, "The laboratories can have anything that is desired—men detailed from any unit, materials at whatever cost, and transportation of the most effective sort."

Equally important, he found some of the personnel who were already at Dijon to be kindred spirits: Hans Zinsser, a distinguished microbiologist from Columbia University; Paul De Kruif, a young immunologist who had trained with Frederick Novy at Michigan; and Jack L. Yates, a thoracic surgeon from Milwaukee, who had previously collaborated with Cannon in his struggles against the antivivisectionists in Wisconsin. Beyond their common interests in research, Cannon found to his delight a stimulating variety of opinions that the group had on social and political issues. "Last evening after dinner," he reported to Cornelia, "we sat around the table and had a long and very lively discussion about democracy, the dangers of labor control, compulsory military service, and the probable effects of the war on the United States . . . We all disagreed heartily. It was almost the best conversation I've heard since coming to Europe, and I was delighted with the prospect of such company. Since the new outlook opened, my whole feeling has changed. I can push ahead now with interest, enthusiasm, and confidence."[11]

Still, when Bayliss wrote him that he had not been able to confirm their earlier experimental results on acidosis, Cannon found himself in a dilemma. "It certainly looks as if acidosis plays a secondary role, if any," he confessed to Bayliss, beginning a cautious retreat from his previous position. But then he added equivocally, "I am inclined to continue to give sodium bicarbonate by mouth to injured men and to reserve intravenous injection of it to the seriously wounded who have to be operated upon before their blood pressure has been restored."[12]

After he received the letter from Bayliss, Cannon concentrated his research more than ever on ideas for treating shock. These included, among others, the best methods of applying heat to the wounded to prevent the onset of secondary shock, the utility of gum acacia in restoring blood volume in the seriously wounded, and the effect of light versus deep anesthesia on blood pressure in a state of shock. While each of the problems was important in its own right, he understood that the ultimate success of his labors depended on maintaining good relations between the laboratory and surgeons at the front. To strengthen those relations, he decided to establish a network of resuscitation teams to undertake the care of shocked soldiers on or near the battlefield.

The project was daunting. Since neither the British nor the French had such units, there were no existing models that Cannon could rely on as a guide. The physicians and surgeons that he wished to train presented further difficulties; some were poorly educated, others had no experience with sphygmomanometers or knew little of blood transfusions. Cannon addressed such problems as they emerged and, with his usual energy and enthusiasm, organized one-week training courses consisting of lectures on subjects such as the physiology of circulation, hemorrhage, anesthesia, and the principles involved in blood transfusion. He accompanied these lectures with a set of demonstrations designed to illustrate the various elements in the experimental production of shock, as well as those involved in the care of patients in shock.[13]

As he worked on the training courses, Cannon found himself wrestling with a problem that stemmed from an honor that had previously been bestowed on him. Before he left London in February, he had received an invitation to deliver the Croonian Lecture to the Royal Society later that spring. Cannon took great pleasure in the appointment. It appeared to him a sign of approval by members of the M.R.C. and put him into the pantheon of distinguished medical scientists who had previously given the lecture—men like John Hunter, Thomas Huxley, Joseph Lister, Emile Roux, Edward Schäfer, and Charles Sherrington. Initially Cannon thought he would lecture on the research that seemed to establish acidosis as the cause of traumatic shock. But six days after his appointment, that prospect had crumbled with Dale's trenchant attack on acidosis during Cannon's final meeting with the Shock Committee.

Casting about for a new topic, Cannon was struck by a vivid memory of wounded soldiers he had seen at Béthune desperately calling for water, and it occurred to him that a lecture on the physiology of thirst might fill the bill. He quickly sent for the notes he had gathered for lectures on thirst to medical students. When the notes reached Dijon toward the end of April 1918, Cannon was busy attending meetings of the Red Cross Society in Paris, as well as training resuscitation teams to treat shock at the front lines. Using what little spare time he had, he began to sketch out his Croonian Lecture.[14]

By mid-May he was using himself as a guinea pig in experiments he hoped would reveal the mechanism that caused thirst. A month later he finished the lecture and, with some trepidation, set off for London. The lecture Cannon gave—essentially a historical and anecdotal account of the various theories of thirst—did not impress the audience, and save for a polite tribute by Sherrington, who chaired the meeting, little was said. "Something was wrong," Cannon confided to Cornelia, "for no other person mentioned the Lecture—not even Dale beside whom I sat down when I finished, nor Bayliss, with whom I have been staying. I should like the approval of my own public—the physiologists—but they seem to have withheld it. And I go back to France disappointed."[15]

Cannon returned to his program of training resuscitation teams, which soon attracted a steady stream of important visitors to Dijon and imparted to the laboratories the aura of a busy beehive. He was delighted with the official attention but dismayed by all of the interruptions he had to endure because "everybody wants to show us off." The visits, regardless of his complaints, did not perceptibly affect the progress of the research he had begun in collaboration with his former student McKeen Cattell on the relation of anesthesia to the development of shock.[16]

Some of their results tended to substantiate Dale's and Bayliss's rejection of acidosis as a cause of shock, and Cannon admitted as much in a letter to Bayliss: "The increase in acidosis accompanying anesthesia which I noted last summer in Béthune is apparently the consequence of the fall of blood pressure and not the occasion for it. You can see that these observations give quite another turn to my thinking and are pertinent to the work which we were doing last winter."

Several days later, he wrote to Dale in the same vein: "The sharp drop in pressure and increase of acidosis which I observed last summer may be accounted for by sensitiveness of the control of the circulation in the shocked animal to ether anesthesia. I recall that you and Richards made a few observations on this point last fall. You can see that somewhat tardily I am drifting towards your view of the relation between anesthesia and the action of acids."

Dale was pleased with the letter and confessed his relief that Cannon was nearing accord with his position—"namely, that low blood pressure produced in any way will lower the alkali reserve, while lowering the alkali reserve does not, per se, produce low blood pressure. Bayliss has for some time past been in agreement with this and it is cheering to know that you have moved so far in the same direction." Best of all, Cannon's changing attitude promised that Dale could write a memorandum on the problem of acidosis and shock that would be acceptable to all members of the M.R.C.'s Shock Committee.[17]

As the resuscitation program progressed, Cannon applied for permission to travel freely to forward areas to inspect the performance of his trainees under conditions of battle. Although Finney and Siler enthusiastically endorsed his inspection trips, Cannon to his consternation found himself thwarted by a jealousy that had begun to manifest itself in a conflict of authority between regular army men and "civilians" like himself. For a time, he had to comfort himself with the hope that what was best for the wounded soldiers would prevail. But as the differences continued to erupt and Cannon's frustration grew, Siler, in an effort to breach the barriers placed in his way, assigned him to duty with the medical unit of the 42nd "Rainbow" Division (an American division so-named because its troops came from all over the country rather than from any one region) in the vicinity of Chalons-sur-Marne.[18]

Within days of his arrival early in July, Cannon found himself in the midst of a fierce battle. He had been under aerial bombardment and shellfire at Béthune, but that experience did not match the savage day-long bombardments of this German drive. During a week of unbelievable horror and fatigue, he somehow managed to keep a hastily-written record of his experiences for Cornelia. All through the battle he worked round-the-clock with the shock teams as the front-line hospital reeled under waves of intense shellfire and a flood of wounded men. When a particularly heavy bombardment all but destroyed the hospital, Cannon kept right on working, "stirred by the consciousness that I was in the midst of a great adventure, the end of which could not be foreseen." As soon as possible, orders came to move everything—the wounded, the staff, the supplies and equipment—to Écury, thirteen miles from the front lines. It hardly proved a respite.

The moment he arrived at Écury, Cannon was directed to the shock ward, which was "already filled with sickening sights associated in my memory with Béthune . . . Men with their bellies torn open, with the sides of their faces ripped out, with brains oozing from skull wounds, with the bladder shot through, with sucking chest wounds—these were the hopeless, pathetic cases left in the shock ward." By the second day, those who had not already died were "likely to be weakened and septic, and restless with approaching death. Cursings, expletives, repeated shouts for 'mother,' or for 'mama' or 'papa', and the stench of septic wounds and dressings, make the shock ward a desperate and depressing place."

Under the stress of battle, Cannon sometimes had difficulty controlling his anger against the hated "Huns" who were responsible for such wanton destruction. Once, when a "very spiritual looking" medical officer visited and asked what was being done for two German prisoners in Cannon's care, he replied stiffly that "there were hundreds of Americans, severely wounded, who were waiting operation, and who were to have first chance." The visitor came again and finding that "the Huns had not been touched, he said, 'It's a shame!'" Cannon dismissed the remark as "contemptible sentimentality," then paused to reflect, "But I suppose I'm no Christian or a very poor one." As the horrors of death and mutilation became fixed in Cannon's mind, he lashed out against the enemy, frequently damning the "Huns" and all their works. After he watched a German plane shot down by anti-aircraft guns and set afire, he admitted "to a feeling of grim satisfaction at seeing the sight." Disturbed by the depth of his own emotion, he asked in wonder, "Have I become totally a savage?"[19]

Cannon gained from his experiences a deeper understanding of the difficulties involved in caring for the wounded in shock. Overall, he was pleased by the performance of the resuscitation officers he had trained. Nevertheless, he saw that their effectiveness was hampered by what happened to the wounded at front-line aid stations. Triage officers had difficulty separating moribund shock patients from those who might yet benefit from evacuation and treatment; some surgeons still did not understand that it was critical to keep the wounded warm and rested in order to prevent the onset of secondary shock. Then, too, he observed that base hospitals needed more than one resuscitation team to care for shock cases during fighting, and that it would be well to establish clinical pathology laboratories at the front lines to send back data for the subsequent treatment of patients at the rear. In sum, he was convinced that the resuscitation program should be expanded.[20]

The period that Cannon spent with the 42nd Division had other consequences. When he got back to Dijon, he was annoyed to find that the resuscitation teams had not been adequately trained during his absence. It became clear that he needed a second-in-command—someone of comparable status who could not only take over the instruction of resuscitation teams when he was away but also work independently on devising new therapeutic methods in dealing with shock.

There was no doubt in Cannon's mind as to the man he wanted to join him. For some time he had been impressed by the work of Oswald H. Robertson on preservation of blood for use in transfusions on the battlefield and in military hospitals. Robertson's investigations led to the development of techniques for preserving red blood cells; he also emphasized the importance of making up the loss of plasma volume by administering fluids by mouth or rectum in cases of hemorrhage. Although Robertson, who had also come to France as a member of Harvard Base Hospital No. 5, had previously expressed an interest in working with Cannon, he preferred to continue his studies on blood volume with Arlie Bock at Boulogne, where he was attached to the British Medical Service. Robertson's pioneer work in setting up what was to become the world's first blood bank had such practical application for treating the wounded and saving lives that he was highly valued by the British. "Do please leave him with us if he can possibly be spared," T. R. Elliott pleaded in a letter to Cannon when he heard of the plan to call Robertson to Dijon. "If he goes, I am bereft and desolate."[21]

Cannon, however, was not to be put off. Some weeks later, with Col. Siler's help, he succeeded in arranging Robertson's transfer to the laboratory of surgical research. When Robertson finally reported for duty at Dijon, he was to prove invaluable in untangling a problem that increasingly troubled surgeons in the field and at base hospitals. It concerned the efficacy of gum acacia in the treatment of shock.

Even before his stay in Bayliss's laboratory, Cannon had been convinced of the utility of gum acacia in restoring blood volume in shock. Later, in his lectures to resuscitation trainees at Dijon, he stressed the importance of its use at the earliest appearance of shock. But during the battle of Chalons-sur-Marne, although he had seen excellent results from infusions of gum acacia, he was taken aback by several cases with unfavorable effects, such as marked shivering and cyanosis, following its use. Even though these symptoms were temporary, Cannon reported his observations to Bayliss.

Bayliss had on occasion seen similar effects in some of his experimental animals and made little of the phenomenon. Instead, he alerted Cannon to a new development of mutual interest. "I saw McNee yesterday," he wrote. "He has been in charge of a resuscitation ward and made some interesting observations, which I have asked him to send a report of to the Shock Committee. One thing is that he is convinced of the very serious result of absorption of toxic products from injured muscle. He finds that removal of the injured tissue or even application of a tourniquet is followed by marked benefit."[22]

This news from Bayliss prompted Cannon to return to his laboratory investigations. "During the past three days," he wrote home early in August, "I have taken up again the experiments that Bayliss and I were doing last winter on the relation of muscle injury to shock. There is no doubt that by injuring muscle, shock can be produced. I am trying to analyze the conditions somewhat; Bayliss and I will publish a paper together on the phenomenon." Each day brought both new and interesting results until, little more than a week later, Cannon submitted a report of his experiments to Army headquarters—in effect, suggesting that a circulating toxic substance released by injury to the muscle might well be the cause of shock. The results of Bayliss's independent experiments pointed in the same direction, and he wrote Cannon, "I quite agree that the results must be due to the chemical product of something like autolysis."[23]

To follow up his assumption that the fall of blood pressure

accompanying tissue injury might be due to early products of protein cleavage, Cannon asked Captain Joseph Aub, one of his former students who had been studying metabolism and venous pressure in shock and hemorrhage, to join him in an experiment to determine if peptone and protease injections could modify shock.[24]

Throughout that September of 1918, Dale apprised Cannon of his progress in preparing the memorandum on behalf of the M.R.C. Shock Committee that laid to rest any claims for the theory that acidosis was the cause of shock. Cannon accepted the memorandum with only minor emendations, and explained his rationale for the use of sodium bicarbonate:

> I saw magical effects from these injections in cases which had reached the stage of air hunger. And I saw also that operation sharply lowered the alkali reserve, so that at the end men were sometimes gasping for breath. It was for the sake of saving them from these conditions that I proposed giving a prophylactic injection of Na-HCO_3. I believe now that the same benefits are obtained better by injecting blood or gum-salt solution at the beginning of operation, to hold up the blood pressure.

Cannon also appended a note of thanks to Dale for saving him from error. Dale, who later characterized the note as "generous and warm-hearted," responded in kind:

> It was quite evident that we who had time to go calmly into the matter in England had an advantage over you who were obliged to form your ideas on observations made in the stress of work at a C.C.S., supplemented by a few weeks of high-pressure experiments in England. The great cause of satisfaction to me is not that you came to agree in the end with our first-formed opinion, but that we are at last all on the same track.[25]

There were, however, questions that remained to be answered—particularly about the relation of muscle injury to shock, as well as the therapeutic efficacy of gum acacia. Late in August Cannon, intrigued with what he had discovered in his experiments on muscle injury, had accepted an invitation from Charles Richet to speak on shock at the Société de Biologie. It was Cannon's hope that a forum with the French might provide information to support his growing conviction that toxemia following muscle injury was a cause of shock.

Although some of the new experiments he conducted in early September gave equivocal results, Cannon was nevertheless heartened by a closer relationship with Dale, who sent him news of his own recent experiments with histamine-induced shock. Early in October Dale wrote, "I think we shall all in the end come to believe that there is more in the effect of retarded circulation than the mere deficiency of oxygen supply. Your further experiments on muscle injury greatly interested the Committee . . . I think we shall end by establishing the absorption of toxic products from ill-nourished tissues as at least *one* factor of importance in shock." A short time later Dale pressed Cannon to put together an account of his latest experiments on muscle injury and added, "We have enough material in sight for what I believe will be a valuable symposium on the relation to shock of absorption of autolytic products from injured muscles."[26]

At the meeting with the Société de Biologie in Paris, many of those present gave strong support to the idea of toxemia. Eduard Quénu, a distinguished surgical pathologist, maintained that shock was primarily the result of contused wounds of large muscle tissues, which set toxic substances free that permeated the circulation. In his experience, shock disappeared after he excised the injured muscles. Some of those present agreed with Quénu that shock could be prevented by amputation or removal of the injured tissues. "Do not wait for shock to pass," one surgeon urged, "operate to suppress shock."

Others, however, were not convinced. A number of surgeons argued that shock was a nervous phenomenon—the result of reflex inhibition—a notion that went back to Weir Mitchell's observations in the American Civil War. Neither Cannon nor Dale, who received an abstract of the discussion, thought the latter position had much merit and largely disregarded it. But almost twenty years later, Cannon would concede in the *Annals of Surgery* that pain, persistent fear, and great emotional excitement might also play a role in the induction of shock.[27]

In the autumn of 1918, surgeons in the field were concerned about the effectiveness of gum acacia in treating shock. Some claimed that blood transfusion was of far greater value; others indicated that their use of gum acacia was followed by ill effects, such as vomiting, chills and urticaria (or hives). Still others asserted that it was responsible for a number of deaths. Chagrined by the complaints, Bayliss sought an explanation in renewed experimentation, but to no avail. "The only explanation that occurs to me," he wrote Cannon, "is either that the solutions which gave chills were made with stale distilled water or that some accidental contamination occurred." The answer to the problem, he concluded, might well lie in the quality of the gum used.[28]

Under pressure from headquarters to get more definitive answers to the mounting tide of criticism, Cannon put Paul De Kruif to work in the laboratory on problems of toxicity of gum acacia. De Kruif, a well-trained bacteriologist, after studying the effect of various samples of gum acacia on guinea pigs, reported that properly prepared gum acacia had no demonstrable toxic effects and was relatively innocuous.[29]

Almost simultaneously, Cannon dispatched Oswald Robertson to forward hospitals to obtain more reliable clinical data. On the basis of observations made at evacuation, mobile, and field hospitals, Robertson reported that the consequences of using gum acacia depended not only on the nature of the case treated but also on the manner in which gum acacia was administered. When gum acacia was used early in cases of shock with moderate hemorrhage, officers in the field recorded favorable results.

The causes of failure, however, were far more complex. Robertson noted that some cases—especially those that were in a state of shock with low blood pressure for long periods of time during evacuation—were invariably beyond the help of either blood transfusion or gum acacia. He also found that patients in shock who were concomitantly infected with gas bacillus did not respond to treatment. At other times failure appeared to be the result of carelessness—the gum was given too rapidly or not sufficiently warmed—but he conceded that it was difficult to trace any causal relation between the gum-salt solution and death of a patient.

Perhaps the nub of Robertson's report was that most judgments about the use of gum acacia were impressionistic and not based on actual data as to blood pressure changes, hemoglobin concentration, etc. "This is due," he concluded, "partly to the lack of apparatus for these determinations (blood pressure apparatus is especially needed) and partly to the fact that methods commonly used for determining the degree of shock and hemorrhage do not give sufficient information as to the fundamental state of the circulation. It is most important to know the amount of blood loss and the total hemoglobin left in the body." But his experiences had convinced him that prevention of shock was critical; and he advised Cannon to train divisional medical officers whose business it would be to "carry the instruction in shock treatment to the front line, and see that the necessary measures are carried out."[30]

Robertson's advice reinforced what Cannon had long been preaching to headquarters. "Ever since early in July," he wrote his wife, "I have been urging by letters, by telephone messages and by complaints, to Colonel Siler, General Thayer and General Finney, the sending here for instruction in resuscitation medical officers from divisions . . . I made ructions again after Robertson returned and reported occurrences that made one weep . . . and at last on Sunday twenty-five divisional doctors appeared and yesterday twenty-six more.[31]

In late October rumors were rampant that the war could end in a matter of weeks or even days; yet Cannon continued his instruction of resuscitation officers much as he had before. When he accompanied Simon Flexner of the Rockefeller Institute and Alexander Lambert of the American Red Cross on a tour of hospitals at the front in early November, it appeared that the war was almost at an end. The booming of artillery could be heard in the distance, but at some of the hospitals, they discovered that no new cases had come in for days, and at others, it seemed as though personnel had gone off duty. Then, as they passed through a small French village on November 11, they learned that the war had ended. "The war, the Great War, was over! The incredible had happened," Cannon later wrote home. "The first shock brought relief, rather than shouting, and then a sense of wonderment as to how events might move."[32]

Although the fighting stopped and there were no longer any wounded, some of the problems of shock remained a mystery. Four days after the Armistice, the British Army in France held a conference at Boulogne for medical personnel to discuss the relative merits of blood transfusion and gum acacia in treatment of shock and hemorrhage. Most of those who attended—generals in the field like Cuthbert Wallace and Sir Anthony Bowlby, or medical consultants like T. R. Elliott, or surgeons serving at casualty clearing stations like J. W. McNee, Ernest Cowell, and Geoffrey Marshall—knew either through personal contact or by reputation Bayliss, Dale, and Cannon, who came as representatives of the British and American scientific establishment.[33]

The criticisms of gum acacia put forth at the Boulogne conference were for the most part subdued or tentative. Few of those whose opinions were reported by General Bowlby allowed that in their experience there were any "bad effects" of infusing gum intravenously. The comments of Colonel George Gask, a respected surgeon and consultant to the British Fourth Army, were typical. Gask indicated that he had never seen any ill effects from the use of gum acacia supplied by the Boulogne Medical Laboratories, but he said nothing of the effect of other gums that might have been used. Colonel Edgar Pilcher, a surgical consultant to the British Third Army, voiced similar sentiments, but then added that he was not in favor of using gum. Casualty clearing station officers who were present at the meeting were equally equivocal. Although some admitted they had seen "chills and rigors" after using gum, they passed them off as "temporary distress." Major J. W. McNee, despite having witnessed several deaths after an infusion of gum, quickly added that he was certain gum was not the cause of death. T. R. Elliott, who earlier had warned Walter Fletcher at M.R.C. headquarters in London, "Don't lay too much stress on gum, it is never as good as blood," said nothing at Boulogne. Bayliss and Dale, too, held their tongues, contenting themselves with keeping a record of comments for future consideration by the Shock Committee.

Cannon took a more active role in the conference. Using Robertson's report as a guide, he delineated the wide range of favorable and unfavorable opinions of American surgeons in regard to the utility of gum acacia, and catalogued the various factors and situations that contributed to its success or failure. The criticism implicit in Cannon's presentation was not about defects in the gum or its purpose, but rather of the lack of judgment on the part of surgeons as to how and when to use gum acacia effectively. He took great pains to emphasize that prevention was of the utmost importance; during evacuation of the wounded from the front lines to the rear, a few simple measures could prevent the onset of shock.

Opinions expressed at the conference on the relative value of the use of gum as compared with blood transfusion were equally controversial. Some believed that when the patient was in obvious shock, blood "ought to be given." Others strongly favored blood transfusion over gum infusion in most circumstances. Still others were in a quandary over how much blood to give. Bayliss made no comment about the use of blood as compared with gum, although when he had given the Oliver-Sharpey Lectures in May 1918, he pointed out that blood was the most appropriate fluid to replace blood. Indeed, after reading a preliminary report of O.H. Robertson's research on blood transfusion, Bayliss wrote Cannon: "I think it would be a pity if there were supposed to be any kind of conflict between the transfusion of blood and the injection of gum. It seems only natural that blood should be used when available and the value of gum seems to me to be rather in the fact that its use is simpler in times of stress and there need be no hesitation in using it, even in apparently hopeless cases, on the grounds of economy." It appeared to Bayliss that the chief thing was to enhance the volume of blood, "no matter how, so long as it is permanent."[34]

The minutes report that by the end of the conference, no definite conclusion was reached about the choice between gum solution and blood in resuscitation work. Actually, it appears that little time was spent on the subject of blood transfusion itself. Nothing was said about the giving of transfusions, or about problems one would have expected to be discussed—such as obtaining donors, matching blood types, preserving red cells, the utility of direct and indirect methods of transfusion, or the availability of transfusion sets.[35] Several months later in Great Britain, on the basis of notes submitted by Bayliss and Dale, the M.R.C. re-evaluated the discussion at Boulogne and came to an entirely different conclusion—namely, "that in all cases of haemorrhage with shock, transfusion of unaltered blood or citrated blood is the best treatment yet available."[36]

The Boulogne conference gave Cannon the opportunity to say good-bye to those he had worked with at Béthune and to enjoy conversations, perhaps for the last time before leaving France, with Bayliss and Dale. One week later, when he participated in an Interallied Surgical Conference at the Red Cross Research Society in Paris where discussion focused on anesthesia, shock, and transfusion, everyone felt "let down and uninterested." "Life is flattened out," he confessed to Cornelia, "and only the excitement of going home again can revive us."[37] Nevertheless, out of a sense of obligation to the Red Cross, he not only attended all of the sessions of the conference but also stayed on for meetings of the Research Society. Nothing new was discussed at either event, and afterward, he was glad to return to Dijon.

In the days that followed, Cannon worked on a history of the surgical laboratories and searched the mail room for his orders to return home. The wait was hard for him to bear. "I've had a troubled, depressing week," he began one letter in mid-December after a visit to Paris. "I've tried distractions. I've looked up literature on shock . . . Yesterday I went to Babinski's neurological clinics at La Pitie and got some insight into the permanent damage of the war—a poilu who had been shaking for two years since a shell burst near him, an aviator who had had a fall and was always dizzy, a young soldier with a wasted right arm—a few trifling examples of the war's wastage in human life." The visit added to the gallery of horrors etched in Cannon's mind.[38]

Several days later he received a letter from Dale that buoyed his spirits, for it addressed a subject that was dear to his heart—the prospect of future collaboration between British and American medical scientists. "May I say once more," Dale wrote,

> how much we have all valued your enthusiastic cooperation. It is especially pleasing to think that, after so much strenuous discussion, we come to what I suppose is the end of our formal association, without any material difference of opinion or attitude. I am sure that the effect of our association will be permanent in strengthening the feeling that we people on both sides of the Atlantic, who speak the same language, think along the same lines, have the same ideals and standards and understand the same jokes, must aim more and more at working together in essential matters like scientific investigation. People who can chaff one another without offence and differ without quarrelling are close colleagues, whatever their spatial or political separation. I am sure you agree.[39]

Cannon barely had time to savor Dale's letter when he received orders to prepare to sail for home on January 11, 1919. Neither he nor Dale could scarcely imagine what the future held in store for them. Through good times and bad, they never lost touch for very long. Because of mutual scientific interests, as well as social and political concerns in common, the two men would move in and out of each other's lives for the next quarter of a century.

3 A Lifeline of Letters

WHEN CANNON went off to war in May of 1917, it was with a sense of purpose, as well as a feeling that he was about to begin a grand adventure. He had been poised to go to France for some weeks; indeed, he was eager to go. For if he were to carry his work on shock further, he would have to leave the safety and calm of his laboratory and go to the front where he would find the large number of human cases that he needed to study.

His wife's feelings, however, were mixed. She understood Cannon's excitement about his research, and she shared his conviction that it might save thousands of lives. "I would not lift a finger to hold him back, of course," Cornelia wrote her mother in St. Paul. "This war means, to me, the great battle for human freedom and world peace." But she understood, too, that America's entry into the war fought "to make the world safe for democracy" could take her husband's life. "Walter thinks it is unreasonable of me to face the worst," she continued, "but I cannot feel it right to let him go until I have accepted the possibility that he may never return."[1]

Before they parted, Cannon and Cornelia agreed to write faithfully at least once a week, and to number each letter so they would know if any went astray. Despite problems with censors and untold delays, surprisingly little of their correspondence was lost, and the letters went back and forth throughout the war. Their sole means of communication for almost twenty months, the letters provided a lifeline of support, filled with love and encouragement, that enabled the Cannons to share the burdens brought on by separation.

Cannon's letters not only served as a record of the progress of his work on shock, they also informed his family and friends about other dimensions of his new experiences in England and France. Cornelia wrote about her own "war work" as well as social and family obligations, and she shared her views on an astonishing range of local and world events. Best of all, she faithfully set down for their father what the children did and said during the weeks and months of his absence.

The children wrote to Cannon as well. Wilma sent him poems that she had written. Marian told him tales of her mischief. Linda carefully described the new places she had seen. Bradford prepared solutions to challenging puzzles that his father sent him (each solution was worth between fifty cents and a dollar). And Helen, little more than a toddler, sent examples of her first efforts to write.[2]

Soon after Cannon's departure with the Harvard Hospital Unit, Cornelia accepted an invitation from her sister, Helen Sommers, to spend the summer with her family. In writing to Helen, for the first time Cornelia hinted of her true feelings: "You do not know what a comfort the thought of coming west to you is to me. It cuts the edge off the loneliness." And when her mother asked how she bearing up, Cornelia replied, "I feel so calm—and so strong. I do not know where my strength comes from, but I feel as if neither life nor death could shake me." Her determination to be strong was one thing; the turmoil of her feelings was quite another. "I took the children early to the voting booth near here to see the young men register," she wrote. "The place was full of beautiful young men in uniform. I stood there, like a fool, the tears streaming down my face—the boys were so matter of course. I suppose they wondered what that old lady was crying about!—so I decided to withdraw as not really fitting the occasion, but it has been a moving day."[3]

Cornelia bravely came to terms with the changed conditions of her life. She sold the family Ford, rented the farm in Franklin, New Hampshire, and left Cambridge with the children for her sister's home in Hudson, Wisconsin. The move proved a welcome change. Cornelia had the solace of an extended visit with her mother and father and sisters, and the children had the pleasure of playing with cousins they had heard about but never seen. In short order they acquired new companions eager to go berry-picking or swimming in a near-by river.[4]

Early in August, wearied by a spell of tending sick children, Cornelia and Helen decided that it was time for a grand adventure of their own. Besides, as Cornelia put it, they had reached middle age and planned to take advantage of that fact by sallying forth to see the world, "as free lances and comrades of the road . . . We shall forget our fondness for soft beds and pleasant foods, to be again the light-hearted and light-footed creatures we were twenty years ago."

They put their children in the care of household help under the supervision of their parents. Then they set out in Helen's car to wander the villages and towns of central Wisconsin and Minnesota, camping out throughout the trip in a tent beside the car. "Though our intelligence tells us that we offer no temptation to the wayward fancy of youth," they recorded, "we carefully secrete ourselves far from the beaten tracks . . . We take a hatchet to bed with us every night to guard against that not impossible behavior." Contrary to fleeting thoughts of possible intruders, Cornelia and her sister were not molested. After ten days of freedom, they returned to Hudson refreshed.[5]

Upon returning to Cambridge, Cornelia wrote her mother that she felt equal to anything. But in fact, during her journey home she had happened to see the headline in a discarded newspaper that unnerved her. Harvard Base Hospital no. 5 in Camiers had been bombed, and she was gripped by a terrible anxiety until she reached home and learned that Cannon was safe.[6] Greatly relieved, she threw herself into the war effort; the myriad volunteer activities in which she became involved soon made their mark on the Cannon household as well as on the Cambridge community.

The wartime project to which she devoted most of her energy was that involving the "radio boys." By the fall of 1917, a radio school established at Harvard to train signalmen had made Cambridge the new home for thousands of Naval recruits. When by chance Cornelia encountered a recruit from Minnesota, she invited him and other Minnesota boys to her home for tea and lemonade. It was the beginning of a new career for her. "The house is full of radio boys," she wrote Cannon, "three have just gone off, three are in your study, three are in the dining-room at work with their books, and one is in the living-room to learn how to knit a sweater. The opportunity to study here away from the turmoil of their quarters seems very welcome to them." Their visits were equally welcome to Cornelia, and to her mother she confessed that the "work with the Radio boys is really saving me, or I should die of homesickness for Walter, now that I am away from you and Helen and back where every stick and stone speaks his absence."[7]

Her efforts on behalf of the radio boys inspired the entire community. At her suggestion, hundreds of Cambridge householders opened their homes to them. A Hostess House, presided over by Cornelia and her friends, was established on Cambridge Common—here the boys could bring their families, and especially their female friends. Harvard professors, eager to help, took the boys on trips to historic sites in and around Boston. One man, a Mr. Wilkey, whom Cornelia had inveigled into providing lunches for 110 boys to take on the trips, told her, "I'm doing this just for you, Mrs. Cannon."[8]

For Cornelia the war was an opportunity to break down social barriers, and she perceived a great rejuvenation of Cambridge social life emerging from the hospitalities offered the radio boys. But not everyone took to the sailors as enthusiastically as she did. After learning that some householders specified they would entertain only "nice" boys, Cornelia bristled, "Perhaps the world would be a barren spot if we could not think of ourselves as superior to someone, but I think to pick and choose among these boys when we are willing to let them die for us without being sure they are 'nice' is a little too much."[9]

Before long, however, the radio boys involved Cornelia and her friends in a social problem. The women became upset when they began hearing tales of thirteen, fourteen, and fifteen-year old school girls who flocked around the radio school, infatuated with the uniform. Since the boys were free to stay out all night, it was inevitable that some of the girls found themselves "in trouble"—thanks to the sailors. Cornelia was deputized by a committee to discuss the matter with the lieutenant and request that he institute a ten o'clock curfew. When he refused, she went to see Harvard president A. Lawrence Lowell, and shortly thereafter, the commanding officer appeared before her Mothers' Club to hear the women out.[10]

While Cornelia was not sure if anything would be changed, she argued that the war did not justify the violation of accepted rules of decency. These matters were very much on her mind when a chance conversation with an acquaintance gave her momentary pause, and she related the substance of it in a letter to Cannon. At a meeting held in Boston ten years earlier, an army doctor had reported "that it was best for a man to indulge once in two weeks," she explained. "When he finished speaking there was utter silence. Everyone looked to Dr. Cabot or some one like him to speak—but no word came. Then a woman in the gallery rose and said, 'I should like to ask the doctor one question. Whose daughters shall these women be?' The effect was profound. If she had spoken an hour she could not have said more—or perhaps as much."[11]

At a meeting of the public health association, her sister Linda told Cornelia, everyone who spoke—army, navy and civilian—was concerned with the subject of "the social evil." The wartime escalation of venereal disease cases added to Cornelia's alarm about promiscuous sexual behavior. Some time later, she sent her husband a pamphlet on sex and asked him about moral conditions in France.

Cannon, though somewhat straitlaced by nature, was interested in the subject of his wife's remarks and had his own experiences to relate. Replying that in Paris open solicitation was unavoidable, he shared with her some of his insights:

> Once I was spoken to three times, on the Boulevard des Italiennes, while walking the distance of one block. "Hello, American boy," "good evening," "will you come with me?"—lisped gently in Anglo-French, usually with the hand placed on the arm, though once I was actually seized so that I had to twist myself free. I felt only disgust for such action. But I can imagine a young fellow, alone here and with few friends, and curious, and venturesome, taking a step that would lead him directly to life-long misery.

He found it difficult to judge the French, who spoke "with the utmost freedom of free-and-easy sex relations, as if such things were as common as possible." Though it might be "more talk than performance," what he had seen for himself was far beyond anything that he'd seen in the United States. "Perhaps we are more covert," he suggested, "that's what makes me hesitate to say the French are more immoral in these affairs; they may be merely less hypocritical."[12]

Cornelia took pleasure in sharing parts of her husband's letters with close friends and neighbors, and they were equally delighted to learn of Cannon's experiences. After hearing her read from a fresh batch of letters, his old friend and fellow member of the Wicht Club, Robert Yerkes, wrote Cannon, "It would have done your heart good, although it might have marred your character, to have looked upon that audience intent on your news, experience and reflections, and to have heard the comments . . . All who know you expressed the feeling of a 'visit.' It seemed as if we had stopped in your tent for a chat."

Mrs. William James invited Cornelia to read aloud from some of the letters in her home. "When I got there, I found the sons and their wives and Major Higginson," she reported to Cannon. "They kept me reading for over two hours. Major Higginson was deeply moved. When you spoke of hoping the Red Cross would finance your researches if the British would not, he said, 'Does he need money? Won't he let me give him all he needs? . . . If he does us the honor of going to France to do this work under my father's name, the least we can do is to see that his work is not halted for a few thousand dollars.'"[13]

After several such readings, Cornelia, prompted by the enthusiastic response the letters had received, began to think they should be read by a wider audience. She sent samples to Ellery Sedgwick, the editor of the *Atlantic Monthly*, to see if he would be interested in publishing them. When Cannon learned of Cornelia's project, he was less than enthusiastic and explained that the material he had written on shock would have to be passed by the British War Office since his research was under their auspices. Furthermore, he had made remarks that were critical of the research on shock of William T. Porter.

Unbeknownst to Cannon, Porter had written several articles based on his own study with Alexis Carrel of shock at the front that had previously been accepted by Sedgwick. At the very time that Cornelia submitted Cannon's letters, Porter's articles were in process of publication in the *Atlantic Monthly*.

When Cannon learned that Porter's articles were to appear in the *Atlantic Monthly*, he was mortified and promptly wrote Cornelia, "If I had known of them, I should have refused to submit anything to Sedgwick."

> Fate seems to have made me a shadow of W.T.P.—often a dark shadow. I was appointed to the position for which he was ambitious,

> and I made a success of the teaching when he failed. He started the *American Journal of Physiology,* and it was while I was president of the American Physiological Society, a position to which he was repeatedly refused election, that the Society took the *Journal* out of his hands.
>
> Then he came over to study shock. Quite by accident, as Chairman of the Physiology Committee of the National Research Council, I was drawn into the shock problem, and the War brought me over here. The results of my work on shock . . . undermine his contentions and demonstrate definite dangers in the method of treatment which he proposes. And now he writes for the *Atlantic* about his experiences and I write about mine! Isn't there an appearance of persecution about this? The articles on shock are sure to upset him—I wish I had not been the agency selected to do the job.[14]

To make matters worse, Yandell Henderson, in conjunction with Porter, had suggested to the British Medical Research Committee that Bayliss and Cannon, when they were about to begin their experiments on shock, undertake on Porter's and Henderson's behalf a series of rebreathing experiments to test their effect on acidosis. At the end of December 1917, Cannon wrote Cornelia that the evidence showed that in an organism with low blood pressure or acidosis, "rebreathing or any other procedure which increases the carbon dioxide (the carbonic *acid*) in the blood has disastrous or unfavorable after-effects . . . I am very thankful that Bayliss is associated with me in this work, for he helps to give the results an 'impersonal' character, which my relations with W.T.P. render desirable."[15]

Cannon's letters never did appear in the *Atlantic.* Later, when an editor at the Appleton Company, who had seen excerpts from his letters in the *Harvard Alumni Bulletin,* asked Cornelia if he could select and edit some of the early letters to her, Cannon erupted again. He cautioned that as he had not written the letters for publication, they would have to be carefully edited if accepted. "I can't do what he wishes done," he wrote home. "I'm not ready to write anything permanent about shock—i.e., anything that is final enough for a book. The idea of mixing up impressions of the war and discussions of a scientific problem is repugnant to me. Further, it has been done by Dr. Porter, and I should refuse in any event to run in his footsteps. And quite as important as these considerations is the fact that I have important work to do for the armies, and cannot take time to write." Ending with the hope that his decision would not be a sore disappointment to Cornelia, Cannon advised that the letters be kept as a record of immediate impressions for their friends and children to read.[16]

Cornelia took Cannon's advice to heart and shared his subsequent letters only with family and intimate friends. As the months of their separation lengthened, she lamented in a letter of March 4, 1918, to her mother, "I feel as if I could never in all my life catch up with the spiritual and mental expansion he has experienced this year." But she need not have worried; although they were worlds apart, Cannon treasured her entertaining and lively accounts of the life that she led in Cambridge.

That life was ever and always affected by the war. On the home front, Cornelia served her country by canvassing for Liberty Loan Bonds, sitting on the Minimum Wage Board, and participating in a wide variety of local efforts to conserve fuel and food. She involved the entire family in making sacrifices. "We have had no sugar or butter for three weeks—we are conservation mad," she wrote her mother at one point. "The children eat oleo on their bread with joy but cocoa sweetened with syrup is odd. [They] have to be keyed up to it by constant references to the Belgian children and suffering Serbia."[17]

She soon became an object lesson for the neighborhood, and it was rumored that Mrs. Cannon was not going "to have a furnace fire this winter," but was going "to dress her children with warm underclothes to strengthen them." Amused, she noted, "A reputation for being 'queer' does lend wings to strange tales!"[18] Nevertheless, her reputation as a conservationist garnered her an invitation from Simmons College to participate in a symposium on food conservation with Graham Lusk, the distinguished nutritionist.

Cornelia welcomed the opportunity to be heard, but to her mother she expressed skepticism that "it is all pretty much words, words, words, for people go right on indulging themselves." Her sister Linda, who was in the audience, wrote home with great excitement: "Saturday afternoon I heard Cornelia's talk on food conservation at Simmons . . . Her idealism and patriotism shot with wit and humor carried the day. Some busily took notes on the hard facts of her talk but all were fired by her enthusiasm. This is the kind of thing she should spend her strength on, not on housework and bedmaking, etc."[19]

Throughout the war, the food shortages continued, but for Cornelia they were the least of her worries. "We are going Hoover one better—we do not use *any* wheat—barley flour, oatmeal and cornmeal are our all. We have given up rye also, as the allies can use that, as well as wheat. I am amazed to see how comfortably we manage."[20]

The weather, unfortunately, could not be managed. By mid-January a persistent and bitter cold spell closed all of the schools in Cambridge because of the lack of coal. Cornelia, who always kept close watch on her children's education, conferred with her friend Isabel Whiting and came up with an ingenious idea: Each day during the crisis teachers should be assigned to take the children on special trips to places of historic interest or museums and libraries that were open—and especially to make use of the wonderful exhibits at the Agassiz Museum. Delighted with their plan, the two friends immediately telephoned the superintendent of schools to urge him to put it into action. When nothing happened, they pushed him to get them an audience with the mayor. After they had made their case, the women were assured by both mayor and superintendent that the idea would be presented to the school board.

But the superintendent played Cornelia and her friend false. He reported their suggestion to a principals' meeting in such a way that it was hooted and jeered at and deemed subversive of discipline. "We are not going to drop it," Cornelia angrily wrote her mother. "It is the inertia and stupidity of the educator of the lower ranks—no mind, no progress. We're going to ask for a hearing on our ideas."

After the better part of a month passed, Cornelia got Charles Whiting, her fellow officer in the Public Schools Association, to appear with her before the school board and present the plan, which was immediately adopted. "The next day," Cornelia triumphantly reported, "the superintendent got busy and now 500 children a day are taught in the Agassiz Museum—trips planned throughout the city, empty schoolrooms used in three buildings, etc." For once, Cornelia had won a round in the continuing battles she engaged in with Cambridge city officials.[21]

While his wife concerned herself with the education of children in the Cambridge public schools, Cannon found time and opportunity to explore the possibility of an educational exchange between students at American and English universities. During the autumn of 1917 on a visit to Cambridge University, he was thrust into a controversy, which was then raging in England over reorganization of the mode in which universities awarded doctoral degrees. Cannon, as well as the officials with whom he met at Cambridge, recognized that after the war many American students were likely to avoid advanced study in Germany, and as a consequence, they would be seeking unversity training elsewhere—particularly in England and France.

He pointed out the difference between Harvard, where the doctoral degree was given when students demonstrated an ability to think independently and follow through with a carefully planned investigation, and Cambridge, where burdensome time constraints were placed in the way of graduate students. And he asked whether the British were prepared to remove rules that stood as an obstacle to American students who had the ability for productive scholarship. Such an action would encourage Americans to come to England, he argued, and their coming would not only increase the productivity of English laboratories but also spread to the United States the spirit and method of English investigators.

Cannon's remarks were especially welcomed by the vice-chancellor of the university, A.E. Shipley, who had previously written an essay on hate in which he made much of Cannon's work on the emotions. Shipley was the leading spirit of the reform faction at Cambridge and in an effort to play fair, introduced Cannon to his opposite number, a conservative named Hobson. The three men plunged into a lively discussion about the inter-relations of English and American universities, after which Shipley requested Cannon to put his arguments in writing for him.[22]

Cannon later regaled his family with the story. "The evening was highly amusing—for I had gone to Cambridge for rest and serenity and got instead total submersion in the biggest whirlpool that has stirred the old place for many a year." When he returned to University College in London the next morning, Ernest Starling was much interested in his experience with the Cambridge conservatives. "He would decerebrate them all, I believe, if he had his way."[23]

Cannon set great store in the American way of getting things done—especially in matters of medical education, medical research and clinical investigation. Throughout the war, his increasing conviction of the importance of collaboration with the British and French only served to strengthen his prewar notions. Once, when he was having lunch in London with Walter Fletcher of the Medical Research Committee, the latter remarked that T. R. Elliott was impressed by the younger American clinicians who came over to serve in British hospitals. Fletcher wanted to know more about the training that had produced such good results, and said he understood that in the United States the best young medical men were encouraged to approach practical work through a discipline in one of the medical sciences and then arrange for free time to continue in clinical investigation.

Cannon saw Fletcher's keen interest as another opportunity to point out the advantages of American training. Stressing that such training encouraged the personal development of the brightest medical men, he took pains to point out that this led to extraordinary clinical productivity, which in turn, was advancing medical knowledge. Impressed by Cannon's remarks, Fletcher indicated that as soon as the war was over, he would make every effort to introduce into English hospitals programs for clinical investigation that provided support for the investigator.[24]

In the spring of 1918, Cannon undertook the opening of the Dijon laboratories in the belief that the research conducted there could fruitfully be applied to the solution of a wide variety of medical problems facing American troops in the field. Each and every achievement of medical research that came to his attention was proof of the superiority of American methods and training, and he began to see these achievements as foundation stones for an important postwar collaboration with England and France. "All these countries need to know each other better," he noted, "and one of the best means is through uncontroversial ways of science and love of learning. I am just beginning to appreciate what France has to offer us in the forms of suggestion and insight, and what we may have to offer her in new enterprises."[25]

New enterprises in the laboratory, however, were being threatened by antivivisectionists. As it became known in the United States that the American Red Cross was supporting experimental medical research in Europe, antivivisectionists in the United States mounted a fierce attack, which for a time undermined some of its fundraising programs. As a result, the Red Cross was forced to pressure its officials in Europe to close down the experimental research programs it supported. When Cannon learned of these events, he wrote directly to Mrs. Abbie Lathrop, one of the leaders of the American antivivisectionist movement:

> When our country found itself confronted by a foe which unscrupulously killed its citizens and seriously threatened the liberties and the democratic ideals of its government, the representatives of the people voted overwhelmingly to send against that foe the young men of the nation. From my own experience, I know that that vote means *human* vivisection, without a whiff of anesthesia, on a scale that is appalling in its magnitude.
>
> The Medical Research Committee of the American Red Cross proposed to employ some of the funds of the Society in attempts to lessen the pain and suffering of our soldiers, in part perhaps by studies on lower animals. American antivivisectionists at once cried out, "Hands off." According to their perverted notions, you may vivisect young men to save us from the miseries of servitude and dishonor, but you must not vivisect lower animals to save these young men from the torment of grave wounds—though the animals are anesthetized and the men are not. The war waged in the laboratories against pain and premature death will last much longer than the Great War, and we who are engaged in medical investigation trust that the good sense of our people will approve conscription of lower animals for purposes of the longer struggles, just as it has approved the conscription of our youth and young manhood in the present emergency.[26]

We do not have Mrs. Lathrop's reply, but given her intransigence in previous debate with Cannon, it is doubtful that his letter moved her.

When the Red Cross Research Committee met in Paris that April, Cannon elaborated on the argument he had advanced to Mrs. Lathrop with telling effect. Mr. Davidson, the president of the American Red Cross who was present, Cannon later reported home, "finally addressed a cablegram to the United States declaring that the work of the Medical Research Committee was perhaps the most important single activity of the Red Cross, that it could not be transferred or given up, but that if persons concerned in raising money would find it more convenient to have a special fund raised for the Red Cross to use for investigation there was no objection. This seems to me a little, but only a little, shifty; the main point is that the Red Cross definitely commits itself to medical research as a humanitarian pursuit."[27]

Cannon did not have to persuade Cornelia as to the value of the American Red Cross. "We have all made contributions to the Red Cross and wear our Red Cross pins with pride," she wrote in answer to his letter. "The organization will get the hundred million they want easily, and may double it. People have perfect faith in the intelligence and integrity of the management." Certainly she did. Cornelia had already been so touched by the Red Cross appeals that her generosity knew no bounds. After the explosion of an ammunition ship in Halifax harbor, she donated Cannon's winter clothes and shoes for its victims. Shortly thereafter, the stories of hardship suffered by the Belgians and people of Northern France wrested from her Cannon's summer wardrobe as well.[28]

By the early spring of 1918 the rush of enthusiasm that accompanied America's entrance into the war had given way to a heavy dose of reality. Her moods fluctuating with every changing condition at the front, Cornelia feared "that the revolutionary debacle in Russia would doom the United States to a long war." In March, after attending a meeting of the wives of men in Cannon's unit, she informed him that there was a perceptible decline in their morale and "they were all depressed by the long stretching out of the war." And at Memorial Day services in Sanders Theatre, she wrote, the audience had been deeply moved when President Lowell at the point of tears said, "This is the darkest day since we entered the war."[29]

In June, when Cornelia found that she was unable to rent the farm, Cannon urged her to set aside her various activities in Cambridge and take the children to Franklin for the summer. Cornelia took his advice. The decision to flee Cambridge helped dispel the gloom that for months had clouded her outlook. Settled at the family farm, the children took turns riding and caring for a new

Shetland pony, and when they recovered from a bout of whooping cough, their mother organized an outdoor production of Jeanne d'Arc. "A marvel of the spectacular," in Cornelia's eyes, it kept them happily occupied and proved to be the first of many theatrical performances in which the entire family, as well as other members of the summer colony, participated during the children's years of growing up.[30]

In the middle of the night of July 19, Cornelia heard the jubilant sound of bells and whistles, and learned that it was in celebration of the reported advance of American soldiers between Soissons and Chateau Thierry. The news from the front seemed to be changing for the better. In reality, unbeknownst to Cornelia, her husband was in command of shock teams in a divisional hospital in one of the fiercest battles of the war—Chalons-sur-Marne. Cannon, on the eve of battle, dashed off a loving note to her as he recalled their honeymoon: "Yesterday was the anniversary of our climbing 'Cannon Mountain' . . . As I looked back, the outstanding thing was not the mountain, but your indomitable spirit! And I saw that you had been showing the same determination and hardihood of soul in these trying days that you showed then . . . I appreciated anew how blessed I have been in your love and companionship through all these years of our life together."[31]

Her husband's tribute as well as encouraging war news made Cornelia look forward to the time when he would be returning home. After her mother wrote advising her to make herself "fit for Walter's return," she replied, "it is more than rest that I shall have to have . . . something is happening to those close up to the horror of [the war] that will set them apart unless we can grow spiritually to meet it . . . We can, at least, respect the silence and reticence of our returned crusaders, when they come."[32]

When she returned to Cambridge in September, Cornelia had news from Cannon that he thought would please her:

> Carrel said that if I did nothing more during the war than the development of the instruction of "shock teams," I should have done a great thing. Major Caplan told me that he had heard from several American sources that the course on shock and hemorrhage was the most valuable service in the army medical organization. And this morning when I asked Col. Siler for something, he remarked, "You can have anything you ask for,—you are doing more important work

> for the American soldier than anybody else in France!" What I do is to discount these statements just about 75 per cent and feel pleased that the attitude towards the work I am doing is a favorable one.

Cannon ended his letter with a word of warning: "Please do not say anything about these matters to anybody outside the family. Anything that is said is sure to be distorted and misunderstood. If there is any truth in what has been said, it will appear in due season and in a proper way. What I have written is purely for *your* consumption—I shouldn't have written it if it had not seemed a rather easy way to make you happy!!"[33] But by the time Cannon's letter reached her, Cornelia was too ill to be happy.

Afflicted with the terrible influenza that ravaged entire communities in a worldwide pandemic (and claimed many more lives than the war), she worried about the care of her children if she did not survive. "The pneumonia afterwards plays havoc so I am taking precious care of myself, for with Walter in France I simply cannot be spared," she wrote her mother. The Spanish flu, as it was called, laid low the Cambridge community and for several weeks, Cornelia's letters were filled with grim reports of those stricken and the appalling number of deaths, especially among young people. Doctors, nurses, undertakers, all were in short supply. Until the disease began to wane, even the thrilling news from France took second place to the influenza epidemic at home.[34]

With October came the first stirring of hope that peace might well be at hand, though several agonizing weeks went by before the hope was finally realized. On November 11, to celebrate that "never-to-be-forgotten day" of the Armistice, Cornelia took the two oldest children, Bradford and Wilma, into Boston to witness the display. "I never saw such crowds, such innocent, joyous, delightful crowds!" she reported. "Boston was transformed, all her inhibitions gone, melted, unself-conscious, with an incredible simplicity. Everyone beamed at everyone else, people in apparent entire forgetfulness of self, were skipping and dancing down the middle of the streets." But inside the Cannon house there was no trace of the patriotic fervor that had marked the family's observance of America's entrance into war—no exuberant singing of the Star Spangled Banner, no reading from the speeches of Lincoln, no salutes to the flag. Instead they celebrated the war's end in relatively subdued fashion by starting up the furnace. "We have been shivering for weeks," Cornelia observed, "it certainly feels good to be warm once more."[35]

Half a world away, Cannon, after hearing the news of the Armistice, made his way with Simon Flexner to Paris and joined a boisterous crowd there in celebration. He described the scene for his wife:

> Here and there a group would form a ring and dance around another ring with much hilarity, for the only way to break out was by kissing! As we stood on an "island" in the street looking over the mass of merrymakers, a nice-looking young woman enthusiastically rushed up and kissed Dr. Flexner on the right cheek. I felt horribly neglected and unattractive—but only for a moment. Another girl suddenly appeared out of the turmoil, asked Dr. Flexner if she might kiss me, and as he instantly gave her permission she seized me and planted a smack on *each* cheek! I was not so unattractive after all—only stern and forbidding! They were nice, exhilarent young things, and we went laughing on our way.

Some days later, he expressed his thoughts on the significance of the war's end and wondered what the future might bring: "What does this all mean, and where may the freed spirit lead us? Here in Dijon the people are dancing through the streets on this, the third night of celebration. But they are catching the joy as it flies, and they and all the rest of us will have to settle down and take serious account of stock before many months have passed." He ended his musing with the hope "that the world has seen enough of murder and wanton destruction to refuse the further spread of Bolshevism" before revealing what was really on his mind, "Of course, the top question personally now is—when can I go home?"[36]

PART II

The 1920s

4 *Postwar Adjustments, and a New Era at the Medical School*

AFTER SEVERAL WEEKS of waiting impatiently in a camp at Sainte Nazarre, Cannon finally boarded the troopship *Manchuria* for his return to the United States. Eleven days of a debilitating rough passage on the Atlantic brought him to Hoboken on January 22, 1919, where he was greeted at the dock by an excited, flag-waving Cornelia. "Life seems all perfect again," she wrote her mother, adding that Walter was going to Fort Dix to be demobilized and she was going with him. "I will not be parted from him again if I can help it," she vowed.[1]

When they arrived home some days later, Cannon was immediately captured by the children, who did not let up except to quarrel about the division of the spoils. "It is incredibly pleasant to be back with my family and to get acquainted with the greatly changed children whom I found awaiting me," he wrote Simon Flexner. "My youngest little girl did not recognize me when I arrived, and apparently thought I was a stranger named 'father,' who was unwarrantably familiar with her mother!"[2]

As the euphoria of Cannon's return abated, Cornelia began to notice subtle changes in his behavior. "Walter is much changed," she observed in a letter to her mother,

> older and graver—his gayety gone, and much of his resilience. He still tells funny stories and laughs at jokes, but the light-heartedness is gone . . . It is too much to expect that one could go through such horrors and be the same—but it makes me sad to see how his whole response is slowed down—as though the grief and agony could never leave him . . . Every thing that love and devotion can give him we can offer, but we shall have to wait for the slow process of time

> to heal the deeper hurts—As I see what this hellish war has done to him and realize that it is repeated a million-fold over the world, I wonder if the devastation of human nature is not worse than all the killing and destruction.[3]

Within weeks, Cannon began to lecture at the Medical School and it appeared to all who knew him that he was his old self. During a lecture in early March, however, he suddenly became faint and found that he could not stand up for more than a few minutes at a time. In the days that followed, he found it difficult to sleep and became depressed. In an effort to help, Cecil Drinker, who had taken over the lectures during the war, urged Cannon to rest and offered to manage the physiology course until the fall. But Cannon refused, claiming he only needed a good night's sleep to recover. On previous occasions whenever her husband had blue moods, Cornelia moved right past the depression as if to deny its presence. This time she met it head on. She went into Boston and bought a Ford car to spare him the strain of traveling between home and the Medical School by streetcar, as well as allowing him to eat lunch at home and take a nap afterwards.[4]

Despite Cornelia's fears, there were few lingering traces of his recent experience to mark Cannon as one of the troubled postwar generation. One telltale sign, however, was a certain restlessness common in many veterans who had seen service in battle. Only months after his return from France, for example, Cannon seriously considered an offer from the Rockefeller Foundation to go to China to teach physiology in the new medical school at Peking. "He is deeply interested in the idea of starting physiology in China," Cornelia noted, "perhaps getting a research laboratory, the first in the Orient, going—and if possible, we shall go." But when the China offer, which had been put on hold, came up again, his former restlessness seemed to have evaporated. "They asked for Walter," she wrote her mother, "but the Dean says the University needs all their men now—to pull the place together—owing to the war demoralization; so Walter stays with the Medical School."[5]

Throughout the spring of 1919, the war and the problem of securing the peace dominated discussions at the dinner-table and at gatherings with friends. Cannon and Cornelia became so annoyed at the *Boston Transcript*'s bitter opposition to the League of Nations that they stopped buying the newspaper. Cannon's particular bane was Senator Henry Cabot Lodge, who led the opposition in Congress against President Wilson's proposal of a League of Nations to secure the peace. Aware that he had little chance of persuading Senator Lodge and others in the Massachusetts congressional delegation to change their minds, Cannon appealed to Senator Frank Kellogg of Minnesota to join with other liberal Republicans in support of the League.

His letter, describing some of the gruesome sights he had witnessed in the shock wards of forward hospitals, reveals that the horrors of war were deeply embedded in Cannon's memory. It was the earnest hope of all thoughtful men in the A.E.F., he told Kellogg, that "the dream of a League of Nations, which should at least make an earnest effort to lessen war, might be realized" so that "the young lives that were lost and the frightful mutilation which men underwent would thereupon be justified."[6]

As true believers in "the war to end all wars," the Cannons looked forward to a new order in international relations. For Cornelia, Woodrow Wilson was the great man of the day with the hope of the world in his hands. "He has a clean slate and bases every thing on principles," she wrote home. "We cannot go back to the old method, [and would] rather perish for the new ideal than consent to the old." But Cannon could not turn his back on the past so easily. When he learned that his sister Ida had as houseguest in Franklin a pacifist who refused to serve in the war, he was outraged. For Cannon, Roger Baldwin, a social worker who spent a year in prison as a conscientious objector rather than be drafted, had exhibited "the last touch of selfishness" by his act of defiance. Though Cornelia tried to defend Baldwin's gesture on the ground of serving some possible future good, Cannon likened him to a man in favor of fireproof construction who let women and children burn up in a wooden house while he advocated steel and brick.[7]

One of the more pleasant associations remaining from Cannon's years abroad were the fond memories and continuing ties of wartime friendships. When he received word that he had been decorated Companion of the Bath by the British government for distinguished war service, his first response was almost flippant: "Companion of *whose* Bath?" he asked tongue-in-cheek. But soon, after cables and letters of congratulations began pouring in from England and Scotland, his irreverent attitude changed to one of humble gratitude. Throughout the summer months, Cannon continued to receive congratulations from friends and colleagues for his recognition by the British and when he learned that General Pershing had been honored as well, he finally gave the award its full measure of respect.[8]

Cannon was no sooner home from the war than he received a request from W. W. Keen that he write a chapter on shock for a new edition of Keen's multi-volume *Surgery, Its Principles and Practice.* "I should want it to be pretty full," the Philadelphia surgeon instructed Cannon. "I want to get in your experimental results from the physiological as well as the surgical side; and that you shall enter rather fully into the treatment of shock, with illustrations as to blanketing, heating, and so on. I would wish particularly that you would take up the question of shock in industrial and civil accidents, as well as gunshot. Moreover, the relation of shock in operation; to the anaesthetic used; to surgical shock in civil life, all of these I should want you to treat." When Cannon later tried to withdraw gracefully from the assignment, Keen, undeterred by his hints of other obligations, blithely asked if he would undertake an additional brief chapter on surgery and the internal secretions. Keen was not to be put off, and after several weeks of considering his options, Cannon agreed to write the chapter on shock.[9]

There were other matters that kept Cannon's attention focused on shock, and not the least a difference that erupted between himself and Samuel Meltzer of the Rockefeller Institute. After returning to Boston, he was surprised to learn that Meltzer some time before had informed a meeting on shock at the Pennsylvania Medical Society that Cannon believed acidosis was the cause of shock. The news distressed him and he wrote Meltzer, "There was a time last winter, as Bayliss and I were working together, when we thought that the diminution of the alkaline reserve might be associated with the production of shock, but those facts were not formally published and I have never myself expressed the idea that there was in the acidosis that developed in shock a primary cause for the low blood pressure."[10]

Meltzer was not persuaded. The evidence for his statement, he informed Cannon, was based on the fact that Cannon had expressly recommended the use of bicarbonate of soda as a means of treatment of acidosis in cases of shock; and also that Bayliss, who had collaborated with Cannon, had publicly changed his mind on acidosis as a cause of shock and tended to look upon its treatment with sodium bicarbonate as of secondary importance.

Meltzer's rejoinder disconcerted Cannon. He not only felt that Meltzer had misinterpreted his position on the relationship between acidosis and shock, he was equally concerned that Meltzer's views (given his status in the scientific community) might be accepted as gospel by other physiologists. In an effort to settle the matter, Cannon wrote Meltzer again, this time pointing out that those of his papers to which Meltzer referred were essentially theoretical:

> I did at one time advocate the use of sodium bicarbonate as a means of avoiding extreme acidosis which may follow surgical operation on shocked men . . . I had seen men who were in extreme distress from air hunger after operation made comfortable by alkaline therapy and it was for this reason that I suggested its use . . . At the time I did this, I was uncertain whether during *anaesthesia* and *operation* the acidosis might not be primary. Later experimental work proved that it was secondary to a fall of pressure and that the main thing was to avoid that. The necessity for alkali thus disappeared.

Some months later, Cannon made a public issue of his differences with Meltzer in a footnote to a paper on shock that he published in *JAMA*. The footnote brought an immediate spirited reply from Meltzer, who detailed again the reasons for the position he had taken concerning Cannon's beliefs of the cause of shock.[11] It was but one of several differences that Cannon would have with physiologists over interpretations of what he said and meant in other papers he had previously published, including those he had written on the emotions and the emergency function of the adrenal medulla.

Of all the news of the Medical School that Cannon received in France during the summer of 1918, none pleased him more than a report from home that David Linn Edsall had succeeded Edward Hinckley Bradford as dean.[12] Cannon was not the only one happy with the choice. When President Lowell called on former President Eliot to help him decide who might replace the ailing Bradford, their selection of Edsall, the Jackson Professor of Medicine at the

Massachusetts General Hospital, was unanimously hailed by the medical faculty.

From his association with Edsall before the war, Cannon knew that the new dean, although a clinician like Bradford, had a keen interest in experimental research and, unlike the former dean, would listen sympathetically to his plans for extending research and teaching in the physiology department. By the time of Cannon's return to Boston, Edsall had spent months quietly making observations, gathering information, and jotting down notes for a comprehensive statement that he could deliver to the medical faculty on his vision of the future of medicine.

Cannon, like every other member of the faculty, looked forward to hearing about Edsall's goals for the School. Most knew Edsall as an excellent physician but wondered how he would comport himself in his new role as dean, and especially what his relations would be with President Lowell. Many were concerned about budgets, salaries, and whether there would be long-awaited increases. There were other questions and they listened attentively while Edsall had his say.

At their meeting on April 7, 1919, Edsall described for the assembled faculty the School as he saw it, and made clear his hopes for its future development as well as his philosophy for how to achieve his goals. He had thought long and hard about problems confronting the medical profession at large in the aftermath of the war, and now proposed a set of guiding principles and policy concepts to deal with the many changes that were certain to come. In so doing, he made a careful comparison of the weaknesses and strengths of Harvard and its most influential rival, Johns Hopkins. And he shared with the faculty his most cherished convictions—among them, the responsibility of the medical profession to a larger community, the undesirability of a rigid and inflexible full-time system in the clinical departments, and the desire to attract and provide for gifted students. One of the credos that would guide him in the future direction of the School was that prevention of disease might well become the dominant side of medicine. Another was the need for a new recognition of public health and of the relation between medicine and social and economic forces.

Edsall devoted the lion's share of his statement to an examination of the laboratory and clinical departments at the Medical School. When it came to addressing the School's financial requirements, his priorities were manifest. "Progress in medical teaching and contributions to medical knowledge have for the past few decades been much more largely through the fundamental medical sciences rather than through the clinical branches," he said, adding that the largest sums were needed chiefly to support research and to develop highly trained men for future teachers, research workers and consultants.[13]

In the discussion that followed, Edsall noted that salaries in the laboratory departments were so low that it was extremely difficult for a man to undertake a career in these fields. Consequently, good men were not attracted; and if there were not good laboratory men, clinical medicine would also suffer. President Lowell joined with the dean in his plea for extending the salaries of preclinical professors.

A debate broke out, in which Harvey Cushing was one of the first to speak up. Cushing declared that this difficulty of lack of money could be overcome by having laboratory work done in the various departments of hospitals—in the surgery department, in the medical department, and so forth. Although Cannon was a close friend of Cushing, he promptly took him to task, saying it would be "a grave error to have all laboratory work done in connection with practical problems." Moreover, he told the assembly it had been brought out by the war that "men in pure sciences have a wider and more penetrating knowledge than those in the applied sciences."[14]

Early in Edsall's deanship, much of the money that came to the Medical School was through donations from outside sources. For instance, in the postwar era the problem of salaries, not only in the Medical School but throughout the University, was a pressing matter. There was rampant inflation, which, added to the fact that salaries had not kept up during the war, created a sense of crisis. To address this issue, Harvard under Lowell's leadership, hired a team of professional fundraisers and began a campaign to raise fifteen million dollars for endowment of salaries.[15]

Almost simultaneously, Edsall applied to various foundations for aid to increase salaries in the Medical School but did not meet with success. When such an application was made to the Carnegie Foundation, Foundation officials replied that their money was reserved for purposes other than salaries for well-established institutions. Undeterred, Edsall, after discovering that the Rockefeller Foundation planned to disperse close to fifty million dollars in order to upgrade salaries, eagerly joined with Lowell to file an application. But again Harvard was turned down, because, the Foundation insisted, large institutions like Columbia, Harvard, Princeton, Yale, etc., unlike smaller ones, had constituencies that could help them out.[16]

At approximately the same time that this search for funding began, some manna from heaven, in the form of an unexpected bequest, fell into Harvard's lap. At the end of 1918, the prestigious New York law firm of Sullivan and Cromwell informed Harvard, Columbia, and Johns Hopkins that one of its clients, Captain Joseph DeLamar, had recently died and, after providing for his daughter Alice, left residuary bequests of approximately three million dollars to each of their medical schools.

DeLamar's will outlined some general requirements including, among others, that the beneficiaries were to devote themselves to the study and teaching of human diseases as well as the effect of food and diet on the human organism. They were also to establish professorships, construct and maintain libraries and laboratories, and ultimately disseminate knowledge of preventive medicine and the conservation of health. Beyond these few requirements, the will put no limitation on the method that the medical schools might in future wish to employ in allocating the funds. It did, however, specify that the beneficiaries should try to keep the principal intact and, by prudent investment, increase their holdings.[17]

Needless to say, news of the gift came as a pleasant surprise to officials at Harvard. Still, it was a mystery. Inquiry at the law firm revealed that DeLamar was not an alumnus of any of the universities. As a matter of fact, his education had begun when he ran away to sea as a boy. At the age of twenty-three, he became captain of his own vessel. After the Civil War he amassed a fortune by raising sunken vessels and selling iron anchors, and then went into trade in Africa. But the debilitating climate made trading on the African coast oppressive for DeLamar, and so he returned to the United States. He studied geology and mineralogy in Chicago in preparation for investment in lead, silver, gold and nickel mines in the far west. At length, he acquired another fortune.

It is not known why Captain DeLamar decided to support medical education. It may be that he followed the example set by other contemporary capitalists like John D. Rockefeller, Andrew Carnegie, and Russell Sage to fight disease and promote health. It may be that he followed the suggestion of Alfred Jaretzski, the young Harvard-trained lawyer at Sullivan and Cromwell who drew up the will, to use his fortune this way. Or it just might have been the preternatural benevolence of a self-made man with a longtime interest in the effect of different foods and diet on health and disease. Whatever the reason, it could not have arrived at a more opportune time.[18]

In 1919 the Harvard Medical School, despite its eminence, was in a tenuous economic condition. Its total endowment at the time was approximately four and a half million dollars, a sum that Edsall believed was not sufficient for the necessary development of medical education and scientific studies. With the approval and cooperation of President Lowell, the dean determined to use the DeLamar bequest to meet the problems facing the preclinical departments at Harvard. Such problems were not singular; they afflicted medical schools throughout the country.

During the autumn of 1919, the National Research Council appointed Dr. Henry Christian, then on leave from Harvard, to travel around the country to make observations about the working conditions in medical schools. In response to a request from Cannon for information about the situation in the preclinical sciences, Christian wrote him a letter early in December, which was, in essence, a preview of a report that a committee of the N.R.C. Division of Medical Sciences was planning to bring out in the spring.

In his letter, Christian suggested that the most serious situation in medical schools throughout the country concerned the preclinical sciences. He pointed out that there were no adequately trained and properly developed men to fill the faculty posts that were available; there was a serious scarcity of assistants in preclinical departments in all of the schools; and the deficiency was considerable, not only as to numbers but, much more seriously, as to quality. For Christian it was critical that as much money as possible be given to the preclinical sciences.[19]

Christian's letter reaffirmed for Cannon Edsall's statement earlier that year about the practical matter of finances in the medical school. A few months later, when Abraham Flexner of the Rockefeller Foundation asked about the organization of the department of physiology at Harvard, Cannon determined to give a full account of the economic problems that he faced. "The chief difficulty," he explained, "has been the small amount of money available to our needs; our budget has never been more than $14,500. During the entire time that I have served as Professor here (fourteen years), I have never had a full time Assistant Professor to help me, and until last year, no full-time secretary."

In Cannon's opinion, the head of a department should be relieved of administrative duty and the routine of running the medical course. These responsibilities should be shouldered by a well-trained assistant professor and one or two instructors—all active investigators. "I am convinced that the preclinical departments are suffering from involvement in a vicious circle," he concluded. "Men will not be stimulated to undertake investigation if they are not brought into relation with active investigators, and they are certain not to be brought into relation with investigators if these preclinical departments are filled by men who are overwhelmed by administrative duties, or are mere stop-gaps temporarily holding the job until they can get into a clinical position."[20]

Not everyone agreed with Christian and Edsall, or for that matter, with Cannon. At a meeting of several organizations held early in March in Chicago to discuss the subject of medical education, Arthur Dean Bevan, chairman of the Council of Medical Education of the A.M.A., gave an opposing view. Bevan contended that professors in anatomy, physiology, and other laboratory branches, as well as laboratory workers in clinical medicine, had been given "too large a part in the task of reorganizing the university medical schools and in developing the medical curriculum." He urged that this tendency be corrected, adding, "We should keep constantly in mind the fact that the first and highest duty of the medical school is to train competent practitioners of medicine." If the training of physicians were done well, it would necessarily result "in training teachers and research workers at the same time as important and desirable by-products."[21]

Bevan's dissenting opinion did not sway Edsall from the course he had set the previous year. In order to determine the manner in which the income from the new Delamar Fund, now becoming available to the School, would be used, he wrote a long letter to President Lowell detailing his plans. The dean began by saying that the three central activities at the Medical School were the teaching of students, the training of advanced workers, and the development of new knowledge—objectives that were well in line with Captain DeLamar's expressed wishes. He went on to address the needs of the physiology, pharmacology, and biochemistry departments, where these goals could be most readily achieved under the leadership of Cannon, Reid Hunt and Otto Folin.

"All three of these men," Edsall related to Lowell, "will need extra instructors; all three of them are thoroughly capable of developing able young men as teachers and investigators, and all are capable of doing very valuable research. All will need increased technical assistance also, and increased amounts of money for research and even for carrying on the activities that the departments are carrying on at present." To meet these needs, Edsall suggested that the DeLamar Fund could not be put to better use than to allot an additional $10,000 to Cannon's department "so that he may be freer to use his real capacity for developing the brilliant students and the brilliant advanced workers." For Folin's department he suggested an additional $8,000, and for Hunt a considerably larger but unspecified sum.[22]

One of Edsall's firmly-held beliefs was that inability to secure or hold distinguished faculty members was the worst misfortune that could befall the Medical School. In the postwar era many of the medical schools, in search of talented clinical or preclinical professors, began to raid other schools that had such men. The dean had heard rumors that various institutions were planning to raid the Harvard Medical School faculty. There were, for instance, rumors that Lawrence Henderson was being courted by Johns Hopkins, and that others were looking at Otto Folin, Reid Hunt, Cecil Drinker, and Walter Cannon. For a time, these rumors remained unsubstantiated, but they were taken very seriously. In fact, Edsall made it a point to discuss his concern with President Lowell and ask his help in dealing with it.

Early in 1920, Edsall learned from L.J. Henderson that he had received a definite job offer from Johns Hopkins of a chair in physiology with a salary of $7500. Although Henderson did not want to leave Harvard, he asked for a comparable salary if he were to stay—"in order to be fair to his responsibility and as a psychological appreciation of his worth." Edsall, informing Lowell about

Henderson's offer, had more bad news. It was expected that Drs. Hunt, Cannon, and Drinker would be approached soon with desirable opportunities elsewhere.

Lowell, on hearing the news about Henderson, lost no time in writing to him: "You must not for a moment consider any offer for leaving here. You know I think you are one of the most valuable people in the University, and you must have what you need to develop your work." Shortly thereafter, all agreed that Henderson would stay at Harvard on his own terms: to teach as much and where he chose, to be free of all committee and administrative work (except as he chose), and to have $12,500 yearly at his disposal from the DeLamar Fund. Henderson spent the following year abroad; after he returned in the fall of 1921, his department became the department of physical chemistry.[23]

No sooner was Henderson's future settled than the dean began to devote himself to making arrangements for Cecil Drinker. Drinker had a close association with Edsall, which had begun in 1917 with a study of manganese poisoning at an ore processing plant in New Jersey. The following year, thanks to the efforts of Dr. Fred Shattuck, money became available to investigate problems of industrial health and hygiene. (Subsequently, Dr. Alice Hamilton, appointed to help conduct studies of public health hazards, defied tradition—and the powers-that-be on the Harvard Corporation—by becoming the first woman on the medical faculty.)

Establishing a place at Harvard for the study of industrial disease was attractive to Edsall, whose abiding interests extended to the economic and social sides of medicine. When Cannon came home from France, the dean transferred Drinker, who had run the physiology department during the war, into a separate department of industrial physiology.[24]

What gave impetus to this change was Edsall's knowledge that Drinker for some time had been negotiating for a post at the University of Pennsylvania. Nevertheless, an offer was not forthcoming until the spring of 1920, when a salary of $6600 a year was proposed for Drinker with an extra $2400 for his wife, Katherine, who was also an industrial physiologist. Harvard wanted to keep Drinker. Not only was he a great favorite of both the dean and the president (Drinker's father was president of Lehigh University and a friend of Lowell's), but he was also a first-rate physiologist.

The problem that confronted Edsall was to find $9,000 to match

Drinker's offer in place of the $4300 Harvard was paying him. Taking part of the money from the DeLamar Fund and part from the endowment funds that had been raised for increasing salaries, he arrived at a figure of $7,000. For the rest, he pointed out that Drinker could easily make up the difference out of funds he received from several firms to tackle some of their industrial health problems.[25]

Drinker, in effect, was in the bag along with Henderson. But before Edsall had time to congratulate himself for fending off the raids on his faculty, he was faced with another instance of raiding. This new threat would soon throw the School into crisis and reverberate well beyond the walls of Harvard.

◆ ◆ ◆

In July of 1920 while visiting his wife's family in St. Paul, Cannon gave a lecture at the Mayo Clinic in Rochester, Minnesota, and had a chat with the recently-appointed professor of medicine, Leonard G. Rowntree. In the course of their conversation, Rowntree proposed that Cannon ought to come to the Mayo Clinic. After Cannon returned to Boston, he received a formal offer from Rowntree to join the staff of the Mayo Clinic at a salary of $15,000 a year—three times what he was getting at Harvard.

When subsequently Donald Van Slyke at the Rockefeller Institute turned down an offer from the Mayo Clinic, Rowntree, interested in the main chance, wrote Cannon that if he would come to the Mayo Clinic, he could choose as head of biochemistry his friend Otto Folin. It would be very nice, he intimated, if Cannon's sister Ida would develop a division of social work at the Clinic as well. Attempting to clinch the deal, Rowntree wrote to Mrs. Cannon, then in Minnesota, and invited her to make a visit to Rochester. In mid-August, he wrote Cannon, "We are extremely anxious to have the Cannons out here. You know that we want you and Miss Ida, and now that we have met Mrs. Cannon, we want the whole family."[26]

The Mayo offer in hand, Cannon went to see the dean, who informed him that money would soon be available from the endowment fund raised to increase faculty salaries at Harvard. By the summer of 1920, some $12,000,000 of the $15,000,000 that the endowment campaign hoped to raise was already certain, and Edsall made it clear to Cannon that full professors in the preclinical sciences at the Medical School would be raised to $8,000 a year. Cannon heard Edsall out and then made an astonishing proposal. In order to stay at Harvard, he said, he would like to receive $10,000 a year and he would like to see all of the laboratory men at the School raised to $10,000 as a signal to the rest of the country of the value of preclinical professors. Cannon's notion not only gave Edsall heartburn, it proceeded to involve Henry Christian, who became an unofficial adviser to Cannon.[27]

Edsall inaugurated negotiations with Cannon in a letter of August 20, by taking up the disparity between salaries of clinicians and preclinicians. "Clinical men, as a group will command higher salaries for their full-time, not because they are more valuable, but because their training makes it possible for them to earn very much more," he maintained. "The real reason that has elevated the clinical salaries here has been that hospitals have taken up the plan of paying their staff. The maximum salary paid laboratory men will now be $8,000. None of the clinical men will get that amount from the school . . . They are paid two salaries for two sets of duties: one for the school and one for the hospital."

The dean said he would be pleased to raise Cannon's salary beyond the maximum of $8,000 "because I think you are worth more than that." But he did not think that the Corporation would approve raising preclinical salaries as a principle because Harvard simply did not have the money to do it.

A few days later, when writing to Christian about his negotiations with Cannon, Edsall admitted that laboratory salaries ought to be raised. But he did not believe that the difference between clinical and laboratory salaries had caused a shortage of men in laboratory branches. Christian, for his part, took issue with Edsall about the salary differences, arguing it was all a matter of supply and demand. He thought there would soon be a change since inadequate salaries were causing the supply of laboratory men to rapidly decrease.

"As to Cannon," Christian went on, "I think we ought to pay him $10,000 because we want him to stay . . . if he goes, in the long run it will cost us just as much because we can't get a new man to come for the present budget and present salary." Then Christian sent Edsall's letter along to Cannon and suggested that he could feel satisfied, even though "the principle of same pay to lab men is not made definite as a stated policy."[28]

Near the end of August, Edsall sounded out Henry P. Walcott, a senior member of the Harvard Corporation and its trusted medical adviser, about his position in regard to retaining Cannon. He wrote Dr. Walcott that with Cannon, Harvard was on the edge of developing physiology and physiologists, and he was most essential to that development. The dean warned of the decline of Hopkins when its medical school had lost several essential men. Stating that he did not desire to establish a principle, he nonetheless recommended a raise for Cannon to $10,000. Walcott's response was swift and short: "My impression is that you have gone as far as we can go." Then he added dryly that he did not believe "the laboratory man" should give himself the trouble of "devising means for lifting the scale of compensation throughout the scientific world."[29]

F. W. Hunnewell, secretary to the Corporation and University comptroller, on the other hand, thought the medical dean had gone too far in his negotiations. He wanted more medical faculty salaries paid from the DeLamar Fund and less from the general endowment. "With over two millions of the DeLamar money already paid in," he wrote Edsall on August 27, "I naturally look upon the Medical School as a wealthy department of the University and should like to see the endowment funds spared as much as possible."

By early September, Cannon acknowledged that Edsall had raised "a number of points which I had not considered in my wish to secure for the medical sciences the same recognition that is granted to clinical departments." At the same time, in a letter to Christian, he noted that a formal job offer to Folin was in the making, but it would not affect his own decision whether to go or not to go to the Mayo Clinic. "It would be a little more comfortable, I must confess, if I were not singled out for special consideration . . . ; and if Folin should be dealt with as I may be, it would ease matters a good deal."[30]

All the while that these discussions were going on, President Lowell had been away in Europe. When he returned and learned where matters stood, he was not happy. By some means or other, he indicated, half of the $50,000 that the medical dean had budgeted for proposed salaries from the endowment funds would have to be withdrawn and the money found elsewhere. Once again, Edsall had to explain all that had happened in his negotiations with

Cannon. He did so in a letter of September 22, 1920, to Lowell, calling the situation "by far the most serious matter that has come up in connection with the Medical School since I have been in Boston."

A week later, Edsall brought Lowell up to date with another letter that began, "I have had a talk with Cannon and, so far as I can see, it would take very little to make him decide to stay here."

> I think if you felt it desirable, some evidence of appreciation from you or the Corporation, by letter for example, would help to convince him of what now occupies his mind chiefly—i.e., where he would make himself most useful. I think I have satisfied him that the Corporation appreciates him and yet could not wisely establish a standard such as was talked of. He grants that there are arguments that he had not thought of that make it at least easy for him not to force that as an issue. I think he is in perplexity whether he has a right, on his children's account, to give up $7,000 additional.

Edsall told Lowell that the Medical School had been in a turmoil, the faculty running in to see him with all manner of suggestions. Christian had talked with Dr. Shattuck who was to see Dr. Bradford, "and the latter may visit you. There is considerable anxiety among them of course over the matter, and everyone urgently wants Cannon retained."[31]

At the end of September, W. S. Thayer, a member of the medical faculty at Johns Hopkins and an influential Harvard alumnus, wrote Lowell about the Cannon problem. Dr. Thayer had heard from an unnamed person deeply interested in the School, who begged him to use any argument which might favor the retention of Cannon. He suggested that perhaps friends of the Medical School might be able to guarantee a sum of money, say $2000, for a certain number of years to be added to Cannon's salary. After talking with Cannon, Lowell replied to Thayer, "Of course, we want to retain him; and, as I said to Cameron Forbes the other day, if there was some means of paying him which did not involve a general raise in salaries—which we can by no means afford—or the discontent caused by raising Cannon without raising others, it would be a very desirable solution."[32]

Matters seemed to be at an impasse, but within a few days Cannon abruptly came to decision. It may be that two personal letters he received helped him to make up his mind. One was from his old friend Robert W. Lovett, the Harvard orthopedist, who wrote:

> I have great confidence in Edsall's policy, and my feeling is that the early part of his administration is a critical time, and that the one thing the school needs most is men like you. I wonder if you realize what an influence you have been here, and how often you are quoted and referred to in matters connected with the school's policy and the school's future.
>
> Personally, I should feel very great regret if I heard you had decided to go . . . because I think in the end, it would make a good deal of difference in the position of the school.

The second letter was from another old friend, the physiologist William H. Howell at Johns Hopkins, who offered Cannon some advice. He did not think the post at the Mayo, no matter how important, could have the dignity or the influence of his position at Harvard. He said he would be sorry to see Cannon make a change because it might be seen as a blow to physiology in the United States. When the full-time system for the clinical departments was inaugurated at Hopkins, he had favored it; but an unfortunate consequence was that it put the scientific chairs at a disadvantage. As Howell put it, the Medical School would have to realize that first-rate physiologists were as valuable as first-class men in medicine, and he hoped that Cannon would be instrumental in bringing about this realization.[33]

On October 4th, Cannon wrote Rowntree and Dr. Will Mayo that he had at last decided not to accept the position at Rochester. Apparently the decision continued to trouble him, for a week later he felt compelled to write an explanation to Dr. Charles Mayo. "As you have clearly seen, there is a great dearth of trained physiologists. It was this fact which really settled my determination to stay on my present job." He was further convinced, he continued, that there was more chance of training physiologists in a university and in association with young men whose careers were not yet clearly determined than there would be at the Mayo Clinic.[34]

When Edsall learned the news, he was delighted. He wrote Cannon that his decision would be "of the greatest moment to the School and the University and I think to Medicine. To me personally it is an intense relief because of my relation to the School and it is even more of a satisfaction, if possible, because of my associations with you and my dependence upon the pleasure and stimulus of contact with you."

In a hasty note to Cannon, Henry Christian, too, expressed his pleasure. "I believe you have already accomplished a lot for the improvement of the situation in the laboratory subjects and that the conditions we want will eventually arrive. Your staying brings me much personal satisfaction, for I would have missed you greatly." He ended somewhat cryptically, "I trust your reward will come sooner than you think."[35]

The meaning of Christian's veiled hint was made clear in a letter he wrote the same day to Alexander Forbes (who as a member of the wealthy Forbes family often gave financial support to the physiology department), arranging to pitch in $200 or more towards the $1,000 being collected for Cannon's salary. The idea, Christian told Forbes, was to send the money to the Corporation on condition that they match it with another $1,000. On Christmas eve, Charles Francis Adams, the Harvard treasurer, formally notified Cannon of the gift received from "certain friends of yours who desire to remain anonymous." Then he added, "May I join, so far as is proper for one who is not of this group, in their expression of appreciation."[36]

To celebrate the occasion of Cannon's staying at Harvard, one of his closest friends, Ralph Barton ("Tom") Perry, composed a poem:

> A professor named Cannon, of dubious wealth,
> was devoting himself to the reason
> of emotion, digestion and hunger and health
> and the glands of internal secretion.
>
> When the plutocrat Mayos got wind of his pay
> and heard of his children redundant,
> and secretly plotted to tempt him away
> by the offers of riches abundant.
>
> They dangled in front of his watering eyes
> all kinds of luxurious uses;
> they promised him shredded wheat biscuits and pies
> to appeal to his gasteric juices.

To Bradford they promised a Sopwith airplane;
to Wilma a studio in Paris;
to Helen a dot and an emerald chain
and the prospect of being an heiress.

To Linda they promised a country estate
with gardens Italian and sunken;
for Marian used Terpsichorean bait
to rival Pavlova and Duncan.

To Cornelia they promised a Rolls Royce sedan,
with a footman, chauffeur and a page,
a permanent wave and an ostrich plume fan,
and a pet Kinkajou in a cage.

Now poor Walter, much moved, half extended his hand,
and began to perceive the sensation
of the super-tax profiteer, fat of the land,
and a natural lord of creation.

The call of pure science grew faint in his ears—
with traditions of pious extraction—
and he almost threw over his uplift career
for the lure of more sordid attraction.

When up spake Cornelia in accents not mute:
"If you think I'll quit housework and slumming,
and the Ford and the farm and the single-piece suit,
why, my dear, you've another guess coming."

"Will you tell me, I ask you, how I can remain
in the class of the poor and downtrodden,
when surrounded by mountains of ill-gotten gain
with my very soul sated and sodden."

Then Walter woke up from his rapturous dreams
(her arguments left him quite dumb).
and turned him again to pale science's faint gleams,
THUS MAMMON TO TRUTH DID SUCCUMB.

In honor of the occasion, there was a dinner for Cannon and Folin, and another for their wives. The men, thirty-five professors in all, went to the Tavern Club, while Mrs. Folin, Mrs. Cannon and sixteen wives went to the Women's College Club. Cornelia described the event for her mother:

> Mrs. Folin made a speech, giving a few of the things that had decided Dr. Folin not to go to Rochester, and then I gave a speech telling my reactions. I did not hesitate to tell funny stories, so that they did not know how bored they were, and I wound up with a proper tribute to the true freedom which Harvard gave to its teaching staff, and the conviction that I had that here Walter found the atmosphere most favorable to the kind of service he had to give. I read Tom's poem, and of course that made a great hit. The whole experience was a very warming one—a new sense of solidarity seems to possess the group, and I think the whole medical school has felt united as never before.

Later, she continued, "Walter came home in a glow. 'I think they like me!' he said, and that meant more than their thinking he knew anything about physiology."[37]

In the final analysis, both Cannon's sense of well-being and the new solidarity that possessed the group at the Medical School owed their existence to the presence of David L. Edsall in the deanship. After years of dissension and dissatisfaction among its faculty, the School at last had a helmsman who would lead it into one of its most fruitful periods of growth and achievement.

5 *The Crucible: Stewart & Rogoff Challenge the Emergency Theory*

IN LATER YEARS Cannon would refer to his research on shock during World War I as "a parenthesis of war"—a digression from the natural progression of his investigations on the adrenals and the sympathetic nervous system. In the numerous publications on traumatic shock that came out of Cannon's work at the front, no mention was ever made of his emergency theory of the function of the adrenal medulla. But the research that he had set aside had been seriously disparaged while he was away. One of his first priorities after returning from Europe was to prepare a response to the challenge mounted by the Cleveland physiologists G.N. Stewart and J.M. Rogoff during the previous three years. Picking up in the laboratory where he had left off, Cannon was determined to vindicate himself.[1]

In order to understand the debate that raged between Cannon and Stewart and Rogoff during the following years, it may be useful here to briefly recapitulate the origins of the differences that had arisen between them.

From the beginning of his research as a first-year medical student on the movements of the stomach in the conscious cat, Cannon, already interested in the emotions, became intrigued by the inhibition of gastric motility induced by fear and especially its persistence after all outward signs of distress had disappeared. He observed that when the animal became frightened, the pupils dilated, the stomach and intestines were inhibited, the heart would beat rapidly, and the hairs of the back and tail stood erect—all evidence of sympathetic activation.[2]

Late in 1910, Cannon with the assistance of Daniel de la Paz, a research fellow from the Philippines, began an investigation to see whether the adrenal medulla, which is innervated by the sympathetic nervous system and whose output mimics sympathetic action, is likewise caused to release a secretion during emotional distress. [Adrenalin(e), adrenin(e), and epinephrin(e) are names variously used for the substance secreted from the adrenal medulla. When Cannon began his work on the adrenals, he followed the English nomenclature by employing the term "adrenalin" for the secretion or extract from the adrenal gland. For many years thereafter, he distinguished between the bodily secretion, calling it "adrenin," and the commercial product that he injected, known as adrenalin. After 1939, he again followed English nomenclature and adopted "adrenaline" as his term of choice, with occasional use of "epinephrin(e)." We have attempted here to reproduce whatever terminology Cannon and others used at the time.][3]

He inserted and advanced a well-lubricated catheter into the femoral vein of an experimental cat (under anesthesia) until its tip reached the level of the adrenal veins in the vena cava. Next he drew a sample of blood to test its effect on a strip of longitudinal muscle of a cat's intestine contracting rhythmically in a small cylinder. The blood did not alter the activity of the contracting muscle. Following this test, Cannon exposed the cat for ten minutes to a barking dog to stress it. He reinserted the catheter to its former position in the vena cava, and then withdrew another sample of blood. This time the blood inhibited the contractions of the muscle strip—a well-known indication of the presence of adrenalin.

In yet another experiment, Cannon and de la Paz removed the adrenals of a cat prior to its exposure to a barking dog. Although the animal showed all the generalized signs of sympathetic stimulation from dilated pupils to the tip of its tail, the muscle strip did not relax.[4]

Some months later, Cannon began a new series of experiments with a former student, Roy Hoskins, to test the effect of asphyxia and sensory stimulation (pain) on adrenal secretion. Since the procedures to be utilized required the use of a general anesthetic, Cannon selected urethane, an agent that did not excite the cats and one that was not attended by increased sympathetic activity and stimulation of adrenal secretion. The techniques used for sampling of blood and assay for adrenalin were essentially the same as those employed in the research with de la Paz.

Again, the blood drawn from the vena cava before asphyxia did not alter the rhythmic contractions of the intestine. Following asphyxia, however, the blood produced a typical inhibition indicating the presence of heightened adrenalin secretion. When Cannon stimulated the central end of the cut sciatic nerve sufficient to maintain pupillary dilatation (such stimulation as would cause severe pain if the animal were not anesthetized), he found it also increased adrenal secretion with consequent inhibition of rhythmic contraction of muscle strips. Given these results, Cannon and Hoskins concluded that the adrenal glands were reflexly stimulated through nervous channels when a sensory trunk was excited, causing the release of adrenalin into the bloodstream.

More experiments followed, and in relatively short order Cannon, with several student collaborators, was able to show that augmented adrenal secretion creates changes in blood sugar concentration, muscle fatigue and coagulation time—all accompanied by signs of sympathetic nerve activation, e.g., cessation of digestive activities, increase of cardiac vigor, and redistribution of blood.[5]

By 1914, Cannon had gathered sufficient data for both an interpretive article, "The Emergency Function of the Adrenal Medulla in Pain and the Major Emotions," in the *American Journal of Physiology* and a book, *Bodily Changes in Pain, Hunger, Fear and Rage* (1915). To crown his findings, Cannon fashioned an over-arching theory of emergency function: "The organism which with the aid of increased adrenal secretion can best muster its energies, can best call forth sugar to supply the laboring muscles, can best lessen fatigue, and can best send blood to the parts essential in the run or the fight for life, is most likely to survive. Such, according to the view here propounded, is the function of the adrenal medulla at times of great emergency."[6]

In the same years that Cannon was publishing the initial results of his studies, George N. Stewart at the Western Reserve Medical School in Cleveland was also studying adrenal secretion. Well known for his research in biophysics and circulation physiology, Stewart had published in 1911 an extensive and highly critical review of the methods used by previous investigators in studying adrenal secretion. He urged that in the bioassay of adrenalin (or epinephrin, a term he came to prefer) at least two test objects should be employed: one that responds to adrenalin by inhibition (e.g., the intestine) and another that is stimulated to contract (e.g., the uterus), thus ruling out the effect of other extraneous agents that act in the same direction on both test objects. A second paper published the following year concerned Stewart's effort to detect epinephrin in the general circulation of patients. Stewart noted that while his methods were adequate to detect epinephrin in the adrenal venous blood of an experimental cat, he was unable to detect the substance in the general circulation of human subjects. Thus, he reasoned, the negative results must indicate the absence of epinephrin or its presence in a concentration below detectable levels.[7]

In an effort to quantify the amount of epinephrin secreted in the anesthetized animal without interference by the adrenal circulation and without dilution from the systemic circulation, Stewart adopted a method shown him by a colleague, J.J.R. Macleod. In Macleod's method, a dog was anesthetized with ether (a drug that unfortunately stimulates the sympathetic nervous system markedly when animals are lightly anesthetized), and catheters were inserted into the carotid arteries. After the abdomen was opened, electrodes were placed on the left and right splanchnic nerves just central to the adrenal glands. Ligatures were then placed in position around the inferior vena cava above and below the entrance of the adrenal veins and around the right renal vein, and left loose while a cannula for collection of the caval blood was introduced through the left renal vein. A sample of caval blood was withdrawn from the cava before the ligatures were tightened so as to make a closed pocket of the cava. That done, a second sample was removed. Later, blood was collected from the closed pocket during stimulation of the splanchnic nerves.

The blood samples were centrifuged, and the serum was assayed on strips of rabbit intestine and uterus. Although the control samples did not significantly alter the contractions of the isolated intestine or uterus strips, the blood obtained from the caval pocket during the splanchnic stimulation inhibited the intestine and stimulated the uterus—suggestive evidence that a marked increase in epinephrin had occurred. Yet when Stewart later utilized Cannon and de la Paz's collection technique and merely inserted a catheter into the undisturbed cava with the tip just anterior to the adrenal veins, he was unable to detect epinephrin in the blood even when the splanchnic nerve was stimulated. He attributed these negative results to the high degree of dilution in the inferior cava coming from the adrenals, results quite at variance with those of Cannon.[8]

For the next few years, Stewart devoted himself to extending his earlier studies on circulation time and blood flow in anesthetized animals. In one experiment, by measuring electrical charges in concentrated salt solutions that he had injected into arteries and veins of various organs, he obtained quantitative data on their transit time through various vascular beds. Later, in a study of the liberation of epinephrin from adrenal glands by stimulation of the splanchnic nerves, Stewart, with his collaborators J.M. Rogoff and F. S. Gibson, used the pupil of an eye sensitized by prior excision of the stellate ganglia to determine that the circulation time of epinephrin from the adrenal gland to the eye in an undisturbed cat was approximately ten seconds. He also found that an injection of adrenalin in the cava at the level of the adrenals induced pupillary dilatation in the same time interval as splanchnic stimulation.[9]

From their inception, Stewart's studies on blood flow and circulation time attracted attention. Early in 1914, William Halsted, chief of surgery at Johns Hopkins, after reading his paper on the determination of blood flow in the extremities, wrote Stewart of his interest in his work: "I wish we could make use of your methods in our clinic. We have enlisted the interest of Dr. Howell and others, but the technical skill required is, I fear, too great for its application except in rare cases." Theodore Hough at the University of Virginia had no such reservations when he wrote Stewart, "I have just finished reading and abstracting your studies of the circulation in the hands and feet. Again I must congratulate you on such solid contributions to our accurate knowledge of physiology and internal medicine."[10]

Soon well-established physicians turned to Stewart for advice on how to deal with patients suffering from circulatory problems. Nor were they alone in their interest. Cannon was so taken with Stewart's observation of the utility of gathering quantitative evidence of adrenal secretion that he tried to use the denervated eye as an adjunct to the catheter method and wrote to Stewart early in

1916 about some difficulties he had encountered. Stewart in reply commiserated and shared some of his own similar difficulties. It was perhaps the last friendly exchange between the two men.[11]

Shortly before the meeting of the Society for Experimental Biology and Medicine that was scheduled to be held in New Haven on May 24, 1916, Cannon received his copy of the program. Much to his surprise, he noted that Stewart and Rogoff were to present two papers on the adrenals. The titles of the papers, "The Alleged Exhaustion of the Epinephrin Store in the Adrenals by Emotional Disturbance" and "The Liberation of Epinephrin from the Adrenals," made him apprehensive. He wrote to Stewart to inquire whether he and Rogoff had evidence of failure of the adrenals to respond to emotional disturbance.

Four days later Stewart replied that his paper basically concerned T. R. Elliott's reports on the content of epinephrin in the adrenal glands, and not the discharge of the substance as reported by Cannon. He indicated that he had been unable to demonstrate any definite diminution in the content of a gland whose splanchnic connection was intact as compared with its fellow whose splanchnics had been previously cut before emotional disturbances were induced. Stewart further maintained that he was not satisfied that what Elliott spoke of as morphine "fright" should be taken as an example of such a disturbance, even though morphine unquestionably produced such an effect as Elliott described on the epinephrin content. But, he continued, "'fright,' 'anger' and whatever other emotions are excited in a dog and cat egged on to vex each other for long periods without the possibility of physical damage being inflicted, do not seem to sensibly affect the content of the still innervated gland in either animal. It would therefore seem best to consider the morphin effect on cats as due to some other action of the drug." He also told Cannon that he had "obtained in dogs, in which there were no symptoms of fright under morphin, very much the same diminution of the medullary content as seen in cats." He concluded that there was no necessary connection between exhaustion or diminution of the epinephrin store and the rate of liberation of epinephrin into the blood.[12]

The reassuring tone of Stewart's letter allayed Cannon's concerns. Shortly after the meeting at Yale, however, another paper by Stewart and Rogoff appeared, entitled "The Spontaneous Liberation of Epinephrin from the Adrenals." Cannon was stunned by the content of the latter publication, since it completely refuted the findings of Cannon and Hoskins and emphasized that no increase in the release of epinephrin was detectable when the sensory nerves (brachial) were stimulated, and if any increase was produced by asphyxia, it was very slight. Since the preliminary note of this paper had been published in the *Proceedings of the Society for Experimental Biology and Medicine* on May 24, 1916, the same day that Stewart wrote to Cannon, and the completed manuscript was received for publication by the *Journal of Pharmacology and Experimental Therapeutics* on July 26, one must assume that most, if not all, of the experiments had been completed when Stewart sent his cordial but apparently misleading letter to Cannon.

Stewart and Rogoff continued to use Macleod's traumatic caval pocket technique in their research—despite the profound alterations it produced in the circulation. The controlled blood pressures that they reported in anesthetized dogs and cats varied from 12 to 140 mm.Hg. and were often in the hypotensive range of from 30 to 40 mm.Hg.—that is, at shock levels. It is curious that Stewart, an experienced cardiovascular physiologist, made little comment on the state of the circulation or, for that matter, that no one else seemed to notice. Notwithstanding these problems with the caval technique, the Cleveland investigators were adamant in their claim of the superiority of their methods over the catheter technique used by Cannon, and for a variety of reasons.

First, they were able to quantify the amount of epinephrin released by measuring the adrenal blood flow and concentration of epinephrin in the pocket. Second, their technique did not necessarily require the withdrawal of blood from the body with possible alterations in the blood samples producing substances that simulated the action of epinephrin. And third, they found that the amount of epinephrin could then be estimated *in vivo* by its action on the denervated eye and on the blood pressure in the same animal.

They did hint in conclusion, however, that with the more extensive operation in their experiments, the spontaneous discharge of epinephrin may have been already so elevated that there was no room for detectable increase by asphyxia, etc., thus accounting for the negative results. But they promptly chose to ignore their own caveat.[13]

Thoroughly convinced of the constancy of epinephrin release and the lack of a reflex response of the adrenal medulla to sensory nerve stimulation, Stewart and Rogoff began a new series of experiments to prove that liberation of epinephrin from the medulla was not indispensable for life or health. In the animals they studied, they excised one adrenal and cut the splanchnic nerve to the other, thus eliminating nervous control of the medulla and lowering the circulating level of epinephrin. They allowed the animals to recover for several weeks before testing them for epinephrin release, at which time they were unable to detect spontaneous epinephrin secretion into the pocket. Even when they employed the more sensitive intestinal and uterine strips for bioassay, the animals behaved in a perfectly normal fashion under a variety of conditions, their very lives casting a shadow on Cannon's emergency theory.[14]

Cannon was deeply troubled by the disparity between his results and those of Stewart and Rogoff. Spurred on by the need to defend the validity of his research, he began, with the assistance of H.F. Pierce, a student, to search for an additional method to demonstrate the effect of asphyxia and sensory stimulation on adrenal secretion. In the course of his search, he chanced upon a 1914 paper by Herbert Gasser and Walter Meek, who had reported that asphyxia in an experimental animal with a denervated heart led to a large increment in heart rate, but after adrenalectomy the rise was small. Given these findings, Cannon reasoned that the magnitude of the increase in heart rate provided a semi-quantitative estimate of the elevated secretion of adrenin.

Before sailing for Europe in May 1917, he fired off a salvo at Stewart and Rogoff—a short note reporting his recent use of the denervated heart in the cat.[15] The method, he noted, was so simple that anyone could use it. It called for isolating the heart from the nervous system by severing both vagi in the neck and removing the stellate ganglia. Cannon also added a word of criticism of the caval pocket technique and warned that a great deal of care should be taken in operating on the abdominal cavity and that manipulations should be minimized. He recalled that such abdominal operations caused continuous discharges of nerve impulses along splanchnic courses. Thus, the adrenal glands would be persistently stimulated—a potent source of error in Stewart's previous work.

Nevertheless, papers highly critical of Cannon's emergency theory continued to flow from the Cushing Laboratories where

Stewart and Rogoff worked in Cleveland.[16] In the month of May 1917, for example, three long, detailed manuscripts were received for publication in various journals, each one buttressed with countless reproductions of kymographic records and tables filled with quantitative data. Throughout 1917 and 1918, during the months that Cannon was at war, a barrage of more than a half dozen similar papers by Stewart and Rogoff followed—all insisting on the continuous output of epinephrin from the adrenal gland with no increase of secretion during sensory stimulation or emotional excitement.

In the early spring of 1919, soon after returning to Boston, Cannon attempted to catch up with some of the many scientific books and papers that had accumulated during his absence. He turned first to those related to his interests in the sympathetic nervous system and adrenal glands. To his dismay, he discovered that J. J. R. Macleod in his widely-used textbook, *Physiology and Biochemistry,* published in 1918, had written that the object of Cannon's investigation of adrenal function "was to test the hypothesis that the reaction of an animal to fear and other emotional states is dependent on hypersecretion of epinephrin into the blood." Macleod noted that a number of physiologists, in the course of their own investigations of adrenal secretion, had found evidence that cast doubt on Cannon's theory of the emergency function of the adrenal medulla.[17]

Macleod's statements startled Cannon, who promptly wrote him, "I have not consciously intended to lead any one to suppose that the existence of fright in an animal and the usual changes in heart rate, blood pressure, pupil, hairs, etc., were dependent on adrenal secretion. The extent of my claim has been that the adrenal glands are stimulated to secrete during excitement and that the secreted adrenin may serve as an aid to sympathetic impulses, possibly augmenting or continuing them." Cannon said he would be greatly obliged if Macleod would tell him where he had been unclear or "perchance led readers astray by some twist of expression."

After re-reading parts of Cannon's book, *Bodily Changes in Pain, Hunger, Fear and Rage,* and several of his papers on the adrenals, Macleod apologized. "I must confess," he wrote Cannon, "that I was surprised to find that you have nowhere made the statement that you believe fear and other emotional states to be *dependent* on hypersecretion of epinephrin into the blood." But he cautioned Cannon that "these writings give the general impression that you believe this dependency to exist . . . I think it is important that you should—as you say you are going to do—state precisely your position with regard to this question." In particular, Macleod promised at the first opportunity to replace the phrase "dependent on" to "associated with."[18]

Given Macleod's pointed statements, Cannon began a careful review of some of his previous investigations on the relations of the adrenals and the sympathetic nervous system. He had hardly started before he was brought up short again, this time by an editorial in *JAMA* that echoed Macleod's remarks:

> The assumption that the physiologic reaction to fear and other emotional states is dependent on a hypersecretion of epinephrin into the blood—an assertion supported by Cannon—has been rendered of doubtful validity, if not actually untenable, by the extensive researches of Stewart and Rogoff and their collaborators at the Western Reserve University School of Medicine. They report that animals in which both suprarenal glands are entirely removed from the influence of the nervous system behave like normal animals when frightened. Therefore, epinephrin appears to have little if any contributory influence in the usual course of events associated with emotional states.[19]

It is not known who wrote the editorial. It may have been Morris Fishbein, the editor of *JAMA,* who had a penchant for bringing current interesting research in medical science to the attention of his readers. Or it might have been contributed by A.J. Carlson, Professor of Physiology at the University of Chicago School of Medicine—a close friend of Stewart, who enjoyed annoying Cannon and other Eastern physiologic cognoscenti from time to time. Given the confident tone of the editorial, it could even have been the work of Stewart and Rogoff themselves, for it clearly rested on the thesis they had developed in their series of articles in leading medical journals ever since 1916.

Whatever its source, the editorial in *JAMA* rudely jolted Cannon back into the dispute he had had with Stewart and Rogoff before the war and signaled to him that the controversy over his emergency theory was far from finished.

Cannon kept Macleod's admonition in mind as he began to fulfill his promise, made in the preliminary note published before he went to Europe, to produce a full-length report on the denervated heart method as an accurate indicator of adrenal secretion. His paper, published at the end of 1919 in the *American Journal of Physiology* as the fifth in his series of Studies on the Condition of Activity in Endocrine Glands, was entitled "The Isolated Heart as an Indicator of Adrenal Secretion Induced by Pain, Asphyxia, and Excitement." Here he referred to his earlier work, which "presented a new view of the function of the sympathetic division of the autonomic system and of the adrenal medulla in important bodily adjustments." This time, he took great pains to communicate in precise terms the results of his research so there would be no misunderstandings or false impressions. In the final section, Cannon reiterated his statement to Macleod; he had never claimed "that there is at any stage a primacy of adrenin in the production of physiological or psychological changes seen during strong emotion."[20]

A large part of Cannon's paper was devoted to skewering Stewart and Rogoff. He clearly stated that in his review of the literature only Stewart and Rogoff's experiments appeared to support the concept of a relatively constant output of adrenin, with no increase of adrenal secretion during sensory stimulation or emotional excitement. Step by step, he presented evidence that refuted their criticisms of the denervated heart technique for measurement of adrenin release. At the same time, he expanded his own critique of their caval pocket method, citing the earlier experiments of Bayliss and Starling, which demonstrated that opening the abdomen produced strong splanchnic stimulation sufficient to inhibit intestinal motility, and by inference, continuously and abnormally stimulating the splanchnic nerves to the adrenal medulla. Thus, Cannon concluded, the pocket method was incapable of yielding any reliable evidence regarding the "normal" secretion of these glands.

Cannon felt quite secure that in the denervated heart he now had a simple, semi-quantitative method for estimating the changes in the rate of adrenal secretion. In fact, he was so confident of the technique that he did not provide any other bioassay data correlating the changes in blood levels of adrenin induced by pain or asphyxia with increments in heart rate. His friend, Joseph Erlanger, professor of physiology at the Washington University School of Medicine, however, after reading Cannon's paper asked for confirmatory evidence. Cannon's reply was surprisingly ingenuous. "That the acceleration of the heart was due to the secretion of adrenin by the glands I had not considered as likely to be questioned," he admitted,

> chiefly because of the absence of any other known agents in the body that will bring about, through the blood stream, such a change as I noted. Furthermore, in 1912, Anrep reported an acceleration of the denervated heart on stimulation of the splanchnic nerve—an acceleration which disappeared when he clipped the adrenal veins and which again reappeared when the check was removed. On exclusion of the adrenals, splanchnic stimulation no longer gave any acceleration of the beat. I repeated these observations but did not mention the results for the reason I have stated above. It seemed to me that Anrep's evidence should be sufficient to show that it is actually secreted adrenin that is accelerating the heart even if the indications of my own experiments were not decisive.[21]

Just before publication of his paper, Cannon wrote to William Bayliss in London:

> After a thorough vacation during the summer, I returned to the laboratory this fall delighted to find that my old zest for work and keen interest in research has been quite restored. The early part of the fall was spent in putting together a paper on adrenal secretion, which will appear in the December number of the American journal. I am particularly desirous of having you read it. It is the first controversial paper that I had ever had to write and the problem of being strict in my scientific criticism and yet personally considerate in my expression required a good deal of care.[22]

While Cannon's buoyant mood was short-lived, his plunge into controversy was not. Further attacks by Stewart and Rogoff would soon engulf him.

In June of 1920, Stewart and Rogoff published a 59-page paper in the *A.J.P.* criticizing the denervated heart method. Although they granted that epinephrin may cause an acceleration of the heart after section of the vagi and stellate ganglia, they questioned whether other factors could be involved, since good acceleration, even as great as that obtained with the intact adrenals, could often be observed after they had removed the adrenal glands.

The Cleveland investigators had criticized Cannon's earlier catheter method for estimating the rate of secretion from the adrenals, but, they noted, at least one could test the blood sample by bioassay; the denervated heart had no such favorable feature and thus "labors under even more serious defects . . . than does the catheter method." They added derisively,

> When Dr. Cannon stimulates the central end of the sciatic numerous reflex effects may be caused, the consequences of which upon the rate of the heart cannot be easily controlled. Among these the vasomotor reflex changes are conspicuous, and it is obvious that in this way great variations may be produced in the pressure in the cavities of the heart, the rate of blood flow through the coronary system and, therefore, the amount of epinephrin passing through the coronaries, any of which may lead to an acceleration of the heart beat without any change having occurred in the rate of output of epinephrin.

To prevent the secretion of epinephrin into the systemic circulation with little trauma to the gland or the splanchnic nerves, Stewart and Rogoff had tied a ligature around the lumbo-adrenal vein distal to the adrenal and one close to the entrance into the vena cava, a "clipping" technique of which Cannon was very critical. In response to his criticism they stated, "Cannon takes exception to 'clipping' because it does not eliminate the heart reaction to sciatic stimulation or asphyxia," suggesting leakage through anastomotic vessels. They continued, in a derogatory tone, that Cannon "quotes, for instance, an experiment in which before ligation of the adrenal veins asphyxia of a certain duration caused an acceleration in the heart rate of 40 beats. After ligation of the veins the acceleration was precisely the same. We should have thought the only possible interpretation of such a result would have been that epinephrin had nothing to do with the reaction."

To provide further evidence that the denervated heart was a poor assay for circulating epinephrin, Stewart and Rogoff compared the response of the denervated eye to that of the denervated heart. Although clipping the adrenal veins markedly reduced the pupillary response to splanchnic stimulation, acceleration of the heart was scarcely diminished, "which shows clearly that the latter reaction cannot be a quantitative test for epinephrin." Moreover, they found similar acceleration of heart rate in all cats tested when the sciatic nerve was stimulated after removal of one adrenal and the cutting of the opposite splanchnic nerve, eliminating nervous control of adrenal secretion, or following bilateral adrenalectomy.

They concluded their searing remarks by labeling as "illusory" Cannon's "so-called proof" of the epinephrin acceleration of the denervated heart. "Our position is simply this," they stated:

> We have investigated the influence of stimulation of the sciatic and brachial nerves upon the rate of epinephrin output by a direct method, correct in principle and free from ambiguity, and have obtained negative results. Doctor Cannon states that by means of an indirect method (the denervated heart reaction) he obtains a positive result. We show that this reaction cannot yield any information as to the rate of epinephrin output from the adrenals or as to changes in that rate, since it is obtainable when the epinephrin output of the adrenals is abolished. And here we are entitled to rest our case, not of course claiming that sensory stimulation *cannot* increase the epinephrin output, but that no increase has hitherto been proved.[23]

Before Cannon had an opportunity to reply, Stewart and Rogoff published a second lengthy and critical review one month later. Whereas Cannon's studies aimed to estimate the concentration of epinephrin in the blood returning to the heart, Stewart and Rogoff insisted that one had to measure the actual amount of the substance secreted per minute. They emphasized that Cannon had no such quantitative data, i.e., measurement of the concentration and flow rate, but only qualitative information. The faster rate of the denervated heart, they argued, might be attributable to a rise of blood pressure and consequent improved coronary flow when the central end of the sciatic, or peripheral end of the splanchnic, nerve was stimulated. In such a case, the amount of epinephrin in the coronary tree could be increased and the heart rate elevated without change in the rate of adrenal secretion merely by a redistribution of blood flow in the body.[24]

To test the hypothesis that "redistribution of blood" accounted for the faster cardiac rate, Cannon, with the help of David Rapport, one of his research fellows, prepared what they called a "reduced" animal with minimal vascular reflexes. Under anesthesia they ligated the carotid, brachial and renal arteries, and the aorta below the kidneys, and then severed the mesenteric nerves. In the abdomen the only viscera that remained innervated by the splanchnic nerves were the adrenal glands. Stimulation of the brachial nerves was not capable of inducing significant redistribution of blood flow, since little of the vascular tree could respond to vasoconstrictor impulses. Now, although blood pressure changes induced by sensory stimulation in "reduced" animals were quite different from those in normal ones as a result of the operative procedures, cardiac acceleration was similar in both groups of animals. These observations did not confirm the thesis proposed by Stewart and Rogoff for redistribution of blood flow.

Cannon, in his paper with Rapport, concluded that Stewart and Rogoff, having employed so-called quantitative methods, believed that their results should be accepted rather than the results of others who relied only on qualitative procedures. "It would be absurd," he insisted,

> "to argue that a carpenter's level is useless and untrustworthy because it does not measure quantitatively the degree of deviation from the horizontal. Its service is to tell whether deviation is present or not. The denervated heart likewise, since unaffected by all except chemical changes brought to it in the blood stream, is, we submit, when used as we have used it, with the adrenals the only innervated abdominal viscera, a reliable and delicate instrument for detecting whether or not adrenin has been added to the circulating blood. There is nothing in the quantitative character of negative evidence that gives it dominance over positive evidence which is quite as accurate.[25]

Since the data obtained from the "reduced" animal had not accounted for the increased rate of the denervated heart during splanchnic stimulation in the adrenalectomized animal, Cannon, with the assistance of Joseph Uridil, a visiting anatomist, and Fred Griffith, a graduate student in physiology, began to search for some other, as yet unknown, factors that might be responsible. This time, he followed the lead of Stewart and Rogoff, who in an earlier paper had suggested that some factor produced in the abdomen was a possibility and raised the question whether the agent could be coming from the liver.[26]

In early 1921, in a brief note in *Endocrinology* entitled "Some

Novel Effects Produced by Stimulating the Nerves of the Liver," Cannon, Uridil and Griffith reported that "After removal of the adrenal glands, splanchnic stimulation will cause a faster beat of the denervated heart. Since this occurs when the inferior cava and the portal vein are closed below the liver, and since it does not occur when the hepatic nerves are cut, the inference is drawn that some agent arising in the liver is the occasion of the faster beat."

Cannon and Uridil expanded on the earlier work and made the prescient observation that although there was similarity of action of the hepatic agent and adrenin in that they both increased heart rate and raised blood pressure, the former did not dilate the pupil, a characteristic action of adrenin. Thus, the chemical released from the liver did not resemble adrenin in all respects.

To provide additional proof that the cardiac factor arose in the liver, they stimulated the peripheral end of the cut hepatic nerves and induced a faster rate of the denervated heart. The results, however, were extremely variable until Cannon hit upon the notion that food intake played an important role. When the cats were fed a high meat diet prior to the experiments, the cardiac acceleration was four to five times greater than that noted in cats that were fasted.[27]

Since the extracts of the adrenal medulla were capable of causing the changes produced by adrenal secretion, it seemed possible to Cannon, now working with Griffith, that likewise a substance might be extracted from the liver that would have the effects seen after stimulating the hepatic nerves. They confirmed that the accelerator agent, appearing when the hepatic nerves were stimulated, must be discharged into the blood stream by "a substance of special and unknown nature." But further than that they were not prepared to go.[28]

As Cannon and Uridil, as well as Stewart and Rogoff, observed, the hepatic factor caused no pupillary change in the denervated iris, indicating that it differed in some pharmacological aspects from adrenaline. Both groups had come close to two remarkable discoveries, one of which, chemical mediation of nerve impulses, would have to wait for another decade, and the second, the duality of the sympathetic mediators adrenaline and noradrenaline, even longer.

Cannon's papers with Rapport, Uridil and Griffith, along with another paper in collaboration with Rapport, were published in the series of Studies on the Conditions of Activity in Endocrine Glands in the *American Journal of Physiology.* It is noteworthy that in contrast to Cannon's pre-war papers, which failed to record the number of experimental trials performed, the protocol used for each experiment, and specific quantitative results, these new reports contained detailed data in the form of figures, graphs, and tables that quantified the results. Even more, they constituted a formidable threat to Stewart and Rogoff's claims for a constant secretion.

Soon after, Cannon, with the aid of Rosendo Carrasco-Formiguera, a Rockefeller fellow from Spain, delivered yet another blow to Stewart and Rogoff's arguments by using their own method, the vena cava pocket, to contradict their opponents' results.[29] Faced with this latest evidence, it seemed only reasonable that Stewart and Rogoff would have to accept Cannon's thesis that asphyxia and sensory stimulation augmented adrenal secretion, or else deny their own experiments.

Stewart and Rogoff did neither, refusing to acknowledge the reports of Cannon's work with Rapport, Uridil, Griffith and Carrasco-Formiguera. Instead, they continued to remonstrate against the use of denervated indicators by the Harvard investigators, as well as by other scientists, averring that "it is not of course incumbent upon us . . . to explain fully why our results differ from those of other workers."[30] Cannon was frustrated by such statements and angry at Stewart and Rogoff for ignoring his rejoinders to their objections, as put forth in his more recent papers. Writing to Roy Hoskins in February of 1923, he accused them of "bringing about confusion instead of elucidation." Torn between pitching into the Cleveland investigators and letting matters slide so he could "go forward with other experiments and not be bothered by this foolish controversy," Cannon was beset by self-doubt. "Am I wrong in calling it foolish?" he inquired of Hoskins. "Should I take a more belligerent attitude and fight them until they give in?"

Hoskins sympathized with Cannon but advised him to bide his time. "As regards the advisability of going in for polemics with Stewart," he replied, "I should personally regret your doing so. My feeling is that with sufficient rope they can be trusted to hang themselves. In the meanwhile, the work no doubt will stimulate a considerable amount of further research and aid in the establishment of the truth."[31]

Hoskins' advice proved to be sound. Within months, more and more evidence began to appear from other laboratories in support of Cannon's work. Frank Hartman, at Buffalo, who had worked in Cannon's laboratory some years before, was one of those who published papers strongly favoring his concept of the emergency function of the adrenal. Using the iris of the unanesthetized cat, sensitized to epinephrin by removal of the superior cervical and ciliary ganglia as an assay of epinephrin release, Hartman and his associates clearly demonstrated that even minor operations increased the output of epinephrin and that ether anesthesia did not prevent an increase in the discharge of epinephrin from stimulation produced by cutting through skin and muscle. In major surgical operations they found that the discharge of epinephrin was greatly augmented, and this finding had a direct bearing upon the caval pocket method used extensively by Stewart and Rogoff. Hartman and his co-workers wrote that they were able to draw but one conclusion: "the 'cava pocket' method does not permit an estimation of the output of epinephrin which prevails in a quiet intact animal."[32]

Based on Hartman's studies, *JAMA*, which over the years had published several editorials critical of Cannon's hypothesis and supportive of Stewart and Rogoff's views, did a turnabout. Its editors wrote that the latest contribution to the subject from the University of Buffalo supported the general conclusion that strong stimulation of many varieties increases the discharge of epinephrin into the circulation, providing further evidence for the emergency function of the suprarenal medulla.[33]

But the crowning blow to Stewart and Rogoff came from Sakuji Kodama, working for several years to perfect their caval pocket method in the physiological laboratory of Professor Y. Satake in Japan. Beginning in 1923 Kodama reported in the *Tohoku Journal of Experimental Medicine* that in his studies using the caval pocket, he observed a distinct increase of epinephrine output during sensory stimulation that contradicted Stewart and Rogoff's observations. When some of Kodama's experiments produced negative results, he questioned whether they might be due to abnormal conditions resulting from the depth of anaesthesia employed, and the trauma inflicted on the animals. He therefore attempted to determine what he called the normal physiological output of epinephrine. In dogs in which the abdominal walls and viscera were previously deafferented, he performed an additional series of caval pocket experiments by cutting the dorsal (sensory) roots from the fourth thoracic to the second lumbar region, hence eliminating the need for anaesthesia for the abdominal manipulations. Stimulation of the median nerve in the unanaesthetized animals now induced an increase in epinephrine in every one of the animals—evidence that anaesthesia does tend to inhibit epinephrine release.[34]

Cannon was exhilarated by Kodama's reports. Although he had been highly critical of Stewart and Rogoff's caval pocket method, he now applauded Kodama's work, in which essentially the same techniques were employed. What mattered to Cannon was that the results were in agreement with his own, and he spread the news of Kodama's observations in his publications as well as in correspondence with his peers. The position, Cannon told one and all, had been changed "from a difference between Stewart and Rogoff and myself to a difference between them and Dr. Kodama." Henry Dale, for one, replied to Cannon's letter in October 1923 that he had not yet seen Kodama's paper, but remarked, "I am glad to hear that you are getting independent support by somebody using Stewart's method. I had a talk with Stewart last year, and he did not convince me in the least."[35]

Rogoff, predictably, was outraged and made cutting remarks about Kodama's publications. It was only too clear, he charged, that Kodama had not adequately mastered the method of assay and therefore his results possessed no value. Some time later, Stewart was even more emphatic in castigating Kodama. He insisted that he and Rogoff had repeatedly found it impossible to demonstrate any increase in the output of epinephrin produced reflexively by stimulation of afferent nerves. He claimed that Kodama's contrary conclusion was based upon an inadequate grasp of the procedure by which he estimated the epinephrin output. The masses of alleged concentrations and outputs with which Kodama filled his paper (and which were much higher than those reported by Stewart and Rogoff) had no relation to reality and, Stewart contended, "it would be very unprofitable to spend time analyzing these imaginary figures."[36]

Kodama did not know of Stewart and Rogoff's criticisms, for he had left Japan to seek further training in Europe. During the next two years, it fell to Sugawara, Watanabe, and Saito, in Satake's laboratory, to investigate the discrepancy between the experimental results of Kodama and those of Stewart and Rogoff. Using the improved caval pocket technique and deafferentation procedure developed by Kodama, by 1926 they demonstrated that anaesthesia (ether or urethane) blunted the response of the adrenal medulla to sensory stimulation and reduced the positive responses by fifty percent. Thus the use of deep anaesthesia may have been in large measure responsible for Stewart and Rogoff's apparent negative results. They also showed that with deafferentation of thoraco-lumbar sensory fibers, stimulation of the median nerve in the unanaesthetized animal induced a larger increment in epinephrine release in all of the cats and dogs they studied, with a rise of four- to five-fold from control levels. The Japanese group concluded that in order to study the effect of sensory stimulation upon epinephrine output in the normal state of the central nervous system, it is essential to avoid the use of narcosis.[37]

The tide had definitely turned in favor of Cannon's emergency theory, but Stewart and Rogoff remained adamant. They stubbornly maintained their firm conviction that since the adrenal medulla is not indispensable for life, it could not play a significant physiological role in humans or animals. Only gradually was it revealed that they had long been involved in studies on cortical function, believing, as they did, that the cortex is the portion of the adrenal gland that must be essential. Indeed, they were among the first investigators to come up with an active cortical extract. Nevertheless, they continued to feed the flames of the controversy. Even after Stewart dropped out (he suffered from pernicious anemia and died in 1930), Rogoff carried on the attack against Cannon, and he never relented.[38]

For Cannon, the impact of the controversy on his subsequent research and career would reverberate for years. Although the annual meetings of the American Physiological Society provided an arena for many lively debates and sharp differences of opinion, the famous argument between Cannon and Stewart and Rogoff had, perhaps, come closest to involving personal animosity and bitterness. Years later, members of Cannon's family, seconded by friends and colleagues, remembered how much he had dreaded and been upset by those confrontations.[39]

It was years later, too, that H.F. Pierce, Cannon's young assistant when he first devised his denervated heart method, reminisced about a talk A. J. Carlson had given at a Federation dinner in the latter part of the 1920s. Carlson, who enjoyed fanning the flames of the controversy to see what sparks he could kindle, told the audience about a dream he had had: He was walking down the gold-paved streets of Heaven when he met Dr. Cannon, who had just arrived and appeared to be in a very melancholy mood. Not having seen Cannon previously in this heavenly spot, Carlson greeted him warmly and inquired why he seemed so unhappy now that he had attained the state of permanent bliss. Alas, Heaven fell short of Cannon's expectations because, he told Carlson, "there is no emergency here."

If Carlson expected to get a rise out of Cannon, he was sorely disappointed. The latter did not attend the dinner but heard the story repeated afterward. Instead of a heated response he reacted with mock *gravitas,* writing a friend, "That showed a good deal of insight on Carlson's part, for I believe that that would be my main objection to Heaven as it is pictured."[40] Cannon could afford to be cool; the dispute that had engaged physiologists with increasing ferocity for more than a decade had been resolved in his favor.

6 *Antivivisection, Another Field of Battle*

THE CLASH with Stewart and Rogoff was not the only controversy that bedeviled Cannon in the years following the war. There were others, and not the least his long-standing struggle with antivivisectionists over the use of animals in medical research. During the spring of 1919, Cannon heard rumbles of increased antivivisectionist activity. In one instance, there was a report that antivivisectionists had succeeded in introducing a bill in the Pennsylvania legislature to prohibit the use of dogs in experimental medical research. Another was an account of a recent meeting of the Washington, D.C., Humane Society in which Franklin Roosevelt, the popular Assistant Secretary of the Navy, had called on public opinion to force the medical profession to stop abuses involved in experimental research.[1]

The reports were not unusual, nor did they appear to be particularly threatening to Cannon. Later, however, after learning from a friend that Senator Henry L. Myers of Montana had introduced a bill in the U.S. Senate to prevent experiments on dogs in the District of Columbia and the territories and insular possessions of the United States, he promptly initiated, as chairman of the A.M.A.'s Committee for the Protection of Medical Research (the new name of the former Defense Council), a campaign to inform physicians, public health officers and medical scientists of the necessity of meeting this new threat to experimental medicine. He also alerted the bureaucracy of the American Medical Association to prepare for a round of antivivisectionist hearings before a congressional committee. A failure to defeat the passage of the Myers bill, Cannon warned, would be disastrous and ultimately lead to the successful initiation of similar bills in state legislatures throughout the country.

The Myers bill was skillfully drawn. Although it seemed to be restricted geographically to the District of Columbia, on reflection it became clear that if the bill passed, it would in effect put at risk experimental research conducted by such government agencies as the Hygienic Laboratory of the U.S. Public Health Service, the Division of Animal Industry, the Chemical Bureau of the Department of Agriculture, the Army and Navy Medical Corps, as well as the laboratories of the Food and Drug Administration. There were other problems. The framers of the bill, by focusing on the use of the dog in research, struck a chord that aroused widespread public sympathy and support for the antivivisectionist cause.[2]

Tales of dogs' exploits during the war in rescuing wounded soldiers from the battlefield added significantly to an already widely-held public conviction of the dog's extraordinary character. Ernest Harold Baynes, a prominent naturalist who had spent time in France on behalf of the American Red Cross, was so moved by the record of the dog's bravery in battle that he began a national drive to establish a fitting memorial to celebrate the dog's faithfulness to man at risk of its own life. *Starry Cross,* one of the leading antivivisectionist journals in the United States, fully seconded Baynes's efforts. "If we had not had animals in the war," it informed its readers, "victory would not have been ours."[3]

The hearings, which both the antivivisectionists and Cannon had anticipated, began on November 1, 1919, before a subcommittee of the Senate Judiciary Committee composed of George Norris of Nebraska, LeBaron Colt of Rhode Island and Henry Ashurst of Arizona. From the outset, they proved to be reminiscent of hearings conducted by Senator Gallinger in behalf of antivivisectionists almost twenty years before. Those who testified in behalf of the Myers bill were for the most part leaders and members of the various antivivisectionist and humane societies of Boston, Philadelphia, New York and Washington. Aside from a panegyric of the dog's virtues at the beginning of the hearings by Edward Clement, former president of the New England Antivivisectionist Society, little more was said about the dog during the hearings.[4]

The first speaker, Mrs. C.F. Farrell of the New York Antivivisectionist Society, set the direction and tone of the testimony. She began with a catalogue of charges: that by definition experimental medicine was cruel, that it was conducted in secret, that physicians who engaged in experimental medicine were different in outlook and behavior from kindly practicing physicians, and that the serums and vaccines medical scientists had concocted to fight disease were useless. Speakers who followed in support of the Myers bill basically sought by their testimony to buttress the various points made by Mrs. Farrell.

Save for the passionate reiteration of their convictions, and the submission of a list of public figures, prelates and writers who shared and supported their beliefs, antivivisection speakers came to the hearings armed with moral righteousness, but with little other proof that bore on the cruelties they claimed were present in contemporary experimental investigation. Mrs. Farrell's deficiencies about the nature and purpose of research were particularly telling. When committee members asked her for examples of cruel practices other than those of the nineteenth-century medical scientists she had spoken of, she cited George Crile's dog experiments on blood pressure. When pressed by Senator Ashurst if she knew the reason for the experiments, she was unable to provide an answer. "I know you are asking me; but there are a great many of these experiments and one cannot tell why they are done," she said.

As the hearings continued, it became apparent that antivivisectionist testimony was not particularly concerned with the use of the dog as an experimental animal. Rather, it was viewed as a wedge for future legislation to ban the use of all animals in medical research. The crux of their position lay in supporting Mrs. Farrell's assertion of the inherent cruelty involved in animal experimentation. Perhaps the most passionate statement in support of this argument was made by Walter S. Hutchins, a former president of the Washington, D.C., Humane Society. "We do not believe," he told the committee, "that any animal should be tortured, not even a mouse. We do not demand—we do want but do not ask in this hall—that mice should be excepted, or that cats, rabbits, or guinea pigs should be excepted. We are asking for the dog now. Later on perhaps we will ask for others." Speaking of the character of medical scientists, he continued, "Probably they have done much good in the world, but we can say that they are cruel in spite of any defense that they can possibly make of their cruelty . . . We think no man has the right, no matter how sincere he may be, to use his superior sagacity, intelligence and knowledge to maltreat helpless creatures."

At no time did Cannon during his tenure as chairman of the A.M.A. Committee for the Protection of Medical Research take lightly any legislative battles with antivivisectionists. In fighting the Myers bill, he determined, in collaboration with George W. McCoy, the director of the Hygienic Laboratory of the U. S. Public Health Service, to recruit a number of medical scientists who worked in various government agencies to testify about the nature and utility of their work as well as of the value of the dog in their research.

The testimony given by Reid Hunt, professor of pharmacology at Harvard Medical School and formerly a member of the Public Health Service, was particularly effective. Hunt spoke directly to the application of animal research in the development of therapeutics. By telling examples, he demonstrated beyond cavil the utility of laboratory experiments on dogs that led to the discovery of new drugs, as well as to establishing standards of strength and purity of various drugs. His testimony about the safety and usefulness of the serums and vaccines developed by medical scientists for treatment of syphilis and diphtheria put to rest the committee's qualms that had been raised by the accusations of the antivivisectionists.

The testimony of Hunt and several government scientists, backed by letters of support from prominent surgeons, physicians and deans of various medical schools who could not attend the hearings, clearly won over the committee. The Myers bill failed, but Cannon knew from past experience that the victory was at best temporary, and that new antivivisectionist bills would be quickly presented again either to the Congress or to one or another state legislature.[5] Although he felt that vigorous steps would have to be taken to counter future antivivisectionist activity, he soon found that some of the new members of the A.M.A. Committee did not share his views.

George McCoy, for example, questioned the advisability of confronting the antivivisectionist forces at every turn. He believed the dangers were two-fold. As he wrote Cannon, "we dignify and give standing to them by meeting their absurd charges and contentions," and by such opposition, one might inadvertently help furnish the diverse groups in the antivivisectionist camp with a common cause. Cannon brushed McCoy's objections aside. "It is wiser," he replied, "to oppose them than to let the audience feel that there is truth in their declarations . . . I do not think that giving them the facts publicly is going to better their situation in any sense."[6]

Even as Cannon wrote McCoy of the need to meet antivivisectionist declarations head on, there was, in the wake of failure to contain widespread epidemics of poliomyelitis in 1916–1917 and influenza in 1918–1919, opposition to vaccination and public health regulation. The lack of public confidence in scientific medicine encouraged peripheral practitioners such as osteopaths, chiropractors and Christian Scientists to increase lobbying various state legislatures for licenses to practice medicine. Joining the search for legitimacy, cultists and quacks in various states organized a Medical Liberty League, and they began not only to argue vociferously for guarantees "to practice the healing arts without discrimination" but to abolish compulsory vaccination as well.

An editorial early in 1919 in the *Starry Cross* called for opposition to medical science and public health regulation: "Insist on your rights to bring up and care for your children free from medical inspection and compulsory vaccination . . . Look with suspicion upon vital statistics prepared by the very people who have a case to prove and a living to make . . . Distrust wire-pulling doctors and arbitrary health boards, which invade homes, close churches, enforce absurd quarantines, and spread fear throughout the community—fight for sanity and sanitation."

Starry Cross's alert found willing allies in the general public. In New Mexico, there was a public protest against the organization of a State Board of Health with a commissioner who not only had to be a physician but also had the power to promulgate as well as enforce all rules necessary to public health. In Montana, the state legislature voted unanimously to reject Senate bills that provided for semi-annual medical examination of all school children and expulsion of those who failed or refused to be examined or vaccinated. In northern California, parents organized a Public School Protective League to fight a legislative proposal to extend the physical examination of school children, arguing that if such a proposal passed, people would find themselves under the domination of physicians exercising state authority.

These were by no means isolated instances. During the fall of

1919, there were public protests in Cambridge, Massachusetts, after the state legislature passed a compulsory vaccination law, and Dr. Samuel Katzoff in Connecticut organized an antivaccination league. Early in 1920, a special officer of the Pennsylvania State Board of Health was sent to a small community to vaccinate the entire population against an incipient smallpox epidemic. When it was learned he had issued arrest warrants for twenty local businessmen who would not be vaccinated, the health officer was driven out of town by an irate mob.

Changes in the political climate—notably the adoption of initiative and referendum laws in various states as well as the extension of the vote to women—had an equally profound effect on the ever-present conflict between antivivisectionists and medical scientists.

In April 1920 Mrs. Rosalind Wright on behalf of the California Antivivisection Society collected over 100,000 signatures and successfully placed a bill prohibiting animal experimentation on the fall election ballot. The Northern California School Protection League soon followed suit with an initiative against vaccination. Chiropractors and osteopaths added their own initiatives to sanction special licensing boards for chiropractic and osteopathic practice. When these new initiatives were added to the ballot, George Hoyt Whipple, the director of the Hooper Laboratories who had long fought antivivisectionists in California, called on Cannon and his A.M.A. Committee to discuss the implications. It was clear, even on cursory examination, that medical scientists could no longer hope to thwart antivivisectionists by lobbying state legislatures to kill prospective bills in legislative committees.

"We have an entirely new sort of activity to engage in with this need for convincing a whole state that our cause is right," Cannon wrote his friend and ally W.W. Keen soon after a conference with Whipple. "The campaign of education which must be undertaken in the fall is going to be expensive. The antivivisection societies of the west are sending a great deal of money to California. Dr. Whipple estimates that already they have spent as much as $25,000 for propaganda in the newspapers." To Cannon, the activity in California was "not a state struggle but merely a new manifestation of the old conflict between light and darkness."[7]

Throughout 1920, the *California State Journal of Medicine* in editorials and special articles constantly alerted regular practitioners about the inherent dangers to medical education and medical practice posed by the various initiatives. During the summer Whipple, with the aid of practicing physicians and public health officials, mounted an energetic campaign to secure the support of leading economic, civic and social groups in California in the coming election.[8] Antivivisectionists, secure in the belief that their efforts to ban animal experimentation could no longer be thwarted by legislative fiat and that women who had just secured the right to vote would certainly join their cause, mounted a fierce unexpected attack on medical research in leading popular magazines that had Whipple and his supporters, as well as Cannon, crying foul.

Little more than two weeks before the election, Albert Payson Terhune, on behalf of the antivivisection movement, published an article titled "Vivisection" in *The Country Gentleman*, one of the foremost agricultural journals in the United States. Terhune, well-known for his dog stories, not only revived accounts of cruel experiments conducted by early 19th-century medical scientists like Charles Bell, François Magendie and Claude Bernard. He also added a farrago of horror stories of the callous behavior of medical scientists that claimed they discarded tortured puppies out of laboratory windows, poured boiling water into dogs' intestines for no rational purpose, and stole kittens from unsuspecting children for "cruel experiments." To cap his account of cruelty, Terhune added horrific observations that Dr. George Starr White, a homeopathic physician in Los Angeles, swore he saw during a visit to one of the largest laboratories in New York City. A week later Terhune, in another article that appeared in the *Ladies' Home Journal*, added a list of horrors he claimed occurred in various medical laboratories in St. Louis.[9]

Cannon was frustrated and outraged by Terhune's articles. Frustrated because he had no information about George Starr White or his standing in the California medical community, and outraged by the sweeping charges that Terhune had made about medical research in St. Louis. Yet, when Dr. Keen asked Cannon to write a reply to Terhune's charges, he declined. "I must admit," he wrote in explanation, "that my attitude is somewhat directed by the fact that I have an overwhelming amount of work before me. There is important research to be done which has been waiting for months, and about eight papers which have long been deferred because of the necessity of finishing the book on Shock."[10]

To Cannon's relief, the problem was resolved in the short run by the outcome of the vote in California. Despite Terhune's sweeping charges, the antivivisectionist initiative to prohibit experimental medical research was defeated by a majority of more than 250,000 of the almost 800,000 votes cast. The initiatives of the antivivisectionists' allies were also defeated but by lesser majorities.[11] Though Cannon was pleased by the outcome of the voting, at the same time he recognized that change was in the air.

The antivivisectionist charges that medical scientists indulged in wantonly cruel animal experiments behind closed laboratory doors, no matter how farfetched and improbable, always found a substantial public willing to believe them. The initiative and referendum laws adopted in California and several other states allowed antivivisectionists to go directly to the public with their demands for the abolition of animal research. With this powerful weapon, Cannon and his allies feared the antivivisectionists could in future invalidate the hitherto successful defense of animal research in hearings before legislative committees.

Not only the attempt to secure hostile legislation by popular vote, but also the extension of suffrage to women, the development of opposition to medical research among Christian Scientists and irregular medical sects, the growth of the League for Medical Freedom—all these were for Cannon instances of a new state of affairs that needed to be addressed. Aware that the effort to protect medical research must in future be conducted on a quite different scale from that which had hitherto been successful, he drew on an idea that had long lain dormant. He proposed to the Committee on Protection of Medical Research, first, that a society of laymen be organized to help the medical profession defend medical research, and second, that the A.M.A. in support of the new group publish a magazine that would popularize how knowledge obtained through such research fought disease and protected and restored health. The A.M.A. for its part promptly accepted in principle the publication of such a magazine but left final acceptance to a later date. Just as promptly, however, the members of Cannon's Committee rejected his companion proposal for a lay society.[12]

Above all else, the situation in California served to underscore for Cannon the necessity of self-policing. "In my mind there is no question that the continued freedom of animal experimentation within the law depends upon control of the practice by the medical profession itself," he wrote Dr. Will Mayo on July 2, 1921.

> For more than twelve years I have strongly advocated this point of view and from all that I have been able to learn we have kept our records clean. I hope that you will be quick to inform me if you know of abuses. I should be glad to put the whole influence of the profession, so far as I can control it, against such abuses and I should not hesitate to invoke the law against cruelty to punish any man who engaged in experimentation with a spirit of carelessness towards the animals operated upon.

Employing a tactic he had adopted in earlier campaigns against the antivivisectionists, Cannon asked the directors of laboratories throughout the United States to post the A.M.A.'s longstanding regulations governing the care of animals used in research.[13] Posting the regulations, he explained to a friend, would not only serve as a behavioral guide to a new generation of medical investigators, it would also reaffirm that laboratories had pledged to police themselves. In addition, he requested the laboratories to publicly adopt an open door policy.

Cannon's first request was unanimously accepted. Obtaining agreement on an open door policy, however, turned out to be more difficult. Authorities at the St. Louis University Medical School insisted that laymen would not be admitted to experimental laboratories unless they had previously witnessed an operation on a human being.[14] Simon Flexner, who had had a series of savage yearly battles with antivivisectionists following their attacks on the Rockefeller Institute in 1908, proved equally adamant.

Although Flexner did not reject the suggestion of an open door policy out of hand, he hedged his acceptance. He would accept such a policy, he wrote Cannon, if a statement were formulated on which all laboratories could agree. Cannon, not to be put off, countered that he was prepared to make public a series of separate statements of those who agreed in one way or another to an open door policy. But Flexner sensed that antivivisectionists would make invidious distinctions between the laboratories that accepted an open door policy and those that did not. Nevertheless, he gave in and reluctantly issued a welcome to visitors to the Rockefeller Institute, providing "they were responsible persons guided by a sincere desire to learn something of the methods and accomplishments of experimental medicine."[15]

As these differences over the open door policy were being aired, Cannon found a most unlikely ally in the person of Ernest Harold Baynes, a prominent naturalist. Baynes, who had a national reputation as a lecturer and writer of animal and bird stories, had interests in animal welfare that drew him to the antivivisectionist movement. Indeed, it was the reports of brutality and torture practiced on animals in the interest of so-called science that encouraged him to undertake an investigation of the charges in antivivisectionist literature. "If these statements could be proved true," Baynes would later write, "vivisection should be abolished at once and I, as a lover of animals, would be among the first to throw off my coat and work for its abolition."[16]

Baynes began his investigation with a visit to Cannon's laboratory at the Harvard Medical School, fully expecting that he would be barred. To his astonishment, Cannon, then in the midst of a study on the sympathetic nervous system, not only made him welcome, he carefully explained the purpose of the research as well as the various operative procedures (such as the denervated heart and adrenalectomy) that he used during the experiments. The visit had a profound effect on Baynes; the care of animals both during and after the experiments that he had observed was certainly different from accounts he had previously read in antivivisectionist journals. In the weeks that followed, Baynes dropped into a number of other laboratories, including those at the Johns Hopkins Medical School and the Rockefeller Institute, each visit adding to a growing conviction that the antivivisectionists had deliberately falsified or distorted the nature of experimental animal research.

In July 1921, Baynes published the results of his inquiry in an article titled "The Truth of Vivisection" in the *Woman's Home Companion.* As soon as the issue hit the newstands, an editorial in the *New York Times* of June 30 called Baynes's article one of the best defenses of vivisection, and perhaps the very best attack on the antivivisectionists, that ever was written. Instead of an exposé of the cruelty of medical research that many had expected, it proved to be a critique of long-held antivivisection verities.

First, it argued that physicians like Edward Berdoe, John Elliotson, Stephen Townsend and Walter Hadwen, trumpeted in antivivisectionist journals as eminent medical critics of experimental medicine, were in fact mid and late 19th-century physicians ignorant of the most recent advances in medicine, surgery and medical science. Then, Baynes directed several well-aimed arrows at another treasured antivivisectionist contention—notably, that callous physicians conducted wantonly cruel animal experiments behind closed doors. These he dismissed as figments of antivivisectionist imagination. Medical experiments were not conducted without purpose or maliciously, he noted, nor in his experience were they done secretly away from prying public eyes.

"I have seen many operations on animals—five within the past month," he explained, "and although I usually visit these places unheralded, I have never seen anything in the nature of cruelty to animals." Instead of the "brutality and heartlessness" he had read about, he found "nothing but kindness and consideration." That said, Baynes ended with a rebuttal of long-held antivivisectionist claims that neither mankind nor animals derived any benefit from medical experimentation. Here he contented himself with a telling catalogue of the contributions of experimental pathology and bacteriology to the practice of medicine and to the elaboration of antisera and vaccines against infectious diseases.[17]

Baynes's article had hardly appeared when Gertrude Lane, the editor of the *Woman's Home Companion,* found herself in the midst of a maelstrom of correspondence. Some antivivisectionists, beside themselves with anger, attacked the author as malicious and full of perversions. Some saw the article as a plot to broadcast support of experimental research in magazines appealing especially to women. And some urged that in the interest of fair play, the magazine publish "an article on the other side." Nor were letters to the editor solely from antivivisectionists. There was in fact a full and enthusiastic measure of support for Baynes from a host of public figures, led by former Harvard president Charles W. Eliot, whose letter was prominently featured. They included, among others, cabinet officers, governors, state public health officials, eminent physicians and surgeons, veterinarians and noted animal lovers.

Baynes sent some of the letters he received to Cannon, who

voiced his enthusiasm at once: "You cannot imagine what a gratification it is after years of labor to make sure that animal experimentation is conducted in the United States in a defensible way . . . , to have a layman and a lay Journal come vigorously to our defense."[18] Eager to keep up the barrage against the antivivisectionists, Baynes began to plan a lecture tour at his own expense to preach the value of animal experimentation. In preparation he asked Cannon to instruct him in both the history and recent developments in medical research. In the weeks of give-and-take that followed, the two became fast friends—Cannon carefully explaining the process and purpose of various medical investigations, Baynes regaling Cannon with stories of his success in converting noted antivivisectionists like Seton Thompson and Owen Wister to become supporters of animal research.[19]

In an effort to drum up lecture invitations, Baynes sent copies of his *Companion* article to various women's group and prominent physicians as well as to local and state medical societies. The publicity garnered him several requests, including one from Dr. Whipple to speak in California before antivivisectionists and their allies began their own campaigns later in the summer and fall of 1922.

From the outset, Baynes's tour of lectures proved successful. "It has stirred the antis—in some cases to fury," he wrote Cannon soon after he began his tour.

> There was plenty of heckling and questioning at all the lectures but one—that was in Santa Barbara . . . On the motion of an elderly gentleman in the audience I was given a rising vote of thanks, and it was unanimous. Besides this lecture I gave two in San Francisco, and one in Palo Alto. I am to give one more, in San Diego and then I shall jump to Chicago where two very important lectures have been arranged by Frank Billings. The first of these will be given before from 500 to 600 prominent women at the home of Mrs. Edith Rockefeller McCormick, who is also financing the second, to be given at Orchestra Hall and which is to be free to the public.[20]

Baynes's letter, along with the newspaper articles he enclosed that described public reaction to his lectures, pleased Cannon. It was the kind of material that bolstered his hope of organizing a lay society in defense of experimental research. Then, too, the spate of invitations to Baynes seemed to promise that more scientists and physicians would take an aggressive stance in meeting antivivisectionist threats to medical experimentation.

Late in 1921, a committee headed by Evarts Graham, professor of surgery at the Washington University Medical School, wrote Cannon that medical schools in St. Louis were facing a crisis because the Humane Society of Missouri had gained control of the local dog pound and instituted an effective embargo on the sale of stray dogs for experimental purposes. Was it advisable, he wondered, given these circumstances, to invite Ernest Harold Baynes to come to St. Louis to give a public lecture on the subject of vivisection similar to his recent article in the *Woman's Home Companion?* There were members of the faculty, he hastened to add, who thought that such a lecture might well be provocative. He asked for an opinion as to whether "it would be better not to stir up too much discussion at the present time or to enter the fight actively in the hope of getting the majority of the public in favor of vivisection before the anti-vivisectionists could convert them."

"It seems to me," Cannon replied, "that both in the east and elsewhere workers in laboratories, physicians, and intelligent laymen have too long put the soft pedal on open discussion of the methods and merits of animal experimentation. We have allowed the antivivisectionists to cry aloud gross falsehoods about our activities without giving a public answer to them." If Graham and his colleagues were feeling the effects of agitation in St. Louis, he advised them to "make use of every opportunity to show that the work carried on is in the highest sense humane and that it is deserving of public support." Cannon believed that Baynes could be considerably effective "because he speaks from the point of view of the man in the street, and as a person who has himself been converted from hostility to advocacy through personal experience."[21]

Emboldened by Cannon's advice to meet the antivivisectionists head on, Evarts Graham with the aid of a committee of concerned medical colleagues drafted a resolution for passage by the Board of Aldermen, which would require the Humane Society as keeper of the dog pound to supply medical schools with impounded dogs needed for the teaching and study of medicine. To gain public approval for the resolution, the committee launched a vigorous political and educational campaign that resulted in the support of every community group in St. Louis—with the notable exception of the American Medical Liberty League.

The Humane Society was not idle, however, and petitioned the Board of Aldermen to defeat the resolution on the ground that it would compel the Society to become a part of vivisection practices. Nor did it leave matters there. When major newspapers in the city announced their support for the resolution, the Society inaugurated a massive letter-writing campaign that appeared to turn the tables. Members of the medical committee were subsequently informed that many women in the city opposed the resolution and that the committee needed a stronger public expression of support if it were to pass. On the day of the public hearings held on March 7, 1922, over five hundred supporters and opponents of the resolution, now in the form of an ordinance, crowded into the hearing room hoping that their shouts of approval and disapproval of the various speakers might in the end affect the vote. Supporters of the ordinance left nothing to chance; they arranged for the aldermen to visit the Washington University Medical School so that they might observe human as well as experimental surgery. Ernest Harold Baynes, summoned to St. Louis by the medical committee, was promptly invited to address the Board of Aldermen in behalf of animal experimentation at a recess before they voted. And then in a spurt of fairness, the president of the Missouri Humane Society was invited to respond to Baynes's talk. One alderman requested that the vote be postponed, but this final objection was defeated and the bill was then passed by a vote of twenty-two to four.[22]

The victory elated Cannon. For well over a decade, the A.M.A. Committee that he chaired had on several occasions attempted to push pound laws through various state legislatures without success. But in St. Louis, a local committee of physicians, with no previous experience in dealing with antivivisectionists and with nothing more than letters of advice to guide them, gained passage of the first bill in the United States that made available by law an adequate supply of animals from local dog pounds for experimental research.[23]

From the beginning of 1922, the lectures of Ernest Harold Baynes and Walter Hadwen served as point and counterpoint in a series of stormy state and local political campaigns designed to persuade an increasingly powerful but scientifically naive electorate of the morality and merits of experimental medical research. Dr. Hadwen, president of the British Union for the Abolition of Vivisection, had been invited by the New York Antivivisection League to undertake a lecture tour in the United States preparatory to various state legislative elections in the fall. As one of the physician-leaders of the British antivivisection and antivaccination movement, he was particularly prized by American antivivisectionists as a keeper of the flame of their ideology. For more than a generation, he had mounted a pamphlet war against experimental medical science. The idea of bacterial etiology of disease, or the use of antisera and vaccines in curing or preventing disease, was anathema to him.[24]

Cannon had nothing but contempt for Hadwen. Yet he viewed the announcement of his lecture tour in Canada and the United States with some trepidation, not the least because of Hadwen's reputation as a public speaker. From the outset of the tour, Hadwen's lectures were an incitement to his audiences. In Montreal, medical students, angered by his attacks on Pasteur, Koch, and the bacterial etiology of disease, interrupted his lecture time and again with catcalls and rowdy shouting. In Toronto, Hadwen fought back by assailing Banting and Best's recent discovery of insulin. Insulin, he told a packed audience, was an unproved, untried claim "that was admittedly dangerous to those who used it." In each of the lectures that Hadwen subsequently gave, audience reaction was the same. Antivivisectionists who were present cheered, while others who could not stomach his views on medical science either jeered or privately regarded them as a joke.[25]

Even though the accounts Cannon received from Canada were critical of Hadwen, he still worried about the impact that the British lecturer's advertised month-long tour in California might have on the state's forthcoming voting. And with good cause. Letters from friends intimated that the defense of medical research in California needed more money and support if it were to meet another attack by antivivisectionists and their allies. To make matters worse, George Whipple, who had long guided the fight there, had recently left to become medical dean at the University of Rochester. Other than a letter from Walter Alvarez containing news that antivivisectionists, chiropractors and osteopaths had again successfully put their various initiatives on the state ballot, Cannon had no inkling of preparations in California to meet the new antivivisectionist threat.[26]

Actually, medical scientists and physicians in California were in better position than Cannon realized. William Ophuls, the professor of pathology at Stanford who had succeeded Whipple, made a special effort to cement previous alliances with the political community as well as to encourage the support of such diverse economic interests as farmers, fruit growers, cattle ranchers, and railroad men. For months, the California Medical Society and the League for the Conservation of Public Health conducted an educational campaign to alert physicians of the scientific and professional implications of the initiatives that the antivivisectionists and their allies had put on the ballot.[27]

Whipple's advice "to do away with dignified neutrality" in fighting antivivisectionists weighed heavily on Hadwen's lecture tour. During a speech in Palo Alto, the audience, in the words of a reporter who was present, "treated him in a decidedly uncivilized fashion." In San Jose, Karl Meyer of the Hooper Institute, before a debate with Hadwen pointedly refused to shake hands with him, leading to an outcry from some in the audience that "even prizefighters shake hands." Nor did Meyer, to Hadwen's discomfort, relent during the debate.

Of all these developments Cannon knew nothing. "Can you explain the utter indifference of the California group to us and our interest in them," he wrote to Dr. William Woodward, the A.M.A.'s new executive secretary. "I have heard nothing whatever of the progress of the campaign there in spite of repeated letters and I have not heard any news of the result of the voting. They seem to be a queer lot."[28] For all his exasperation at being kept in the dark, Cannon was reluctant to join with a local group in Colorado when a similar campaign was undertaken there.

Late in 1921, when he learned of initiatives in Colorado reminiscent of the beginnings of the antivivisectionist campaigns in California, Cannon had alerted Henry Sewall, one of the leaders of the Colorado Medical Society, of the imminent threat and urged him to seek counsel from Dr. Whipple on how to meet it.[29] In the spring of 1922 Sewall, following Whipple's advice, organized a Colorado Association to Protect Public Health and, with the aid of some of his colleagues in Denver, began to plan a campaign to educate businessmen, farmers and ranchers of the hidden danger to the health and well-being of both the human and animal population of Colorado in the initiative proposed by the antivivisectionists.

When Sewall subsequently asked Cannon for help in organizing the campaign, the latter bluntly replied that while he was willing to share with him information and experiences in fighting the antivivisectionists, the responsibility for conducting the campaign ultimately had to be assumed by local physicians and other organizations. Sewall, who had expected more from Cannon, appealed directly to A.M.A. headquarters for help. But he had no better luck.[30]

Dispirited by the apathetic and negative response of the national group to their needs, Sewall and his colleagues aired their frustration and anger in an editorial in *Colorado Medicine,* the journal of the Colorado State Medical Society. Using Wellington's cry of impatience at the lack of promised reinforcements during the battle of Waterloo, "Oh! for Blucher or night!" they cried out once more to the A.M.A. for help. Colorado, they concluded, "has sought and expected timely aid which has not yet arrived. It does not want night. It wants to continue the battle, but—where is Blucher?"[31] The editorial hit the mark. Following its appearance, Cannon agreed to join with Dr. Woodward from A.M.A. headquarters in an on-site investigation of the Colorado campaign against the antivivisectionists.

To their great surprise, Cannon and Woodward discovered that the prospects for success were better than they had expected. In the weeks before their arrival, the Colorado Association had hired an executive officer with considerable expertise in advertising and uncanny powers of persuasion—qualities which injected new life into the struggle against the antivivisectionists. No less important, Ernest Harold Baynes, during the course of a visit to lecture on medical experimentation, had managed to convince the editor of the *Denver Labor Bulletin,* a supporter of the antivivisectionist campaign, to introduce in his paper material favorable to medical research.[32]

Although Cannon kept informed of the progress of the Colorado campaign, except for intermittent advice and his site visit in September, he took no direct part in it. Throughout the antivivisectionist struggles in the West, he focused his efforts for the most part on containing Hadwen. As Hadwen travelled East, Cannon pursued him with a series of unsigned editorials in *JAMA* warning physicians of the dangers to medical development implicit in his attacks on medical science. The editorials on the whole were businesslike, but a week before the election, Cannon took the gloves off and published a personal attack on Hadwen. "No evidence is forthcoming," he wrote,

> to indicate in any way that Dr. Hadwen is a distinguished physician or surgeon except as he may be said to be distinguished from the rest of the medical profession by the manner in which he allows his medical degrees to be exploited in the interests of an unscientific propaganda, thus lending to his bizarre utterances a weight they would not otherwise have. Like most faddists he does not confine his advocacy to a single fad. He not only is an ardent antivivisectionist, an antivaccinationist, and a vegetarian but also has been a proponent of movements connected with temperance, premature burial, and almost every agitation against modern science and modern medicine.[33]

In contrast to his rather passive role in California and Colorado, Cannon took both an active and visible part in the election campaigns against antivivisection and antivaccination bills in some of the eastern states. Keenly aware of the importance of the woman's vote, he appeared before the Massachusetts Federation of Women's Clubs for a resolution to the state legislature in support of medical research through animal experimentation. The need for such support was particularly manifest in Massachusetts, where the American Medical Liberty League in cooperation with antivivisectionists had from 1921 launched a series of harsh attacks against smallpox vaccination. These attacks outraged not only practicing physicians throughout the state but laymen as well. Taking note of the educational campaigns conducted by the League for Conservation of Public Health in California against antivivisectionists and their allies, Thomas Barbour, a zoology professor at Harvard, and others organized a lay group called the Committee for the Protection of Animal Experimentation to advance a similar struggle in Massachusetts.[34]

In the November 1922 elections, antivivisectionist initiatives were defeated in California by a three-to-one majority, and in Colorado by a five-to-one majority. The Massachusetts Medical Society, with help rallied by Cannon, Harvey Cushing and Elliott Cutler and in collaboration with lay allies, defeated a bill against animal experimentation and vaccination in their state that had been proposed by the American Medical Liberty League. In New York, where there was no initiative and referendum, Simon Flexner, after a hard fight with antivivisectionists in hearings before a state legislative committee, succeeded in having a bill killed in committee. Although it would seem that medical scientists and physicians achieved total victory in these elections, they did in fact suffer one subsidiary defeat. In California, the initiatives proposed by osteopaths and chiropractors that they be examined by their own licensing boards passed, despite vigorous effort to defeat them.[35]

Throughout his lecture tour, Hadwen had given the impression of a triumphal reception at every stop he made. At the end of the tour, however, he confessed that he had encountered some difficulties as well. To Dr. Keen, an old adversary who wrote him of the defeat of the antivivisectionists in California, he replied gloomily, "Never were the forces of evil put forth with greater malignity and determination, or with more unscrupulous fanaticism than during the California campaign, and I was quite prepared for defeat." Later, writing to the editor of *JAMA*, he grumbled about the editorials that Cannon had written and demanded proof of "one single great discovery made by experimental medicine for the benefit of mankind."[36]

Cannon was unperturbed by the demand; he knew it was useless to engage in further debate with Hadwen. Although he savored the triumph over Hadwen and the antivivisectionists, his elation was tempered by qualms about the ability of physicians to take responsibility for defending medical research in the future. Convinced that the strategy previously employed to thwart antivivisectionists would no longer suffice, Cannon proposed that the A.M.A. Protection Committee help organize a national lay society to educate the public of the value of experimental research to humanity and animals.

Nine years before, when Cannon had made a similar proposal, it was promptly rejected. This time a group of members of the Committee including, among others, William Welch, Simon Flexner,

Frederic Lee, and Joshua Sweet, accepted the proposal. Indeed, they gave Cannon a mandate to breathe life into the new society. First, he wrote a constitution for the new organization and then, after much effort, persuaded Thomas Barbour of the Massachusetts Committee in Defense of Medical Experimentation to become the director of the national group, which was named the Society of Friends of Medical Progress.[37]

Cannon's desire for a lay organization to take up the cause was not only a reflection of the growing complexity of fighting the antivivisectionists. It was also a sign of his weariness in trying to balance the demands of defending medical research with his duties in the physiology department of the Medical School. More than once during 1922, he indicated that he would like to resign from the chairmanship of the A.M.A. Committee for the Protection of Medical Research. Following the adoption of the pound law in St. Louis, for example, he complained to Victor Vaughan that the defense of research had become so time-consuming and onerous that he was unable to carry on his research with the attention and care it deserved. When Vaughan said that he did not see how he could get along without Cannon and begged him to delay even thinking about resigning until the fall campaigns were over, Cannon reluctantly agreed.[38] But after the elections, he again hinted that he wished to give up his chairmanship.

"I feel that the time has come for a greater activity than I have been able to carry on in connection with the A.M.A.," he wrote W.W. Keen on September 29, 1922.

> Not only the antivivisectionists, but also the Christian Scientists, the League for Medical Freedom, the Chiropractors, etc., are uniting in efforts to check modern methods in medicine . . . It is going to require more time than I can give to meet their attacks in various parts of the country—that is, if I am to continue as an investigator, for already for a number of years I have been deeply disturbed at the amount of time that I have had to give to outside interests at the expense of time for investigation. The years are beginning to be relatively few that remain for research and I have still a long program to carry out.

Despite his dilemma, Cannon still could not bring himself to resign from the A.M.A. Committee. There was simply too much at stake, both for him and for the community of his fellows who engaged in animal experimentation to further the cause of medical knowledge. Although he had already done yeoman service for the protection of medical research, several further battles with the antivivisectionists lay ahead of him before he would succeed in turning the leadership over to someone else.

7 A Spirit of Fellowship: Ties with Pavlov and Yerkes

AFTER HIS SERVICE in the war had ended, Cannon's prime obligation to the community of his fellows was to protect the freedom of experimental research. Increasingly, however, in the belief that the scientist as a citizen has an added responsibility to the society in which he lives, he began to take part in further battles against, as he put it, the true enemies of mankind—war, ignorance and disease. In the process, he became adept in developing a panoply of political skills that he would be called on to use in future.

One example of the proliferation of Cannon's social concerns was his role in providing medical relief to Russian physicians caught in the maw of revolution, famine and epidemic disease in the years between 1920 and 1923. The initial spur to that activity was a letter that appeared in the French magazine *Les Dernieres Nouvelles* in October 1920, which reported that the great Russian physiologist, Ivan Petrovich Pavlov, was starving in Petrograd and would probably be unable to survive the approaching winter. Cannon was aghast at the news and promptly wrote Carl Tigerstedt at the University of Helsingfors for more definite information about Pavlov's circumstances. "Do you know," he inquired, "whether there is any possibility of our coming to his assistance? Can you name how aid might best be given? I feel sure that American physiologists would be glad to do anything in their power to ease his condition and to render it possible for him to undertake active investigation again."[1]

Not content to rest there, Cannon also wrote William Howell at Johns Hopkins and Simon Flexner at the Rockefeller Institute about Pavlov's difficulties, and then he alerted officials at the American Red Cross, who promised to bring Pavlov's problems to the attention of the State Department. The Red Cross subsequently announced that Simon Flexner had invited Pavlov to come to the United States, but Flexner denied the report, writing Cannon that he felt it would be a mistake for Pavlov to come to America. "I know of no means by which he could be supported, and he would find here no Russians of his class with whom to associate." Cannon agreed with Flexner and came to believe that it would be wiser if Pavlov were to leave Russia and settle in Helsingfors, Stockholm or Copenhagen, where he would be near friends and not so far from Russia in case conditions developed which would permit him to return.[2]

By the spring of 1921, Cannon had collected $2,000 from friends and colleagues, which went to Prof. Tigerstedt to help meet Pavlov's needs. Although he later received information from the Red Cross that Pavlov was being cared for, rumors of the Russian physiologist's difficulties persisted. One foreign news report announced that Pavlov had been removed from his chair by the Soviet government because he would not accommodate himself to the new regime. Another claimed that he had been sentenced to death because of the tone of a lecture he had delivered at a popular meeting that was hostile to the government. Still another reported, in the most imaginative terms, that he was dead, claiming that "the poor gray savant" had suffered greatly, could bear the misery of cold and hunger no longer, and was taken to a hospital in January 1921 where he died after a few days.

Cannon, in an effort to scotch the rumors, wrote *JAMA* that summer that Tigerstedt had been sending consignments of food and clothing twice monthly to Pavlov through the Finnish Red Cross Commission, and that by all recent accounts he was alive and well.[3]

Save for the stories of starvation and political repression that swirled around Pavlov, Cannon actually knew little about daily living conditions in Petrograd or, for that matter, in the rest of the country. The first inkling that there was substance to the rumors of famine and epidemic disease in Russia occurred when the Soviet government in July of 1921 officially asked the United States to send food and medical aid. When the Soviet distress signal reached

Herbert Hoover, Secretary of Commerce, he ordered Col. William Haskell, one of his close associates, to organize a Russian Relief within the American Relief Administration (A.R.A.) to conduct a preliminary survey of food and medical needs.

By the end of August, Haskell, supplied with $9,000,000 for a child feeding program and $3,000,000 donated by the Red Cross for medical relief, began the survey. Within weeks, the data he collected indicated such overwhelming shortages that unless large amounts of food, clothing and medical supplies were sent to the Soviet Union at once, the Russian people faced a massive number of deaths brought on by starvation and disease. By year's end, Congress, prompted by Hoover's persistent lobbying, appropriated $20,000,000 in food and grain and $4,000,000 in Army medical supplies to begin relief operations. Congressional action opened the flood gates, and soon a host of private charitable organizations ranging from the American Friends Service Committee to the Jewish Joint Distribution Committee began to collect funds for Russian relief as well.[4]

Horsley Gantt, who was posted to Petrograd early in 1922 by the A.R.A., later reported the conditions he found on his arrival in Russia, where "millions were starving and nearly all the population was seriously undernourished. Dogs, cats, horses, camels, and even rats were eaten . . . In some places cannibalism existed then; quite frequently dogs and men could be seen in the streets of Leningrad fighting over the carcass of a horse. On many occasions I have seen people eating roots, wild berries, the barks and leaves of trees, and 'bread' made of dirt, grass and dung."

Chief among other horrors that Dr. Gantt observed were the rampant epidemics of typhus and relapsing fever, with mortality estimated at three to six million people. As if that were not enough, he reported concurrent epidemics of malaria, typhoid, cholera, tuberculosis, syphilis, and dysentery. Those who fell ill and needed medical attention found that physicians and hospitals could give little help.

During the years of war and revolution, hospitals had deteriorated markedly and all too often lacked basic equipment and supplies. Surgeons despaired over the lack of anesthetics, hemostats and catgut, as well as the condition of their deteriorating and worn out surgical instruments. Physicians could do little for their patients, much less for themselves. "I saw hundreds of physicians," wrote Gantt, "pale, wan, and haggard, living on black bread, herring, and perhaps some potatoes, and huddled with their families into one or two rooms, still trying to carry on their practice or research, and I saw many actually starving."[5]

In the spring of 1922 the A.R.A., concerned that the deteriorating physical condition of Russian physicians put at risk the implementation of various relief programs they had devised, appealed to Americans to send food and clothing for the physicians directly to special A.R.A. depots in Russia. In New York, Mrs. Henry Villard and some of her friends among the socially elite, together with the help of such luminaries as Charles Mayo, Lewellys Barker, Haven Emerson, Harvey Cushing, Jacques Loeb and William Welch, organized the National Campaign of Physicians and Surgeons in Behalf of American Medical Aid to Russia.[6]

In Boston, Cannon was so moved by the plight of Russian physicians that he and a group of colleagues at Harvard began a drive to ask the local community to contribute to medical aid for Russia. "This is an appeal [made] solely on the basis of broad humanity," they informed the public. "It has nothing to do with any theory of government or of international politics. You do not have to believe in the present Russian government to heed and answer the appeal of noncombatants. We as physicians ask you to do what you can to help us alleviate the physical suffering of millions of helpless people."[7]

During the winter and early spring of 1922, Cannon again became anxious about Pavlov, in part because he had not received replies to any of the letters he had written him months before, and in part because of continuing reports of the desperate living conditions in Russia. In late April or early May, however, a letter from Pavlov, addressed to "My dear Friend," finally arrived.

Pavlov was in Helsingfors for a short stay, "looking over the psychological literature for the last five or six years," and from Professors Robert and Carl Tigerstedt he learned for the first time who had come to his aid.

> I beg of you first of all to receive my most sincere thanks and also to give them to all my American colleagues. Science, literature, art unite the whole world, not the lower instincts, the standard of the present Moscow III internationals, who in the words of Lenin first

> promised Terror and Violence to all nations. Yes, science has given and is giving me not only the fullest satisfaction and joy of my life, but also is giving me and my whole family the greatest support, both moral and material, in the form of world-wide sympathy and different kinds of help in our very sad circumstances.

Pavlov told Cannon that work was keeping him going, for without it, he would be "continually depressed by the conviction that my Fatherland was falling to ruins."[8]

Cannon was happy to hear from Pavlov. But he was soon to become troubled by matters relating to relief and medical assistance for beleaguered Russian physicians—particularly, the activities of the National Campaign of Physicians and Surgeons, which he had joined at the behest of his friend Haven Emerson. In one instance, Mrs. Villard and her friends at the New York office pledged to maintain a hospital in Moscow for one year at a cost of $200,000. "Do you not agree with us that this comparatively small sum . . . can be raised from a limited group of contributors?" the executive secretary of the campaign subsequently wrote Cannon. "You, my dear Dr. Cannon, will readily think of persons whose names are not known to us. Some of these are to be found among your personal acquaintances and your clientele . . . [and] we are begging you, with all possible earnestness, to reach them for us."

Cannon was annoyed and irritated by the appeal letter. Annoyed because he thought that a policy of supporting a single hospital was ill conceived, given the tremendous need of so many Russian physicians for relief and medical aid. And irritated not only by the notion that $200,000 was a comparatively small sum of money, but even more, by the allusion to Cannon's wealthy medical clientele as a source for such funds. "Unfortunately," he retorted, "I have almost no acquaintances among wealthy people. Indeed, I have lived as a hermit in laboratories all my professional life. In other words, my position is very different from that of Dr. Barker, Dr. Mayo and Dr. Cushing."[9]

Several months later, Cannon's misgivings were reinforced by a report given at a medical meeting by members of the National Information Bureau's Commission on Russian Relief. The Bureau, which had been investigating the conduct of various charitable organizations, charged that the policies of the National Campaign of Physicians and Surgeons overlooked the major problems of the health situation in Russia—namely, the need for food and clothing to protect the health of individual physicians and nurses and for instruments, drugs, medical supplies and literature, to make their work effective.

Cannon, alarmed, wrote Haven Emerson, who served on the executive committee of the campaign, about the report, explaining: "Last night I heard a rather doleful account of the way in which the organization to which our name was attached, the American Medical Aid for Russia, had been mismanaged. Apparently, no large amount had been obtained, and such as it was, it was used for foolish purposes." If aid were warranted, he continued, "it should be undertaken on a large scale and not in patches in this or that city." Only a few days ago, he had heard that most of the $2,000 he had collected for Pavlov "was still in the bank at Helsingfors, and that all that Pavlov was asking for was bonbons! Personally, I want more information before I take further action . . . If you have reliable knowledge of the medical needs in Russia, I would be glad to know of them."

Emerson took pains to mollify Cannon, promptly replying that "readjustment" of the National Campaign policies and methods of procedure were already underway. "I am astonished at the reports concerning Pavlov," he commiserated, "and do not wonder that you are critical and conservative with regard to further moves for relief without adequate information."[10]

Within weeks of Emerson's response, Mrs. Villard wrote Cannon that the National Campaign had changed its policies and would devote its efforts to providing general medical relief to Russian hospitals as well as supplies and supplementary food rations to Russian doctors and nurses. All funds and supplies that the campaign collected, she assured him, would in future be turned over to the American Friends' Service Committee for distribution. In a postscript Mrs. Villard asked for help to organize a program to supply medical books and periodicals to Russian physicians.

It was a request that Cannon could appreciate. He had come to believe that educating Russian physicians about recent progress in medicine was necessary for the success of programs to deal with the ongoing famine and epidemics in Russia. Indeed, he had for months been collecting and shipping medical literature to Horsley Gantt, whose work as district medical officer for the A.R.A. daily brought him into contact with medical personnel in and around Petrograd.[11]

Gantt not only sent Cannon requests for particular medical books and periodicals, he also sent him information about Pavlov, who had returned to Russia. "Dr. Pavlov is a very interesting man, active and enthusiastic and daily working in his laboratories," he wrote in mid-April 1923. Gantt supposed that Cannon was familiar with the present work of Pavlov on conditioned reflexes and reported, "he is working in three laboratories, lecturing in three universities, and he has about sixty doctors working under him." Almost as an aside, he imparted the news that Pavlov expected to visit physiologists in the United States in early August after attending the forthcoming International Congress of Physiology in Edinburgh.

Cannon believed that Pavlov's visit would bring much pleasure and stimulation to his friends in America. But he thought August was a poor time for the Russian's tour, for he would find practically all of the laboratories closed and workers away on vacation. If only Pavlov would defer his visit until the autumn, he wrote Gantt, "he would find hosts of friends wherever he might go." Cannon wrote a similar letter to Pavlov, who reversed his travel plans in order to visit America before going to Edinburgh.[12] Accompanied by his son Vladimir, Pavlov early in June attended the Pasteur Jubilee in France, and at its conclusion sailed to New York to stay with his friend Phoebus Levene, who worked at the Rockefeller Institute.

Some days after his arrival in America, Pavlov decided to visit the physiological laboratories at Yale University. The trip turned out to be a disaster. Pavlov's son had hardly gone to check their luggage aboard the train to New Haven when some thieves in the station set upon Pavlov, purloined his wallet, and made off with approximately $800. Outraged by the assault on his person and desolate at the loss of his travel funds, Pavlov went straight to the Rockefeller Institute and told Dr. Levene that he wished to return to the Soviet Union at once because he feared for his life in the United States.

Levene, attempting to set matters right, succeeded within hours in raising $800 from Simon Flexner and F. S. Lee to enable

Pavlov to continue his travels in the United States. But Pavlov, as a matter of principle, refused to accept the gift. Levene persisted, and after some hours of talk managed to convince Pavlov that it was reasonable to take the money as an honorarium for future lectures. "Of course," Levene wrote Cannon the next day, "he would need an interpreter. The hope that such . . . might be arranged cheered him up, and he let himself be persuaded to go to Boston."

Although Cannon made haste to arrange for Pavlov's lectures at Woods Hole and the University of Buffalo, they did not prove to be necessary. In the interim, the Rockefeller Foundation solved the problem not only by inviting Pavlov to visit some American laboratories, but also asking him to attend the International Physiological Congress in Scotland as an official of the Foundation.[13]

Cannon was eager to meet Pavlov in person, for their previous contacts had been exclusively through correspondence. As part of the visit, he hosted a luncheon at the Harvard Club in Boston so that the Russian physiologist could also see and speak with Harvey Cushing, George Parker and Robert Yerkes. During lunch, Pavlov regaled Cannon and the others with stories of life in Russia and of his arrest after the revolution by the dreaded Soviet secret police. While being interrogated, he told his enthralled listeners, he drew out his watch and to the astonishment of his captors announced that he had to give a lecture at the medical school; without further ado he left the police station. In retrospect, it appears that Pavlov embroidered the story, perhaps to emphasize for his audience his antipathy and scorn for Soviet authorities.[14]

The visit to Boston cemented Pavlov's friendship with Cannon. Both men discovered that, in addition to their common interest in physiology, they also shared the credo that experimental research was central to the development of knowledge and ultimately to the advancement of humanity. Although the meeting was relatively brief, Cannon retained a warm memory of the occasion. Years later when a young California physician asked for information about Pavlov, Cannon remembered him—the man rather than the scientist—at this, their first meeting:

> They [Pavlov and his son] spent a day with me in Cambridge. As we left the house in the evening (it was in the summer and my family were in New Hampshire) he asked, "Where is the watchman?" When

I told him there was no watchman he said, "Someone will enter the house." When I stated that there was no danger of that, he said that a thief would take my fine automobile (it was an old Ford), and when I discounted that danger he threw up his hands and remarked, "What a profound difference there is between the morality of New York and the morality of Cambridge!"

On the way to Woods Hole from Boston he conditioned a word for me. The large headline on a newspaper held up by a man sitting in the seat in front of us included the word "fizzle." Pavlov turned to me and asked, "Was meint das Wort *Fitzel*, Fiasco?" I have never seen the word "fizzle" since then without thinking of Pavlov's pronunciation of it.[15]

On Cape Cod, Jacques Loeb took the Pavlovs in hand. He insisted that they stay at his house and shepherded them through the various laboratories at the research station. The enthusiasm with which Pavlov was received at Woods Hole was repeated at other laboratories he later visited in Baltimore and Washington.

Unfortunately, Pavlov's trip to the United States ended as it had begun: with an incident that upset him—this time by an act of the British Consul in New York who refused to give him a visa to go to Scotland on the grounds that he was a Bolshevik. Although Phoebus Levene tried to persuade the Consul that Pavlov was neither a Bolshevik nor a supporter of the Soviet regime, his testimony fell on deaf ears. In the end, Pavlov went back to France where a more understanding British Consul, who knew something about his work, expeditiously granted him a visa to attend the meetings of the International Congress of Physiology in Edinburgh.[16]

Cannon did not attend that Congress, nor did he attend the one held in Stockholm in 1926. But it would be at future physiological congresses that Pavlov and Cannon were to meet for the second and third times—one held in Boston in 1929, the other in Leningrad in 1935.

Whenever Cannon allowed himself to be drawn into the politics of science, one of the contributing factors was likely to be his willingness to come to the aid of his friends and scientific colleagues. We now look at another example of the expansion of his activity in the public realm. Here, instead of joining in an effort to send emergency relief to physicians caught in a disastrous situation in Russia, the object was to supply funds and opportunities to medical scientists for biological research that could be applied to the solution of social problems.

Soon after the establishment of the National Research Council (N.R.C.) in 1916, Cannon had written William H. Welch to express his views on the potential opportunities for support of the sciences. In his letter, he stressed the need for recognition of important movements already in existence, "which might be correlated with the efforts of the Council to direct attention to the values of science for social welfare."[17] Cannon's words would prove prophetic. Five years later, one such movement was taken under the N.R.C.'s wing when it established under the auspices of its Medical Division the Committee on Research in Problems of Sex—a committee that was to function admirably under the direction of Robert M. Yerkes for the next quarter of a century.

Yerkes and Cannon had been professional colleagues at Harvard before the war, with Yerkes, a comparative psychologist, in part working in Cambridge on laboratory studies of animal behavior in the psychology department, and in part working in Boston doing clinical psychology at the Psychopathic Hospital, then directed by Ernest Southard. All three men had been members of the Wicht Club—an association of young academics with a common interest in scholarly inquiry and scientific ideas. In addition to these collegial ties, Robert and Ada Yerkes and their children were cherished members of the Cannons' circle of friends, especially when both families were in residence at their neighboring farms in Franklin, New Hampshire.[18] Although Yerkes left Harvard when he was offered a post at the University of Minnesota, the two families kept in touch and Cannon and Yerkes continued to share mutual professional interests.

Yerkes never filled his appointment at Minnesota; the war intervened and he spent 1917–19 as a major in the Surgeon General's Office, where he conducted intelligence tests on Army personnel. At the same time, a letter of commendation from Cannon to George Ellery Hale, the prime mover in the creation of the National Research Council, was instrumental in securing a place for Yerkes not only as a protégé of Hale, but also as an officer of the N.R.C. in various capacities.

The need for scientific advice in time of war had led the National Academy of Sciences to establish the National Research

Council before America entered the conflict. As the war progressed, the medical and allied groups within the N.R.C. were reorganized into a new Department of Science and Research of the Council of National Defense, and Yerkes was subsequently appointed vice-chair of its Division of Medicine and Related Sciences. He also served as chairman of the N.R.C.'s Psychology Committee, just as Cannon acted as chairman of the Physiology Committee.[19]

On March 10, 1918, Yerkes wrote Cannon, then in France, to bring him up to date about the reorganization:

> In recognition of my efforts to organize the psychological service, I was made vice chairman of the Division. And so it comes about that I meet with the medical men and participate in their interesting discussions. Our group includes Flexner, Welch, Vaughan, Pearce, Howell, Mayos, Dakin, Davenport, etc. . . . Needless to say, I enjoy my association with this group and am inspired by the varied businesses in hand. The N.R.C. under Dr. Hale's leadership is doing wonders . . . It is splendid to be working with and for him. I am now a member of the executive board of the council and am therefore familiar with various important matters with military bearings.

When the war ended, Hale, an ambitious man who had built the world's largest telescope on the summit of Mount Wilson, had equally ambitious hopes and plans for the National Research Council. To help him achieve his goals, he recruited Yerkes for an important position as head of the Research Information Service. Although Yerkes had difficulty obtaining the funds needed to develop this service, he went on to assume the chairmanship of two grant-making committees—one, a Committee for Research in Problems of Sex and the other, a Committee on Problems on Human Migration. Based in Washington, D.C., at the headquarters of the National Research Council, Yerkes, according to one of his biographers, "thus spent the period of the war and afterward in the heart of the American scientific establishment, helping to make policy and channel funds . . . he served on committees and advisory groups and exerted great influence on the institutions and programs of American science in the 1920's and 1930's."[20]

Without a doubt, the most successful and prolonged venture that emerged out of Yerkes' association with the N.R.C. was his chairmanship of the Committee on Research in Problems of Sex. And yet, although the Sex Committee (as it was known) became a model for later committees that sponsored research in a special field, the original idea for it did not come from Yerkes.

In 1920, Earl Zinn, who had done graduate work in psychology under G. Stanley Hall, came up with the idea that a special effort should be made to encourage research in the field of sex. After talking it over with Max Exner, then organizing a program of sex education for the American Social Hygiene Association, and Katherine Bement Davis, director of the Bureau of Social Hygiene, Zinn spent a year with Dr. Davis so he could seek the advice of competent scientists and work out the means to establish a research program. He then drew up a proposal for the National Research Council—which called for the attention of Robert Yerkes in his capacity as chief of the Research Information Service.

Yerkes immediately recognized the importance of the proposal placed before him by Zinn and tried to carry the plan forward, but his efforts were not well received. After further discussions involving biologists, physiologists, psychologists and physicians, the project became more widely understood and appreciated. And it received the blessing of Victor C. Vaughan, who had recently become chairman of the Division of Medical Sciences. The next step was to schedule a special conference, with Vaughan as chairman, and Yerkes, Zinn and representatives of the N.R.C. and its divisions in attendance, to discuss the feasibility of a research program. Among others invited to participate in the meetings held on October 28–29, 1921, were two members of the Harvard medical faculty: Alice Hamilton, assistant professor of industrial medicine, and Walter B. Cannon, professor of physiology.[21]

George Corner, in his history of the Sex Committee, has recorded more about this extraordinary event. In 1921, he noted, it took a good deal of courage for a group of professional biologists and psychologists to organize a committee for research on the problems of sex. Previously in the United States, only a rare physician like Robert L. Dickinson or a psychiatrist like Adolf Meyer was seriously working on human sex biology. "Such men were few, however, and they had to work against a deep reluctance, even in the medical profession," wrote Corner. "These occasional exceptions emphasized the need for the dignified sponsorship of research on sex problems that was to be supplied through the National Research Council."

There were other antecedents as well. Following a sensational investigation of "white slave traffic" in 1910, a group of influential persons headed by John D. Rockefeller organized the Bureau of Social Hygiene, which undertook a comprehensive study of prostitution. The American Social Hygiene Association, founded in 1913, became the agency for change of public attitudes toward prostitution. Before long, these efforts caught the attention of the women's movement, which campaigned for a single standard of sex morality. "Without doubt," Corner commented, "the feminist campaign and the birth control movement contributed to the growing demand for greater knowledge of the physiology of reproduction and the psychology of sex."[22]

The 1921 conference, with funding from the Bureau of Social Hygiene, was a great success and ended with whole-hearted endorsement of the proposal that the National Research Council establish a committee to sponsor sex research. The impetus for this decision was led by Cannon, who as chairman of a temporary subcommittee, drafted a resolution stating the problem and its difficulties, and in so doing, shaped the attitude and the aims of the program for the next twenty-five years. "The impulses and activities associated with sex behavior and reproduction," the resolution read,

> are fundamentally important for the welfare of the individual, the family, the community, the race. Nevertheless, the reports of personal experience are lacking and the relatively few data of observation have not been collected in serviceable form. Under circumstances where we should have knowledge and intelligence, we are ignorant. To a large degree, our ignorance is due to enshrouding of sex relations in a fog of mystery, reticence and shame. Attitudes toward the subject have been fixed by moral teaching, medical instruction and social propaganda, all based on only a slight foundation of well-established fact. In the presence of this secrecy and prejudice scientific investigation would be difficult. The committee is convinced, however, that with the use of methods employed in physiology, psychology, anthropology, and related sciences, problems of sex behavior can be subjected to scientific examination. In order to eliminate any suggestion that such inquiry is undertaken for purposes of propaganda, it should be sponsored by a body of investigators whose disinterested devotion to science is well recognized. For these various reasons, the committee recommends that the National Research Council be advised to organize and foster an investigation into the problems of sex.[23]

An organizing meeting was held soon after, at which a committee of five scientific advisers was recommended, Cannon among them. At the first meeting of the new Committee held in January 1922, Yerkes was elected chairman and Earl Zinn was appointed executive secretary—a salaried position that he held until 1929. A good deal of Zinn's time during the first few years was devoted to surveying the state of scientific knowledge of sex, as well as the workers actually engaged in sex research. In the fall of 1925, he was sent to Europe to survey the work in progress there.[24]

To begin with, the Sex Committee devoted itself to the task of working out a plan of research, dividing the field of study into the anatomical and physiological, anthropological, psychological, and sociological aspects of sex. Two outlines were produced: one by Frank Lillie covered problems in biology, the other by K.S. Lashley (at the suggestion of Yerkes) covered work in neurobiology and psychobiology. While Cannon found the outlines impressive, he confessed to Yerkes "that the elaboration of a large scheme of research is difficult for me. I am an experimentalist and prefer taking the obvious next steps rather than attempt to evolve the whole system of arrangements."[25]

Meetings were held at first bi-monthly, then quarterly, at which committee members gathered in Washington to discuss and vote on grant applications. The quality of the chosen investigators was very high, with three of the early recipients—Lillie at the University of Chicago, Charles Stockard at Cornell, and Herbert Evans at the University of California—running laboratories that in time emerged as centers of research in the biology of reproduction. Later, funding was provided for several other centers of research, such as those that evolved at Columbia under Philip E. Smith and at Johns Hopkins under Philip Bard.[26]

Usually, the Sex Committee adhered to a policy of giving more substantial grants to existing investigations in a limited number of institutions. When some of the research groups grew too large in number, strength and needs for its resources, the Rockefeller Foundation and other organizations took over their financial support. The question of how to give generous support to significant projects without making unwise commitments for indefinite periods of time was a major problem for the Sex Committee throughout its existence. By the time of Cannon's resignation in 1945, however, it devoted a large part of its funding to a single project: the case-history study of human sexual behavior under the direction of Alfred C. Kinsey at Indiana University.[27]

When Cannon was first asked to become a member of the N.R.C. Committee for Research on Problems of Sex, he wrote Dr. Vaughan that he hesitated not because he failed to appreciate the serious import of the work but because of "the encroachment on research that this administrative work will necessarily involve." He did, however, tentatively accept the invitation in order "to get the work started on profitable lines." Even so, he added a stipulation. If requirements of time and attention proved too exacting, he wanted to feel free to "step out without being regarded as slacking or deserting."[28]

A year later, Cannon again hesitated about whether he would serve on the Sex Committee but finally agreed to accept reappointment. "I recognized, as strongly as you did," he wrote Yerkes, "the importance of keeping the personnel of the Committee undisturbed. I hope, therefore, we may continue to work together at least until the new organization is organized."[29] Still, it must be remembered that these were years when Cannon was pressed hard with the labor of starting a new lay society for the promotion of medical research as well as opposing legislative campaigns waged by the antivivisectionists. Whenever he felt swamped by these extra responsibilities, to say nothing of the obligation of running the physiology course, revising his laboratory manual, or the myriad commitments to investigations underway in his laboratory, he would ask to be relieved as much as possible from the demands made on him by Yerkes or Zinn. In the years that followed, he even tried several times to resign from the Committee, but Yerkes always managed to dissuade him. And Cannon ended by serving for twenty-three years—just two years short of Yerkes' own term of office.

The Sex Committee served Cannon in turn, for it gave him an opportunity to learn the ways and means of grantsmanship (no small part of the politics of science). The experience and skills he acquired in the process of evaluating applications were put to good use when in future years he would guide a number of other committees entrusted with seeking opportunities for medical investigators and arranging for financial support of their research projects.

One such committee followed close on the heels of the founding of the Sex Committee when the Ella Sachs Plotz Foundation was established in 1923 at the Harvard Medical School. The Plotz Foundation, administered by an executive committee consisting of Cannon (chairman), Francis Peabody (secretary), Bernard Sachs and Paul J. Sachs, awarded small grants for research directed toward the solution of problems in medicine and surgery, or in branches of science bearing on medicine and surgery. Another was the Medical Fellowship Board of the National Research Council, to which Cannon was appointed in 1927. Although the sums of money that such committees doled out may seem paltry by later standards, they often made the difference between life or death of a given research project—or even of a career.

Cannon's service to the Plotz Foundation at Harvard, as well as his management of the Elizabeth Thompson Science Fund there, gave him an enthusiasm for small foundations in the service of medical research. He believed that their grants in aid, added to the customary funding available in universities and research institutions, made many important investigations possible, which would not have been able to secure help from the larger foundations that concentrated on major projects. Letters from students and scholars the world over who had received such awards brought Cannon not only warm messages of gratitude but also confirmation of his faith in the value of small sums as an element in the development of scientific research.[30]

When Yerkes early in 1918 wrote Cannon about his multifarious wartime projects, he was clearly enthralled with the brave new world he had so recently entered. There was a new appreciation, he rejoiced, that the war must be won by brains and not by brute strength or arms. Even more to his liking, he joined with a group of other psychologists who believed that their ideas and attitudes would help to decide the conflict. What with the rapid expansion of psychology, he sought to launch a psychological offensive "which will put the submarine to shame" and "show Teutonic enemies the significance of American behaviorism."[31]

Yerkes' special contribution to the war effort was as a guiding force in the conduct of intelligence tests on officers, enlisted men and draftees for the U.S. Army. By the time the conflict ended, he believed that the army testing program had been a great practical success and that it provided a rich lode of data on the heritability of intelligence. It would be difficult to find anyone today to agree with that view. Historians and scientists have long since published detailed critical accounts of the program showing that the tests, at best, lacked rigorous standards and, at worst, were bad science. But during the early part of the 1920s, there were many educated and articulate people to be found, and not least in the Boston and Harvard communities, who held eugenical views and found sustenance for them in the Army intelligence tests.[32]

In the summer of 1921, Cannon's wife Cornelia wrote Robert Yerkes that she had been reading two books—McDougall's *Is America Safe for Democracy?* and *Army Intelligence Tests* by R.M. Yerkes. After calling McDougall's book "an interesting and suggestive book but the kind that a diletante might have composed," she proceeded to express her admiration for Yerkes' book: "*There* is a real contribution, something that will stand as a permanent addition to man's knowledge about man . . . That book of yours is an extremely important publication and . . . the accumulation of data from which this is culled will prove a gold mine which should surely be more fully worked."[33]

The books that Cornelia had read were merely the tip of an iceberg. A postwar literature that swept the country spread the news derived from the Army test data: America was a nation peopled by inhabitants with an average mental age of thirteen to fourteen years; and the foreign-born tested even lower in mental ability than native Americans. Throughout the early 1920s, articles by Yerkes and other commentators on the scientific and social aspects of the Army I.Q. test score results appeared in popular magazines such as *The Independent, Scribner's* and the *Atlantic Monthly*.[34] One of the most controversial of these was published as the lead article in the February 1922 *Atlantic Monthly*. It was called "American Misgivings" and it was written by Cornelia James Cannon.

"I read yesterday for entertainment on the train your article in the February Atlantic," Yerkes wrote Cornelia soon after publication.

> If I were a debating society I am sure I should be seriously overworked for the indefinite future by the question evoking statements of your paper. I am happy indeed to see that Mr. Sedgwick prized the discussion so highly as to give it first place.
>
> It occurs to me that you, the Atlantic Monthly, or both, will be

> called on to defend various statements in your article and to offer additional information and explanations. I wish to offer you, and you may pass the information along to Mr. Sedgwick if you like, the facilities of our Information Service in this connection.

Cornelia's reply to Yerkes was grateful: "Your letter certainly showed feminine insight—I had already referred one gentleman from Iowa to you as the source of all wisdom on this subject. I am certainly in for brick bats but I depend on *you* to back me for after all you got me into the row. Mr. John G. Brooks came over with some confirmatory evidence by J.S. Frazer and Galton for he said he wanted me not to weaken when the storm broke. I am sending your letter to Mr. Sedgwick for his use."[35]

Brushing aside the brickbats, a fearless Cornelia went on to write more articles that dealt with provocative and controversial subjects. But not Yerkes, who began a strategic retreat that eventuated in a return to his career as a comparative psychologist. He had long had hopes of heading an institute of psychobiological research, with emphasis on primate studies. After James R. Angell, a fellow psychologist, became president of Yale, Yerkes' hopes were realized—though it took some time before funding could be obtained. In 1924, through the actions of Beardsley Ruml at the Laura Spellman Rockefeller Memorial, Yale was promised $200,000 for an Institute of Psychology, with Yerkes in the fold, and the following year, the Rockefeller Foundation granted $40,000 outright to him for anthropoid research.[36]

Yerkes' work with monkeys and apes stemmed from a half year spent in 1915 with G.V. Hamilton's primate colony in Santa Barbara, California. In preparation for his return to the field of primate studies, he acquired a pair of young chimpanzees, who were given the names of "Chim" and "Panzee" by his and Cannon's friend, the philosopher Ralph Barton Perry. When Yerkes brought the chimps to his farm in Franklin, all of the children in the summer colony were fascinated with their new playmates. As Cannon's son, Bradford, later remembered:

> They were a friendly pair, walking hand in hand with us children, and were agile in the birch trees, instinctively making nests by folding branches together. My father was friendly with both until, on one occasion, Chim was being a nuisance and he pushed him aside. Chim became angry, showing his teeth, stamping on the

> ground with all fours, and making threatening sounds. Doubtless blood pressure and pulse rate also increased, and his adrenal medulla poured out adrenaline. Apparently, Chim did not find that all humans looked alike and, much to my father's distress, Chim reacted with similar anger whenever they met, even weeks and months later.[37]

Yerkes spent the remainder of his career at Yale, where, in James Reed's opinion, he was a much more successful investigator in dealing with animal subjects than he had been with human beings. In 1929, he published, with his wife Ada Watterson Yerkes, his monumental book *The Great Apes*. That same year, he founded an experimental station near Orange Park, Florida, which later became the Yerkes Laboratories of Primate Biology. Still later the animals were moved to Atlanta to a facility at Emory University named the Yerkes Regional Primate Research Center, one of several such centers throughout the United States.[38]

8 Homeostasis and the Sympathico-Adrenal System

FOLLOWING THE COMPLETION of his experiments with Rapport, Uridil, Griffith, and Carrasco in 1921–22, Cannon felt that he had fully met Stewart and Rogoff's criticisms of his research. He also believed that in the process he had vindicated his use of the denervated heart to determine the output of secretion from the adrenal medulla under conditions of asphyxia, reflex stimulation of sensory nerves, and emotional excitement. Bolstered by the results of his investigations, he thought that the only option Stewart and Rogoff had left was to retire gracefully from the controversy.

However, Cannon misjudged his opponents. They paid little attention to his latest conclusions and merely repeated their old assertions that the caval pocket method they had used to measure adrenal secretion was far superior to any method that Cannon employed. They also continued to maintain that during the course of their experiments they found adrenal secretion to be constant and unchangeable, and indeed, that it was impossible to judge whether this secretion was increased or decreased unless one actually knew the rate of blood flow.[1]

Nor were Stewart and Rogoff the only ones to attack Cannon's research on adrenal secretion. Gley and Quinquaud in France, for example, appeared unable to comprehend his meaning or, for that matter, to keep up with his latest findings. These criticisms and underappreciations of his work so nettled Cannon that in December 1922, he complained to Carrasco-Formiguera, who had returned to his practice of endocrinology in Spain, that physiologists in Europe had overlooked the results of the experiments they had performed together. "Why do you not make a demonstration before the Society in Barcelona and show them what the denervated heart will do in the way of proving adrenal secretion to a degree that it affects structures within the body?" he suggested. "If anybody then attacks you I will gladly enter the fray myself. There are too many interesting new things that are calling for my time to make it possible for me to undertake those old controversies at the present time."[2]

Despite Cannon's growing exasperation with the continuing doubts about his research that he believed he had laid to rest, the criticisms of his opponents prodded him to exploit several promising avenues for extending the research he had begun through use of the denervated heart. The success that he achieved in responding to Stewart and Rogoff and others, and how much he learned in the process, may be determined by comparing two very different statements made at different times.

In 1921, when contributing an editorial to *JAMA,* Cannon solemnly told clinicians, "Unfortunately, we are still deeply ignorant of many important aspects of endocrinology. The conditions in the body which control the glands of internal secretion are still almost wholly unknown." A mere four years later, when delivering an address to the Congress of American Physicians in May 1925, Cannon could claim that "advances in our knowledge of the relations of internal secretion to metabolism have lately been so great and so various that an attempt to present an encyclopedic survey in brief compass would be futile."[3]

For Cannon, the decade of the 1920s would prove to be an extraordinarily active and productive one in the laboratory. Before it ended, he not only would put the quibblings of his critics into retreat, he also would contribute two novel and important ideas to the fields of biology and medicine. One was homeostasis, the conception with which Cannon's name is most frequently associated today, and the other was a new explanation for a theory of the emotions. We begin first with an examination of homeostasis.

Most commentators, when discussing the development of Cannon's concept of homeostasis, begin with Claude Bernard's observation of the *milieu intérieur*—that is, the body's natural ability to stabilize its internal environment. After all, early in his definitive discussion of "The Organization for Physiological Homeostasis"

(1929), Cannon credited Bernard with first pointing out that in animals with complex organization, the living parts exist in an internal environment of blood and lymph which bathe them—"what we may call the *fluid matrix* of the body." He went on to describe the mechanisms or agencies that "are automatically called into service which act to bring back towards the mean position the disturbed states." And yet, even though Cannon referred to Claude Bernard and his concept of the *milieu intérieur* in his lectures to Harvard medical students as early as 1921, it would appear that he was unaware of its pertinence to his own work until he was well along on the path towards the development of his idea of homeostasis.[4]

"Whatever the spark was which set off this series of experiments," F. L. Holmes, a student of Cannon's work, has stated, "it required no drastic jump from previous work. The techniques were much the same, involving sympathetic and other stimulation of the adrenal glands, adrenalectomies, and the use of the denervated heart as an indicator for adrenalin." Time and again, Cannon himself explained that his research followed a logical sequence of events that only gradually and imperceptibly led to his concept of homeostasis. In the Preface to *The Wisdom of the Body* (1932), he traced the steps in the natural sequence of ideas that had become apparent to him in the progress of his career before he was able to arrive at the subject of the book, homeostasis. "Indeed," he revealed,

> not a few researches on the service of the autonomic in providing for stability of the organism had been completed and published before the connection of that system with regulatory arrangements was clearly understood. We found that we had long been working on the role of the autonomic system in maintaining steady states without realizing that we were doing so! Then facts already discovered took on new significance.[5]

Cannon's disclosure told only part of the story. An omnivorous reader with a knack for making insightful correlations even on fragmentary evidence, he was able to recognize and willing to test interesting ideas of others that came to his attention. Here and there, luck favored him with unforeseen possibilities which his prepared mind was keen enough to follow up—the "gains from serendipity" he so candidly acknowledged.[6]

One influence may have come from the investigation of the effects of adrenalectomy on metabolism by Joseph Aub, Jonathan Forman and Elizabeth Bright, all working in Cannon's laboratory at Harvard. From 1920 to 1922, Cannon collaborated with Monroe McIver from the department of surgery and Sidney Bliss, a doctoral candidate in physiology, in an investigation of the relation of adrenin to metabolism. He then encouraged and helped McIver and Bright to test the effects of reflex stimulation of adrenal secretion on the rate of metabolism. These experiments established the adrenal system as one of the regulators of short-rate changes in metabolism, a finding of importance because many internal conditions are affected by metabolism.

Perhaps another early stage in the development of homeostasis may be traced to Cannon's efforts to produce a volume summarizing the war work on traumatic shock (written in 1920 but not published until 1923), which gave rise to David Rapport's experiments on the relation between shock and adrenal secretion. By constricting the blood vessels of non-vital regions, epinephrin helped to maintain a constant blood pressure in the vital organs when a loss of blood threatened to depress it. This work, together with the studies by Aub and his colleagues, may well have suggested a general relation between adrenal secretion and the maintenance of certain conditions in the face of factors tending to alter them.[7]

Cannon's output of scientific publications was extraordinarily prolific in 1922. The year brought to light no less than nine papers relating to the endocrine glands and internal secretions. Some of these have been described in a previous chapter, but we may also note here the research reported by Cannon and Philip E. Smith, which demonstrated that the thyroid responded to stimulation of the cervical trunk of the sympathetic nervous system and appeared to provide clear evidence of nervous control of thyroid secretion. Publication of Cannon and Smith's study, begun in 1919 and reported to the Society for Experimental Biology in February 1920, had been delayed because attempts to obtain further verification were not successful.[8]

All of the above may be viewed as steps along the pathway leading to Cannon's concept of homeostasis. Still another, of special significance, was his attendance at a meeting in New Haven of the

American Physiology Society at the end of 1921. This was the meeting at which Frederick Banting presented the results of the work he had done with Charles Best in Toronto on the internal secretion of the pancreas of the dog and made a preliminary communication regarding their isolation of insulin. Banting was not a polished lecturer, and as a result, many in the audience were skeptical of the importance of his conclusions. Cannon, however, was very impressed with the work of Banting and Best. Moreover, he immediately began to formulate ideas about how he could use insulin as a tool to study the inter-relationship of the autonomic nervous system and the regulation of blood sugar levels—a move, in fact, that put him on the highway to homeostasis.[9]

Cannon began to follow the burgeoning literature of diabetes and insulin and was intrigued by the descriptions of the symptom complex induced in diabetic patients given excessive doses of insulin, with marked reduction in blood sugar levels. He noted the similarity of the response in hypoglycemia to that seen with increased sympathetic activity—that is, sweating, pallor, pupillary dilatation and tachycardia. Cannon also noted that Wilder and Boothby at the Mayo Clinic had reported that, with the first subjective evidence of hypoglycemia induced in a diabetic patient given an excessive dose of insulin, there might be a spontaneous outpouring of adrenalin that was responsible for the tachycardia and other symptoms. However, the Mayo investigators did not provide any experimental evidence to prove that adrenal secretion was stimulated.[10]

To test the hypothesis concerning the inter-relationship of hypoglycemia and adrenalemia, as well as the effect of altered blood sugar levels on the normal heart beat, Cannon, assisted by Monroe McIver and Sidney Bliss, began a series of experiments in animals with denervated hearts, a preparation which, as he had demonstrated, offered a sensitive measure of the rate of adrenal secretion. He was fortunate enough to obtain a supply of insulin from Dr. George Clowes, the director of the program producing insulin commercially at Eli Lilly Company.[11]

Since adrenin is a counter-hormone to insulin with opposite effects, Cannon, McIver and Bliss surmised that insulin serves to store sugar in the liver while the secretion of adrenin liberates it. After the adrenals had been inactivated, the injection of insulin resulting in a lowering of blood sugar might have disastrous effects.

They were able to prove quite conclusively that when hypoglycemia was induced by a large dose of insulin in cats with denervated hearts, tachycardia was also induced. But if, on the other hand, the adrenal glands had previously been removed, or if one had been removed and the other denervated, a fall of blood sugar below the critical level was not accompanied by an increased heart rate. The cardiac acceleration, therefore, was not due to a direct action of insulin on the heart or on the adrenal gland but rather to increased adrenal discharge in response to nervous impulses.

When they made a full report of their findings, Cannon, McIver and Bliss stated that the mechanism protecting the body from dangerous hypoglycemia probably operates in two stages—a preliminary stage in which sympathetic activity with adrenal secretion occurs and mobilizes sugar from the liver; if this proved inadequate, a secondary stage takes place in which the activities of the first state are intensified and augmented in convulsive seizures. "The mechanism here described," their 1924 report concluded, "is another remarkable example of automatic adjustments within the organism when there is a disturbance endangering its equilibrium."[12]

Along with the accumulation of other evidence, this work required Cannon to alter his emergency theory in favor of a broader view that explained the body's ability to make adjustments in time of stress or special need, in conjunction with its built-in mechanisms for maintaining a steady equilibrium. He had long been interested in glycosuria induced by emotional factors, beginning with his studies with Shohl and Wright in 1911, and up until this time had stated that adrenin produces glycosuria and raises the blood sugar. Starting with the paper published with McIver and Bliss in 1924, however, he held that hypoglycemia produces increased adrenin, which in turn produces increased blood sugar. Or, to put it another way, Cannon in 1914 had ascribed to adrenal secretion the mobilizing of sugar into the blood to aid in extreme muscular effort; now, ten years later, he conceived that the role of adrenin was to maintain the constant level of blood sugar under any circumstances.[13]

News of these investigations on the role of the adrenal gland in hypoglycemia spread. James B. Collip, the Canadian biochemist who had worked with Banting and Best on the discovery of insulin, later expressed his appreciation thus: "Professor Walter Cannon and his colleagues in their brilliant presentation have shown that the suprarenal glands may be activated in a number of ways to discharge adrenin; chief among these were pain, hunger, fear, and rage. A good example of this emergency function of the suprarenals was recently reported to me." And he continued,

> A diabetic patient taking the insulin treatment realized one morning walking down the street that he was developing a hypoglycemic reaction. Finding that he had forgotten to provide himself with a chocolate bar, he proceeded to the nearest drug store. By the time he reached the store his gait was unsteady and his speech incoherent. He tried to explain to the druggist what he wanted, but the latter, fully convinced that he was dealing with a drunken man, threw him into the street. The patient, still conscious and terribly enraged at being so treated, promptly recovered and proceeded to another store unaided, made his wants known, and continued on his way. Obviously we have here an example of the activation of the adrenals as a result of anger, leading to the release of enough adrenin to cause an increase in the patient's blood sugar sufficient to restore his equilibrium and his powers of speech.[14]

There was still work to be done, however, before Cannon could reach the point that allowed him to elaborate his innovative ideas in public. Yet another process that he sought to investigate was the body's regulation of its internal temperature. The erection of hairs, constriction of peripheral vessels, and increase in blood sugar—all characteristic of emotional excitement in animals—suggested that the same mechanism might be involved in the regulation of body temperature, either during exposure to cold or in the presence of fever.

Working with Arie Querido, a research fellow from the Netherlands, Cannon investigated the effect of cold by calculating the heat debt in an animal whose heart had been denervated, discovering in the process that the measurement would be more precise if cold water were introduced into the stomach. The "heat liability" thus produced could be nicely adjusted for the purpose of the experiment. When it was more than 1,000 calories per kilo of body weight, acceleration of the denervated heart (indicating increased adrenal secretion) was almost universally attended by shivering. When the heat liability was 900 calories, shivering rarely occurred or was of short duration, and the faster heart beat showed that the adrenal factor was operating. But if the adrenal glands were inactivated and a heat liability of 900 calories was established, shivering almost universally occurred and was of long duration.

A second study that Cannon undertook with Jayme Pereira, a Rockefeller fellow from Brazil, proposed to determine whether fever was also associated with more abundant adrenal secretion. Again using the denervated heart, he produced fever in experimental cats by injecting a suspension of dead typhoid bacilli supplied by Hans Zinsser (whom Cannon had been instrumental in bringing to Harvard as professor of bacteriology). When the adrenals had been rendered inactive they found that injection of the bacillus, instead of causing first a rapid rise in temperature followed by a second slower rise, caused only the second slow rise. Hence, they reasoned that in fever the adrenal glands were stimulated to greater activity. In their report, Cannon and Pereira concluded, "It may be that the response of the adrenal medulla during toxic fever is serviceable to the organism." Cannon would be compelled to come back later to the subject of temperature control.[15]

One other study conducted at this time may be noted in relation to the development of Cannon's concept of homeostasis. Assisted by two Harvard medical students whom he had taken into his laboratory, the Linton brothers, James and Robert, Cannon determined to examine the discrepancy between the positive and negative effects of muscle metabolites (carbon dioxide and lactic acid) on the denervated heart in intact animals. Their study, reported late in 1924, corroborated Johansson's long-standing observation that the denervated heart beat accelerated when muscle metabolites were set free in the circulation. But when, after the adrenal glands had been eliminated, acceleration failed to occur, the authors concluded that "the muscle metabolites have no direct effect on the heart rate, but act by stimulating the nervous control of adrenal secretion."[16]

Between 1919 and 1925, Cannon frequently made mention of the sympathetic control of adrenal secretion, especially in the 1922 papers that signalled his move in the direction of homeostasis. Nevertheless, he continued to refer to the "emergency theory of the adrenal medulla." It was in his lecture on "Some General Features of Endocrine Influence on Metabolism" that he gave the first indication in public of the seismic shift in his thinking. Although he did not then use the words "homeostasis" or "homeostatic," he enunciated six propositions he had formulated regarding physiologic factors that maintained steady states. "Some recent research done by McIver and Bliss and myself," he stated, "has revealed a sympathico-adrenal mechanism as one factor in the balance, protecting the body against the dangers of too low blood sugar."[17] Cannon's first use of the term "sympathico-adrenal" here clearly signals a shift in his emphasis from medulliadrenal secretion to a sympathico-adrenal (or sympathoadrenal) system.

The address that Cannon delivered to the Congress of American Physicians in May 1925 indicates that he had arrived at some definite conclusions, based on his recent series of investigations. Right from the start he made his thesis clear: "A fairly constant or steady state, maintained in many aspects of the bodily economy even when they are beset by conditions tending to disturb them, is a most remarkable characteristic of the living organism." The existence of "steady states" (a term commonly employed in physics) in the body had long been recognized, he continued. He now laid emphasis on these states because of an interest in "some tentative general considerations with regard to physiological factors regulating them" and in their possible bearing on internal secretion.

Cannon then advanced six propositions, suggesting that they might have value in determining lines for future research as well as in providing a basis for critical judgment. In his words, they are as follows:

1. In an open system, such as our bodies represent, complex and subject to numberless disturbances, the very existence of a poised or steady state is in itself evidence that agencies are at hand keeping the balance, or ready to act in such a way as to keep the balance.

2. If the state remains steady, there is an automatic arrangement whereby any tendency toward change is effectively met by increased action of the factor or factors which resist the change.

3. Any factor which operates to maintain a steady state by action in one direction does not act at the same point in the opposite direction.

4. Factors which may be antagonistic in one region, where they effect a balance, may be cooperative in another region.

5. The system of checks which determines a balanced state may not be constituted of only two antagonistic factors; on either side

there may be two or more, brought into action at the same time or successively.

6. When a physiologic factor is known which can shift a steady state in one direction, it is reasonable to look for a physiologic factor or factors having a contrary or counter-balancing effect.

These six propositions Cannon expanded on and illustrated with examples from his recent experiments and their findings, including a description of the sympathico-adrenal factor that works to maintain blood sugar balance and to protect the organism from the danger of hypoglycemia. He also discussed factors affecting protein equilibria and factors affecting the endocrine glands in maintaining temperature equilibrium. In studying the value of protein for bodily welfare and function, he related that he had been led to believe the liver is of central importance for both protein and carbohydrate metabolism, and its reserves could be called on by sympathetic stimulation and adrenal secretion in order to maintain the body's steady state.

Even though his address contained much that sounds like homeostasis, Cannon was not yet ready to perfect a doctrine. "In my opinion," he concluded, "an attempt at present to develop an elaborate and intricate conception of the inter-relations of the endocrine glands would be altogether premature and might be quite misleading." Acknowledging that the necessary proof was not yet at hand, he nevertheless insisted that endocrine factors play an important role in establishing and maintaining physiologic equilibria. It was an idea that was reasonable and moderate in its scope, but it would not involve "a far-reaching projection of theory into the unknown."

After he had finished his talk, Lewellys Barker, who chaired the meeting, commented that he was glad to hear Cannon make these generalizations. In Dr. Barker's opinion, too often investigators were afraid to speculate about their observations and consequently never did more than gather facts. "The great harm comes from assuming that hypothetical constructions are established principles; but there is no reason why we should not make judicious use of hypotheses. They are extremely useful in directing experimental work, and in clinical work they are a help in controlling observations and in keeping records."[18]

In May 1926, exactly one year after Cannon delivered his talk to the Congress of American Physicians, a jubilee volume honoring the French physiologist Charles Richet was published in Paris. Cannon, invited to contribute to the festschrift, submitted a paper (brief like the other papers in the volume, it is only two and a half printed pages) titled "Physiological Regulation of Normal States: Some Tentative Postulates Concerning Biological Homeostatics." Basically a boiled-down version of his six propositions, it has the distinction of containing Cannon's first use in print of his newly-coined word.

He began with a quotation from Richet, who stated in 1900, "The living being is stable. It must be so in order not to be destroyed, dissolved, or disintegrated by the colossal forces, often adverse, which surround it." Then Richet elaborated that the living being is, in a sense, "stable because it is modifiable—the slight instability is the necessary condition for the true stability of the organism." Cannon followed with a statement from Claude Bernard, who even earlier than Richet had observed, "All the vital mechanisms, however varied they may be, have only one object, that of preserving constant the conditions of life in the internal environment." Here we have another first: Cannon had not previously mentioned Bernard and his *milieu intérieur* in the context of his own evolving concept.[19]

It is possible that Cannon's reacquaintance with Bernard's thinking on this subject may have occurred while he was looking into Richet's life and work when preparing his paper for the jubilee volume. The historian Donald Fleming has argued in a recent appraisal of Cannon's work on homeostasis, on the other hand, that Bernard's impact on the Harvard physiologist's thinking came about in a different way. Fleming suggests that Cannon's association of Bernard's concept with his own work on bodily regulation occurred when, in searching for clues regarding antagonistic agents in the body which through their opposing forces bring about equilibrium, he read, or more likely re-read, J. S. Haldane's *Respiration* (1922). Cannon himself stated that there was no better example of the autonomic regulation of bodily processes than Haldane's work on respiration.

Haldane was one of a number of English physiologists, most of them Cannon's contemporaries and some his colleagues during the late war—including T. R. Elliott, C.S. Sherrington, E.H. Starling, and Joseph Barcroft—whose work closely paralleled Cannon's own.

Fleming further observed that whereas Bernard had specifically commented on the total lack of knowledge in his day about the function of the adrenal glands, Cannon ("ideally equipped to add the perspectives of another fifty years to Claude Bernard's own") became one of the principal elucidators of the effects of adrenaline, never wavering in maintaining T. R. Elliott's association of it with the sympathetic nerves. Among the empirical researches that Cannon incorporated into his concept of homeostasis, Fleming rightly notes, the most important category of discoveries made since Bernard's time had been the hormones in general and adrenaline in particular.[20]

Aside from the connection with Claude Bernard and the first use of the word "homeostasis," Cannon's contribution to Richet's jubilee volume is also noteworthy because it lays bare the resolve that held Cannon in its grip. "It is clear," he maintained, "that an examination of homeostatic conditions in the body and the agencies controlling them is of very great biological and medical interest and importance. It is a field which has been too little cultivated."

The examination of homeostatic conditions was to occupy Cannon for some months, and was carried on in tandem with his study of the nature and seat of the emotions. Increasingly, the two major strands of his research became inextricably woven together, sometimes separating, sometimes interlocking. Although it is evident that homeostasis remained very much in his thoughts, not until 1929 would Cannon be prepared to issue a comprehensive account.

❧ ❧ ❧

It was not unusual for Cannon to have several lines of inquiry going on in the laboratory at one and the same time. Some of these delved into questions that could be answered with a continuous series of experiments. Others uncovered puzzles that perplexed him, mysteries that defied solution, which had to be put aside until a more propitious time. And then there was one particular problem that stubbornly nagged him, its secrets eluding him for a period of years.

While the Harvard physiologists were conducting researches to determine what factors might control steady states, an old problem appeared out of the blue to trouble them. After the work on the denervated heart in 1921 and 1922, Cannon and his associates had been satisfied that they could eliminate factors which previously interfered with its efficacy as an indicator of adrenal secretion. (That is, by removing the stellate ganglion they obviated any thyroid influence, and by cutting the hepatic nerve they prevented the liver from causing an acceleration in the heart beat.) In the summer of 1924, however, they experienced a serious setback.

Observing a cat in which they had inactivated the adrenals six months previously, they found that when they stimulated the animal's brachial nerve, its heart rate increased by forty-six beats per minute. This startling phenomenon caused some anxious moments in the laboratory because, unless it could be explained and its effects excluded, it would render unreliable all of the conclusions that had been based on the denervated heart.

The resulting search for the unexpected trouble-maker dragged on interminably and involved many of the workers in the laboratory. While Cannon, assisted by his postdoctoral fellows, J. T. Lewis and Sidney Britton, carried the work through, several others were involved in this lengthy endeavor as well. Since the unaccounted-for acceleration occurred some time after operation, the group were able first to rule out the effect of anesthesia as a causative factor. Next they considered the advisability of denervating the heart in stages more closely spaced together than previously, and soon succeeded in completing the denervation in two operations, one closely following the other. But the acceleration persisted, being at first small and increasing with each passing day.

By February 1925, after several experiments, Cannon was convinced that they had not cast much light on the cause of the disturbance. Was it possible, he wondered, whether the denervated heart might have become sensitive to changes in blood pressure after a few days? He also questioned if metabolic products might not in some way have affected the heart rate. But on investigation, he and his associates found that neither of these factors could be held accountable for the acceleration. They even tried atropine to block the vagus action in the event that the augmented heart beats might be due to some agent that blocks vagal impulses—but again, they came up short.

When all these efforts provided no answers, the frustrated workers sat down together and tried to systematize the remaining possibilities. Faced with a baffling situation, the only idea that occurred to them was that there might be some accessory accelerator fibers going to the heart from the sympathetic chain of ganglia below the stellate ganglia, which had already been removed in the denervation process. They had considered such a possibility some months earlier, but had hesitated because removal of these portions of the sympathetic chain was difficult and involved a long and tedious procedure. Furthermore, the cats often caught cold, became sick and died, thus creating more delay and wasted effort. Still, they saw no other way to proceed.

In early operations, although they removed the thoracic sympathetic chain down to the third intercostal space on the right side and to the fourth or fifth space on the left, the acceleration still occurred. At last, in September 1925, they began removing the sympathetic ganglia down to the ninth intercostal space. By December, when the increase in heart rate fell from 106 beats per minute to only four, they believed they had solved their exasperating problem. Cannon and his associates conducted several more experiments to confirm their results, and by March 1926 they concluded that the known acceleratory factor was due to nerve fibers connecting the heart and the thoracic sympathetic chain below the stellate ganglia.

Nevertheless, even after all these factors had been eliminated, there remained a minor increase of four to fourteen beats per minute when the cats were excited—an acceleration that still could not be explained. But Cannon considered this so slight, compared to the increases that resulted from adrenal, hepatic or thyroid secretions, that it would not affect the reliability of the denervated heart as an indicator of endocrine activities.[21]

In the meantime, Cannon had returned to his investigation of temperature regulation as a component of his notion of homeostasis and reconsidered the role of adrenal secretion in the chemical control of body temperature. Sometime in 1925 or 1926, aided by Arie Querido, Sidney Britton and Elizabeth Bright, he set about to determine more thoroughly and more specifically whether secretion from the adrenal medulla was augmented when cold caused a discharge of sympathetic impulses. The group also examined the function of shivering as well as the role that metabolism played in the process, using the Benedict apparatus, or spirometer (as modified by Roth), to measure the rate of oxygen utilization in humans before and after exposure to cold. Repeating many of the experiments that Cannon and Querido had conducted earlier on cats, both with the adrenals present and absent, they used as their indicator the denervated heart prepared with additional sympathetic ganglia removed. In January 1927 they reported their findings: that a disturbing heat loss brought about an automatic increase in medulladrenal secretion; that the extra output of adrenin, by hastening combustion, served to protect the organism from cooling; that heat debt brought about a rise of metabolism to produce the same effect; and that, when shivering occurred, it too caused a marked increase in metabolic rate.[22]

Later in 1927 Cannon, Lewis and Britton published a preliminary report on the dispensability of the sympathetic division of the autonomic nervous system, which picked up where they had left off the year before. They stated that the apparent innocuousness of the surgical procedure reported previously had led them to attempt complete removal of the sympathetic system from the superior cervical ganglion to the united ganglia of the two lateral chains just below the rim of the pelvis. For more than a year animals thus operated on had lived in the quiet of the laboratory, apparently in perfect health. They ate well and maintained their weight or grew fat. Nevertheless, in some ways they did not behave like normal cats.

For instance, their hairs did not rise when they were excited, and they lacked control of the surface blood vessels that would permit regulation of heat conservation and heat loss by shifting the circulating blood. As a result, in cold weather they were more likely to seek a warm place than normal animals. Neither did their rate of metabolism change nor their blood sugar rise in consequence of excitement. In other ways, they were surprisingly normal. One female cat had even given birth to kittens and was nursing them naturally. It was noteworthy, too, that the experimental animals, although deprived of that part of their nervous system which was characteristically aroused during great emotional excitement and also lacking control over adrenal secretion, exhibited all of the typical signs of aggressive feeling (except erection of hairs) when approached by a barking dog.[23]

Not until 1929 did Cannon publish a full report on total sympathectomy. By that time, Lewis and Britton were no longer at Harvard and Cannon had a new set of collaborators, including two young surgeons, H.F. Newton and R.M. Moore. In this paper, he and his collaborators reviewed previous experiments on removal of the sympathetic division of the autonomic nervous system in the Harvard physiological laboratory, as well as those reported by Meltzer in 1920 and Montgomery in 1926. While they had seen effects that might well account for the high mortality rate described by Meltzer, the procedures that they had followed eliminated such fatal outcomes. "Meltzer's intimation that the fatal effects of superior cervical ganglionectomy might be due to removal of some important internal secretion associated with the ganglia," they maintained, "receives no support in our observations."[24]

The authors then described the effects of sympathectomy, including among others, effects on growth and development, on female functions in reproduction, on emotional conditions occurring under stress, and on metabolism. They reported among their findings that following bilateral sympathectomy, animals had lived in the laboratory in good health and performed routine functions for many months, thus disproving prior evidence that the superior cervical sympathetic ganglia or the mesenteric nerves are essential to life. Also following bilateral sympathectomy, emotional excitement caused no erection of hairs, no consistent increase in blood sugar, no polycythemia or increase in mononuclear cells, and no marked rise in blood pressure.

Sympathectomized animals were found to be sensitive to cold, and lost heat more rapidly than did normal animals. Although basal metabolism usually fell somewhat after sympathectomy, the drop was not more than ten per cent. These slight effects resulting from sympathectomy led to the question, What is the function of the sympathetic nerves? Cannon's answer was that when considered in the light of natural conditions, the sympathico-adrenal system, "dispensable in the protected conditions of the laboratory, finds its great service at times of critical emergencies when it adjusts the internal organs of the body for use of the mechanisms responding to external exigencies."[25]

Attempts to separate the development of Cannon's ideas on homeostasis and the bodily effects of strong emotions may at times seem almost arbitrary. For example, two articles published in 1928, which do not deal with homeostasis *per se*, are significant in showing not only the direction of his thinking but also how he connected the assortment of scientific facts that came out of his research. They are noteworthy, too, because their titles contain words that reflect

Cannon's incorporation of new knowledge as it developed. We have already made note of his steady use of "sympathico-adrenal" (or "sympathoadrenal") as a system, beginning in 1925. By the late 1920s, study of the adrenal cortex and the search for a cortical hormone by Stewart and Rogoff, Frank Hartman, and others had quickened to the point where Cannon increasingly found it necessary to use "medulliadrenal" in referring to his work on adrenal secretion.[26]

"Emotional Polycythemia in Relation to Sympathetic and Medulliadrenal Action of the Spleen," a study that Cannon conducted with J. J. Izquierdo, a Rockefeller fellow from Mexico, examined previous research by Ferrari on the influence of emotional excitement on the blood count and by Lamson on the role of the liver in emotional polycythemia. Between 1923 and 1927, new light was thrown on Lamson's data and new knowledge of the spleen was developed through experiments conducted by Joseph Barcroft, who explicated the spleen's service to the organism as a means of quickly increasing the number of circulating erythrocytes and then storing them away again.

Izquierdo and Cannon sought to determine whether the spleen, instead of the liver, could be the source of emotional polycythemia, and also if the adrenal glands played an essential role, as Lamson had maintained. They conducted a series of extensive experiments that pointed to the spleen as the immediate cause of emotional polycythemia. Though forced to admit that the results were equivocal, Cannon clearly believed the work to be important, for in his second edition of *Bodily Changes in Pain, Hunger, Fear and Rage* (1929), he devoted an entire chapter to "Emotional Increase of Red Blood Corpuscles."[27]

The other 1928 paper was published in the *Ergebnisse der Physiologie,* then edited by Professor Leon Asher of Berne, Switzerland. Asher, an admirer of the emergency theory, was engaged in physiological research closely paralleling Cannon's work on the adrenals. After a visit to Boston in 1924, he had written to inquire if the Harvard physiologist would contribute a review paper on the functions of the adrenals for his journal. Cannon declined with the excuse that since he was a party in continuing debate on the subject, his viewpoint might appear one-sided. When Asher persisted and assured him that he felt nearer to him in the controversy than to his opponents, Cannon promised to reconsider the invitation at a later time. At length, he sent on his article, which Asher's wife, Else, translated into German: "Die Notfallsfunktionen des Sympathico-Adrenal Systems" (The Emergency Function of the Sympathico-Adrenal System).[28]

In all likelihood, Cannon's change of mind was prompted by his continuing belief that his work was being overlooked or misunderstood by physiologists in Europe, and he hoped that publication of a comprehensive statement in Asher's *Ergebnisse* might help to remedy the situation. His article was devoted entirely to his emergency theory—but rather than follow with "of the adrenal medulla," there was a telling adjustment to "of the sympathico-adrenal system." It went over the history of his early researches, as well as the researches and reactions of Gley and Quinquaud and of Stewart and Rogoff. Then it reviewed his subsequent work with the group at Harvard on the role of the adrenals in emotional excitement, on muscle action, on hypoglycemia, on asphyxia, in hemorrhage and traumatic shock, and under the influence of cold and infection and fever. Only the ending, a statement on the biological significance of the sympathico-adrenal system, hinted that Cannon was doing some related thinking on homeostasis concurrently, for his article made no reference to his evolving concept, nor did it mention homeostasis by name.[29]

We have no clue as to the exact timing of Cannon's full-blown realization of his doctrine of homeostasis. There probably was no "Eureka!" moment, as there had been on January 16, 1911, when he exclaimed, "This is a hallelujah day!" and rejoiced over the clear evidence he and de la Paz had obtained of an emotional production of adrenalin. Nor are there diary notes or protocols extant to tell us how he determined the order in which to issue public reports of his dazzling array of researches. That his conception of homeostasis was essentially complete by early 1928 is evident in an address that Cannon delivered to the Middlesex South District Society in Boston with a title that would appeal to practitioners, "Reasons for Optimism in the Care of the Sick."[30] Although he said little or nothing about the sympathico-adrenal system and refrained from even mentioning homeostasis in his talk, the overriding concept was, in a general way, all but there.

Beginning by asking what he, a mere laboratory man, could do for physicians who daily confronted the practical problems of their patients, Cannon humbly answered that he could only offer suggestions that might be useful for laying a basis of optimism in medical practice. When looking back over more than a quarter of a century of physiological investigation, he had noted a common thread that ran through many of his studies, of which he had not been aware until a few years before. "Then I began to realize," he revealed, "that a considerable number of the researches which had been carried on in the laboratory were connected with factors which make for stability of the organism"—mechanisms that were in readiness to hold the organism steady and keep it from oscillating widely from a median position.

In surveying the variety of mechanisms for maintaining steady states, Cannon related, he had repeatedly seen agencies that tip the organism in one way or another so as to make adjustments and prevent disturbances from going too far. "In other words," he explained, "our bodies are built to take very effective care of themselves, in many ways which we have become aware of only in recent years." He referred in passing to this process as "the wisdom of the body," and acknowledged Meltzer's contribution in emphasizing the factors of safety in the animal organism. After citing numerous processes in the body as examples, he asked another question—What then is the function of the physician?—and again he provided the answer. His services are called for when the mechanism is overwhelmed and threatened unless external aid is given, such as administration of insulin for diabetes, thyroxin for myxoedema or cretinism, antitoxin for supporting the body's line of defense, and the like. Thus, the physician fulfills his role by helping nature to make effective and self-regulating adjustments of the organism that had become disarranged or in need of reinforcement. Aside from that, the great service that the physician renders to his patients in bringing good hope and cheer justifies his presence.

❧ ❧ ❧

By 1929, three years after he had initially published his tentative principles on steady states and the body's self-regulating mechanisms, Cannon was ready to share his doctrine of homeostasis with the world. He did so in several articles, the most definitive being "Organization for Physiological Homeostasis." Addressed to his fellow physiologists and biologists, this first offering constitutes an all-encompassing statement on the subject and has come to be looked upon as Cannon's classic paper on homeostasis—and, for that matter, one of the classic papers in the history of physiology.[31]

His paper began by acknowledging those before him who had been impressed with the ability of living beings to maintain their bodily stability, stretching from the *vis medicatrix naturae* of Hippocrates to the self-regulatory arrangements of Richet, but especially, the *milieu intérieur* (or in Cannon's words, the fluid matrix) of Claude Bernard. After crediting Bernard for giving these ideas a more precise analysis in his 1878 report, Cannon referred to the highly developed living being as an "open system" that interacts with its environment. When changes in the external surroundings excite reactions in this system and internal disturbances are produced, such disturbances are normally kept within narrow limits by automatic adjustments acting within the system that maintain its constancy. The word "equilibrium" had come into use to designate these constant conditions, but it had exact meaning with regard to simple physico-chemical states in closed systems that were known to be balanced (a nod to the work of Cannon's Harvard associate, L. J. Henderson). Instead, Cannon proposed that "homeostasis"—a word he had coined in 1926 from the Latin to signify a condition that may vary but is relatively constant—be substituted in its place.

"According to Bernard," Cannon continued, "the conditions which must be maintained constant in the fluid matrix of the body in order to favor freedom from external limitations are water, oxygen, temperature and nutriment (incuding salts, fat and sugar)." New insight gained since Bernard's time, however, required more ample classification. In consequence, Cannon suggested "a serviceable grouping of homeostatic categories," which operate as a means for the organism to achieve stability, and he named a variety of materials for cellular needs: glucose, protein and fat; water, sodium chloride, and other inorganic constituents such as calcium, oxygen and internal secretions. In addition, osmotic pressure, temperature and hydrogen-ion concentration are environmental factors that affect cellular activity.[32]

Cannon went on to discuss the two general types of homeostatic regulation involving supplies and processes. The storage of supplies he subdivided into storage by inundation and storage by segregation. Storage by inundation he likened to a bog or swamp, into which water soaks when the supply is bountiful, and from which it seeps back into the distributing system when the supply becomes meager. Storage by segregation, subject to more complicated control, commonly involves changes of physical state or of molecular configurations and appears to be subject to nervous or neuro-endocrine government; but, he freely admitted, large gaps exist in our knowledge about these matters. Among the examples he gave was sugar, which is stored in the liver as glycogen and is controlled by the opposing agencies of insulin and adrenin.

Turning to regulation of processes, Cannon discussed maintenance of neutrality in the blood and uniform temperature. In a section on the role of the autonomic nervous system, he employed the terms "exterofective" for the voluntary or cerebrospinal system and "interofective" for the involuntary or autonomic system. Then he gave a complete list of his six propositions concerned with physiological factors maintaining steady states in the body. Before ending, Cannon admonished his readers that the statements in his presentation should not be taken as conclusive. Nevertheless, he closed with the firm conviction that "regulation in the organism is the central problem of physiology" and that further research was desirable "into the operation of agencies for maintaining biological homeostasis."[33]

Another of the 1929 papers on homeostasis was addressed to a different audience, for Cannon had previously read it before a meeting of neurologists and psychiatrists in New York. Essentially covering the same ground as his "Organization" paper, but more briefly and with less technical detail, it was titled "The Sympathetic Division of the Autonomic System in Relation to Homeostasis." Cannon pointed out that the concept of homeostasis did not contravene his emergency theory of the sympathico-adrenal system; "rather it is an extension of that theory from a restricted relation to major disturbances, to a general relation to minor disturbances of the organism."

A third paper was delivered before a joint meeting of the Institute of Medicine and the Chicago Society for International Medicine in October 1929. Titled "Functions of the Sympathetic System in Maintaining the Stability of the Organism," it reported Cannon's experience with sympathectomies and how they contributed to his understanding of the nervous control of the body.[34] One can imagine its reception: a talk on stability given at the time of the great stock market crash.

By the end of the decade, the primary change in Cannon's view of the self-regulating nature of the human organism, aided by his observations of sympathectomized animals, was accorded to the role of the autonomic nervous system as the most important mechanism of homeostasis. As the word passed into the mainstream of the English language, the concept was eagerly embraced by biologists and social scientists. Indeed, Fleming has ranked homeostasis just a notch below evolution through natural selection in proving to be one of the most influential of all integrating concepts in biology.[35] Though Cannon went on to explore other workings of the sympathetic nervous system, his core conviction that physiological functions are admirably designed to fulfill a purpose would never be more cogently expressed than in his doctrine of homeostasis.

9 An Alternative Theory of the Emotions

ALL THE WHILE that Cannon was exploring the functioning of the sympathetic division of the autonomic nervous system and formulating his principles of homeostasis, he held firm to the ideas he had expressed in his emergency theory involving the physiological effects of emotional excitement. As new and important findings came to light and illuminated his understanding, the two lines of research on which his mind had become concentrated increasingly converged and fed one another.

In his long-standing search for a deeper comprehension of the bodily expression of pain, hunger, fear and rage, Cannon's earlier focus on the manifestation of emotional excitement had been stimulated in part by a reading of Charles Darwin's *Expression of the Emotions in Man and Animals*, as well as William McDougall's *Introduction to Social Psychology.*[1] His research later led him to turn his attention to the results of C. S. Sherrington's pioneer experiments on the nature of reflexes and T. R. Elliott's investigation of the operation of the sympathetic nervous system, as well as the clinical research of neurologists like Henry Head. Ultimately, Cannon's application of these ideas would have a direct bearing on his investigations of the emotions.

Cannon had an opportunity to speak with Sherrington when he stopped at Harvard in 1907 on his way to give the Silliman lectures at Yale based on his recently published monograph, *The Integrative Action of the Nervous System*—a pioneer study of reflexes. He later became interested in Sherrington's work on an experimental animal in which he had disconnected the cortex from the brain stem (decerebration) and, after cessation of anesthesia, stimulated an afferent nerve that triggered a cascade of bodily reactions—among others, opening of eyelids, snarling, moving the head from side to side, protruding of claws, etc. These reactions Sherrington termed pseudaffective (or pseudo-affective) reflexes. His study had an added value for Cannon: it introduced him to the research of Sherrington's teacher, Heinrich Goltz, who during the course of early experiments had decorticated a dog.

As Goltz demonstrated, the centers normally under the inhibitory control of the cerebral cortex are capable of energetic response when conditions demand urgent and impulsive action. If the cortical government was set aside, the subordinate activities became prominent. Basal ganglia freed from dominance of the cortex might function in a manner which would exhibit many of the physiological phenomena attending intense emotional disturbance. After cessation of the operation, the animal, without particular stimulation of sensory nerves, showed vigorous signs of sympathetic discharge and lived for some months in the laboratory subject to Goltz's observation.[2]

The work of T. R. Elliott also influenced Cannon's direction. Almost from the beginning of his investigation of the operation of the autonomic nervous system, Cannon became fascinated by Elliott's study of adrenalin and particularly his postulation that when the adrenal medulla poured its secretion into the blood stream, it proved capable by itself of reproducing, or mimicking, visceral alterations of the action of sympathetic impulses. In time, Elliott's research, along with Sherrington's study of pseudaffective reflexes, would prove of equal importance to Cannon's prospective investigation of the expression of the emotions.[3]

In preparing for his future research, Cannon paid special attention to neurological studies of emotional expression. He was particularly taken with Henry Head's suggestion that the nervous system is organized in a neural hierarchy, and when the cortex was set aside, subordinate activities were given over to the diencephalon for control in its absence. At such times, even slight stimulation was able to produce an extraordinary exhibition of rage. This seemed to indicate, as Head observed, that the neural apparatus for emotional expression is set and ready for energetic discharge and, if the superior control is weakened or inhibited, that appropriate stimuli elsewhere evoke an intense and powerful response.[4]

In an address to the New York Neurological Society late in

1921, Cannon informed his auditors that evidence placed the control of internal secretions in a part of the primitive (or archaic) nervous system. After citing the work of Woodworth and Sherrington, as well as Goltz, on emotional expression, he continued,

> Head and Holmes because of clinical studies were led to the conclusion that the thalamus in man is concerned with the affective side of sensation. It appears, therefore, that the emotional reactions of animals are not managed in the neo-pallium but are ingrained in an archaic portion of the nervous system. Possibly it is for this reason that the expression of emotions are so similar in widely different animals, such as the cat or the dog, and man. The fact, however, that such is the case permits us to use the lower animals in an analysis of conditions accompanying emotional outbreaks and justifies the expectation of finding similar conditions and a similar analysis in human beings.[5]

Thus it may be seen that Cannon was mapping out the territory he was going to explore in his search for the control of emotional expression. At the same time, he was letting the clinicians know that his work on experimental animals would be useful to them in the long run because it had applications for their work on humans, who have similar reactions to those seen in animals.

On similar occasions Cannon took the opportunity to speak to groups of practitioners at medical meetings to weave together the various ideas he was mulling over in his mind. Not only did he defend the results of his research that had been challenged by Stewart and Rogoff, he also used such presentations to announce plans for future investigations.

In May 1922, when giving an address that reported evidence for nervous and humoral control of internal secretions before a joint meeting of the sections on the practice of medicine and on pharmacy of the A.M.A., Cannon tailored his remarks to an audience that included pathologists, pharmacologists and physiologists. In speaking of the endocrine glands, he did not hesitate to use dramatic imagery to capture the attention of his auditors. "There appears to be a tendency just now," he told them, "to pay these glands not only respect, but reverence and homage and a blind worship. The endocrine organs are referred to as 'regulators of metabolism' and as 'regulators of personality,' as if they occupied an imperious position in our internal economy, and the other organs were mere servants to do their bidding." This led him to question, "Is that the mode of organization of the body? Is our internal government autocratically ruled by a czar or a kaiser or by groups of kings and princes, represented by the ductless glands, or is there a more democratic arrangement whereby influences pass to and fro among the various parts of the organization and, after many mutual adjustments, finally express themselves in function?"

In remarking that activity of the sympathetic system may be varied, he referred to the work of T. R. Elliott (although not citing him by name) to express the trend of his thinking:

> Quite possibly, therefore, conditions may exist in the body, characterized by either greater or less sympathetic tone than normal; but such states would not be determined by independent dominance of suprarenal secretion, for . . . lack of any secretion at all does not lessen the efficacy of sympathetic impulses, and extra secretion is ordinarily dependent on the effect of nerve impulses. The condition of the nervous system, not of the suprarenal medulla, therefore, is of primary moment.

Concluding that he had merely hinted at some of the signs of cooperation among certain endocrine glands, Cannon acknowledged that much work was still to be done before the mechanisms of control would be recognized.[6]

In yet another lecture in 1922—this time before a meeting of the American Psychiatric Association—he discussed new evidence for sympathetic control of internal secretions. But he laid his emphasis on the emotions, stating that "we must know about the natural history of the emotions, the occasions for their onset, their modes of expression, their attendant visceral effects, and the possibility of modifying them, before we shall be in a position to judge completely the character and degree of abnormal emotional disorder." Here, in a sense, Cannon set out one of the purposes of his future research, not as a physiologist but rather as a clinical investigator assessing patients' problems.

All emotions are divided into three parts or stages, he continued. First, there is a subjective state, namely a feeling of tone of emotions. Second is a state of bodily changes, some with typical expressions of the face and posture of the trunk and limbs, others with profound disturbances of the internal organs beyond any government of will. And third, there is the instinctive behavior, the fight in case of anger, the flight in case of fear. In the years since William James had expounded his theory of the emotions, Cannon said, physiologists had learned of a considerable number of new features of the bodily changes that attend great excitement. Especially was this true for the location of the nervous mechanisms involved and for the extent and variety of the visceral disturbances.

"We now know," Cannon told his audience, "that the substance, adrenin, poured into the blood stream by the adrenal medulla, is capable by itself of reproducing the visceral alterations characteristic of the major emotions." And he expressed his belief that since the glands of internal secretion could be brought into action under emotional excitement, as well as under asphyxia and reflex stimulation (pain), it became a matter of importance to secure evidence that they actually secrete their peculiar products when sympathetic nerve impulses are discharged.[7]

Cannon proposed that the similarity of bodily changes in widely different states, such as rage and fear, did not permit them to be used as the source of feelings that were experienced: "We may conceive these patterns as quite different for joy and for sorrow, for anger and for terror. And when the nerve impulses flash through these ready, but unworn, pathways, the effects may be quite as strange and startling and as rich in feeling tone as if the impulses arose from the altered viscera or the tensions of facial and skeletal muscles."

Bodily changes, he contended, may be induced by appropriate stimulation without evoking a specific emotional attitude, and thus he saw the emotion as a nervous *pattern* that lies embedded in the archaic part of the neurone system. The physiologist, leaving matters of dissociation to those more expert, finds a fascinating field for research in the simple reactions, emotional and instinctive, that occur in an ancient portion of the nervous organization of the body —the basal ganglia, the medulla and the sympathetic system—and that bring into operation the ductless glands so fundamentally important for the organism. He noted in conclusion that this was a field of interest lying in the twilight zone where physiology and psychology, pathology and psychiatry are likely to meet.[8]

When finally in 1924 Cannon laid out his program for studying the emotions in a thorough and detailed way, he was gratified by the help he had on hand in conducting his experiments. He now had a varying group, no longer consisting chiefly of Harvard medical students but with a good measure of graduate students and trained research fellows, who came from far and wide. These workers were capable of doing independent research or, if they wished, could work collaboratively with him.

Sidney Britton, a recent graduate of McGill Medical School who came to Harvard as an N.R.C. fellow in 1924, was one of Cannon's most productive associates. Along with J. T. Lewis, a Rockefeller fellow who came from B.A. Houssay's laboratory in Argentina, Britton had the distinction of working with Cannon on several lines of research, including homeostatic control and the emotions. He never continued those interests after he left Harvard but rather took up study of the adrenal cortex.[9]

In their first investigation together, Cannon and Britton set out to study those bodily reactions attending great emotional excitement—reactions involving changes regarded as instinctive adjustments to critical situations (self-preservation) that dominated behavior. Administering anesthesia to their experimental animals (cats) was a problem that at first harried the investigators because it promptly obliterated the responses that were the object of the study. But here Cannon turned to a study conducted in 1904 by Woodworth and Sherrington, from which he learned that the central mechanisms for emotional expression could be artificially activated in the decerebrated animal.

It was Woodworth and Sherrington's experiments that produced the knowledge that the thalamic region plays a role as the center for emotional expression. Sherrington had observed that truncation of the brain of a mammal at the mesencephalon annihilated the neural mechanism for which affective psychosis is adjunct and left fairly intact the reflex motor machinery "whose concurrent action is habitually taken as an outward expression of inward feeling." He and his colleague employed the expression "quasi-emotional" for reactions induced by stimulating an afferent nerve—such as a snapping of jaw, lowering of head, dilatation of pupils, snarling, and increase of blood pressure.

Cannon and Britton, drawing on the clinical findings of Hughlings Jackson and Henry Head, as well as the experiments of Goltz and others on chronically decorticate animals, attempted to obtain a preparation that would exhibit many of the physiological phenomena attending intense emotional disturbance. They made use of the Sherrington decerebrator to bring about a destruction of the cortex without destroying the basal ganglia. The decortication that resulted was done very rapidly so that the period in which the experimental animal was under anesthesia would be as short as possible. In these circumstances, there was usually a good showing of Sherrington's "quasi-emotional" phenomena, which Cannon, coining his own expression, called "sham rage." One of the things Cannon and Britton wanted to determine was whether the phenomena relating to sham rage induced secretion from the adrenal medulla. Accordingly, they used the denervated heart method to ascertain whether adrenal secretion was actually occurring. They also experimented on fasting animals to make sure that the hepatic factor would only play a temporary role in causing cardiac acceleration.[10]

The evidence Cannon and Britton gathered (reported in a paper on "The Pseudaffective Medulliadrenal Secretion" in 1925) indicated that there was not only increased secretion from the adrenal medulla but that secretion was accompanied by indications of sympathetic activity. In sum, they inferred that the adrenal medulla participates in the general widespread discharge of sympathetic impulses while the animal is in the pseudaffective state and pours adrenin into the blood stream to such a degree that the denervated heart is driven for a considerable period to maintain a rapid beat.

They also found, first, that there was a true emotional secretion from the adrenal medulla. Cats, excited by new experiences in the laboratory, had a smaller adrenin content in the medulla than did quiet animals. Secondly, that if the splanchnic nerves were cut previously on one side, the adrenal medulla on that side had a larger load of adrenin than did the other gland after the animal had been excited by morphine. And third, that if the superior cervical ganglion was removed, a paradoxical dilatation of the pupil occurred when the animal became angry—a reaction that did not occur after both adrenals were excised. Cannon also pointed out that many of the physiological accompaniments of great emotional excitement could be produced without anesthesia in an acute experiment and thus could be readily studied for hours.[11]

Cannon's study with Britton was followed by another one on the influence of motion and emotion on medulliadrenal secretion, in which they used the lasting preparation of the denervated heart and had the collaboration of J. T. Lewis and A. Groeneveld. This time they observed minor movements of cats with little or no affective accompaniment of emotion; emotion with little or no movement of skeletal muscles; and a combination with movements made vigorous by excitement. They noticed that with changes in activity, the rate of the denervated heart was augmented in each of these various states—that is, with little, moderate, or great emotional excitement. In their discussion, they concluded that the output of adrenin is greatest when emotional excitement is combined with violent struggle.

> That all these changes may occur to some small degree . . . indicates that what takes place on a large scale in emergencies takes place on a small scale in the common behavior of the organism. We can hardly conceive that slight efforts are given a critical importance because the heart has been denervated . . . [T]he emergency theory would have to be altered insofar as it might imply that the sympathico-adrenal mechanism is called into action *only* at times of violent emotion. According to the evidence now in hand, the greater the emergency, as measured by intensity of excitement and struggle, the more is that mechanism utilized.[12]

In 1927, Cannon asked Philip Bard, one of his doctoral candidates, to search for the mechanisms responsible for the bodily changes that accompanied the emotional behavior of fear and rage observed in experimental animals. A native Californian, Bard had followed the advice of Walter Alvarez, whom he had known from boyhood, that he direct his attention to the study of biology and medicine. After graduation from Princeton, he applied for admission to the Division of Medical Sciences program at Harvard to work for the doctorate in physiology under Cannon—an association that became a close friendship and, as he later declared, the basis for "whatever success I have had as a physiologist."

Bard, like Britton, arrived at Harvard in the fall of 1924, but unlike Britton, had first to undertake preliminary coursework. He became immersed in several introductory courses of the graduate curriculum, as well as in a special course that Cannon had organized, which had as a feature laboratory exercises repeating various pioneer experiments in circulation physiology. In his second year, Bard took a course in pharmacology and began more advanced laboratory work that included learning how to denervate the hearts of cats. Under Cannon's direction, he also began an investigation of increased sympathetic activity that occurred during muscular exercise. Even more, Cannon introduced Bard to studies he had underway with Britton and others in the laboratory in a search for conditions that induce vigorous sympathico-adrenal medullary activity.[13]

Bard could not have begun his doctoral research at a better time. At the outset of Cannon and Britton's experiments in 1924, they had used decerebrated animals to evoke pseudaffective reactions, but found their reactions for the most part disappointing—in Cannon's words, "they never amounted to an effective action of attack or escape." Decorticated animals, on the other hand, proved more useful since their widespread display of rage with its astonishing intensity—lashing of tail, arching of trunk, protrusion of claws, attempts to bite, very rapid panting, accompanied by large increases of heart rate—showed extraordinary emotion which, to Cannon, appeared clearly directed to the preservation of the individual. That the reaction was primitive and energetically purposeful further suggested to him that the reaction was dependent on the older subcortical divisions of the nervous system.

Bard's thesis topic was directed toward the delimitation of that part of the brain stem responsible for the sham rage of the decorticate cat. It was believed that not only would this elucidate the neural basis for this emotion. It also might yield some evidence for the existence of a predominant central mechanism so clearly implied by the tendency of the sympathetic to discharge vigorously and as a unit under conditions of stress.

If Cannon's investigations with Britton paved the way for Bard's research, the techniques that Bard adopted to decorticate his experimental animals differed markedly from those that Cannon had used. Cannon entered the cerebrum by way of the orbits of both eyes to decorticate his animals. However, Bard removed the vault of the animal's skull to expose and discard the cortex of both hemispheres of the brain. That done, he began to transect the brain stem at different levels until the sham rage failed to occur.

All forty-six of Bard's experiments were acute. A number of them were devoted to an exploration of the zone of transition between the diencephalon and the mesencephalon. Bard found that the sham rage occurred after ablation of both hemispheres, the corpora striata and the cranial half of the diencephalon; but that it failed to develop after transecting the caudal half of the diencephalon or the cranial portion of the mesencephalon.

Although Bard's work in connection with his 1927 doctoral dissertation was supervised by Cannon, Bard's name stood alone on his publications.[14] Nevertheless, his research would provide much of the basis for Cannon's emerging concepts, not only of the thalamic theory of emotion but, to a lesser extent, of homeostasis as well.

While the work of Britton and Bard fit nicely into his emerging concepts, some of Cannon's ideas about the effects of strong emotions on the organism remained essentially unchanged from his pioneer elucidation of the emergency theory. Throughout the years, one of those with whom he shared an interest in the bodily effects of emotions was the gastroenterologist, Walter C. Alvarez. A letter that Cannon wrote to Alvarez in the summer of 1927 might well have been composed more than a decade earlier:

> When we consider the very great rôle which fear and anger have played in the history of the race, as essential factors in self-preservation, there is little wonder that they are involved in physiological processes of a most varied and extensive character, directed towards the mobilization of the bodily forces for perhaps supreme physical struggle. In such contingencies the digestive process can not appropriately claim attention from the circulatory apparatus—it is set aside for more urgent matters. Such is our racial inheritance. And now, as we fear operations, or the consequences of moral error, or threatening financial disaster, the physiological preparations for action occur, as they have occurred in fear for myriads of generations in the past.
>
> The trouble is that the causes of fear are not so likely to be *temporary* as they were with our remote ancestors,—we can not run away, or go out and kill or be killed. The causes persist and there is little to *do* . . . These persistent causes can result in establishing neural paths—habits of reactions as real as any other, as real as buttoning and unbuttoning. No pathologist could find a change to account for such trouble. This is what I should call a *functional* disease or disturbance.[15]

But in the same year, 1927, Cannon also expressed his most recent ideas on the emotions when he published a critique, or modified view, of a theory that William James had formulated—known in a slightly altered form as the James-Lange theory of the emotions. In this paper, Cannon proposed an alternative theory and pointed to the thalamic region of the brain as the central station of nerve impulses and the coordinating center of emotional reactions. His evidence was based in large part on the research that he and Phil Bard had been carrying on concurrently, and on the work involving total sympathectomy as well.[16]

In 1914, a year before publication of *Bodily Changes in Pain, Hunger, Fear and Rage,* Cannon had been invited by the American Psychological Association to give a talk on his prospective volume. To the disappointment of his audience, he made no mention of the James-Lange theory. And indeed, when the book came out, there was again no special mention of James's theory of the emotions. Other aspects of James's work were discussed but not his theory of the emotions. Cannon in essence had paid no attention to it.[17]

In 1922, when Cannon spoke at a meeting of the American Psychiatric Association, he pointed out that James in his theory of the emotions attributed subjective states to impulses aroused by bodily changes. But after noting that many new facts had come to light since James's time which did not accord with that view, Cannon abruptly left James and his theory to concentrate on the glands of internal secretion. Five years later, Cannon published a lengthy paper in the *American Journal of Psychology,* in which he finally had his say about the James-Lange theory.[18]

James had expressed his views on the emotions first in an article published in 1884 in the recently-founded journal *Mind,* and afterward elaborated them in a lengthy chapter in the second volume of his *Principles of Psychology,* published in 1890. Around the same time, in 1885, the Danish physiologist Carl Georg Lange had published a small pamphlet setting forth his own theory, which was similar to that of James but elaborated more narrowly, and focusing on changes in the circulatory system.[19] Lange not only held that stimulation of the vasomotor centers aroused the emotions by whatever means, he also maintained that the symptoms of the major emotions were all referable to vascular changes, and distinguished between one and another emotion on the basis of different bodily accompaniments. That said, we can dispense with Lange and concentrate on the more fertile ideas of William James.

James hypothesized that bodily changes follow directly the

perception of the exciting fact, and that our feeling of the changes as they occur *is* the emotion. He believed the following sequence to be incorrect: "Common-sense says, we lose our fortune, are sorry and weep; we meet a bear, are frightened and run; we are insulted by a rival, are angry and strike." The more rational statement, for James, was that "we feel sorry because we cry, angry because we strike, afraid because we tremble, and not that we cry, strike, or tremble because we are sorry, angry, or fearful, as the case may be."

When discussing what he called the vital point of his theory, James wrote, "If we fancy some strong emotion, and then try to abstract from our consciousness of it all the feelings of its bodily symptom, we find we have nothing left behind, no mind-stuff out of which the emotion can be constituted, and that a cold and neutral state of intellectual perception is all that remains." He contended that "for *us*, emotion disassociated from all bodily feeling is inconceivable."

James was certain that of the two things concerning the emotions, one must be true. Either separate or special centers are their brain seat, or else they correspond to processes in the motor or sensory centers that are already assigned or not yet known. He held that the latter possibility was all that his theory demanded. Although certain emotional reactions could be explained, James admitted, there remained many others that could not be explained at all. Among these were the effects of the viscera and internal glands; the dryness of mouth and diarrhea and nausea of fear; the urinary secretion of sanguine excitement and the bladder-contraction of apprehension; the various sweatings of the skin and its flushings. He supposed that even the change of blood pressure and heart beat during emotional excitement might, instead of being teleologically determined, prove to be purely mechanical or physiological outpourings through the easiest drainage channels—"the pneumogastrics and sympathetic nerves happening under ordinary circumstances to be such channels."

From the fall of 1921, when Cannon announced his intention to search for mechanisms that he believed might cast light on the expression of the emotions, until 1924, when Britton and Bard came into his laboratory, he pondered the James-Lange theory[20] and discussed his musings with his close friend, the philosopher Ralph Barton Perry—to whom he later dedicated his book *The Wisdom of the Body.* "Tom" Perry was then in the midst of a study attempting to bridge the gap between common sense and science; like Cannon, he was a former student and ardent admirer of William James. Both men were eager to clarify their ideas on James's thoughts on the emotions. Unfortunately, we cannot eavesdrop on their conversations, but an important exposition of Perry's notions may be found in his book, *The General Theory of Value,* published in 1926.

"The history of James's doctrine," wrote Perry at one point, "is not unlike that of Darwin's theory of natural selection. It is so strongly fortified by proof and so repeatedly confirmed by experience that it cannot be denied its substantial truth. In spite of elaborate refutations it shows no signs of obsolescence." "Nevertheless," Perry admitted, "like most daring and innovative theories, it requires both refinement and supplementation. The objections which have been raised against it have built upon it and extended it but cannot undermine it."

There obviously had been give-and-take in the discussions between the two friends, for at the end of his section on the James-Lange theory Perry noted that Cannon, because of his observations on visceral and glandular changes induced by emotionally exciting stimuli, had raised objections to the theory. And yet Perry's interpretation was that Cannon's results, "in so far as they emphasize and report in much detail the profound and pervasive character of these disturbances, and their biological connection with the circumstances under which emotions may arise, may be said to corroborate and clarify the James-Lange theory."[21]

Perry notwithstanding, when in 1927 Cannon finally published his views, he went far beyond anything his friend might have imagined; for he was highly critical of James's theory, and in its stead advanced his own physiological explanation for the basis of the emotions. Even the title of the paper, "The James-Lange Theory of Emotions: A Critical Examination and an Alternative Theory," promised to deliver fireworks.

After a historical review, Cannon launched into lengthy argumentation, remarking that the processes of the thoracic and abdominal organs did not add flavor to experience—that is to say, did not greatly affect the emotions—as James had suggested. Rather, they functioned to adapt the internal economy so that, in spite of shifts of outer circumstances, the even tenor of the inner life would not be disturbed. And in recent years, he maintained, all observations, both experimental and clinical, "point to the optic thalamus as a region in which resides the neural organization for the different emotional experiences." In making this statement, Cannon fired the first shot of a volley that was to follow in support of the thalamic theory of the emotions.

Cannon paid special attention to the visceral factors so highly favored by James in accounting for the richness and variety of emotional feeling. He pointed to total transection of the spinal cord and vagus nerves of dogs by Sherrington in the early 1900s, which destroyed any connection of the brain with the heart, lungs, stomach and other abdominal organs and thus served to "isolate all the structures in which formerly feelings were supposed to reside." He also cited recent work in his own laboratory, which succeeded in keeping cats alive and healthy for many months after removal of the entire sympathetic nervous system had abolished all vascular reactions controlled by the vasomotor center. These extensively disturbing operations had little if any effect on the emotional response of the animals—despite the fact that the secretion of the adrenal medulla could no longer be evoked, the action of the stomach and intestines could not be inhibited, the hairs could not be erected, and the liver could not be called on to liberate sugar into the blood stream.

James and Lange, on the one hand, had affirmed that if organic sensations were removed imaginatively from an emotional experience, nothing was left. Sherrington and Cannon, on the other hand, had removed the sensations surgically, abolishing all visceral disturbances through sympathetic channels (the channels for nervous discharge in great excitement) and found that their animals acted with no lessening of the intensity of emotional display. In one of Sherrington's dogs, "her anger, her joy, her disgust, and when provocation arose, her fear, remained as evident as ever." In Cannon's sympathectomized cats, all superficial signs of rage were manifested in the presence of a barking dog: "hissing, growling, retraction of ears, showing of teeth, lifting up the paw to strike—*except* erection of the hairs."

Continuing with his arguments against previously held notions, Cannon went on to explain that the viscera are relatively insensitive structures, despite the common belief that the more deeply the body is penetrated, the more sensitive it becomes. On the contrary, he wrote, we are unaware of the contractions and relaxations of the stomach and intestines during digestion, of the rubbing of the stomach against the diaphragm, of the squeezing motions of the spleen, of the processes of the liver—"only after long search have we learned what is occurring in these organs." Furthermore, visceral changes are too slow to be a source of emotional feeling since the viscera are composed of smooth muscle and glands, which are known to respond with relative sluggishness.

If the emotions were indeed a consequence of visceral changes, in accordance with the James-Lange theory, Cannon observed, we should expect them to follow these changes in all cases. He then introduced clinical evidence to show that the injection of adrenalin in a large number of patients or research subjects had brought about a large variety of experiences, contrary to previous expectations. Numerous events occurring in the viscera in consequence of great excitement, as detailed in his own research, "have been interpreted as supporting the James-Lange theory," he noted, in reference to Perry's statement. But, he cautioned, "it should be clear that that interpretation is unwarranted."

In final analysis, Cannon argued, because visceral processes are not a considerable part of sensation and since even extreme disturbances in them yield no noteworthy emotional experience, we can understand why such pronounced and differing emotions as fear and rage, or chilliness, asphyxia, hyperglycemia and fever, though attended by these disturbances, or even total exclusion of visceral factors from emotional expression—why all of these make no difference in emotional behavior. "It is because the returns from the thoracic and abdominal 'sounding-board', to use James's word, are very faint indeed," Cannon stated, "that they play such a minor role in the affective complex."[22]

The remainder of Cannon's paper, equal in length to what had gone before, was devoted to an exposition of his thalamic theory and intended to prove that emotional expression resulted from the action of a subcortical center. As early as 1887, W. Bechterev had reported that after removal of cerebral hemispheres from various kinds of experimental animals, appropriate stimuli evoked corresponding responses of an affective character. Since these effects disappeared when the optic thalamus was removed, Bechterev concluded that this organ plays a prominent role in emotional expression. The investigations of Woodworth and Sherrington in

1904 on the pseudaffective (or sham) response appeared to lessen the importance of the thalamus as a center, but recent findings by Cannon and Britton (1925) and by Bard (still unpublished) again emphasized its dominance. These results, to Cannon's mind, clearly pointed to the thalamus as a region from which, in the absence of cortical government, impulses are discharged that evoke an extreme degree of "emotional" activity, both muscular and visceral, and he cited clinical and pharmacological evidence in support of his contention.

In light of this accumulated information, Cannon proposed a modification of the section in William James's discussion of his theory headed "No Special Brain Centres for Emotion." Accordingly, he suggested that the neural organization for an emotion is as follows: "An external situation stimulates receptors and the consequent excitation starts impulses toward the cortex. Arrival of the impulses in the cortex is associated with conditioned processes which determine the direction of the response. Either because the response is initiated in a certain mode or figure and the cortical neurones therefore stimulate the thalamic process, or because on their centripetal course the impulses from the receptors excite thalamic processes, they are roused and ready for discharge."

Cannon continued that the reaction patterns typical of the different affective states prove that the thalamic neurones act to combine in a given emotional expression. These neurones do not need innervation from above in order to be driven into action. Rather, *release* for action is the primary condition for their service to the body, and they then discharge precipitately and intensely. "Within and near the thalamus the neurones concerned in an emotional expression lie close to the relay in the sensory path from periphery to cortex . . . when [they] discharge in a particular combination, they not only innervate muscles and viscera but also excite afferent paths to the cortex by direct connection or by irradiation."

In consequence, Cannon hypothesized, "*the peculiar quality of the emotion is added to simple sensation when the thalamic processes are roused.*" It was a theory that appeared to fit all the known facts. "When the thalamic discharge occurs, the bodily changes occur almost simultaneously with the emotional experience." The coincidence of disturbances in muscles and viscera with thrills, excitements or depressions had misled James because "with the rôle of the thalamus omitted from consideration, the obvious inference was that the peculiar quality of the emotion arose from the peripheral changes." Indeed, contrary to James's view that sensations from the peripheral changes are "pale, colorless and destitute of emotional warmth," the evidence Cannon presented showed that "the thalamic disturbances contribute glow and color to otherwise simple cognitive states."[23]

The following year, Cannon published two other papers in which he argued that the neural arrangement for emotional expression was near the optic thalamus. The first was written as a chapter for a volume on *Feelings and Emotions.* Here he confined his remarks to the emotion of rage to best illustrate his thesis, noting that the complex of bodily alterations that appeared in rage had many of the features resembling the simple reflexes, such as sneezing, coughing and sucking.

He asked if rage had its locus in the cortical neurones or in the more primitive parts of the nervous axis, and observed that in lower vertebrates lacking a cerebral cortex, a group of centers carried on the elementary functions for maintaining existence. In higher forms, these centers, although normally held in check by the dominant cortex, were capable of energetic responses when conditions demanded urgent and impulsive action; and, when the cortical government was set aside, the subordinate activities became dominant. In Goltz's and Rothmann's hemisphereless dogs and in Dusser de Barenne's cats deprived of the neopallium, even slight disturbances had proved an effective stimulus for evoking emotional responses.

Interpreting these instances of a typical rage-reaction as examples of "release phenomena" (in Hughlings Jackson's phrase), Cannon described his experiments with Britton on animals deprived of the cerebral cortex but with the base of the brain left intact, which exhibited on recovery a sort of sham rage with all of its accompanying physical activities (lashing of the tail, arching of the trunk, snarling, attempts to bite, and so forth). But animals that had been cut through the mid-brain and decerebrated, leaving the diencephalon and all parts of the cerebrum anterior intact, exhibited disturbances of an otherwise fairly continuous rigidity.

This clearly proved, in Cannon's words, that the "part of the neural organization which *directly* sends forth the rage-response lies posterior to the section through the mesencephalon, i.e., in the remnant of the mid-brain or in the medulla." The difference in the behavior of the two preparations must thus be due to some part of the brain-stem lacking in the decerebrate but present in the decorticate animal. To put it another way, he reasoned that responses in a typically widespread outburst of rage appeared to be controlled by a superior center. Since Britton and he left untouched almost all of the basal gray matter of the anterior brain-stem, he asked, "Where among these ganglia does the dominating center reside?"

Actually, this very problem had been the subject of an investigation by Philip Bard, who reported that the dominating center was situated near the base of the diencephalon, probably in the subthalamus. "Thus in the diencephalon," Cannon wrote, "in a part of the old brain, which is common to all members of the vertebrate series from the fishes to mammals, is localized the neural apparatus for integrating the complex reaction system of rage, not only the external expression, but also the internal mobilization of the bodily forces for the violent physical efforts in which rage typically culminates."

Cannon observed that visceral alterations accompanying fury were profound and widespread, evoked by discharges through the sympathetic division of the autonomic nervous system. The neurones of this division were so arranged as to discharge impulses diffusely to smooth muscles and glands in all parts of the body, in cooperation with the adrenal medulla liberating adrenin at the same time and acting in every part of the body with the same effect as the nerve impulses. Although the sympathetic system was called into operation in various circumstances, it was activated under quite different conditions. The physiological organization that established the reflex figure of rage might serve as a prototype for other primitive emotional responses—fear, joy and grief, for example—which, good evidence showed, might also lie subcortically, and, specifically, in the thalamic region of the brain.

In his chapter, Cannon touched briefly on the subject of cognitive consciousness. Since cognitive consciousness is associated with the functioning of cortical neurones, it followed, as a corollary, that the neural mechanism for the primitive emotions operates in a region outside the range of such consciousness. It also explained some of the most characteristic features of emotional experience in terms of a sudden and powerful domination of the bodily forces by subcortical neurones.[24]

In the second of the two 1928 papers, which Cannon presented as the annual discourse before the Massachusetts Medical Society on June 6, he again discussed the mechanisms of emotional disturbance in bodily functions. Reporting that experiments on lower animals had demonstrated that the neural arrangement for the emotions was near the optic thalamus, he found the evidence consistent with indications that, in man, emotional expression was also managed by parts of the brain below the cortex—specifically, by centers near or in the optic thalamus.

Ever mindful of providing information that would be of use to practitioners, Cannon told his audience that his research had implications for practical bearings on the care of patients with emotional disturbances. Still, he urged them to proceed with caution. "If an objective cause for a patient's complaint is not found, nothing is easier than to attribute the difficulty to nervous factors . . . The assumption that emotional agencies are causing mischief in the organism should be a last resort." Even when doctors suspected that fear or rage, or some other strong feeling, was at the root of the trouble, proof "should be carefully sought both at the source of the trouble and in the effect of appropriate therapy." He ended his discourse with a direct appeal to the physicians:

> I hope that I have convinced you that interest in this realm of medicine should not be relegated to cults, mental healers and the clergy. The doctor is properly concerned with the workings of the body and their disturbances, and he should have, therefore, a natural interest in the effects of emotional stress and in the modes of relieving it. The field has not been well cultivated. Much work still needs to be done in it. It offers to all kinds of medical practitioners many opportunities for useful studies. There is no more fascinating realm of medicine in which to conduct investigation. I heartily commend it to you.[25]

Meanwhile, Bard, in a series of papers anchored in the work he had undertaken and accomplished for his doctoral degree, continued to reinforce Cannon's views on the workings and control of bodily emotions. Much of an article on the diencephalic mechanism that he published in the *American Journal of Physiology* in 1928 was devoted to reporting in succinct form the experiments he had conducted to locate "the seat of the emotions" in the inner brain. The next year, Bard published another article on the central representation of the sympathetic system, which, like the first, was derived from his dissertation. Both papers reiterated many of his mentor's prior assertions in attributing the emotions to the regulation of the sympathetic nervous system, adding that it was, in turn, controlled by higher subcortical processes.[26]

After four happy and fruitful years in Cannon's laboratory, in 1928 Bard accepted an offer to return to Princeton for a post in the biology department. Before he left Harvard, however, Cannon received a letter from Carl Murchison at Clark University, asking him to contribute a chapter on the emotions for his forthcoming textbook in psychology. Though Murchison told him, "you are the one man that American psychologists think of in connection with the chemico-mechanical aspects of emotion," Cannon declined the invitation. When Murchison asked him to recommend another author to undertake the task, Cannon instantly named Bard, because he knew "thoroughly the work that has been done here." Bard not only wrote well, Cannon maintained, he could also present the physiological side of emotional behavior quite satisfactorily.

Murchison was pleased with the suggestion and asked only that Cannon endorse the manuscript after it was completed. But Cannon made haste to assure him that Bard "knows his stuff" and "would make the work stand on its own feet without any authoritative support from me." Cannon was not mistaken; when the *Handbook of General Experimental Psychology* came out in 1934, Bard's chapter on "The Neuro-Humoral Basis of Emotional Reactions" was one of its highlights. That same year, Bard published a two-part paper on emotional expression after decortication before turning his attention to other aspects of the subject.[27]

During the decade of the 1920s, Cannon made great strides toward advancing knowledge of the nature and workings of the emotions based on what physiological research could reveal. His studies had even emboldened him to challenge the abstract tenets of his former teacher, William James, who remained in the eyes of many, and particularly of the psychologists, a sacred cow, to be worshipped, and left untouched. Despite Murchison's flattering remark to Cannon, not everyone in his profession was inclined to swallow the thalamic theory whole.

Some of them felt obliged to pitch into Cannon with personal letters of complaint or refutation. But in 1930, three psychologists at the University of Kansas, E. B. Newman, Theodore Perkins and T. H. Wheeler, went public with so detailed a critique of Cannon's theory that he was compelled to reply to them in print. After the trio from Kansas had reviewed all of Cannon's criticisms of the James-Lange theory and his offer of a physiologically based explanation in its stead, they concluded that emotion is a function of a unified neuromuscular system and refuted Cannon's alternative view. Worse, they asserted that his evidence for localizing all emotional experience in the thalamus was full of inconsistencies.[28]

In reply to these critics, Cannon contrasted his views with those of James and Lange with even more care than previously and showed by means of diagrams how his alleged nerve connections and pathways differed from theirs. To elucidate his notion that the thalamic region was the coordinating center for the so-called emotional processes, he defined in a footnote the expression "thalamus" or "thalamic region" to include not only the sensory stations in the various parts of the diencephalon but also the motor centers "(e.g., the hypothalamus)." To shore up his argument, he had occasion to use a reference he knew about but overlooked when he had written his earlier article.

A 1921 paper by the New York neurologist, C.L. Dana, described the case of a paraplegic who had suffered almost complete paralysis from the neck down but nonetheless displayed the emotions of grief, joy, displeasure and affection. In lieu of the James-Lange theory (which had "never been satisfactorily proved or convincingly refuted"), Dr. Dana had proposed a new theory that "emotion is centrally located and results from the action and interaction of the cortex and thalamus," and that the centers controlling the bodily sensations accompanying emotion are mainly in the brain-stem. In presenting his conclusions, Dana noted that Cannon had "presented evidence to show that the center for sympathetic discharges occurring on pain-stimulation lies in the brain-stem as high up as the anterior corpora quadrigemina."[29]

Cannon's 1931 defense of his thalamic theory ended by pointing out that in 1884 William James had appealed to asylum physicians and neurologists to confirm or disprove his hypothesis, "for they alone have the data in their hands." Since then, eminent and experienced specialists in nervous disorders had invariably denied its validity. Likewise, Cannon abrogated the claim of Newman, Perkins and Wheeler that organic and kinesthetic processes were "necessary" for emotional experience, since human witnesses who suffered damage to their nervous systems were unable to provide expert testimony. Inasmuch as both clinical observers and physiological experimenters had been forced by the facts to abandon the idea that the emotions have a peripheral source, Cannon reiterated his proposal that the thalamic theory offered an alternative view to the James-Lange theory and appeared "to be in accord with the facts now known."[30]

Where, we may ask, does Cannon's understanding of the emotions and his thalamic theory fit into later views of the emotions and the underlying anatomical centers and physiological forces that govern them? What is its value in terms of hindsight and the long stream of history? To answer these questions, we must jump ahead a few years to when James W. Papez, a neuroanatomist at Cornell, proposed a more elaborate theory, which not only delved into Cannon's research but took his conclusions a giant step further.

In a paper published in 1937 entitled "A Proposed Mechanism of Emotion," Papez presented a simple explanation that, in Melvin Konner's words, "was as revolutionary as it was elegant, and it set the tone for thought and research about emotional physiology ever since." Papez located the emotions and their expression not merely in a center in the brain, as Cannon had, but in a circuit of interrelated structures that in time became known as the limbic system (so named by Paul MacLean in 1949). He credited the work of Cannon, Bard, and others for greatly advancing knowledge of the functions of the hypothalamus, and claimed that in light of these researches, "the connections of the hypothalamus to the medial wall of the cortex gain a new significance." His discussion presented anatomic, clinical and experimental data relating to the hypothalamus, the gyrus cinguli, the hippocampus and their interconnections. "Taken as a whole," Papez wrote, "this ensemble of structures is proposed as representing theoretically the anatomic basis of the emotions."[31]

The limbic system includes not only parts of the thalamus, but parts of the cortex as well, and, most important, the hypothalamus which, situated in the center of the head below the thalamus, is the organ through which the brain regulates the hormones of the pituitary gland and, in turn, those of the body. There was, by the second half of the 1930s, strong evidence to support a clear relationship between this circuitry and the emotions. Indeed, Papez himself later disclosed that he had written his paper quickly, in a fit of spleen, after reading about a large grant that had been made in England for study of the brain's role in emotion. "I was mad," he confessed, "because the English proposal seemed to ignore what was already known about this subject"—one that he had been investigating at least since the early thirties.[32]

In lieu of a central locus for governing the emotions, Papez proposed a theory involving three "streams" of impulses corresponding to three broad categories of psychic life, all of them beginning with the sense organs and each one involving some part of Cannon's thalamus. The first one, the "stream of movement," relays the sensations through the thalamus to the corpus striatum, a large structure forming the central mass of each cerebral hemisphere. The second, the "stream of thought," relays the sensations through the thalamus to major portions of the cerebral cortex. And the third, the "stream of feeling," relays the sensations through still other thalamic regions to the hypothalamus and other parts of the cortex located between the two hemispheres near the core of the brain, which causes the sensory excitations to receive their emotional coloring.[33] Papez's contribution, building on the foundation of Cannon's work, stands as an outline of what is understood today about the neuroanatomy of the emotions.

Others have added to that understanding. A recent commentator, Sherwin B. Nuland, in reporting the present state of knowledge, notes that the various anatomic elements of the limbic system (so called "because it is in a sense in limbo between heaven and the nether regions"), i.e., the cerebrum and brain stem, lie within the deepest parts of the cerebral hemispheres: the amygdala, the hippocampus, and several communicating structures.

> Because of their many connections to both the older and lower parts of the brain, which are anatomically below, and the newer and higher parts, which are anatomically above, the limbic system plays a vital role in learning, memory and emotional life. It connects areas of the brain that are instinctive with those that consciously respond to incoming stimuli. The limbic system receives sensory input from voluntary and involuntary centers and sends motor output back to areas where it can be acted upon. Connections between the hippocampus and cortex, particularly, are vital to memory.

In both the structural and functional sense, Nuland points out, the limbic system is associated with autonomic activity, so that some neurologists regard the hypothalamus to be included in both, or at least to serve as the gatekeeper. Thus, the hypothalamus may be viewed as the highest part of the brain stem, as well as the lowest part of the limbic system.[34]

While Cannon may have been wide of the mark about some of his specifics, there is no denying that he furthered the course of historical development that was to enhance our knowledge of the neural and humoral control of the emotions. Given the state of knowledge at the time Cannon and his colleagues were working, this is all the more remarkable. Many discoveries that would be crucial to later studies—instruments like the electron microscope and the electroencephalograph, as well as techniques and developments in biochemistry, pharmacology and the neurosciences—were unavailable or newly emergent in the years that he was conducting his research on the glands of internal secretion and the sympathetic nervous system. In imagining a functional hypothesis full of architectural detail, Cannon took an important step forward. In so doing, he transcended James and Lange by proposing an alternative theory, which located the seat of control for the emotions not in the peripheral organs, but in the brain.

10 *Friends and Associates of Medical Progress*

AFTER THE legislative battles of the early 1920s, Cannon did not again take a prominent role in hearings or elections to combat antivivisection or antivaccination bills. There was still a good deal for him to do, however, and he continued as a valued adviser whose recommendations for the protection of medical research were customarily followed. If from time to time he had differences with officers of the American Medical Association on matters of policy, those conflicts were invariably settled amicably.

One of Cannon's abiding concerns was that accurate medical knowledge be presented to the public at large. After more than two years of discussion, the A.M.A. accepted his suggestion to publish a magazine aimed at lay readership, which would promote the value of experimental research in advancing preventive medicine and health care. During the spring of 1923 the first issue of *Hygeia* appeared, its contents a cornucopia of popular articles on an array of medical topics ranging from a sketch of Pasteur's achievements, a brief history of tuberculosis, and an exposition of the growing role of women in medicine to accounts of patent medicine fraud, the use of music by American Indians in treatment of the sick, and the fly as a carrier of disease. As a tribute to the role of laboratory research in furthering medical knowledge, the editors included a popular account by Cannon on the organs of internal secretion and their diseases.[1]

There can be little doubt that Cannon was gratified when *Hygeia* finally broke into print. Indeed, he saw it as an opportunity and pressed the A.M.A. to help foster the growth of the newly-incorporated lay society, Friends of Medical Progress. Initially some members of the A.M.A. argued that such activity would surely increase the cries of "Medical Trust" broadcast by the American Medical Liberty League. But after discussion with Cannon, they agreed in the fall of 1923 to publish an editorial in *JAMA* praising the purposes of the Friends and promised to print notices of the organization's upcoming activities in the *A.M.A. Bulletin*.[2]

With Thomas Barbour serving as acting director, it took the better part of a year for the Friends of Medical Progress to get off the ground. Once launched, the society under Barbour's leadership thrived. He energetically raised funds, instituted membership drives and, to obtain a public image, contributed publicity about the Friends to exhibitions at local health shows. Louis Dublin of the Metropolitan Life Insurance Company was so impressed by the activities of the Friends that he declared the society "a Godsend to the health officers and others who were working for individual and community welfare."[3]

From the beginning, Cannon found himself continually trying to smooth over differences between the lay members of the Friends and its medical advisory board. One of the earliest issues related to Ernest Harold Baynes. From the publication of his article on vivisection in the *Woman's Home Companion* in 1921, Baynes had at his own expense given popular lectures throughout the United States on the value of medical research, often at the request of local and state medical societies engaged in bitter fights with antivivisectionists. Not only that, but through his wide acquaintance with members of the press, he had been able to secure added space in newspapers and popular magazines for favorable stories and editorials about experimental medicine.[4]

When Cannon learned via the grapevine that Baynes's lectures in behalf of animal experimentation had practically dried up all income derived from his popular lectures on animal life, he wrote Simon Flexner in support of Baynes's stated wish for a post as field secretary for the Friends of Medical Progress. Flexner, who believed that anyone involved in the permanent organization of the society should have a background in science, put Baynes's appointment off, giving lack of funds as an excuse. Only after pressure from Cannon and W. W. Keen did he later agree to give Baynes the appointment.[5]

The issue of scientific background of staff would not go away, and at times reappeared in a number of guises to threaten the harmony of the society. One such instance involved a letter that Mary Lee Thurman, an assistant secretary of the Friends of Medical Progress, wrote to the Health Commissioner of Baltimore for statistics on the number of stray dogs in the pound that had been put to death by the local humane society.

When an antivivisectionist early in 1924 brought the letter to the attention of the Friends' officers and advisory board, Dr. Flexner immediately asked Cannon for an explanation. Miss Thurman's query, Cannon replied, was part of a fact-finding mission for a future effort to obtain impounded animals for the purpose of medical teaching and investigation. "If the Friends of Medical Progress," he explained, "are able to promote 'progress' in the direction of educating people as to the value of animals in medical schools, and in the justification for using impounded animals, which are to be put to death in any case, as a means of advancing medical knowledge, they will be most helpful to us."[6]

Cannon's answer did not satisfy Flexner; he thought the advisory board of the society should be consulted about "large, general, and aggressive policies of a controversial nature." Indeed, he felt so strongly about it that he threatened "to wait a convenient season and then draw out, if I am expected to assent to actions of a comprehensive and belligerent nature, of which I have not had any information in advance." To add point to his complaint, Flexner told Cannon, "As you know, I appear annually before the New York Legislature. If I should be asked this year whether I approve of turning over animals from the public pound to the medical schools, I am in grave doubt as to whether I should consider it wise to answer 'yes.'"

Although Cannon tried to reassure Flexner that no steps would be taken to implement matters of policy without consulting the advisory board, he was not about to give way to threats. "If I were placed in the position you may be placed in by a question regarding the use of animals from the local pound for medical teaching and research," he bluntly advised him, "I should state as vigorously as possible that in my opinion that was the only rational arrangement to make . . . the local humane society insofar as it places obstacles in the way of men who are engaged in advancing medicine, is a foe of medical progress and ought, I believe, to be so characterized."[7]

Flexner was not the only adviser who balked at some of the initiatives that the Friends of Medical Progress had undertaken. George McCoy, on more than one occasion, made vigorous complaint to Cannon that the society's efforts to testify against the bills that antivivisectionists perennially presented to congressional committees interfered with the successful strategy he had devised for dealing with them. Cannon politely acknowledged the complaints, but made little of McCoy's discontent.[8]

McCoy was equally disapproving when Cannon called on the U.S. Public Health Service for help in challenging an article that Fred Kelly, a syndicated columnist, had recently published in *Current History,* a magazine put out by the *New York Times.* Kelly's article provocatively titled "Is Better Health Due to Doctors?" maintained that the trend in declining death rates for many diseases would have occurred irrespective of anything doctors and health authorities might have done. The core of Kelly's article was based on Raymond Pearl's comparative analysis between the rate of decline of mortality for several controllable diseases (tuberculosis, diphtheria, typhoid and the like, which health agencies had been active in fighting) and several uncontrollable diseases (such as bronchitis, paralysis without specific cause, and softening of the brain, which health agencies had not fought as successfully).

Convinced that Pearl's premises were shot through with a variety of defects, Cannon was startled and upset when McCoy, in reply to his request for help, stated that the Public Health Service was not interested in refuting the claims in Kelly's article. "If the Public Health Service is not concerned with the effects of public health activities or the effects of medical treatment in changing the mortality figures," Cannon acidly queried him, "please tell me to whom I might apply for authoritative information." McCoy was unyielding, and refused to budge from his position. So far as the Hygienic Laboratory was concerned, he answered, it was not worth their effort "to point out the inaccuracies or to in general controvert the statements Pearl has made." Furthermore, he suspected that it could lead to "a rather unprofitable" controversy. Then he added a conciliatory afterthought, "You know what almost any clever man can do with figures and Pearl does take the curse off his figures by some kindly things he says in the text."[9]

Cannon did not pursue the matter further with McCoy, but he worried that antivivisectionists in future would draw the same conclusion as Kelly from Pearl's statistics and adopt it for their propaganda that medical scientists used animals experimentally to no useful purpose. So he turned to Simon Flexner for advice in dealing with Kelly's article and Pearl's statistics. Flexner proved to be about as much help as McCoy. He told Cannon that his contacts with Pearl on the use of a particular kind of statistics had not been very reassuring. "He has a fine disdain for those of us who are chary of some of them. I see little or nothing to do beyond possibly getting at Dr. Welch who is I think little likely to act . . . I should hate to be assigned the job of restraining Pearl."[10]

Even before the founding of the Friends of Medical Progress, Cannon emphasized that the most important matter before the society would be to remain a lay organization; and while it was right and proper for its members to seek counsel from the medical advisory board on technical matters, he urged that they guard against domination by professional interests. The very activities that Drs. Flexner and McCoy had opposed gave Cannon cause to turn over to the Friends a share of his duties as chairman of the A.M.A. Committee for the Protection of Medical Research in regard to monitoring the various meetings and publications of the antivivisectionists.[11]

One obligation Cannon kept entirely for himself, however, that of monitoring the research reports of medical scientists. He took great care to insure that the investigators clearly stated the purpose of their research and that the animals they used were protected against pain—in other words, that their work was defensible against possible antivivisectionist charges of cruelty or claims that it was undertaken to satisfy "an idle curiosity." Sometimes publications were attacked that had passed muster with Cannon. In these cases, he rose to their defense, often with passion. At other times, if a report gave the least intimation that the research might have involved painful experiments, Cannon did not hesitate to censor parts of the paper or reject it outright—even on grounds as vague as his feelings about the "character" of the author. Two incidents illustrate Cannon's *modus operandi*.

Early in 1924, Bishop Alexander Mann (a member of the board

of directors of the Friends of Medical Progress) received a complaint about a paper that Carl J. Wiggers had written in 1918 on experimental shock, charging "that Wiggers was curious to note the effect of intolerable pain on the central nervous system and blood pressure, etc., of experimental animals without regard to the suffering of the animals during his research." After a time, the complaint reached Cannon, who indignantly called the charges direct falsehoods, not based on statements in the paper. "I wish to state at once," he nevertheless cautioned Miss Thurman, "that Dr. Wiggers' paper is far from being pleasant literature to read."

Having said that, Cannon proceeded to defend Wiggers, who had been a member of the National Research Council committee that undertook an investigation of wound shock during the war.

> The purposes of his work were directed towards the relief of human suffering and, as I read his paper, his work was done with due precautions to avoid unnecessary pain. The implication in the letter to Bishop Mann that there was an attitude of idle curiosity on Dr. Wiggers' part is quite unwarranted.
>
> I wish that Dr. Mann's correspondent had seen service in a shock ward after a battle as I have had to do repeatedly. The beds are filled with horribly wounded men—men with arms and legs torn to shreds, with bellies opened and the liver and other organs protruding, or with jaws shot away . . . These human beings have been brought to such distressing plights without the whiff of an anaesthetic. Nothing even remotely approaching the horrors of such a place ever occurs in a laboratory of medical research. It was to help men, wounded in these frightful ways, that Dr. Wiggers was working.[12]

Another time, George H. Simmons, the editor of *JAMA*, asked Cannon to review a paper on the surgical treatment of burns that A. Murat Willis, of Richmond, Virginia, had submitted to him for publication. In search of informed opinion, Cannon turned to Harvey Cushing and his young assistant, Elliott Cutler, both respected surgeons, for advice. Although Cushing and Cutler believed that the paper, which featured experiments on dogs, was important and ought to be published, Cannon held off recommending publication. "Personally," he wrote Dr. Simmons, "I should want to know something more about the reliability of the man before taking his word for what he says. It is a difficult judgment, which you can perhaps postpone until after you have seen what sort of man Dr. Willis is and whether he is willing to make any changes you may suggest."

Following the receipt of Cannon's letter, Simmons refused publication, his explanation resonating with Cannon's oft-stated concern about cruelty. "These experiments," he wrote Willis, "must be regarded as the extreme of cruelty. An anesthetic was used in the initial work, of course, but the after-suffering in such experiments must be great. What we are thinking of is the effect on the average reader—even on the medical man. If published in its present form, the report would give the ordinary layman the horrors." Above all, Simmons feared that Willis's report "would, of course, be utilized by the antivivisectionists, and would arouse against animal experimentation many who otherwise would not be unfavorable to it."[13]

Late in 1924, Thomas Barbour, recently appointed Curator of the Agassiz Museum at Harvard in addition to his post as Professor of Zoology, resigned as director of the Friends of Medical Progress. Cannon was sorry to see Barbour go. Not only had he relieved Cannon of the need for keeping track of antivivisectionist actions throughout the country, he also had a congenial relationship with the boards of directors and advisers who respected him as a scientist. Finding a replacement with Barbour's qualifications as well as his interest and energy in developing the work of the society seemed an almost impossible task.

After a diligent search, however, Cannon by good fortune came upon Benjamin Gruenberg. A graduate of Columbia University with a splendid record as an innovative biology teacher in New York City high schools as well as experience in educational promotion for the United States Public Health Service, Gruenberg appeared to be made for the job. Best of all from Cannon's angle, Gruenberg firmly believed that the best way of dealing with antivivisectionists was to develop a varied program of public education.[14]

As the new director of the lay society—now removed to New York and renamed the American Association for Medical Progress, Inc.—Gruenberg, in furtherance of his views, embarked on a public relations tour of the United States. He not only addressed state and local medical societies and lectured at universities about the goals of the Association, he also spoke to high school students and teachers as well as to women's clubs, labor unions, insurance companies, social workers and lay guilds of churches—indeed, he spoke to anyone who would listen to him. Moreover, he set out for the advisory boards his philosophy of administering the society, telling them, "I am strongly impressed with the need for widespread education, especially among adults regarding the foundations and methods of modern science. There is particular need for such education regarding organic processes and the applications of our knowledge to preventive medicine and hygiene." He was inclined to give antivivisectionist propaganda public notice only where it involved legislation or obstruction of public health administration. "Instead of engaging in controversial activities calculated to refute the claims and charges of the antis," it seemed to Gruenberg "more important to develop positive educational materials and methods."[15]

He then suggested a series of ambitious programs—among others, that the members of the Association publicize their activities at meetings of state and local medical societies, that they prepare and distribute leaflets and pamphlets about the nature of disease and public health, that they organize material for exhibitions, and especially that they work through other agencies, such as newspapers, teachers' associations, women's clubs, labor unions, church groups and insurance companies. Gruenberg's display of intelligence and energy in developing the programs of the Association was a joy for Cannon to behold. He enthusiastically responded to a report of Gruenberg's first year as managing director: "It shows an extraordinary grasp of the situation throughout the country and a great deal of imaginative insight into possibilities. I heartily commend your emphasis on educational methods, rather than attacks. I am sure that attacks merely serve to fix mental attitudes and do not lead to changes of conviction."[16]

While Gruenberg opposed the annual antivaccination and dog exemption bills presented to Congress and various state legislatures, in most matters he tried to avoid direct confrontation with the antivivisectionists; rather, he relied on indirect means to get his points across. One such occasion involved the body builder and health-food advocate, Bernarr MacFadden. He was also proprietor of the *New York Graphic*—a tabloid newspaper that titillated the public with tales of sex, mayhem and murder.

During the winter of 1925–26, MacFadden, an ardent antivaccinationist, launched a series of blistering editorials against compulsory vaccination as well as the use of antitoxin against diphtheria. Gruenberg countered MacFadden's various assertions by gathering information from the deans of forty-five medical schools as well as hospital superintendents in New York. "There is of course general agreement that nothing is to be gained by giving MacFadden and his dupes further publicity by taking them seriously in an open controversy or refutation," he wrote. "It seems worthwhile, however, to make available to educators and to the profession the actual facts in such cases."[17]

In an effort to contain the damage of MacFadden's editorials, Gruenberg mounted a campaign of public education with the aid of a vice-president of the Metropolitan Life Insurance Company. District managers and insurance agents of Metropolitan Life, armed with pamphlets on smallpox and diphtheria prevention supplied by the Association, carried the message of medical research and disease prevention throughout the city to clients of the company and their neighbors. As far as is known, MacFadden did not continue his diatribe, and returned to the more profitable schemes of promoting physical culture and publishing lurid stories and editorials guaranteed to sell newspapers.[18]

Of greater matter was an internal controversy that Gruenberg soon became enmeshed in, which would involve Cannon as well. Early in March 1926, Margaret Sanger, on behalf of the American Birth Control League, called on the Association to support two proposed amendments to the U.S. Penal Code—the first to allow physicians to teach women how to avoid conception, and the second to allow the mails to be used to distribute contraceptive information and supplies. Gruenberg, who sympathized with Mrs. Sanger's views, went to the board of directors for advice on how to handle her request, and asked whether it would be permissible to give the Birth Control League the Association's membership mailing list. Gruenberg's action brought forth a storm of protest. Cardinal Denis Dougherty, a vice-president of the American Association for Medical Progress and a member of the board from its inception, announced his intention to resign. Other board members wrote Cannon that Gruenberg had "made himself so obnoxious and useless as to warrant a demand for his resignation."[19]

Cannon was appalled at the vehemence of the reaction. "It seems to me," he pointed out to W. W. Keen, "that the effective and also the Christian act on the part of Cardinal Dougherty . . . would be to stand by the organization and see that his advice counts strongly in what he regards as the right direction . . . We had a very difficult time to find a person of Gruenberg's abilities and training. He was the only candidate in the field who at all fitted our needs. It would be extremely difficult to replace him, but of course we also want Cardinal Dougherty . . . I do not see any good reason why we should not have both men in our service."

Gruenberg weathered the storm, but his willingness to even consider matters relating to birth control was deemed by the directors to be so grave a heresy that they felt it necessary to issue a reprimand. "The Board of Directors," they resolved, "wish most vigorously to point out that this inquiry was unnecessary and ill-advised inasmuch as the objects of the American Birth Control League are wholly and completely foreign to the interests and purposes of the American Association for Medical Progress." As if that wasn't clear enough, they went on to resolve that the Association had "no interest in the methods and aims" of the Birth Control League and that "in future no further communication be held with said League."[20]

With Cannon's department at Harvard growing by leaps and bounds and his research becoming ever more complex and engrossing, he again yearned to give up his post at the helm of the A.M.A. Committee. At last, certain that the various programs he had fashioned had taken root and were functioning smoothly, and that Gruenberg could successfully advance the activities of the lay society, he rendered a final report and turned in his resignation. "Some weeks ago I sent to the secretary of the A.M.A. my resignation as chairman of the Committee for the Protection of Medical Research," he confessed to Dr. Keen early in 1926.

> My reasons for doing so were that I had served in that position for seventeen years and had begun to feel the burden of it, and also that it seemed to me desirable to have a younger man being trained to take over the job while I was still at hand to contribute such advice and information as would be useful . . . I do not expect to be any the less interested in our campaign, but my move will relieve me of responsibility for organization and for close watching of the situation over the country.[21]

Finding a replacement for Cannon, however, was no easy matter. Indeed, it was not until early 1928 that he learned that his former student, Elliott C. Cutler, had agreed to head the Committee. In the meantime, A.M.A. officials took Cannon at his word and continued to call on him for advice in dealing with the antivivisectionists. An early call for help came after an attack on some of the experiments conducted by his longtime adversaries, G.N. Stewart and J. M. Rogoff.

In November 1927, Sarah Cleghorn, a prominent antivivisectionist, charged in *The Nation* that Stewart and Rogoff had removed the adrenal glands from some pregnant dogs for no apparent useful purpose. Within weeks, the Vivisection Investigation League reprinted a leaflet, "Torturing the Pregnant Dog," with a selection of grim excerpts from a paper entitled "The Influence of Pregnancy upon the Survival Period in Adrenalectomized Dogs," published by the Cleveland investigators earlier that year, so as to illuminate Miss Cleghorn's charges.[22] The resultant letters of protest to editors of various newspapers and magazines served antivivisectionists well.

When the Vivisection Investigation League leaflet came to Cannon's attention, he sent a copy to Olin West, the new secretary of the American Medical Association. "I have been for many years in a controversy with Stewart and Rogoff and am not the proper person to inquire regarding the facts of this case," he candidly informed West. "Should I do so, it might seem like an escalation of personal feeling. I think that it might be well to get from them the facts in the case, as well as to point out to them the jeopardy into which they are placing research by careless experiments."[23]

Cannon's remark about the "careless experiments" shows how deeply his acrimonious debate with Stewart and Rogoff had cut. He was well aware of the Cleveland investigators' pioneer work on the adrenal cortex and of their report that pregnancy exerted a protective influence in adrenalectomized animals. They had, in fact, suggested a probable connection between adrenal function during pregnancy and certain toxemias of pregnancy. Hence, the experiments were not done, as the antivivisectionists claimed, "to satisfy an idle curiosity at the price of suffering purposely inflicted." Rather, they were part of a search for knowledge of the role played by the adrenals in the pregnant state as well as in other conditions.

Dr. West, fearing that the pamphlet might persuade members

of Congress to pass the perennial Dog Bill that antivivisectionists hoped would halt animal experimentation, took Cannon's advice and wrote to Stewart and Rogoff himself. He stated that the language used in their article provided the antivivisectionists with "telling material" and asked that descriptions of experiments be more guarded in future lest they "add fuel to the flame" of their movement.[24] Rogoff, who had been in combat with antivivisectionists throughout his career and felt that he needed few reminders from the A.M.A. and its Committee on how to conduct and report his research, took West to task. "It is . . . deplorable that in this age it should be necessary to disguise a scientific report or to subject it to a particular sort of censorship in order to evade malicious distortion and misrepresentation on the part of organized cranks, liars or ignorant fools," he lashed out. "The possibility of introduction in Congress of a bill to interfere with scientific research is a colossal monument to the inertia and lethargy, in such matters, of our scientific, educational and professional organizations."

Though Rogoff admitted that some of the descriptions in the paper gave sustenance to the antivivisectionist attacks, he ardently defended the right of the scientist to express himself unequivocally. "Trope and metaphor do not readily lend themselves to the description of the phenomena observed by the scientist," he acidly commented. "Such observations as the occurrences of vomiting bile and bloody materials, or the passing of bloody fecal matter, are rather difficult to describe in lyric verse. Yet they are very important clinical observations and must be recorded, even at the risk of offending the aesthetic sense of the antivivisectionist or the sensitive admirer of good English."[25]

West, although agreeing with most of what Rogoff had written, nevertheless pointed out that it was necessary to provide a statement to the effect that animals used in experiments had been anesthetized and treated in a humane fashion. Then he sent copies of his correspondence with Rogoff to Cannon.

There is no denying that Rogoff's letter, with its sharp attack on the inactivity of the A.M.A.'s Committee, stung Cannon. Referring to Rogoff's "curious psychology" in thinking that the introduction of a bill into Congress meant inactivity of scientific, educational and professional organizations, he defended the committee he had chaired, which had "repeatedly managed hearings in Washington at which the case of the investigators was so effectively presented that the bill never appeared outside of committee."[26] When the flames of anger finally died down, Rogoff acknowledged to West that he had been indiscreet and promised that he would act differently in the future. For Cannon, it was at best a small satisfaction.

In the weeks that followed, Rogoff, not waiting for support from the A.M.A. but with the aid of a number of practicing physicians who admired his research, conducted his own defense against the unrelenting attacks of the antivivisectionists. Despite these efforts, it became clear, from the continuing flood of letters to the editors of newspapers and magazines, that the public generally accepted the antivivisectionist charges. Even more serious, the Western Reserve Medical School found it increasingly difficult to obtain animals for purposes of teaching and research. The dean for administration, W.G. Leutner, recognizing that Western Reserve faced a crisis, invited Benjamin Gruenberg to attend a conference with Drs. Rogoff and Wiggers and Stanley Wilson, the university's legal counsel, to discuss ways of promoting legislation that would give the school access to strays in the local dog pound.

Gruenberg was openly skeptical of Leutner's approach. Even if legislation were to be enacted, he told the dean, it would still be necessary to educate the public about the spectacular applications of medical research as well as the methods by which scientists obtained their results. He underscored that such a strategy would require not only utilizing existing educational institutions in Cleveland, but also obtaining the cooperation of innumerable lay groups.[27]

The value of educating a lay audience to an understanding of science and the scientific method was an article of faith with Gruenberg, and he worked hard to achieve it. In 1927, he edited a volume in cooperation with medical scientists, physicians and public health leaders titled *Modern Science and the People's Health* that received encomiums from reviewers. The following year, Gruenberg founded and edited *Medical Progress: A Journal for Laymen* that contained a variety of popular articles on medical science. Initially, Gladys Dick, the co-discoverer of scarlet fever vaccine, submitted an account of the research leading to the development of the vaccine. Gruenberg contributed sketches of the scientific achievements of Karl E. Baer, John Hunter and William Harvey, as well as articles reviewing legislation that affected medical progress and notes on therapies relating to tuberculosis, blood transfusion and pneumonia. A letter from Rogoff about the antivivisectionists and a review of Pavlov's *Conditioned Reflexes* rounded out the first issue.[28] It was a promise of things to come.

Unfortunately, that promise was never fulfilled. In the spring of 1928, George McCoy, then engaged in the latest round of Congressional legislative battles with antivivisectionists, wrote Cannon that he desired the lay society to stay out of the fray. McCoy reminded Cannon that he had always been "somewhat less than lukewarm to the Friends of Medical Progress, feeling that organized opposition to the Antis would really do them more good than harm, no matter what influence it might have on our side of the fight."

McCoy went on to quote from a recent issue from *Living Tissue:* "In the period from 1921 to 1927 there have been two particular events which have helped our society's growth. The first was the violent attack upon anti-vivisectionists by our opponents—particularly the Friends of Medical Progress . . . This stimulated interest and added to our membership." Cannon politely thanked McCoy for his letter, then added, "It looks very much as if your wisdom should have prevailed in the judgment as to formation of a lay society. Gruenberg informs me that probably the American Association will go out of existence for lack of funds."[29]

Cannon said nothing about his own disappointment on hearing the news of the Association's demise or, for that matter, about the apparent vitality of the antivivisectionists. Some months previously, he had received a letter from Dr. Keen, who had done battle with the antivivisectionists almost from the start of their campaigns in America in 1883. Keen held out little hope that the vitality of the antivivisectionist movement would be short-lived:

> The more I see of attacks upon experimental research the more decided I am that all our setting forth of the benefits of experimental research are practically useless. There is but one thing that we want to correct, and it is going to be a very difficult thing to do, and that is to persuade the entire community that experimental research is not cruel. That should be practically our sole aim at the present time . . . I like to meet them on this ground, that the success of their efforts to prevent research and the medical, surgical and obstetrical progress that result from it, is the greatest cruelty to human beings and

> animals that can possibly be devised, for it insists on allowing both human beings and animals to continue to suffer and to die by hundreds and thousands from diseases that we might ameliorate or even abolish by experimental research.

Although no reply from Cannon has been found, in a re-examination of his struggles against the antivivisectionists made late in life, he found a satisfaction that had eluded Keen. "It is gratifying to report," Cannon wrote in his autobiographical memoir, "that thus far the antivivisectionists have not been able in the United States to interfere with medical investigators in their efforts to increase knowledge of disease and to discover effective ways of treating it. With due precautions against the infliction of pain, they are still as free as the experimenters in physics and chemistry to project their explorations into the dark range of ignorance."[30]

11 One Man's Family

FRIENDS AND ASSOCIATES never failed to remark how essential Cannon's wife and family were to his well-being. Henry Dale, for one, visiting from England not long after the war, "found himself welcomed to a household glowing with happiness and affection, and resounding with friendly chaff and merriment." He later noted that "any one who had not seen Cannon as a father had missed a very important and attractive side of his character."[1]

Happy, too, were the memories of the Cannon children. As adults, their descriptions of growing up in the large rambling house at No. 2 Divinity Avenue (a street of Harvard landmarks featuring the Semitic Museum, the Germanic Museum, and the Museum of Comparative Zoology), with idyllic summers at the beloved farm in Franklin, New Hampshire, evoked any number of coming-of-age stories. Who can resist the images of a (surely not typical) academic family in the 1920s, which Marian Cannon Schlesinger has brought to life in her engaging memoir, *Snatched from Oblivion?* The nuclear family extended to nine permanent members—the five children and four adults, including two maiden aunts in the persons of Cannon's sisters Ida and Bernice—who all got along amazingly well. With pet names lovingly formed through infant lips, then adopted by the entire family, "Dada" and "Beecie" continued to be cherished members of the household well into the latter part of the 1930s, so that the children proudly boasted they had three mothers.

Visits to Ida's office in the Social Service Department of the Massachusetts General Hospital or to Bernice's in the Personnel Department of Filene's Department Store were red-letter days for the children. This was in marked contrast to their father's laboratory at the Medical School, which was sacrosanct and strictly off limits to them. But, as Marian tells us, plenty of traffic flowed from the other direction, for a steady stream of Cannon's visitors, often distinguished and interesting personages from around the world, passed through the doors of the Cambridge house or the Franklin farm.[2] Beginning in 1920, the Cannons instituted the practice of having at-home teas for his students. For the family, the company of young people was stimulating and fun. The students in turn felt privileged to be there and especially appreciated the antics of impish Marian, who was inclined to amuse the guests by turning handsprings in the kitchen.

The rhythm of the household revolved around the children's school schedules and the adults' working life. Evenings and weekends for Walter and Cornelia were filled with a dizzying round of social events ranging from informal dinners and meetings with old friends to gala celebrations and special events. More especially, life centered on ties to Harvard and Radcliffe, with everyone participating in a rich program of university activities and a full calendar of academic events. Although Cambridge and Boston were at the core of the Cannons' universe, it was during vacation times in New Hampshire that family life was most cohesive. That was when Cannon, undisturbed by outside distractions, was able to spend more time with the children.

George and Florence Pierce, Tom and Rachel Perry, as well as the Whitings, the Crothers, the Jacksons, the Southards, the Yerkes—all friends from the early years of their marriage—had remained part of the family's intimate circle and were there to welcome Cannon when he came home from the war. As the years passed, this circle of friends dwindled, and those who remained became even more precious. One of the first losses occurred in February 1920, when Ernest Southard, who had gone to New York for a meeting and lecture, came down with a virulent pneumonia and died there within forty-eight hours. The loss of Southard was followed within a relatively short period of time by the deaths of Allen Jackson's wife, Bess, and Maria Baldwin, the revered black principal of the Agassiz School. Cornelia took Bradford and Wilma to Miss Baldwin's funeral service, which was held in "a big church packed with white and colored people paying tribute to a noble personality."

The passing of dear friends alerted Cornelia to the frailty of her parents, then in their seventies, and she increasingly took care in her letters to tell them of her love and devotion. Returning home from a trip out west in the summer of 1922, she learned that Mrs. William James was dying; by October, she was gone. Cornelia felt the loss keenly and wrote her mother, "I wish all of us who love each other could go together . . . Instead, each generation has to tear itself away from the next generation and suffer the pang of letting the previous generation slip away first."[3]

Throughout the decade she made it a point to visit St. Paul and attend family occasions like her parents' fiftieth wedding celebration. And members of the James family visited her—not only her sisters and their families but also her mother and father. In between visits, she continued to strengthen the bonds between the James and the Cannon families by bringing Cambridge to St. Paul through anecdotes and vivid accounts in her letters that made everyday happenings come alive.

Fortunately, it is through Cornelia's letters to her family that we have a record of the Cannons' private lives during these years. How else would we know, for instance, how thrilled she was when in November 1920 she exercised her newly-granted power to vote in a presidential election? After she and Cannon cast their ballots for Cox, the Democrat (because he was for the League of Nations), she wrote her mother, "I do not know when I have been so moved. It was with difficulty that I could keep the tears back." Or that Walter was peeved when one of her former "radio boys" brought his bride from South Dakota to visit them in the autumn of 1922 and the visitors lounged about the house for days. He was "neither amused nor pleased with Main Street at his front door," she reported. A few days later she went with him to a meeting where W. W. Keen, 86 years old, spoke on the sixty years of surgery he had seen, and was congratulated by President Eliot, 88 years old. "It was a sight to remember," she reflected, "the two wise men and leaders of a generation that is gone still showing the qualities of leadership and farsighted vision. It was like looking at the elder statesmen, the sages of a great era."[4]

Cornelia recorded the milestones as well. When Cannon was awarded the Distinguished Service Medal late in 1922, the whole family attended the ceremony at the Army Base in South Boston. In later years, she would recall that he was pleased by the honor chiefly because he could see his reputation with his children rising. "He had to go through the humbling experience after his return from France of seeing his status as hero diminish. But this honor helped to reinstate him in the children's minds as something out of the ordinary." At the time, however, she noted that "Walter keeps thinking of all the other people who deserve this more than he, but just the same he is awfully happy over it—that his own people, as well as the British, had thought his work deserving of credit." She added that later in the week Cannon preached in church on the ideals of the man of science, and marveled, "That was a week—lambasted by the antivivisectionists on Monday, decorated by the government on Friday, and preaching in church on Sunday!"[5]

In the fall of 1923, much to Cornelia's joy, her mother paid a visit to Cambridge. Mrs. James was a semi-invalid and generally spent her mornings in bed; but during afternoons, indomitable in spirit, she sallied forth with Cornelia to attend lectures, teas and parties, visit museums, and enjoy one another's company. Blessed with a ready wit and shrewd insight, Mrs. James kept her family in St. Paul informed about the Cannon household as well as her "wild careering" with her daughter. At one time, she described attending a lecture on Manchuria in the Fogg Museum, where "the audience was composed of people out for culture—and as I looked around at so many dull, heavy faces, I wondered if that was the way I looked! It certainly was not a cheering thought." Another time, she divulged that Cornelia wanted to send Bradford to the state university in Colorado next year, but "there was a lot of rebellion" both from the boy and from his father.

The entire family took great delight in Mrs. James's presence, and not the least her son-in-law. "Walter said last night that he could not get over your eager, searching mind," Cornelia wrote her mother after she had left Cambridge. He had marveled at the way she wrestled with the theory of relativity or sympathized with the hardships of the Pilgrims, all the while being helpful about Wilma's art and Bradford's turbulent adolescence and Helen's whines and Marian's impishness and Linda's quiet power of producing irritation—but ready at any moment to drop it all for pure conversation.[6]

Everyone, adults and children alike, depended on Cornelia to keep the household running smoothly and on an even keel. She, like Cannon, admitted that she rather liked to run the machine. Not only that, but she could be counted upon to dispense affection and care to each and every one without fail or favorites. During an era when live-in maids were a commonplace, she managed with the help of only a day worker—never one of the Irish "Bridgets" so ubiquitous in Boston and Cambridge, but rather a procession of "colored" girls until the cherished Hortense came into their lives and stayed with the family for years.[7] They, in turn, were supplemented at the Franklin farm by local girls—mostly daughters of the Polish immigrants who had bought up derelict Yankee farms and were making them thrive.

Not satisfied with being merely a "stay-at-home" mother, Cornelia devoted herself to an increasing array of good works and social reforms that were important to her. Though at times she may have seemed like one of "these volatized women who run about in motors and can't keep still" (as one doughty New Englander put it), there were two causes to which she remained steadfast and true. One had to do with educational matters, the other with birth control.

As the Cannon children moved through the grades in the Cambridge school system, their mother sustained her crusade to improve public education for them and their school-mates. From about 1922 to 1924, as secretary of the Cambridge Public Schools Association, she contributed a regular column to the *Cambridge Chronicle.* But she frequently met with frustration. The elected and appointed officials running the Cambridge government and school system—like those in Boston, often corrupt "machine" politicians and administrators who represented the large number of Irish Catholic immigrants—dominated the city's institutional and social life. Without the approval of this powerful ethnic group, no reform that she advocated was possible, and these men remained the bane of her life.

Mrs. Cannon's running battles with Mayor Quinn, Mr. Hurley, and their like became the stuff of legend, sometimes making newspaper headlines. Such was her fame that the *Cambridge Chronicle,* noting in June 1923 that Cannon had received an honorary degree from Yale, reported, "To most people in public life, Prof. Cannon is perhaps best known as the husband of Mrs. Cornelia James Cannon, . . . whom everyone not only loves and admires for her engaging personality but for her constructive and sympathetic work for public schools and in other civic lines."

Officials at City Hall, however, were not quite so captivated by her charms, and she was known there as "that woman" or "the critical Mrs. C." At meetings of the school committee, she presented ideas that gave her nemesis, Charles Francis Hurley (later the governor of Massachusetts), apoplexy. He roared and bellowed at her until one day, she could stand it no longer. "My hair may be graying but my hearing is good—don't yell at me, Mr. Hurley," she shouted back as she rose and walked out of the chamber. Despite winning a skirmish in public now and again, Cornelia more often than not had to settle for asserting her conscience against "the Irish clannish spirit" in private.[8]

In November 1921, after yet another school election was lost to her opponents, Cornelia, smarting from the defeat, escaped to New York, on what Wilma called "Mother's sabbatical." There she attended an international conference on birth control organized by Margaret Sanger, the leader of the movement in America. Finding the occasion extraordinarily interesting and stirred by events that she attended, she kept a running account for her mother. "The 'lunatic fringe' has apparently fallen off and a finer, more serious, right-minded lot of people I never saw," she wrote after attending one session. "I am proud to be one of the pioneers in so important a movement."

Pride turned into outrage when, during another session the New York City police, at the instigation of Cardinal Hayes, rushed in and arrested "the beautiful Mrs. Sanger." To Cornelia's way of thinking, the arrest was a clear-cut attack on civil liberty and freedom of speech. "That is what our land of freedom has come to under Irish Catholic control," she furiously scrawled on the train going home. "The blood of my Protestant forebears is boiling in my veins!"

On her return home, she was still burning with rage, and fired off an editorial to the *Woman Citizen* about the incident, calling it a blow to the very foundation of the Republic. Walter, for her forty-fifth birthday, gave her Prof. Zechariah Chafee's little book on freedom of speech so she could relate her "flame of indignation to other flames of indigation against this attempt upon our liberties." Within a few days, though, Cornelia got over her "liberty frenzy" and calmed down enough to see that the birth control battle "will have to be fought to all eternity and probably never fully won." Soon after, when Ernest Harold Baynes, Cannon's ally in the defense of medical research, came to dinner and remarked that his talks had exposed the lies of the antivivisectionists, she observed with feeling, "It takes courage to stand up for an unpopular cause."[9]

In December, Margaret Sanger visited Cambridge and spoke to a group in the Cannons' home that included representatives of the medical profession. When Cornelia told her mother afterward that the doctors had expressed concern for "the saving of the better strain," she could hardly contain her annoyance. "I must say in face of the individual suffering woman, I find it hard to think solely and steadily of the *race*. What if the race does die out? I cannot see how life is of any particular value when it is worry, pain, disappointment, and slavery."

A few years later, when Mrs. Sanger returned to the Boston area, Mayor Curley threatened that if she proposed to address a birth control meeting, he would block her by seeing that the hall had its license revoked. Cornelia was livid; Cannon, already the veteran of too many conflicts, was alarmed. He recognized that his wife's passion might well push him into the vortex of yet another controversy, and he had neither the zeal nor the inclination for it. "I want to get into the fight," she explained to her mother in the spring of 1924, "but Walter shows a grave reluctance to having me take part in anything inviting such unpleasant notoriety. Yet it is the one cause worth being a martyr about today." Cannon may have agreed that birth control was a worthy cause, but he was not willing to martyr himself for it. Nor was he willing, just then, to have Cornelia do so. For the time being, at least, she bowed to his wishes and stayed away from the battle for birth control. But she would come back to fight another day.[10]

Not that Cornelia was any stranger to controversy. Good conversation was meat and drink to her—the more copious and audacious, the better. She enjoyed nothing more than the high-spirited and argumentative debates in which she engaged with friends at Discussion Club meetings or around the dinner table. Out of this intellectual ferment came no end of topics that she could, and often did, commit to writing. When George and Florence Pierce presented her with a new Corona typewriter, she crowed with delight, "I do not know when I have been so happy over anything." Then she added that the Pierces had "decided that I was destined for a literary career and did what they could to help me on my way."

Without betraying a shade of self-consciousness or a particle of reserve, Cornelia was capable of giving expression to her every thought, feeling or reaction with absolute ingenuousness. Beginning in 1920, she offered her ideas and views on an astonishing array of subjects to properly oriented readers of such middle-class magazines as the *Atlantic Monthly, North American Review* and *Woman Citizen.* In her first published article, "Can Our Civilization Maintain Itself?," she discussed the economic slavery of low-wage workers and asked whether the habit of material possession in the nation might be renounced so that the struggle between the haves and the have-nots would not go on forever. When the article elicited a flurry of letters to the editor, she was heartened and burst out, "Let the wise and the foolish alike chew their cuds." For her, asking the questions in the first place was the point.[11]

The following winter, the editor of the *Atlantic,* Ellery Sedgwick, wrote Cannon to say how much he had appreciated Mrs. Cannon's contributions and that he wished to encourage her to develop further her topics. Before long, Sedgwick invited Cornelia to come and see him to discuss future contributions; he told her his problem was to find people with ideas. "You know, more and more, as I assemble my mental processes," she wrote her mother in May, "am I impressed with the fact that intellectually I am like you—I react in the same outrageous way to the facts of the universe, and so you have only yourself to thank for my indiscretions and audacities!"

Both Sedgwick and Cornelia expected "to get a lot of comeback" from her next article for the *Atlantic,* "Philanthropic Doubts," in which she pondered the putative success of charitable organizations as well as the failure of public support for aid to the poor. They were not disappointed. When it was published in September 1921, Cornelia's provocative piece elicited such an outpouring of letters and rejoinders that she planned to follow it with another essay called "More Philanthropic Doubts."[12]

But it was her article on "American Misgivings"—which reviewed publication of the Army Intelligence Tests and gave praise to the work of its author, her friend Robert Yerkes—that became her most controversial "think piece." Published in the February 1922 *Atlantic,* it is still taken note of today by historians of the eugenics movement. While attending meetings of the A.A.A.S. in Cambridge with her husband, Cornelia was brought up short when she heard herself being defended by a professor who "extolled my paper on 'American Misgivings'" and "annihilated my critics among the educators." "Wasn't that amusing?" she remarked with feigned insouciance. "I did not know I was being fought over, but apparently the educators are as mad about that article as the philanthropists were about the 'Doubts.'"[13]

Throughout 1923, the provocative nature of Cornelia's ideas and views continued to bring a storm of protest down on her head. When she spoke on "The Real Opportunity of the Philanthropist" to a group of social workers in Boston, she met an audience "with eyes blazing and their axes sharpened for me." Her article, "Can Our Public Officials Be Our Leaders?," published in the *North American Review,* elicited pungent remarks from Felix Frankfurter. "Why do they not let a poor lady have her ideas without being so severe about it?" she appealed to her mother. "If I had to prove every statement I make, I could be stricken dumb for life." Another article, "The Crabbing of Youth by Age," in the *Atlantic,* brought the criticism close to home when Wilma, with pointed irony, asked her mother, "Why, when you say it is so awful for the old people to crab the young, do you keep right on crabbing us?" Even Ellery Sedgwick, who solicited her articles for their freshness and vigor of treatment, could not agree with one of her essays on education; although he accepted and paid for it, he held off publishing it.[14]

All of this Cannon regarded with amused detachment. When his friend E.G. Martin at Stanford wrote him that he and his wife usually argued rather heatedly about Cornelia's "very suggestive contributions," he replied, "You apparently are having the same sort of reaction that most people have to what Mrs. Cannon writes. She almost invariably receives both brick bats and bouquets through the Editor's office after she has put out a paper, and, curiously enough, the arguments, one way or the other, are very warmly stated." Nevertheless, Cornelia, when she was given more brickbats than bouquets, put on a brave face and played the misunderstood prophet with gusto.

All the same, controversy did have its compensations, for there is no denying that Cornelia had come into her own. The published articles brought her invitations to speak before community organizations, women's clubs and educational groups (her fees for these engagements often went to the Radcliffe Endowment and Alumni Funds). A modest celebrity was now hers, a measured success; and she confessed to her mother that she rather liked it: "Do you know, this writing in the magazines, while it is a mild little affair as far as achievement is concerned, has very pleasant human associations. People think more highly of me than they did and nice things come my way that did not before, although I was as noble a woman before my name appeared in print as now."[15]

But she never said it was easy. By disciplining herself to work late into the night, her output of articles on social, political and cultural issues soon mounted up. Still, the effort took a psychic toll on her. Feeling herself pulled and hauled in every direction, she expressed a conflict of emotions. In one breath she would exclaim, "As far as the family is concerned, they all groan when they see the typewriter come out for they mark it as the end of Mother." And in another, "I sometimes wonder if the children will remember me as a Mrs. Jellyby when they grow up. But I tell you my heart is all aglow when the instant the front door opens, the call of 'Mother' always echoes through the house. It is lovely to be wanted in this world."[16]

Though she sometimes admitted to feeling irritable while she was writing, at other times Cornelia could be engagingly whimsical, and knew how to use her pen for pure fun. By writing and performing monologues and humorous skits, she added another string to her bow and, at the same time, found an outlet for her repressed talents as an actress. She even wrote poems in the style of Amy Lowell or Walt Whitman. In January 1921, Cannon described for his mother-in-law a Stunt Party where Cornelia had held the floor: "The event of the evening was Cornelia's burlesque of Amy Lowell. The audience went into peals and waves and torrents of laughter . . . and realized that they had been listening to and watching perfect art—nothing too much or too meager and so subtle a travesty of the New Poetry, in form and content, that the delight was permanent."

What had begun as hilarious imitations at private stunt parties developed into star acts on a public stage when she began giving monologue afternoons for the benefit of Radcliffe. On a trip to Seattle and Spokane with Cannon in 1922, while he spoke to medical men about the physiological effects of the emotions and his latest research, Mrs. Cannon entertained their wives with pieces she had written on "Bradford's Cello Lesson," "The Breakfast Scene," and "The Diet of the Future." Whenever and wherever she performed, she presented a picture full of absurdities; indeed, absurdity was the essential condition of her art.[17]

In 1924, Cornelia's writing career had a setback when a volume of essays she had put together under the title of "Social Builders" was judged unsuitable for publication. Instead of dwelling on the disappointment, she promptly decided that she must write in a lighter vein. Despite her resolve, she continued to contribute serious pieces to magazines on subjects ranging from issues of public health or patriotism to a daughter's decision to attend Radcliffe.[18] But the following year, when she visited Arizona and New Mexico with her sister Helen, she stumbled into a whole new way of expressing herself.

Cornelia fell in love with the Southwest. A visit she made to the Acoma pueblo in Arizona was a revelation that would have special consequence for her. Returning home, she had scarcely unpacked her suitcase before making a beeline for Harvard's Widener Library and Peabody Museum, where she plunged into the chronicles of sixteenth-century Spanish Conquistadors as well as the studies of Pueblo Indians published by the Bureau of American Ethnology. Absorbed by her new interest, she imparted her excitement to her mother: "Have you ever read about Coronado's expedition into New Mexico? . . . I am reading all the old diaries of the explorers I can get hold of and the children say that New Mexico and Coronado are the only words they hear from morning to night."

As Cornelia's excitement grew, she could hardly wait to get to New Hampshire to begin writing again. Only this time it would be a children's book. As soon as she had the young Cannons (and any number of cousins) settled on the farm for the summer, she moved her typewriter and notebooks into "the chalet" to begin work on her "Acoma" book (later to become *The Pueblo Boy*). When she finished the first chapter, she gathered the children together to read it to them—"and got their unqualified disapproval." Helen, seeing her mother's crestfallen face, tried to soften the blow by telling her, "If you have a lot of nice pictures, no one will mind what you have written." Nevertheless, Cornelia listened well to her critics' comments; duly humbled, she prepared further chapters "according to their prescription and succeeded better."[19]

During Cannon's absences from Franklin, she established a strict routine and disciplined herself to stick to it. Leaving the family to its own resources, her orders were that "nothing but sudden death" be allowed to interfere with her morning work. Though she missed her husband when he found it necessary to be away, his absences allowed her complete freedom to throw herself into her work. And the work progressed by leaps and bounds. By early July, she was jubilant and exclaimed, "I have never found a more congenial task." She even felt a little ambivalent about Cannon's return. "I do not see how Mrs. Humphrey Ward wrote her books, brought up her children, and retained her husband's affection," she wrote her mother.

She needn't have worried. Cannon not only contributed to her efforts by carefully going over her drafts, he also took the finished manuscript to Houghton Mifflin for Ferris Greensleet to read. From Boston, Cannon reported that Greensleet had said Cornelia's story of the conquest of New Mexico, as far as he knew, "was virgin territory for children's writing and since virginity is always charming, he would read it with pleasure and have the editor of the juvenile department do likewise." To the delight of all, Houghton Mifflin accepted the book. When it was published in 1926 (the first in a series of four Pueblo books for children), Cornelia was already immersed in her next project.[20]

Cornelia had scarcely finished her Acoma book when she wrote her mother that she was planning to tackle a novel. "Why not?" she asked. "Stupider people have done so. But shall I ever find such a stimulus as wonderful Acoma?" She did find one—in her native Minnesota—for what would prove to be her greatest literary success, a novel called *Red Rust*. Upon her return to Cambridge in the autumn of 1925, she haunted the Harvard library for historical and factual material and, settling on a formula that mixed romance with realism, came up with a story about Matts, a Swedish immigrant farm boy who devotes his life to an effort to develop a strain of wheat resistant to rust disease.

Drawing inspiration from her childhood experiences, she consulted people like her father and Uncle Jake in Elba who knew the

Swedes, experts in the Department of Agriculture who knew the science, and Otto Folin, Cannon's colleague at the Medical School who knew both. She wrote her mother that she was reading a lot of literature about the pioneers ("It seemed so odd to sit in the Widener Library stacks and come suddenly upon old friends") and that "I spent last evening with Dr. Folin, going over his early memories of Minnesota for my projected novel." In the grip of a creative surge, Cornelia bent to her task and managed to work steadily. By the summer of 1926, she was able to report, "Walter is sitting in front of the barn, reading the first nine chapters of my novel (40,000 words) while I sit trembling within waiting for his verdict. He came up from the chalet after reading the first chapter, enthusiastic, so I am full of hope." Later that fall, she gave her manuscript to Louise Jackson to type and sighed, "It is an inexpressible relief to have that burden off my shoulders."[21]

Early in 1927, Cornelia entered a preliminary draft of "Matts" (as she then called her book) in two literary competitions, one at Harper's, the other at the Atlantic. Although she had mixed feelings about the novel's ending, her uncertainty vanished when Walter and Ida said they liked it. "Walter was thrilled over the last chapter and said any scientist would be, for it was recognition of a worthwhile work and left him saying to himself, 'That was a fine life. Any man would have been satisfied to have done that' . . . I hated to have Matts die and shed tears over his demise myself. My next book shall have a happy ending."

The entire household was caught up in the thrill of the contest and everyone had an idea of how to spend the prize money. Cannon "wanted the paint around the doorknobs washed," while the children indulged their fantasies. As for herself, Cornelia wished only for music lessons, books and travel. She recognized, however, that the greatest financial reward would come from royalties and even began to think of motion picture rights, exclaiming "Think of Mary Pickford as Lena!" But her greatest delight lay in the children's excitement and interest in every step of the process. She took particular pleasure in Bradford's remark, "Well, Mother, you're quite a women"—even though she thought it pure flattery.[22]

When she had to choose between the two contests, Cornelia decided to withdraw her novel from Harpers. Although *Red Rust* came in second at the Atlantic, she had the consolation of seeing it become a bestseller when it was published in 1928—only Hugh

Walpole's *Winter's Moon* and Thornton Wilder's *The Bridge of San Luis Rey* outsold it. One critic called it a "remarkable first novel"; another asserted it "really deserves the Pulitzer Prize." In *The Nation,* a reviewer wrote that it was "an unpretentious . . . tale, relieved by honest beauty in the description of the million meticulous details of nature by one who knows her Minnesota."

Not all of the reviews were favorable. The *New York Herald Tribune* complained that the novel was saccharine—the hero's life "is one of unrelieved noble sweetness." Other detractors noted that the author's ideas got "in the way of both the characters and the story"; that it was "an example of strong material in weak, if bravely well-meaning hands." Even though *Red Rust* was a huge popular and financial success, holding steady on the bestseller list, two of Cornelia's daughters also expressed reservations. "It's awfully queer to read a book by your mother," observed Wilma; "I've already come across six words I never heard of before and I never realized my mother knew anything about 'love.'" Linda let her mother know that she thought it dangerous to write impassioned love scenes because "you might put ideas in our heads!"[23]

By June 1927, Cornelia was already into her second novel, which she titled *Heirs* (thus providing her family the opportunity to shamelessly pun, "Now that Mother has a bestseller, she's putting on Heirs"). Still torn between the creative urge and her family responsibilities, she brooded that "poor Walter has no time or attention from me" and repeated her conviction that "the real problem in careers for women is, as I have always said, the divided mind." The shock of her mother's death the following month and subsequent events contributed to her sense of turmoil, and work on the new novel was often sporadic. Cornelia's spirit may have wavered but it never failed. Although the loss of her mother was a terrible blow, she went on to finish *Heirs,* which was published in 1930.

The story of *Heirs,* like that of *Red Rust,* takes place in a setting that Cornelia knew well. Marilla, a school teacher in rural New Hampshire, gives up her career to marry Seth, an enlightened mill owner. She is unable to bear children and her infertility almost wrecks the marriage—just as an economically impoverished New England is almost destroyed by an exhausted racial stock of Yankees. Seth contracts polio and Marilla devotes herself to caring for him—just as the hardy Polish immigrants who work in his mill devote themselves to tending the rockbound farms that will transform the face of New Hampshire.

During the 1920s, Cornelia worked intermittently on yet another novel, tentatively titled "The Clan Betrays." Again, it features autobiographical details, is set in a locale that she knew well, and reflects her intense sociological preoccupations. Utilizing one of her favorite subjects, America as melting-pot, the story is about a young couple who meet in a Hostess House during World War I and, after the war, enter into a difficult ethnic and religious intermarriage. The novel develops the theme that the war, for all its horror and violence, served as an instrument to break down class barriers and defy social convention. While the rhetoric may seem radical for its day, Cornelia's storyline follows tradition. Despite adversity, the marriage holds, the heroine prevails, love conquers all.

The publication of *Heirs* was hit hard by the deepening Depression in the United States, for in such desperate times only writers of the stature of Pearl Buck or Willa Cather would sell well. Cornelia's second novel failed to make the mark that *Red Rust* did, and "The Clan Betrays" would remain unpublished.[24]

Life with the Cannons was not always harmonious. There were days when Walter or one of his sisters (coming from a melancholy family) was given to sick headaches and everyone else had to tiptoe around a darkened house. At other times, the male members of the family, living in a household where they were roundly outnumbered by the distaff side, displayed evident signs of stress. Bradford found an outlet for his emotions in open warfare with his four strong-minded sisters; Marian, the most contentious, was his favorite adversary. His father, more mild-mannered and undemanding, either retreated to the peace and quiet of his third-floor study or, when he was discouraged with the way his work was going, gave in to blue moods.

As the children grew into teenagers, the stress-line became more evident. There were outright differences in style between the father of the family, whose anxiety and pessimism could make moral issues out of normal adolescent behavior, and the mother, an incurable optimist who was able to take the turmoil of budding youth in stride. "What is the matter with fathers!" Cornelia exclaimed at one point. "They will not let the plant grow quietly but are always jerking it up to see if the roots are secure." Notwithstanding her exasperation, a shared sense of humor generally saved the day, and all of the children grew up to be a joy and comfort to their parents, as well as a source of pride.

High school gave the Cannon daughters ideas and they became more independent, Marian later recalled. When they began to wear high heels and lipstick, their father "was horrified . . . , thinking we were no better than fallen women, and went into a depression for days."[25] If Marian thought that she and her sisters had given their father fits, it was nothing compared to the heartburn he suffered over Bradford.

Cornelia, believing that Walter had never got over the loss of his mother when he was but a boy of ten, countered a possible inheritance of depression in her children by heaping praise upon them. She noted, in particular, that Bradford was in many ways like his father. Once, in an interview she gave to the *Boston Sunday Globe*, she advised, "If your boy is lazy, don't nag him. He may be a genius in the act of thinking." In the fall of 1925, after Bradford had entered Harvard College, his parents were apprehensive. "This terrible self-consciousness of the introspective Cannons curses him," Cornelia complained. "Walter says he can remember being just like him and as a matter of fact his life has been dogged by that pressure of convention."

Cornelia's concern for the boy colored her correspondence. "Our problem now," she divulged to her mother, "is to inspire the diffident Bradford to launch forth. He is like Walter in his lack of faith in himself and fear of being conspicuous." Before long, however, Bradford appeared to be gaining in confidence—and in a manner that alarmed his parents. His tales of "sweet mamas and honeyed women" over the dinner table may have fascinated his sisters but they sorely tried his father. Even his mother, who listened to Bradford's stories with amused tolerance ("he is convinced that his career would be that of lady-killer"), had to concede that his interest in young women was far more intense than his interest in his studies.[26]

By the winter of 1927, Bradford was in despair over his grades, and his mother wrote her family, "There has been nothing but talk of suicide or going away to work but no suggestion of pitching in and studying hard!" Caught in the middle between Bradford's anguish and Walter's attitude that the children were lost souls if they didn't get A's, Cornelia wryly commented that their father "gets up a lot of emotion over artificial standards of measurement." When Bradford was named manager of the Harvard basketball team after an eight-week competition, the family "made much of him." Still, the friction between father and son intensified as the boy's German examination approached.

The story has passed into the annals of Cannon family history: It seems that Bradford had been unable to pass the reading examination in German, which was essential for graduation. After numerous failures, Cannon took up the task of tutoring him in the subject. The boy would sit glassy-eyed across the table from his father, trying to take in what was being said, or dropping off to sleep from intense boredom. Finally, after many trials at the examination, he came home glowing with pride. "I passed," Bradford announced gleefully, "I got a D." Cannon looked at him glumly. "Bradford," he said, "that's the first D I ever got in College."[27]

While Cornelia may have made light of these incidents and seemed to attribute Cannon's depression to an inflated sense of duty as a parent, there were also hints in her letters that he was suffering from physical problems. For example, writing to Bradford, who was spending the summer out west in 1925, she told him that "Father has given up eating meat or fish or eggs and seems cheerful and happy as a result." In typical Cornelia fashion, she wanted to spread the news to "all wives of despondent men to cut out the meat and try starvation." Though he kept to his meatless diet for some weeks, by the end of the year Cornelia wrote Wilma, then attending art school in New Mexico, that "poor Father has not been sleeping well for a week and is in the depths of despair about himself as a result." After that he appeared to improve, but subsequently developed a pyelitis that required him to be admitted to the Brigham Hospital for treatment in October 1926. He received a clean bill of health afterward, but had a difficult time getting back to normal.

Cannon himself was inclined to attribute his illness to overwork and lack of a proper vacation. By the end of the academic year, he looked forward to spending the summer quietly at Franklin, writing, resting and getting into fit shape. That seems to have done the trick, for there is no inkling that he suffered from ill health again during the remainder of the decade.[28]

Cornelia was always fond of travel, and though both she and Cannon treasured their summers in New Hampshire, she sometimes was able to pry him away. In 1922, for instance, worried that her husband was "being driven to death," she managed to convince him that they could combine a round of speaking engagements in the northwest with family visits and an opportunity to repeat their honeymoon tour of Glacier National Park. By the time they left in June, she wrote Ida that Walter was completely relaxed and as pleased as she was to have this holiday together. They sent Bradford off to summer camp, Wilma to Alexander James's art school in Dublin, New Hampshire, and the other children to Bernice's farm in Pondville, Massachusetts. Then they departed for St. Paul to attend the wedding of Cornelia's sister Linda to William Bennitt, and went on to spend their twenty-first wedding anniversary at Glacier National Park.

Although they had hoped to climb Mount Cannon (the high peak formerly known as Goat Mountain that had been renamed in their honor) and reclaim the note they had left in a bottle on the summit more than two decades earlier, the fates were against them. They found the hiking harder the second time around. The views were as heavenly as they remembered, but they suffered the debilitating effects of high altitude—the result of the softening of muscles that accompanied their advancing years. Coming down from their hike, Cornelia wrote in a family letter on July 6: "We came into Spokane yesterday in our tramping clothes, Walter with a three-days growth of beard, carrying two loaded rucksacks. The doctor sent to meet us overlooked Walter and when we finally pressed ourselves upon him, laughed 'til he wept, saying he took him for an old prospector. But now that he is washed and in his legitimate clothes, he passes for a great man."

The trip was a rousing success, except that they were homesick and eager to see the children. When they got back to the farm that August, they found a change awaiting them. "Two more Poles have bought places near us," Cornelia reported. "We expect to be a Polish settlement finally. They make quiet hard-working neighbors so far, though we are told they will bring all their relatives in close juxtaposition, and develop tenement house grouping on the old farms. But what a chance for Americanization!"[29]

One summer away from his family in Franklin and his laboratory in Boston proved enough for Cannon. Even the lure of an International Physiological Congress in Edinburgh in 1923, and another in Stockholm in 1926, could not budge him from his routine. Cornelia, on the other hand, traveled—usually with one of her sisters—more frequently as the children grew older. By 1928, with Bradford at Harvard, Wilma at Radcliffe, and the vista of an empty nest looming ahead, she arranged a family vacation out west.

A Buick touring car (purchased with some of the earnings from *Red Rust*) took all seven of them to New Mexico, the Grand Canyon, Yellowstone, and other natural wonders, and finally to Glacier Park in Montana where, in their eyes, Cannon Mountain took pride of place. Whenever they had to take substandard secondary roads, Bradford's superiority as a chauffeur and his ingenuity in times of emergency combined to save the day. Near the end of the trip they put Linda on a train for the first leg of her journey to Constantinople, where her mother had arranged for her to spend a year at a school run by a former Radcliffe classmate. For Cannon, a major focus of the trip west was to show the children various sites of their parents' history, and he took them to visit places from his youth around St. Paul, Elba, and even Prairie du Chien.[30] For Cornelia, the trip was just a taste of things to come. Travel would become a regular feature in the lives of the Cannons.

Bradford was not exempt from the wanderlust that appeared to infect his mother and sisters. In the winter of 1929, as a senior at Harvard College, he cast about for something to do after graduation. Linda was finishing her year in Constantinople, his father had been offered an exchange professorship to France for 1930, and everyone in the family except him seemed to be making plans for going places. His first thought was to offer himself as a teacher of chemistry to Robert College in Constantinople. But when he found that nothing short of a three-year contract would be required, he gave up his travel plans and looked around for something else.

Without anything especially appealing in sight, he next decided to amble over to the Medical School and look into the possibility of enrolling as a student there. A pleasant talk with Worth Hale, the assistant dean in charge of admissions, brought a resolution. Instead of traveling, Bradford was to begin medical school in the fall as a member of the Class of 1933. His father and mother, who had not been privy to Bradford's plans, were pleased with his choice.[31] After commencement, his father welcomed him as one of the initiated into his laboratory—a place in which Bradford had scarcely set foot up to that time. His mother was ecstatic. "Just think of such a chance for that boy," she wrote in a letter to St. Paul, "to work with one of the most inspiring research men in the country. He has his chance now, if he will only improve it." His father was more restrained, but hopeful. "It is a great pleasure to feel that Bradford and I are coming closer to each other," he wrote Cornelia from Boston early in July. "I really think the prospects of the medical course attract him and that he will rejoice with next year's experience."

During the summer months Bradford worked on two research projects; for one he did a study of blood pressure in dogs, for the other he made implanted electrodes for wrapping around nerves. Both eventuated in publication—his first and second published papers. As the weeks flew by, Cannon spoke proudly and happily about the young man's work. Cornelia gathered from his accounts that "Bradford's own attitude changes as he works, not with 'Father' but with a distinguished physiologist—and respect for the scientist takes the place of suspicion of the paternal autocrat."[32]

Cornelia's words were more than wishful thinking; the strained relations between the two were indeed a thing of the past. Cannon's struggles with the adolescent Bradford gave way to a mutual respect and bonding between father and son that would become more highly prized with each passing year.

From left to right: Captains John Fraser and A. N. Hooper, Royal Army Medical Corps, and Captain Walter B. Cannon, U.S. Army Medical Service, at Casualty Clearing Station No. 33, Béthune, October 1917

Cannon and his collaborators, early 1920s: W. B. Cannon in center; clockwise from left: P. E. Smith, J. E. Uridil, F. R. Griffith, D. Rapport, S. W. Bliss, M. A. McIver, R. rasco-Formiguera.

Top: Cat prepared for fight or flight, showing effects of stimulation of sympathetic nervous system; drawing from Charles Darwin, *The Expression of the Emotions in Man and Animals* (1872).
Bottom: Bronze mask by Prof. R. Tait Mackenzie, showing facial expression of violent effort; illustration from A. E. Shipley, "Hate," *The Cornhill Magazine*, September 1915.

Cannon and his collaborators, late 1920s: W. B. Cannon in center; clockwise from top J. T. Lewis; S. W. Britton, H. F. Newton, R. M. Moore, V. Menkin, J. J. Izquierdo, Linton.

International Congress of Physiology, Boston, 1929: Top, I. P. Pavlov and W. B. Cannon; Bottom, from left to right: E. D. Adrian, C. S. Sherrington, Alexander Forbes.

I. P. Pavlov with the Cannon family in Franklin, New Hampshire, summer of 1929: Linda, Marian and Mrs. Cannon are at the left; Grandfather James and Dr. Cannon are on either side of Pavlov; Helen and Wilma are seated in front, and Bradford is at the far right.

Some workers in the physiology department, 1920s/1930s: Top, Alfred Redfield, Cecil Drinker, Katherine Drinker; Center, Alexander Forbes, Hallowell Davis, Philip Bard; Bottom: Elizabeth Bright, Bradford Cannon, Magnus Gregersen.

Cannon and his collaborators, 1930s: W. B. Cannon in center; clockwise from top left: Z. M. Bacq, A. Rosenblueth, K. Lissak, H. B. Friedgood, H. Haimovici.

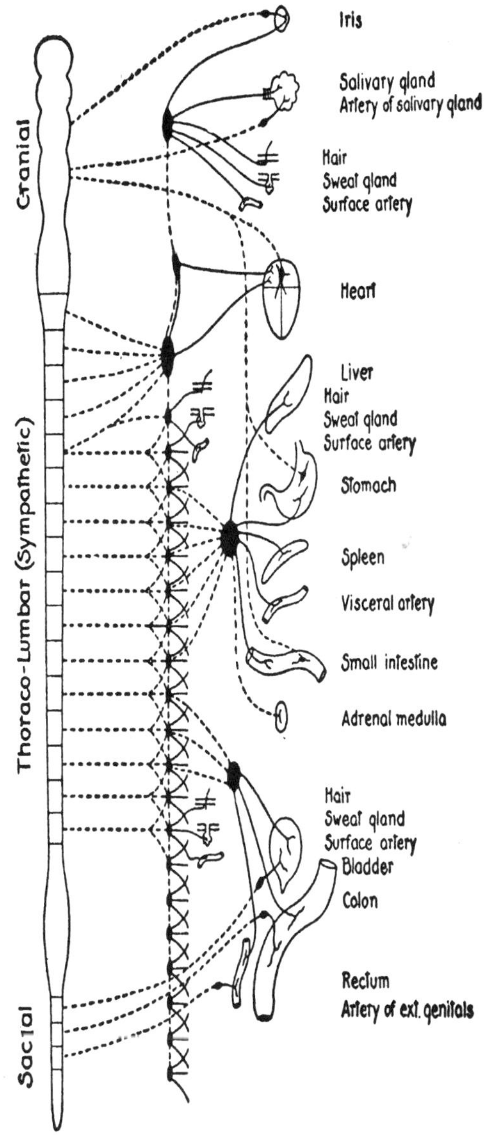

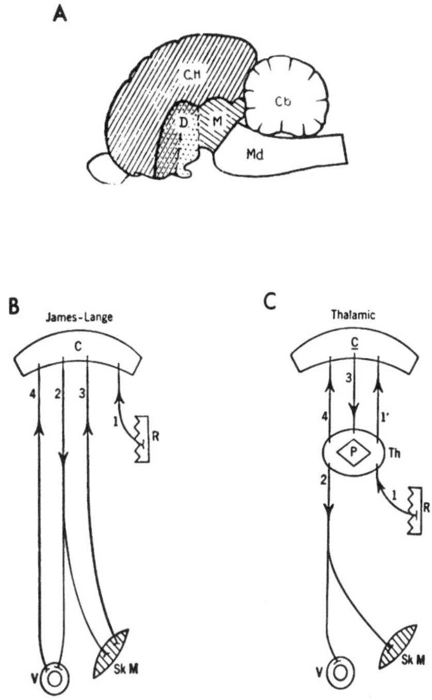

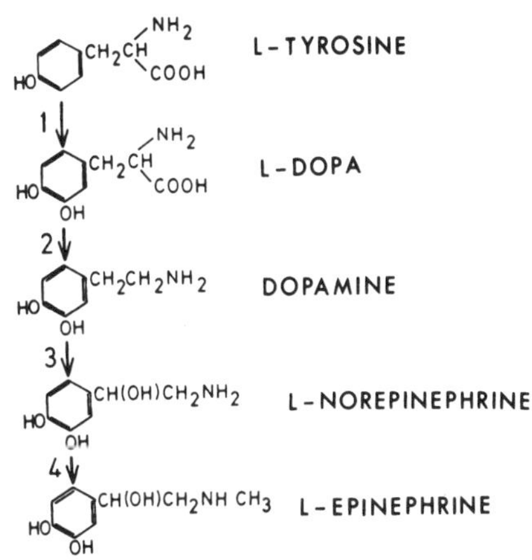

Left: diagram of the general arrangement of the autonomic nervous system [from *The Wisdom of The Body* (1932)]; Right: diagrams of the median section of the brain and of the James-Lange and Thalamic theories of the emotions; synthesis of norepinephrine and epinephrine [both from *The Life and Contributions of Walter Bradford Cannon* (1975)].

Grandfather and grandson: Top, W. W. Keen; Bottom, N. E. Freeman.

12 *The Physiology Department: An Extended Family*

WHEN, IN 1926, Cannon spoke to a group at the Medical School about the history of his department, he devoted a large part of his talk to the era of Henry Pickering Bowditch and the programs of instruction and research in physiology that were developed under his guidance. Dr. Bowditch, much to his credit, had created an environment which served as a magnet to attract a diverse group of talented investigators, Cannon told his auditors. For all his success in building a flourishing laboratory of physiology, by the end of his career a sadness came over him when he saw that new methods were replacing the old ones he knew so well. What remained, however, was a rich inheritance of traditions and influences that had come down to Cannon as Bowditch's successor.

Those traditions and influences came from near and far. "I think of myself," Cannon reflected, "as being one of Dr. Bowditch's sons, and a grandson of Ludwig on one side, and of Jeffries Wyman on the other, with perhaps a great uncle in Claude Bernard." Dr. Forbes and others, he added, had brought in the English tradition through study abroad with Sherrington and Foster. Now new generations of workers in physiology were passing through, and would continue to pass through the Harvard department, thus perpetuating the image of family. It was a large and ever-growing family, and at its head stood Cannon as paterfamilias.[1]

Another feature that had been relayed down through the years to the laboratories under Cannon's direction was the impressive range of studies underway. "I am sure that if Bowditch were alive today and could see how greatly that variety has been extended,"

Cannon declared, "—how the two attic rooms have developed into a laboratory of Comparative Physiology, a laboratory of Physical Chemistry applied to physiological problems, and a laboratory of Physiology in relation to Public Health—he would be deeply gratified." Indeed, as Alfred Redfield had reported a short time before Cannon gave his address, the laboratories of physiology at the Harvard Medical School, which had recently "doubled in number," comprised "an institute of physiology superior in its scope to any individual department." The four laboratories now gathered under one roof, wrote Redfield, "brings together a larger group of physiologists representing a greater variety of special interests within this field than is to be found elsewhere in this country."[2]

From the time that Cannon had been named head of the physiology department in 1906 until he went off to war in the spring of 1917, his staff was made up of a few young instructors like E.G. Martin and Alexander Forbes, and a meager assortment of teaching fellows, hardly more than half a dozen in number at any given time. Of this group, only Forbes remained in the department, where he was joined by Percy Stiles and Cecil Drinker.

When Cannon and Forbes went into military service, a depleted physiology department was left in the hands of capable but inexperienced assistants. While Stiles had charge of teaching physiology to dental students, Drinker took over the teaching of medical students, assisted by another instructor, Alfred Redfield, and a teaching fellow. They performed their duties so well that Drinker was promoted to assistant professor and Redfield was regarded as a valuable man to hold onto.[3]

In the fall of 1919, with Cannon fully returned to running the department, there were changes in the status of his assistants. Cecil Drinker was appointed associate professor of applied physiology and head of a laboratory engaged in research on problems in industrial hygiene and public health. In 1923, he became professor of physiology in the newly-established Harvard School of Public Health, a post he was to hold for the remainder of his career. When Alfred Redfield decided to pursue further training in Canada and England, Cannon gained the services of three new instructors: Joseph Aub, Wallace Fenn and Stanley Cobb. Aub and Fenn worked at times with Drinker in his rapidly growing subdepartment, while Cobb worked in part with Forbes and in part with the department of neurology. All would go on to distinguished careers, Cobb in neurology and psychiatry and Aub in clinical medicine at Harvard, and Fenn in physiology at the University of Rochester.[4]

In 1922–23, Fenn left the department to pursue further training in England and was succeeded as instructor by Fred Griffith, who had collaborated with Cannon and Uridil on studies of hepatic function. Shortly thereafter, Aub moved over to applied physiology to work on problems of lead poisoning, and he was replaced by Redfield who had returned to Boston and took charge of the laboratories for Cannon. Before long, however, Redfield was pulled in two directions, for he was asked to undertake, in addition to his other duties, a new educational experiment by serving as tutor in physiology. Although Cannon's name is frequently linked with the tutorial program that was inaugurated at Harvard, in fact it put him on the horns of a dilemma. Writing to Fenn about Redfield's new assignment, he confessed, "Naturally, it is not yet clear that the experiment will succeed. At present, however, and permanently hereafter, in case the experiment does succeed, Dr. Redfield will be engaged in that sort of special teaching. The consequence is that I am left without an Assistant Professor."[5]

In considering the matter of reliable help, Cannon had not given up his wartime vision of an educational exchange of young scholars between England and America. To be sure, several members of his department did go to England for further training, but he had a difficult time getting the traffic to flow from the other direction. In the spring of 1923, he wrote to his friend Joseph Barcroft about the possibility of having F. W. Roughton come to Harvard from Trinity College, Cambridge, for an instructorship in physiology. Hoping to attract the young Englishman, Cannon described a custom he had brought back from his war experience in London:

> Every Tuesday afternoon at 4:30 the group in the building, including Henderson, Cohn and their collaborators, Drinker and those associated with him, as well as Forbes, Redfield and myself, with others in the department, have tea in the library, and some one of the group tells about recent work in which he has been interested. This, of course, is not a required exercise, but a source of interest and stimulation. I think that any one coming here would find that the investigations which are going forward are so diverse and yet so centered on Physiology that he would not find the experience uninstructive."[6]

Much to his disappointment, Roughton decided not to come, and Cannon was left to look for another assistant.

In an effort to obtain the help he badly needed, Cannon called on Hallowell Davis, a former student, to return from postdoctoral work with E.D. Adrian in England in order to assist in running the physiology course. When later Davis, too, was drawn into joining Redfield in operating the tutorial program, Cannon again expressed feelings of frustration. While he had not interfered with their taking on these duties, he lamented to Dean Edsall in 1925, it meant that he was without the full-time services of both an assistant professor and an experienced instructor; consequently he was left to carry the burden of teaching physiology to the medical students himself. In spite of Cannon's fretting, in the years ahead a good part of the administrative burden of running the department would fall on the capable shoulders of Hal Davis. Davis would also become a favorite collaborator of Alexander Forbes, whom he later characterized as "the single individual most important in shaping my career as a neurophysiologist."[7]

Although Forbes did no teaching, his value to the department was unmistakable. A pioneer in applying electrical recording to the central nervous system, he made important improvements in physiological recording techniques. In 1914–15, as an early user of the string galvanometer, he had made the first electrical recordings of central reflex phenomena. In 1919–20, he developed the first vacuum-tube amplifier for physiological research and used it to record action currents in nerves. A few years later, Forbes and his collaborators utilized the amplifier to make the first recordings of sensory impulses from muscle stretch receptors.

Not only did Cannon testify as to the great importance of Forbes's contributions to knowledge of the physiology of reflex action, he told Dean Edsall that there wasn't anyone anywhere "who has equalled his technique or who has a more thorough acquaintance with the field in which he has been working." After Forbes's 1922 review article on "The Interpretation of Spinal Reflexes in Terms of Present Knowledge of Nerve Conduction" received widespread recognition, he was promoted from instructor to associate professor in one fell swoop. Not that the salary increase he received mattered. As a man of independent wealth, Forbes frequently helped to finance projects in the department or to provide the salaries of some of the laboratory assistants, secretaries and technicians who might otherwise have been let go in times of budgetary crisis.

The team of Alex Forbes and Hal Davis, both highly expert investigators concerned with analytical methods in their ongoing studies of nerve impulses, beautifully complemented Cannon's focus on synthetic, or integrative, physiology in studying the responses of the total organism. "The variety of our endeavors in solving physiological problems," wrote Cannon retrospectively, "was an additional advantage to our young collaborators because it gave them direct demonstration of the virtues of the two research methods."[8]

In the latter part of the 1920s, Cannon's permanent staff was relatively stable, with Forbes, Stiles, Redfield and Davis at its core. After the Rockefeller Foundation gave a generous grant to support the School of Public Health, Cecil Drinker moved his laboratory of applied physiology out of Building C, thus making way for plans that had begun to take shape for a new laboratory of electrophysiology run by Forbes and Davis. (Besides, Edsall divulged to Cannon, there was another reason for Drinker's move into the School of Public Health: people coming there tended to get a bad impression when they saw how little was going on in the building that bore the name of the School.)[9] Philip Bard, while earning his doctorate in medical sciences, served as an instructor in physiology for two years from 1926 to 1928, and was followed by Harold Blum and Magnus Gregersen. Last, but not least, at the end of the decade Roy Hoskins, Cannon's very first graduate student, returned to the physiology department as research associate to run the newly-established Memorial Foundation for Neuro-Endocrine Research.[10]

Before World War I, Cannon had drawn his laboratory assistants, for the most part, from the ranks of Harvard medical and doctoral students who were attracted to physiological research. After the war, Cannon's teaching and research assistants were of a different stripe. Some, drawn from all over the country, came as newly-minted physicians interested in physiological problems or as postdoctoral physiologists eager for more experience in research. Or again, some came from foreign countries as visiting fellows supported by grants from the Rockefeller Foundation and other special agencies both here and abroad.

During the years from 1919 to 1929, no less than twenty-five young investigators collaborated with or worked under the direction of Cannon—and perhaps there were even a few more. Some were physicians and surgeons like David Rapport, Monroe McIver and Harlan Newton, who were graduates of Harvard Medical School, or like Joseph Uridil, Sidney Britton and Robert Moore, who were graduates of other North American medical schools. Some were medical scientists trained at Harvard and elsewhere like Philip Smith, Fred Griffith, Sidney Bliss and Philip Bard. Some were National Research Council (N.R.C.) fellows who had either medical or doctoral degrees, like Homer Smith, Raymond Zwemer and Simon Dworkin. A few were Harvard medical students whom Cannon took under his wing, like James and Robert Linton and Valy Menkin.

The largest single group, however, were the research fellows who with foundation support came from abroad to seek postgraduate training. During the 1920s, Rockefeller fellows came to Cannon's laboratory from Spain, Brazil, Argentina, Mexico, the Philippines, and Austria. Arie Querido and Arnold Groeneveld came from Holland supported by funding from both the Rockefeller and Netherlands-American Foundations.

Cannon welcomed these aspiring young men into his laboratory and often into his home. He enjoyed their associations, believing that he gained as much from those he taught and supervised as he gave. There is no denying that he came to view them as an extended family and cared deeply about what happened to them. Many became good friends with whom he enjoyed contacts at meetings, exchanged visits and letters, and kept in touch for years afterwards.

In return for the time and effort he put into guiding his fellows, Cannon expected a high level of performance from them; he could even be obdurate if he felt their behavior breached his sense of propriety. When Harry Veach, one of his teaching fellows, approached Cannon to say that he would require additional funding, he was sternly lectured that "no student has any claim whatsoever on any fellowship" and offered little hope of further aid. But when Veach went on to receive the degree of Doctor of Medical Sciences, Cannon mellowed and praised him for sailing through one of the best examinations by any candidate he had ever known.

Though Veach remained in the department for several years, he never collaborated with Cannon, who later remembered him as an odd, gloomy fellow always seeming to seek shelter under a black umbrella. Perhaps it was Veach whom Cannon had in mind when he set forth one of his personnel policies: "Personal qualities, apart from real difficulties of cooperation, seem to me of minor importance in the presence of outstanding ability. We are constantly overlooking defects of personality in men whose skill and insight and devotion to their chosen work make these peculiarities of personality seem minor affairs."[11]

❧ ❧ ❧

Cannon's work day customarily began with a lecture at 9 a.m. While he was teaching, one of his assistants would set up equipment and prepare animals for whatever experiments he had planned. After he left the lecture hall, Cannon would then spend the rest of the day in the laboratory except for going to his office, whenever opportunity allowed, to dictate letters to his secretary. He seldom missed getting home for dinner, and despite an active social life, spent many an evening in his study at the top of the house, reading, writing, and going over the day's work. According to his colleagues, he kept careful notes of what he was doing and of ideas that occurred to him.[12]

Relations within the laboratory were relatively informal. Part of Cannon's success in the department was that he was an enjoyable person to be with; his enthusiasm seemed to infect those who worked with him. In 1922, soon after leaving the department, Wallace Fenn acknowledged his debt to Cannon:

> I well remember talking with you one evening at your house about the Physiology Department and the glowing account which you gave me of the working conditions there. Every word has turned out to be true. I could not imagine a happier three years in more congenial or instructive surroundings and I shall always be proud of that connection and proud to have been associated with you.[13]

When more than a score of years later he came to write *The Way of an Investigator,* Cannon would draw on his own experiences in describing the role of the head of a laboratory as mentor to his students and collaborators. The activities he believed to be important, among others, included maintaining high standards, supervising publications, suggesting new directions or setting limits, being "generous in the care and treatment of those with whom he is working," and aiding in providing sources of funding and support when they went out to their own laboratories.

Though Cannon did not drive his students as hard as he drove himself, he expected them to be interested in the work and he insisted on quality. He justifiably took pride in his ability to write accurate and readable prose, and he insisted that his students do the same. "It is essential," he later stated, "that a scientific observer should be able to write a clear and definite report of his aims, his methods, his results, and his conclusions. Flowering language and dramatization are out of place in scientific exposition. Likewise the art of persuasion—i.e., special pleading—must be excluded, for the facts presented should be convincing without an appeal to feelings. The prime requirements are clarity and brevity." But then he added that it was tempting to strive for the occasional imaginative phrase—if only one could find the words that "nicely fit the thought" and were also "piquantly picturesque."[14]

Cannon put a great deal of care into writing letters of recommendation. For example, two of the Ph.D. candidates in the Medical Sciences program were women getting their degrees through Radcliffe College. In the spring of 1923, he lent support to a request for scholarship aid in behalf of Theobald Smith's daughter Dorothea, pointing to her very high record and work on the poisonous effects of lead, in which she had shown "keen insight, devotion and technical skill." A few months later he recommended C.S. Minot's niece Anne for a post teaching physiology at the Woman's Medical College of Pennsylvania. "She had one of the highest records ever attained in the course in physiology given to our medical students," he wrote. "If it were possible to appoint women to positions in the Medical School here I should not hesitate to ask her to become a member of the Department of Physiology." In following years, as old restrictions slowly (very slowly) eased, Cannon would welcome into his laboratory still other talented young women who were accepted for work in physiology as full-fledged fellows and collaborators.[15]

Even in the prosperous times of the twenties, it was not always easy to help certain students obtain jobs and funding—especially if they were Jewish. Cannon asked several people to be on the lookout for a position for one of his collaborators, David Rapport, who worried that he might have to go into practice if he could not soon find a place in academic medicine. Another case in point was

Valy Menkin, who worked his way through medical school as a student assistant in physiology. Late in the decade, Peyton Rous asked Cannon for information about Menkin, who had inquired about the possibility of working in Rous's laboratory at the Rockefeller Institute. Cannon, in reply, characterized the young man as hard working, intelligent and alert, but felt compelled to add, "He is a Hebrew without the pronounced features of his race and is inclined to be timid rather than aggressive. I am still somewhat doubtful about his originality. He greatly desires to go into laboratory work."

Rous was clearly troubled. He responded that it would not be easy to make a place for Menkin because his laboratory was adequately filled. Not only that, "but to take him on with what this would mean later in the way of responsibility to place him,—married, impecunious, without great initiative and Jewish,—is a deal too much." Then, in a burst of sympathy, Rous lamented, "What is one to do nowadays with the young Jews that one might train up,—such nice and able fellows as they so often are?"[16]

In his annual report for 1923–24, the medical dean pointed with pride to the considerable number of young men at the School who were on fellowships financed by outside organizations. In the subject of physiology alone, he observed, there were several units that included on their staffs exceptional men who were capable of guiding visiting fellows in diverse and important lines of research. While it would seem that Edsall painted a rosy picture, he acknowledged to his friend Richard M. Pearce, head of the Division of Medical Education at the Rockefeller Foundation, that he had problems as well. There were indeed unexcelled opportunities in physiology that presented themselves to the fellows who came to Harvard for training; but if the men were sent in the belief that they were to work with Cannon, they might be disappointed. Since there was a tendency for Cannon to become overburdened with the various students coming from various sources, Edsall suggested to Pearce that the fellows might work sometimes with Cannon and sometimes with other men, like Lawrence Henderson, or Cecil Drinker, or Alexander Forbes, and thus get particularly advantageous results.[17]

A consequence of all this activity was that it put a premium not only on Cannon's time but also on the space that was available to his department. Less than twenty-five years before, when Dr. Bowditch presented the plans of the Longwood Medical School buildings to President Eliot, the president could scarcely believe their projected size and expressed the opinion that the medical faculty would "rattle around in this great plant of ours for years to come." Now, with all the growth and expansion that Edsall was actively encouraging, there was a great competition for available space, and none more than in Building C where the physiology department was housed.

Cannon's concern about overcrowding was expressed as early as 1922, when he wrote John Warren of the Space Committee that a working environment for the department of physical chemistry, established the previous year, needed to be set up. Office and laboratory space was required not only for L.J. Henderson, who headed the department, but for his assistants Edwin Cohn and Ronald Ferry as well. After proposing several arrangements, chief of which related to the rooms on the top floor of Building C assigned to W.T. Porter for his department of comparative physiology, Cannon concluded, "As I see the situation, it is a matter of securing temporary convenience for Dr. Henderson until Dr. Porter surrenders his space, and gives us all release from a considerable amount of pressure in this wing."

In March of 1925, Cannon again voiced his concern about space problems, this time to Edsall: "Already the resort here by Fellows of the Rockefeller Foundation and National Research Fellows has crowded our quarters. Of course, if Dr. Porter should relinquish the rooms which he is holding up stairs our difficulties would vanish—at least for a time—but just now I feel that we must be careful about our hospitality because of the crowding."[18]

In truth, Porter's retirement date, which was set to take effect when he reached the age of sixty-six, was anticipated by Edsall and President Lowell long before it actually took place. As early as 1920, after he was left out of the round of salary raises because it was charged that he had not been productive, Porter was forced to round up testimonial letters, both from members of the medical faculty and from outsiders, to substantiate his worth to the president and the dean. Although this maneuver did manage to gain him a 20% increase, it did not strengthen his position with the administration of the Medical School. In the many instances where

Edsall called attention to the great "School of Physiology" being built at Harvard, Porter's name was invariably left out.

At last, in the spring of 1928 the time came for Porter's retirement, and the Space Committee received an official request to turn over to the department of physical chemistry the rooms he had occupied for more than twenty years—but with the understanding that the space be used for purposes of the whole building and perhaps even the whole School. When Porter vacated the premises that summer, his quarters on the top floor of Building C were assigned to Edwin Cohn.[19]

Throughout the decade, what with departments at the Medical School growing by leaps and bounds, Dr. Edsall made it a priority to seek outside financial support for their expanded activities. In the spring of 1927, soon after returning from a trip to China undertaken for the Rockefeller Foundation, he was invited to become a member of the governing boards of the Rockefeller enterprise. Before doing so, he decided to make one last effort to obtain further funding for the Medical School from them.[20]

To begin with, Edsall sent Richard Pearce a preliminary statement of his over-all needs in coming years. He asked Pearce, in a letter marked personal, to keep in mind that a large proportion of the departments at the Medical School were working on low budgets compared with those in other institutions that had obtained aid from the General Education Board (G.E.B.) and the Foundation. Although the Harvard departments had accomplished a great deal, they were underfunded and running "on a rather narrow ledge."

His reason for writing privately, he told Pearce, was based on his previous experience (and the more recent experience of President Lowell and Dr. Henderson) with Abraham Flexner. Flexner, if reasonably possible, "would block giving any considerable sum to this institution as he has repeatedly before blocked me in my requests on grounds that he stated, and then has been responsible for giving to other institutions that he is especially interested in when precisely the same grounds for declining were present in even a more marked form than here."

Edsall complained that a year and a half before, Flexner had encouraged him to file an application for financial support and actually led him to believe that all would be smooth sailing. Although he had followed instructions to the letter, after months of inaction Flexner said it would be impossible to bring his request before the Board until after Edsall came back from China, "and that he doubted what their feeling would be about the matter." To add insult to injury, Flexner claimed the dean "was quite mistaken in thinking he had ever given me any definite encouragement, that he only wished to express a friendly interest in the matter."

Despite his complaint, almost immediately Edsall made formal application for aid in "A Memorandum of Details of Needs and Costs for Harvard Medical School" dated April 26, 1927, and sent it off to both Pearce and Flexner. It is a lengthy document, only a small part of which is devoted to physiology. Edsall wrote that the department very badly needed more money for technical services and supplies, and perhaps some small stipends for further research. Cannon had put the prospective cost of doing what he had in mind at $2500; "he always asks for the very smallest amount possible," Edsall commented, but he was quite sure that less than $5,000 would not really serve the purpose. He then asked for a like amount for the department of physical chemistry, citing the importance of Arlie Bock's work with Henderson as well as Cohn's work with George Minot on pernicious anemia.

In his cover letter to Pearce, Edsall expressed the hope that he had not sounded prejudiced concerning the treatment Harvard had previously received from the Rockefeller Boards, as compared to Hopkins. Although he realized it was a principle for the Boards to give money to more needy institutions, it was also a fact that those Boards together had given Hopkins as much as or more than the total endowment of the Harvard schools of medicine and public health combined. Edsall was frank to admit that he was "left a little cold" as to an explanation for this treatment:

> If Hopkins had been for the last ten or fifteen years a superior place to this, as it was in the years before that, I should feel that it was quite justified. When, however, it is true, as it is undeniably true, that for the past ten or fifteen years this place has been recognized very widely indeed as being in most ways and in general effect decidedly superior in its contributions to medical progress, in the number of men that it has trained for the profession and for advanced positions in academic medicine, that seems to me still further to emphasize the partisanship in the matter.

Edsall told Pearce that he had talked with Flexner (at Flexner's behest) about an idea for developing certain definite subjects so that they might produce more highly trained men for professorial jobs. Flexner especially mentioned physiology, "which proposition he said came originally from you."[21]

Flexner acknowledged Edsall's memorandum almost immediately. He wrote that he would need to confer with Drs. Pearce and Rose and examine the proposal in regard to the sum still remaining unexpended from Mr. Rockefeller's gift, for that gift was rapidly vanishing. He cautiously stated, however, that he might be able to get action on financial aid for physiology, biochemistry and physical chemistry. "These three departments require, if I understand your memorandum, additional income to the extent of $15,000 per year.

Now Flexner's letter infuriated Edsall, and he shot off another personal letter to Pearce to complain that Flexner was fencing so as to avoid giving the Medical School anything except a trivial sum of money. Edsall vowed that if the officers of the General Education Board were unable to support his general plan, he would request an entire withdrawal of the proposition because he did not feel there was any advantage in taking up minor details of it. As he put it in a nutshell, "I am quite unwilling to accept a little sop and let Mr. Flexner or the General Education Board feel that they have absorbed thereby any conscientious feelings they have that they ought to aid us." When the Board subsequently met and discussed Edsall's proposal, its members decided to postpone taking any action.[22]

In the fall of 1927, Cannon, desiring to ascertain the status of support for the Harvard laboratories, asked Pearce about it. Pearce reminded Cannon that the plan to support selected laboratory departments involved not the Rockefeller Foundation but the General Education Board. When Flexner had approached Edsall about aid for the three departments at Harvard, nothing came of it because at the time, as Pearce understood it, "your Dean was interested in a larger program and disinclined to consider minor projects. May I suggest that you ask Dr. Edsall for the details."

After receiving Pearce's reply, Cannon reported his disappointment to Edsall: "I fear that there is nothing that can be done now to secure the money for this year. I should hate to think, however, that the Department of Physiology at the University of Chicago is receiving a subvention of considerable amount whereas we, who have taken care of Rockefeller fellows for years and are now doing so, are being neglected."

The upshot to all of this was that Edsall wrote Flexner a confidential letter, in which his tone was greatly moderated. Referring to earlier negotiations between them, he asked for help from the G.E.B. to develop surgery at the Boston City Hospital, and then, as if by the way, digressed momentarily. He would also be grateful if the Board could see its way clear to provide aid for physiology and physical chemistry:

> We can, as I said, provide what I think is necessary for Biochemistry. The total amount necessary for Dr. Cannon and Dr. Henderson would be about $10,000. There is a little research money unemployed that might cover about one-third of this. I might possibly be able to stretch that somewhat further, but not much. It would be a pleasure, and I think a recognition, of the very important work those men have done if that could be accomplished.

Edsall's letter must have hit the right note. At the next General Education Board meeting, it was voted that an appropriation of $22,500 be given to Harvard Medical School over a period of three years for developing the department of physiology.[23]

The following year, Edsall became chairman of a committee appointed to prepare a report that could be used as a basis for the restructuring of the Rockefeller Boards. In October 1928, Cannon responded to a request from Edsall, who was then gathering material for his report, by sending him three remarkable letters. The first was a statement of research activity in the department of physiology over the past twenty years. Cannon began by noting that the department had continued the traditional diversity of interests that characterized its early days under Dr. Bowditch's guidance. Such variety, he said, could be found in the work of five of his associates: the valuable researches on the physiology of nervous conduction and of the central nervous system by Dr. Forbes and Dr. Davis; the important papers on the metabolism of surgical shock and on the role of adrenal secretion in control of body temperature by Dr. Aub; the interesting investigations on the effect of radiation on cells and an elaborate study of the comparative physiology of the blood by Dr. Redfield; and the thorough examination of the activities of bone marrow as a blood-forming organ by Dr. Drinker.

Cannon saved his own achievements until last. Laying down a progression that began with his x-ray investigation of the motions of the alimentary tract, he pointed to his studies of the bodily effects of strong emotions, and of the functions of the sympathetic nervous system and the adrenal medulla, as well as to his wartime work on wound shock. His most recent researches were concerned with the conception that the sympathetic system acts in crises to preserve the stability of the organism—"the so-called *emergency theory* of the sympathico-adrenal system."[24]

A few days later, Cannon sent Edsall two further statements comprising a catalogue of men who had received training in physiology under his direction. One was a list of those who received their first doctorate or had important experience in research and teaching in the department before going on to academic positions. The other was a list of students or recent graduates of the Medical School who in most instances had their first research experience in the department and who since had gone into productive scholarship in medicine. Although by no means complete, Cannon's catalogue contained a total of forty names—a number that Edsall found quite astonishing.

With three deft strokes of his pen, as it were, Cannon delineated the teaching and research efforts in physiology that had flowered for more than a score of years under his direction. At the same time, he reinforced the image he had presented a few years before when he talked about maintaining Dr. Bowditch's ideal—that is, a large and growing family of laboratory workers cooperating to work on a varied complex of physiological interests.[25]

On March 11, 1929, Richard Pearce had dinner with David Edsall in Boston and they discussed future aid for physiology. Pearce suggested that the proper course for Edsall would be to make application for renewal nine months before the end of the three-year grant period to allow the Rockefeller Boards to take up the matter on its merits. Almost immediately Edsall began gathering the information he would need for ammunition in his renewal application.[26]

In July, Cannon sent Edsall an update of the myriad projects in his department, noting that he would have opportunities to promote the development of new ideas and new methods at two events: the upcoming Physiological Congress in Boston and his approaching exchange professorship to France. In August, Edsall drafted for President Lowell a tentative memorandum to be used as a basis for the renewal application in support of the work of Henderson and Elton Mayo at the Fatigue Laboratory, and of Cannon and Cohn at the Medical School. It would be difficult, in his opinion, to find another such group that had already done such brilliant things and that bid fair to do further valuable things. Once Lowell bestowed his approval, Edsall planned to give his memo to a committee consisting of Henderson, Dean Wallace Donham of the Harvard Business School, and himself. He would next get Pearce and Edmund Day of the Rockefeller Foundation to study it and come to Boston for a site visit no later than early October, so that any amendments could be made before the application went to the Board in November.[27]

But matters did not move quite as fast as Edsall expected. Pearce, seeking additional information about financial details, arranged to have a personal talk with Cannon at the Harvard Club in New York early in December. Together, the two friends projected Cannon's needs as head of the physiology department at Harvard until his retirement, and discussed not only a program for utilizing him to attract and train men but also ways of conserving his best efforts as the principal leader of physiology in the U.S.A. Only a slight modification of Edsall's proposal was necessary to meet his needs, Cannon told Pearce; what would help him most was relief from administration and routine. Moreover, he expressed enthusiasm for a proposal that would give a year for research free of instruction to hand-picked students from the medical course. Indeed, he thought such an arrangement could probably be more effective than graduate fellowships given under the National Research Council.

Soon after their meeting, Cannon, in response to Pearce's request, drafted an outline of his general program for the next ten years. He noted that he had been in excellent health and expected to continue in vigorous prosecution of his research, and he maintained that the more collaborators he had, the more useful he could be. Pearce's reply arrived on the eve of Cannon's departure for Europe. Cannon had given him exactly what he wanted and he hoped the matter would be settled well in advance of his return to Boston.[28]

Early in 1930, Pearce prepared a memorandum for an office conference at the Rockefeller headquarters with the title, "Aid for a Leader: Support of Physiology, including Physical Chemistry, for Dr. W.B. Cannon of the Harvard Medical School." Recommending a continuation of the three-year G.E.B. grant, he called Harvard "probably the outstanding center of physiology in the world today," and then noted: "It is desirable to take care of the remainder of Dr. Cannon's active life and therefore the period proposed is *ten* years." In elaborating on Cannon's strengths, it is of interest that Pearce stressed the fact that he had recently developed an organism wholly unknown in the history of the universe—an animal without a sympathetic nervous system.

After the officers of the Foundation agreed to present the proposal to the Trustees at their April meeting, Pearce wrote Cannon (c/o Dr. Alan Gregg in Paris) that he had every reason to believe it would be received favorably. Before the meeting could take place, however, Pearce died. Edsall, writing to Cannon again, broke the news about Pearce's sudden death. While he was deeply grieved and startled by his passing, he was not surprised, for he had feared that Pearce was working too hard. Edsall took pains to assure Cannon that he did not believe the loss of Pearce would interfere with the plans underway.[29]

The resolution of the Rockefeller Foundation, passed on April 16, 1930, awarded a sum of $175,000 for researches in physiology under W.B. Cannon and for physical chemistry for a period beginning in July 1930. If Dr. Cannon should for any reason cease to head the department of physiology at Harvard, the Trustees stated, the situation would then be reviewed by the officers of the Foundation.

Edsall took undisguised pleasure in sending Cannon news of the award. Officially informing him that there would be available for his use for the next seven years $15,000 per year, as well as $10,000 for Edwin Cohn, the dean exhibited a sense of personal gratification as well. He called the action "a very happy tribute" to Cannon, since it was the first time that the Foundation elected to give financial support to a leader in medicine. There had been instances before of support for particular investigations under distinguished men, to be sure. But "the proposal in this case was to make a definite example of the policy to support a man who had made himself a leader," Edsall told Cannon, "and the matter was presented to the Board on the docket with the statement that it would be difficult to find as good an example as yourself in the field of medical science in this country."[30]

Edsall had good reason to be pleased, for Cannon's award put the crowning touch on his efforts to obtain financial support from the Rockefeller interests. Despite the dean's complaints that Harvard had been left out of the running, under his leadership the medical campus became a major beneficiary of Rockefeller grants, with awards going to the School of Public Health, the Lying-In Hospital, new laboratories at the M.G.H., the departments of psychiatry and neurology, and the tutorial system, among others.

PART III

The 1930s

13 *Scientific Outreach at Home and Abroad*

AS CANNON'S LABORATORY attracted an ever-widening world of young medical scientists and physicians looking for the very best training they could find, so, too, did world leaders in physiology and medicine flock to the Medical School, which was rapidly taking its place in the international community. Some came to Boston for meetings and conferences, and some made Harvard a stopping place when visiting similar institutions along the eastern seaboard. Still others, men at the very pinnacle of their careers, came in response to specific invitations.

In 1923 a fund established by Mrs. Edward K. Dunham in memory of her husband (a member of the H.M.S. class of 1886) for an annual series of invited guest lectures added to the Medical School's prestige. Known as the Dunham Lectureship for the Promotion of the Medical Sciences, its purpose was "to bring to this country men who are distinctive contributors to medical science or lines of scientific work closely related to medicine." To administer the lectureship, a committee of six was appointed. Chaired by Cannon, it had Drs. Aub, Folin, Wolbach, Zinsser and Edsall as members.

When precarious health prevented Ramon y Cajal from accepting an invitation, Willem Einthoven of Leiden was selected to deliver the inaugural lectures in 1924–25. While in Boston that October, Einthoven heard that he had been awarded the Nobel Prize for his discovery of the electrocardiogram—thus adding an unforeseen lustre to what promised to be an exceptional lecture series. Dunham lecturers for the remainder of the twenties were Ross Harrison of Yale (1925–26), Richard Willstätter of Munich

(1926–27), C.S. Sherrington of Oxford (1927–28), Louis Lapicque of the Sorbonne (1928–29), and Joseph Barcroft of Cambridge (1929–30).[1]

Although the committee had a large number of distinguished candidates from which to choose, as well as a handsome honorarium of $2500 to offer prospective lecturers, its deliberations were not without problems. From the beginning, when its members learned that Cajal would not be able to initiate the series, they had difficulty obtaining their first choices. This often resulted in a scramble for a speaker that made a juggling act out of scheduling and availability—not to mention ability to speak English. The unfortunate pattern continued during the early years, with several first choices disappointing the committee for one reason or another.

For example, Heinrich Zangger, Professor of Legal Medicine at the University of Zurich, chosen as Dunham Lecturer for 1925–26, had to withdraw because of illness. In a frantic search for a last-minute substitute, the committee was thankful when Ross Harrison, who only had to come from as far away as New Haven, answered their call for help. Or again, after it was learned that Dr. Brumpt had to decline appointment for the 1926–27 lectures, an invitation went out to Karl Pearson of the Galton Laboratory at the University of London; but, he, too, sent his regrets, pleading previous commitments to postgraduate students. Sir Charles Sherrington, on the other hand, was lined up well in advance to coincide with the delivery of his Herter Lectures at Johns Hopkins, so his lectureship in 1927–28 went smoothly and gave the committee a respite from their labors. But the following year, when William Bate Hardy, a pioneer in colloidal chemistry at Cambridge University, declined because he had no new research that he believed would be appropriate, the "first-choice jinx" reverted to pattern.

Despite these snags, in a period of five short years the Dunham Lectureship not only served as a magnet to draw the world's scientific elite to the Medical School, it also had become a plum. The honor of receiving an invitation was such that when Dr. Hardy regretfully turned it down, he wrote Cannon that so distinguished a lectureship "is not for the tribe who 'read for Metaphysics under the letter M, and for Chinese under the letter C, and combine the information.'" Cannon waxed enthusiastic when he wrote Mrs. Dunham about the possibility of adding clinical investigators to the list of candidates and told her, "The lectureship has already proved to be an important means of bringing together the visiting investigators and their friends and associates here and offering a valuable opportunity for exchange of ideas."[2]

During the course of the first few years, as arrangements for the Dunham Lectureship settled into a routine, Cannon developed a form letter that he sent to prospective guest speakers. "It is expected that the lecturer will give two or three lectures, or perhaps four if he cares to do so," he explained, "and that he will spend two weeks here, becoming acquainted with the investigators and their investigations in friendly conversations and conferences." The series could be scheduled at any time during the academic year between October and May, he continued, usually at the discretion of the speaker. But after Boston was selected as the site of the 1929 International Congress of Physiology, Cannon seized the opportunity to secure two of the most eminent guests as Dunham Lecturers—one for late in 1928–29, the other for early in 1929–30. When family illness forced a last-minute cancellation by Professor Otto Warburg of Berlin, the committee considered itself fortunate to obtain the services of Dr. Lapicque, who, together with Barcroft, rounded out the decade.[3]

The thirteenth International Congress of Physiology convened in Boston on August 19–23, 1929. More than two years previously, William H. Howell of Johns Hopkins, in behalf of the national committee, had accepted an invitation from the Harvard Medical School to hold the Congress in Boston, and he proposed that Cannon be named chairman of the local committee. Cannon, in turn, chose Alfred Redfield and Edwin Cohn as his secretaries and put them in charge of making the necessary arrangements.[4]

The Congress attracted approximately 600 members from some forty foreign countries, together with 1,000 registered from the United States, for a total attendance nearly three times that of any previous one. It had been feared that the expense of the journey would prevent a large number of foreign members from coming, but a fund-raising effort in their behalf proved spectacularly successful. When the *S.S. Minnekahda* offered a special rate to the delegates, every berth was taken by physiologists coming to the Congress. They had such a good time and such good fellowship on their voyage that (it was reported) they declared all future congresses should be held on shipboard. On arrival, all of the visitors, together with their families, were housed in Harvard dormitories in Cambridge and Boston or became guests in private homes.[5]

The choice of Boston for the meetings was a signal honor. Not only was it the first time the Congress gathered outside of Europe, it also marked the 50th anniversary (more or less) of the beginnings of physiology as an independent science in America. "We look upon this meeting of the Congress at Boston as a recognition that American physiology has come of age," Dr. Howell greeted the delegates, "and has been admitted to full membership in that old world group to whom we have so long looked for guidance and inspiration." To underscore his message, each of the delegates received a commemorative bronze medal of William Beaumont, accompanied by a facsimile edition of Beaumont's pioneer study, *Experiments and Observations on the Gastric Juice and the Physiology of Digestion* (1833), and Sir William Osler's essay on Beaumont, "A Pioneer American Physiologist."

On August 19, Howell as president of the Congress gave the opening address at Sanders Theatre; that evening August Krogh of Copenhagen delivered an oration on the progress of physiology. During the next four days, sessions in six amphitheatres were held simultaneously, with scientific communications limited to ten minutes each because an overwhelming number of papers had been submitted. There were also demonstrations given in the laboratories and hospitals of the Medical School, as well as special events that included a concert under the stars on the Longwood quadrangle by the Boston Pops Orchestra, led by its first-year conductor, Arthur Fiedler.[6]

Cannon and his local committee, run ragged by the demands of preparation, were kept just as busy during the whirl of the Congress; but the success of the meetings compensated them well for their labors. Bradford Cannon, about to become a medical student, volunteered as an assistant to Edwin Cohn and manned the Congress Bureau located in the Administration Building at the Medical School. "Bradford is the keystone of the Congress office," his mother wrote her family. "Dr. Cohn told Walter that he was 'magnificent,' 'the best helper we have, knows where everything is and manages the entire place.' Walter swelled with pride."

A few days later, she had more news to impart. "Prof. Pavlov and his son arrived unexpectedly for breakfast and Bradford breezed down with his 'Guten Morgen' as if reared in the land of the Teutons. Prof. Pavlov is the most lovable man . . . We all sit at his feet." Pavlov at the age of 80, she explained, was entitled to a little peace and would be staying with the Cannons. Their other house guests were Henry Dale's daughter Alison and Rudolf Magnus' daughter Gretl.[7]

Ever since their meeting in 1923, Cannon and Pavlov had kept up a correspondence with one another. Some letters were warm exchanges of greetings on the occasion of holidays, others were brief comments on the progress of their ongoing research. At one point, Pavlov, concerned with the welfare of two of his students who had immigrated to North America, Boris Babkin and Vasily Boldyreff, brought Cannon's attention to their research in the hope that he might in future be of some assistance to them. Early in January 1929, Pavlov sent Cannon an inscribed copy of his recent Croonian Lecture. Cannon, upon receiving this latest expression of Pavlov's friendship, promptly invited him (and his son, Vladimir), to stay at his home when he attended the International Congress in Boston later that year.[8]

If the Congress symbolically stood as a celebration of American physiology, it was no less clear that Pavlov, in attendance as a Soviet representative, was its chief hero. Although he could communicate adequately in German and French, he knew no English and was dependent on one of his former students, G.V. Anrep, to act as his translator during the Congress. Anrep was more than equal to the task, especially on the two occasions when Pavlov spoke in public.

The first of these took place spontaneously. One afternoon, Cannon held the attention of a small audience with an exhibit of a series of animals from which the sympathetic nervous system had been completely removed—"a surgical feat which has shed important light upon the functions of this system and its relation to emotions," John Fulton later reported. After Cannon had finished demonstrating his experimental surgical skill, he invited an audience of between forty and fifty men into a room adjoining his laboratory to listen to Pavlov deliver some informal remarks. Harvey Cushing, who was present, later left a description of the Russian physiologist "serving up his latest ideas of inhibition in relation to neuroses, hot from the griddle":

> Vivid, alert, gesticulating, the old man poured out his Russian phrases, like a mitrailleuse never missing fire, directing his attention meanwhile chiefly to Anrep, who sat calmly alongside smoking innumerable cigarettes; Pavlov would suddenly stop and point menacingly at Anrep who possibly would ask him a question or two to make sure of his ground—indeed even interrupt him. Pavlov, moving his watch and chain along about six inches farther on the table in front of him, would slump down in his chair, shifting his ischial tuberosities to one side or the other—whether because the chair was hard or because this was one of his reflexes, I am not sure. Anrep would then begin, always composedly, and give a most brilliant and concise presentation in English of what had gone before, Pavlov then picking up the thread again and continuing. This went on for an hour and, except for the intrusion of a few belated guests who crowded into the room, one could have heard a pin drop.

When it was over, Cushing was dismayed to find that no one had recorded Pavlov's lecture, or even used a camera to take pictures of him in action. When he asked Anrep if he could write out Pavlov's remarks as given, Anrep replied that not by any possibility would he be able to do it again, that throughout he had been acting under the hypnotic influence of the old man.

Later that week Pavlov gave a scheduled paper at a scientific session, this time to a much larger gathering. As soon as he appeared, Fulton related, the entire audience rose spontaneously to its feet amidst thundering applause. Anrep, again equal to the occasion, translated Pavlov's paper as it was given, and fielded the questions from the audience and Pavlov's answers rapidly and with ease.[9]

At one point during the Congress, Pavlov was invited to witness an operation Cushing was to perform on a left-handed patient with a large tumor in the left hemisphere of the brain. Cushing thought the case would be of interest to Pavlov since the patient did not at any time present clinical symptoms of aphasia, and also because Pavlov too was left-handed, indeed came from a long line of left-handers in his family. The operation was successful, and Fulton, also in attendance, later wrote, "Pavlov was so much interested in the procedure that he all but put his prominent whiskers in the operative field." To commemorate Pavlov's visit, Cushing, who had a fine sense of history, produced a large slab of beef and had Pavlov inscribe his name on the beef with his knife—and later preserved the specimen in his Brain Tumor Registry.

The afternoon of the final day of the Congress was given over to history. Léon Frederique, who had served as president of the Second International Congress in 1892, reminisced about the early days of the Congress. In the evening, Cannon, serving as toastmaster at the banquet, added to the historical atmosphere by sketching the contributions that each nation had made to the development of physiology and then called upon distinguished representatives to make a response on the achievements of their countrymen. It was a gala affair, ending in Harvard Yard with dancing by the light of a silvery moon aided by the glow of Japanese lanterns. Not even the curse of Prohibition, which obliged the physiologists to drink the toast "To international good feeling" in water instead of champagne, could mar the occasion.[10]

Frederique had nothing but praise about the conduct and results of the Congress. "I have been astonished at all that has been reported in America from the point of view of biologic thought," he wrote Sherrington some time later. "It was admirable to see the large number of young investigators and the spirit that animates them."

After the meetings ended, the delegates travelled to Woods Hole and various other parts of the country. When Cannon learned that Pavlov intended to go to Montreal before returning to the Soviet Union, he offered to drive the Pavlovs, father and son, to the family farm in New Hampshire, which would take them half way on their journey. The diversion, he assured Pavlov, would spare them the tedium of the long train ride to Montreal.

On their arrival in Franklin, Cannon took great pride in introducing Pavlov to his daughters, his sisters Bernice and Ida, and his wife's father, Henry C. James, who was visiting from the midwest. "My father," Mrs. Cannon later recalled,

> was Pavlov's age and through interpreters they found they agreed about the essentials of life. Pavlov said, looking at the circle of relatives, that there was nothing in life more wonderful than family love and family life and my father nodded an assent, both men having experienced it in the eighty years of their lives.

In the few hours that Pavlov remained at Franklin, she continued, the children never left his side, "as he talked to them in French and German, which one or another child would try to translate to the others amid laughter and a true fellowship of youth and age." When it was time for him to leave, they all waved a sad goodbye. "To the children he remains the Glory of Russia; to the adults, the great scientist."[11]

One evening late in 1928, Cannon came home and casually mentioned that he had just turned down an offer of an exchange professorship to France for the following year. "The stunned silence of the family," his wife later recalled, "was succeeded by a chorus of reproaches for a father who would let slip such an opportunity to introduce his family, relatively painlessly, to the wonders of Europe." He was finally persuaded to telephone Dr. Edsall, who had made the original offer, to ask if the appointment was still open. Since the request from France had specifically been for a medical man, the dean, after Cannon refused, had written to another doctor offering the appointment to him. But being a Saturday afternoon, he had not waited to sign the letter and it was probably just as he had left it on his desk. Edsall promised Cannon that if it was still there Monday morning, he would have a chance to reconsider. "Sunday was an anxious day for the family," Mrs. Cannon continued, "especially for the college girls, who were thumbing their Baedakers to decide where they wanted to go to round out their art courses."

On Monday came the welcome news that the other doctor had been cheated of his opportunity and the Cannons were destined for a year in Europe. When Cannon decided that he could only be away from his laboratory for half a year, there seemed no reason why the four girls and Cornelia should not go on ahead and see Italy first. In the succeeding months, while five "potential landscape artists" were getting ready for their grand tour, Cannon worked on his lectures and had them translated into French. In the autumn, he practiced reading them to Marcel Florkin, a Belgian graduate student who coached him and corrected his pronunciation.[12]

Soon after the close of the Physiological Congress in Boston, Cannon accompanied his family to New York in a seven-passenger Cadillac (the Buick having been traded in for a beautiful 1926 model, purchased second-hand) that he had customized with locked boxes on the running-boards for their clothing and art supplies. Seeing them off on the *S.S. Roma*, he exercised his political powers so skillfully with the ship's doctor that they were later given first-class cabins on their second-class tickets. Cornelia's sister, Margaret Burt, joined them when they landed at Genoa, and for the next several months they motored adventurously down the boot of Italy to its toe, then on to Sicily.

At the suggestion of Aunt Margaret, the entire management of the trip was handed over to the girls and a division of labor was promptly made. Linda was the main chauffeur, Wilma art guide and interpreter; Marian had charge of the car and handled all the money; Helen read maps, acted as road guide and selected accommodations. Cornelia and Margaret read guide books and made suggestions. "The only real problem," Cornelia later wrote self-mockingly, "was presented by the mother of the party. After having been boss of a large family for twenty years, abdication was extraordinarily difficult for her . . . the strain of being an equal in a world where she had been a superior so long told on her."[13]

While Cornelia and the girls were in Italy, Bradford began his first year of medical study, and, since their Cambridge house had been rented out, both he and his father lived in Vanderbilt Hall, the new dormitory of the Medical School. In October, Cannon accepted an invitation from Perrin C. Galpin, Secretary of the C.R.B. Educational Foundation, to visit Belgium and lecture at four universities there, in addition to his lectures on the emotions and homeostasis in France. All of the Belgian universities—Brussels, Louvain, Liége and Ghent—when offered a choice between semi-popular or scientific lectures, asked for scientific lectures to be delivered in French, one on the role of the sympathetic system in maintaining stability of the organism, the other on the neural basis for emotional expression.[14]

A few days before Christmas Cannon arrived in Paris, where he found the family staying at the Hôtel de l'Université and Wilma's beau John Fairbank, a Rhodes scholar at Oxford, visiting. The hotel, on the east bank, was an expensive one, but took them in at a reduced rate because the stock market crash in October "had sent all their American patrons scurrying home." Besides, the prospect of six long-term tenants was not to be sneezed at. They spent the holiday in the French manner with attendance at Mass in a Catholic Church instead of with carols and gaily-wrapped presents around a Christmas tree.

For Cannon, Paris brought back old memories and afforded new opportunities. He immediately got in touch with Alan Gregg, a former student then in charge of the Paris office of the Rockefeller Foundation, who took him to tea to meet members of his group working in public health and offered the services of his office to Cannon. He also contacted André Mayer, his old friend at the Sorbonne, and other medical scientists in Paris. "They extended their charming hospitality to us," Cornelia noted, "even at times entertaining the entire family."

On one occasion, there was a luncheon with Charles Richet, and on another, a dinner party given by Prof. and Mrs. Lapicque, in return for the Cannons' kindness when Lapicque had delivered the Dunham lectures in the spring. After the dinner all of the students of physiology at the Sorbonne and the Collège de France joined the party and, Cannon wrote Bradford, "we had a grand time talking French . . . Mother did valiantly—but apparently used certain French words that are not admitted into polite society!" Another dinner party was given by the French physiologists who had attended the recent International Congress—fifty-five of them in all—and retained pleasant memories. Eugène Gley made a flattering speech and presented Cannon with an inscribed medal of Claude Bernard as a token of appreciation for his hospitality to them in Boston.[15]

Yet another social event offered a different sort of experience. One day early in 1930, Cornelia took the entire family to tea with Gertrude Stein, whom she had known as an undergraduate at Radcliffe. As Wilma later described the event to her Aunt Ida,

> We six wholesome Cannons were almost too much for the exotic atmosphere she struggled to maintain in a large dimly-lighted studio hung thickly with Picassos . . . Miss Stein sat in a low chair with her long woolen underwear showing and laid down the law in a very determined voice to Mother, who countered in her usual brusque fashion. Mother maintained hotly that no expatriated American can understand America truly. Gertrude Stein said she had it all *within* her.

Wilma heard no more of their conversation because she was sent off into a corner to speak French with a young man, a protégé of their hostess. But when she happened to look out of the corner of her eye, she saw her father glaring across the room at her and the "degenerate," as he afterward described the exotic poet.[16]

To prepare for his lectures, Cannon was given a study in the Sorbonne in which to practice reading them aloud in French, under the tutelage of a pair of Lapicque's graduate students. During February, he gave two series of lectures, first at the Ècole de Médecine of the University of Paris and then at the Sorbonne, on homeostasis and the bodily changes brought about by emotional effects. At his inaugural lecture, he was marched into the Vulpian Amphitheatre with great pomp and circumstance by an escort of academic and medical dignitaries. The lecture was "a severe ordeal, but I managed to live through it somehow," he wrote Bradford. Alan Gregg, who was in the audience, assured Cannon that he had come through with flying colors; his French was understandable, his subject matter interesting, and the audience exceptionally attentive—for a French audience. When Cannon repeated the lectures at the Sorbonne, he was astonished to see some of the professors there who had attended the first series, including the dean and Georges Dumas, the eminent French psychologist.[17]

The professors spoke kindly of Cannon's French, which he used freely albeit with a certain amount of trepidation. By the end of his stay in Paris, however, he felt more at ease and when he gave a talk to the French Psychological Society on the removal of the sympathetic, followed by moving pictures of his cats, his wife recorded that the "whole thing was a veritable triumph" and that he answered questions in French "with grace and facility." Best of all, the psychologists had shown real enthusiasm for Cannon's research. "After the film," he wrote Bradford, "Professors Lapicque and Gley both spoke of the significance of the work, and Pierre Janet, one of the foremost psychiatrists . . . came to me all aglow and told me of his intense interest in what he had heard and seen."[18]

In March Cornelia accompanied Cannon to his Belgian lectures, taking the Cadillac and leaving the girls in Paris with their Aunt Margaret. After he had shown her Béthune, where he had been stationed during the war, and fulfilled his lecture commitments in Belgium, they returned to Paris and prepared for a motor tour through southern France and Spain. It was Cannon's first experience of traveling with the entire entourage, and it took some adjustment on his part to fit in with the party. In the beginning, he tried to take charge of the maps and even the driving, but before long his daughters inveigled him to take a back seat (just as they had their mother), and the trip was more relaxed and peaceful after that.

That is to say, it was more relaxed and peaceful for them. But not for Cannon, who began to suffer from headaches and stomach upsets. Nevertheless, he was able to deliver the lectures he had contracted to give at the provincial medical schools in Lyons, Montpelier and Marseilles. He had fine audiences and was presented with a medallion in commemoration of his visit to Montpelier—the fourth one bestowed on him since his arrival in France. Although he was suitably impressed with the great age of the medical school in Montpelier, Cannon dryly remarked, "There is much in the past which they point to with pride, but very little in the present or recent past that they can boast of." Marseilles he found old but "indescribably benighted," with not a gleam of intellectual interest to be seen—very different from "dear old Cambridge."[19]

They drove over the Pyrenees to cross the frontier into Spain; then down from the mountains and over the plains until they reached Barcelona, where a party of Spanish physiologists was waiting to greet Cannon. Rosendo Carrasco-Formiguera and Jayme Pi-Suñer, former workers in the physiology laboratory at Harvard, led a delegation who were most enthusiastic about their distinguished visitor, calling him "the greatest living physiologist." The Spaniards took them all on interesting side trips, gave them elaborate dinners, and couldn't do enough for them. Cannon, deeply touched, wrote Bradford, "I think that one of the greatest compensations in the life of a teacher is found in the respect and devotion of former students . . . I really feel that I have friends wherever old students of mine are settled."

After a welcoming banquet at the Ritz with speeches and toasts and warmth of feeling, Cannon lectured the next day at the medical school on the sympathetic in relation to homeostasis and showed his film. "The medical students were a young lot," he later observed with mellow tolerance,

> showing the traditional characteristics—making the sounds of cats when the sympathectomized cats were shown in the film and bow-wowing when the dogs were shown. They were stumped by the monkeys—or otherwise it might have sounded quite like a talking movie![20]

Dr. Carrasco told the family at dinner one night that he had never known such kindness as Cannon had extended to him when he worked in his laboratory, and he showed them moving pictures taken by one of the Spaniards at the Boston Congress. "And lo and behold," Mrs. Cannon wrote Bradford, "here in far Barcelona whom should we see walking on the terraces at the Medical School with the crowd but you! We screamed with joy and excitement . . . these people have the most radiant memories of the Congress. It is like a fairy dream to them all and they feel as if they could not do enough for us to show their gratitude."[21]

On leaving Barcelona, the Cannons pushed on to see as much of Spain as they could possibly squeeze in. On their arrival in Madrid for Easter, the doctors once more turned out to greet them. There, as in Barcelona and elsewhere in Spain, they heard much political talk, for just a short time before, Primo de Rivera, the dictator of Spain, had been driven from office. Many spoke freely and hopefully about the coming of a Republic. Chief among them was Juan Negrin, the professor of physiology at the University of Madrid, who, Cornelia noted in her travel letter, "is a charming man with true scientific ideals—one of the coming men in Spain, I think." A few days later, she described Dr. Negrin again: "a real man, a republican like all of the people we have met and anti-monarchist." Marian later remembered that "the fervor and idealism of the intellectuals whom my father saw in Spain made a profound impression on him."

In Madrid they also met Georges Marañon, the endocrinologist, who had spent a month in prison for speaking out against Rivera. He found it a most interesting experience, however, for he had used the time to study the bodily effects of emotions. The physiologists in Madrid, like those in Barcelona, were kindness itself to their guests. After Cannon had given a demonstration of the removal of the sympathetic chain as well as a lecture, they not only gave a dinner in his honor, they also paid the family's hotel bill, thus allowing them extra money for sightseeing.[22]

In May Cornelia accompanied Cannon on a quick trip to England, where he was due to deliver the Linacre Lecture in Cambridge on the autonomic nervous system. There they stayed at the home of Joseph Barcroft, who had stayed at their home previously (Cornelia had marveled "how he got on in our hit and miss household"). Following an enthusiastic introduction by Sir Humphrey Rolleston, the Regius Professor of Medicine, Cannon gave his lecture to a packed house. Even though he soon discovered that his film had been wound wrong and reading the captions backward was no easy matter, he maintained his self-possession, took it all good-humoredly and, in the words of his wife, made a great hit. At the conclusion, when asked to say a few words more, Cannon told the medical students in the audience how pleasant it would be for them to find wherever they went in the world "fellow medical men animated by the same ideals and working for the same ends—the relief of man's estate."

Henry Dale came to Cambridge just to hear the lecture, which pleased Cannon very much, and that evening, at the "feast" at St. John's College, he and Barcroft sat across the table from Dale, while Sir Donald McAllister ("a hardy old medical man, who knew Oliver Wendell Holmes") and J. J. Thomson, the renowned physicist, were to his left and right. They all had a jolly time gossiping and swapping stories. Before they left England, Mrs. Cannon had luncheon with Lady Bayliss and Mrs. Dale in London while Cannon visited Dale's laboratory at the National Institute for Medical Research. "It is an interesting place," he noted, "and Dale has some extraordinarily good work going forward there."[23]

For the next few weeks, save for an occasional lecture or demonstration given by Cannon, he and Cornelia toured central Europe in the Cadillac, the highlight being visits to World War I battlefields and burial grounds dotting the landscape from Rheims through the Argonne and round Verdun. Retracing the trail of Cannon's journeys during the war, they took the girls to Chalons-sur-Marne, where during a major drive by the Germans in the summer of 1918 he had tended the wounded in a field hospital, and observed a Memorial Day service at an American cemetery just south of St. Mihiel. It was a moving experience for all of them.[24]

Shortly after Cannon visited his friend Leon Asher in his laboratory at Berne and took the family to tea with Dr. W.R. Hess in Zurich, the trip hit its first real snag. In the second week of June, with Wilma driving, they had a minor car accident. Although the damage was very slight, a policeman insisted on taking depositions. "Walter was pale with agitation," Cornelia later wrote home, "very indignant with Wilma, and generally upset far beyond any requirements of the situation." They were told they had to go to the police station, where there was endless discussion, papers to sign, Wilma's name entered on the police blotter, etc., "all accompanied by excessive emotion on Walter's part." When they finally got away, they stopped for a picnic lunch, at which Cannon drank half a bottle of beer.

On their way again, he suffered an attack of hematuria (blood in the urine), became anxious about his symptoms, and decided to seek medical attention when they reached Ragaz. The doctor in Ragaz pronounced the trouble hemhorragic nephritis brought on by the beer, which he said came from a batch of green hops that year. Cannon was inclined to believe it the result of the excitement raising his blood pressure; nevertheless, he took the doctor's advice to remain quietly in bed for some days until his condition improved. While the rest of the party went on to Munich, Cornelia stayed to nurse Cannon. At last he became impatient with the enforced bed rest, told the doctor he was leaving and said goodbye.[25]

Reunited with the girls, they motored through Germany, then down the Rhine and into Holland, where they met Bradford, who arrived in Rotterdam at the end of June. En route, Cannon suffered another attack of hematuria, which cleared up by itself, but a third attack persisted. Finally, upon reaching Brussels he consulted Dr. Govaerts, whose examination revealed that the trouble was not kidney but a small papilloma—what Cornelia called "a sort of internal wart"—in the bladder. Govaerts advised Cannon to seek treatment when he reached London and the party continued on its way.[26]

In London, Cannon was seen by a Harley Street surgeon, Sir John William Thompson Walker, who cystoscoped him and confirmed that the bleeding was coming from what appeared to be a papilloma at the orifice of the diverticulum in the left lateral wall of the bladder. Told that an operation would be necessary, he was advised by Harvey Cushing, whom he happened to meet by chance in London, to return to Boston where he would have more personal care than in an English nursing home, and also be among friends. With Cushing's words in mind, Cannon arranged by cablegram to have the operation done by Dr. William Quinby at the Peter Bent Brigham Hospital. He and Cornelia then booked passage on the first boat available, leaving their five children with Aunt Margaret to see England without them. When Bradford later became ill with what turned out to be jaundice, these plans unfortunately went awry as well.[27]

At operation on Cannon on July 31, Quinby removed the papillary neoplasm in its entirety. Burt Wolbach's pathology exam reported "small islands of tumor cells that give a very definite picture of malignancy." Though the findings were indicative of more rapid growth than the average papilloma of the bladder, Wolbach ventured to make a completely favorable prognosis. Not only Quinby and Wolbach but also David Cheever and Cecil Drinker took pains to reassure Cornelia, and Ida Cannon, too, that the operation had been a great success and that they expected no further trouble. Nevertheless, as a precaution, Cannon was to be examined every two months for several years to see if there was any possibility of recurrence.[28]

By the time the children came back from England, their father had fully regained his strength in Franklin and could hardly wait to get back to his laboratory. If he thought he had bid farewell to the international community when he left Europe, Cannon was mistaken. He found the world on his doorstep in the persons of two young research fellows waiting to work with him in the fall of 1930. One was Zenon Bacq, a C.R.B. fellow from Belgium who was eager to resume the investigation he had begun with Cannon the previous year. The other was Arturo Rosenblueth, a newly-appointed Guggenheim fellow from Mexico who would become a favored collaborator of Cannon for the remainder of his career.

In November of 1930, Cannon received a visit from W. W. Norton, the head of a publishing firm in New York. This was not the first time Norton had come to see Cannon. In the spring of 1929, while Cannon was negotiating with the French universities to give a series of scientific lectures on homeostasis, Norton had proposed drawing up a contract for a book based on his lectures. Cannon, who had become unhappy with his previous publishers, Appleton & Company—and especially so in his dealings with them while preparing the second edition of *Bodily Changes in Pain, Hunger, Fear and Rage*—was interested in Norton's proposition. But he put him off because he had not yet written the volume and did not know if it would turn out to be the semi-popular book that Norton wanted.

When Norton brought the matter up again, Cannon explained that he had not done anything with his lectures because his surgery during the summer left him in a condition in which he was unable to carry out his plans. But soon after the visit, the publisher sent a formal proposal which Cannon found so attractive that he intimated he might find time the following summer to put a book together. Norton continued his pursuit of Cannon, visiting him again in the spring of 1931, until on May 9 a contract was drawn up for a book, tentatively titled "Factors of Stabilization," which Cannon duly signed.

Early in August, Cannon wrote Norton that he had finished the first draft of sixteen out of nineteen chapters of his book—an expansion of his lectures on steady states in the body. "Such is the result of pegging away in a routine manner daily for a month and a half," he added. He was puzzling over a title and felt "How We Stay Normal" might be best, though he also considered, among other possibilities, using a one-word title like "Stabilitas" or "Homeostasis." Before closing his letter, Cannon asked Norton if he had any suggestions for Mrs. Cannon, who was having trouble finding a publisher for her travel book, "Art Awheel in Italy." While her Boston publishers had commended it, they told her the book trade was "in the doldrums" and that a book of travel had little chance to sell.[29]

By year's end Cannon had ready a finished manuscript of his book, which he now called "The Wisdom of the Body." But he was holding onto the last chapter relating biological to social homeostasis (written at the suggestion of his wife), because he wanted to submit it to experts in economics and sociology for comment. "Of course it is a little hazardous for a physiologist to enter into the economic and sociological field in any case," he wrote Norton with some trepidation, "and perhaps it is a mistake for me to put forward the ideas that I have suggested in pointing out analogies between the body politic and the body physiologic."

Norton's reply left no doubt of his feelings. He bluntly told Cannon not to "dare let any conservative economist or sociologist . . . scare you out of having your little party in the last chapter. By no possible stretch of the imagination could the parallels which you draw be considered radical." As to whether it was "politic for a physiologist to indulge in such parallels," it seemed to him that Cannon had done so "with such regard for the scientific method as to disarm any possible criticism on that score."

It took Cannon another month before he sent in the final chapter—now called an Epilogue "in order to distinguish it from the main biological part of the book." Meanwhile, production had gone ahead and Cannon was "very agreeably shocked" on receiving the news in February 1932 that his book was already off the press. And when Norton learned that Cannon would be delivering the Herter lectures at New York University, he invited him to stay at his apartment and meet his wife, who was Dr. Christian Herter's daughter.[30]

As a means of promoting sales, Cannon told Norton that he thought his book would tie in well with a book Wallace Donham, dean of the Harvard Business School, was publishing that spring. Some time earlier, after hearing Donham deliver a talk on the social and financial problems produced by the Depression, Cannon had sent him a reprint of his article on physiological homeostasis with the comment, "It seems possible that the mechanisms for stabilization which are used in the body physiologic might be serviceable also in the body politic." When Donham's book, *Business Looks at the Unforseen,* was published, he acknowledged his indebtedness to Cannon for many of the concepts.

Cannon was pleased with Donham's acknowledgment. And he was delighted when *The Wisdom of the Body* was named a selection of the Scientific Book Club. Almost at once, rave reviews appeared in the *New York Herald-Tribune* by John Fulton and in the *Saturday Review of Literature* by Yandell Henderson. By May, sales were so brisk that Norton ordered a second printing. He was also negotiating with Kegan Paul to put out an English edition.[31]

In Cannon's scrapbook there are several pages of newspaper and magazine notices of *The Wisdom of the Body.* The volume, like *Bodily Changes in Pain, Hunger, Fear and Rage,* was widely reviewed and highly praised for its stimulating substance as well as for its clear, admirable style. Graham Lusk, for one, in a review in *Science,* marveled at Cannon's use of lucid language, without resort to complicated mathematical formulae. His straight-forward exposition of "the exquisitely sensitive regulatory mechanisms which maintain the units of the body in a state of nearly balanced equilibrium" thus rendered physiological phenomena comprehensible to biologists and general readers alike.

A nameless reviewer in *Nature,* for another, lauded the author for being "careful to show where he is stating ascertained fact and where making use of conjecture, thus at once attracting the layman by the candour of his science and suggesting fresh fields of experiment to the biological worker." He found Cannon's use of the title of E.H. Starling's Harveian Oration especially apt and picturesque in describing the content of the book, and ended his review, "It is science, but it reads like a poem." One of Cannon's friends, after reading the review, wrote him jocularly, "What am I to say when the pater-familias of all scientific journals apparently relishes the idea that science should read like a poem? And what am I to say to a scientific investigator who is flattered by the idea?" Indeed, he fully expected when next he saw Cannon in Boston to find him with "hair grown long, a Windsor tie upon his bosom, and wearing a smock" instead of a lab coat.

English-language reviews on both sides of the Atlantic were published not only in a variety of medical, scientific, and popular journals, but also in a large number of city and town newspapers. Several reviewers noted the influence on Cannon's concept of homeostasis of Claude Bernard, who had lately been rediscovered and whose broad principles of thought handed down to posterity were at last vindicated. Recommended as a "must-read" for readers of all stripes, most reviewers called attention not only to the book's popular appeal but also to the timeliness of the Epilogue. *The New York Times,* on March 6, 1932, devoted nearly a full page to its review, with a bold headline proclaiming "How That Going Concern, the Body, Maintains Stability" and as a sub-heading, "Dr. Cannon looks for the key to equilibrium, whether individual or social, in the processes of the physical organism."[32]

A year later, when Cannon gave a speech to the alumni at M.I.T. titled "Biocracy: Does the Human Body Contain the Secret of Economic Stabilization?" his proposal that biocracy be used as a substitute for technocracy in treating the ills of mankind stirred up an enthusiasm that took him by surprise. "I have told most persons who have consulted me," he wrote Norton in February 1933, "that they would find the full story in the Epilogue in 'The Wisdom of the Body.'"

In the very depths of the Depression, Cannon began his lecture: "You will perhaps pardon me if I take as my text, 'Civilization is in a hell of a mess'; the expression seems justifiable because it is so exactly desciptive." He went on to say that a new group appropriating the slogan "Technocracy" had taken hold and seized popular fancy; "they promise an overthrow of the price system, and ultimate abundance for everybody. What all these efforts intend is a replacement of the present hell of hunger, nakedness, demoralizing idleness and fear with the plenty and security of the kingdom of heaven. I wish to suggest that the kingdom of heaven is within us!"

Cannon, a liberal Republican who had not voted for Franklin Delano Roosevelt, proceeded to propose some rather un-Republican concepts. However, he knew better than to expect that a real biocracy would arrive soon or, for that matter, that technocracy would have the influence its promoters desired. "The technocrats and the biocrats must wait," he concluded, "because the Democrats have the vote!"[33]

Cannon's address at M.I.T., like his book, received international attention. An especially entertaining account in the British newspaper, *Nottingham Evening Post,* must surely have tickled his fancy. "Technocracy, the new economic doctrine that was to save the world, is dead," it began. "But although technocracy failed, and failed badly, to impress the world, that fact has not deterred another American scientist from putting forward an entirely new theory of world organization, the theory of biocracy."

After explaining that Cannon's concept of homeostasis was rooted in the principles of biology and that it was being newly applied to social and economic problems, the reporter concluded that biocracy would probably have received scant attention but for one thing. Inasmuch as F.D.R.'s various schemes and "general seizure of powers far beyond those normally enjoyed by a president savour strongly of biocracy," he believed that Cannon's proposal could well be established as political doctrine—in which case Mr. Roosevelt would go down in history as the first practical Biocrat.[34]

14 Chemical Mediation of Nerve Impulses:

I. Searching for the Mysterious X-factor

IT IS RECOGNIZED today that nerve impulses are transported along complicated multineuron pathways from nerve cell to nerve cell over minute gaps called synapses. Since each neuron or nerve cell is a distinct cellular unit anatomically separated from all other neurons, each synapse represents a connection between one neuron and another that has information to convey; and after the information is integrated, it is sent along as a nerve impulse to an adjacent, or postsynaptic, cell. Transmission of the nerve impulse along a nerve fiber is an electrical process; at the end of the fiber, however, the impulse is transmitted to a neighboring cell by means of a chemical mediation.

When a specific chemical is synthesized by the nerve cell, it is stored at its axon terminal and, upon arrival of the impulse at the nerve ending, is released. The transmitter diffuses across the synaptic cleft (i.e., "jumps the gap") and produces a change in the permeability of the postsynaptic membrane. This, in turn, results in changes in the postsynaptic cell that may be excitatory or inhibitory, hence causing muscles to contract or relax and, as the case may be, other organs to accelerate or become depressed. After the neurotransmitter is liberated, it is destroyed by an enzyme or reabsorbed into the nerve terminal, so that its effect is localized and of short duration.[1]

Contrast this current knowledge with the view held at the end of the 19th century, when nerve pathways were little understood. Physiologists then generally believed that there is a direct connection between one nerve cell and another, and that a nerve impulse in a preganglionic fiber is conveyed directly to the postganglionic by means of this physical continuity. It was C. S. Sherrington who, in 1897, first advanced the idea that nerve cells are anatomically discontinuous, although physiologically continuous, and that "a feature of the concatenation of neurons more probably explicative of modification and delay of nerve impulses is the synapse." Just how transmission is stimulated to take place remained unknown, although the cause was thought by some to be electrical in nature.[2]

The notion that nervous effects might be transmitted by release of a specific chemical stimulant was first suggested by T. R. Elliott in 1904. While a student of J. N. Langley in the department of physiology at Cambridge University, Elliott had worked out the details of the interaction of adrenalin with the sympathetic nerves. He observed that long after the sympathetic nerves had been cut or had degenerated, structures previously innervated by them responded in a characteristic manner to the direct application of adrenalin—e.g., smooth muscle contracting or relaxing as sympathetic impulses formerly had provoked it to do. On the basis of this observation, Elliott reasoned that whenever a sympathetic impulse arrived at a smooth-muscle cell, it might liberate adrenalin locally in the cell, and afterward he was led to propose that "adrenalin might then be the chemical stimulant liberated on each occasion when the impulses arrived at the periphery."

It took Elliott's teacher, Langley (who, more than anyone else, was responsible for elucidating the nature and structure of the autonomic nervous system, dividing it into sympathetic and parasympathetic divisions), to explain the fact that sympathetic nerve impulses, as well as injections of adrenalin and even electric shocks, bring about contraction in some cells and relaxation in others. In 1905, Langley hypothesized that in reacting cells two sorts of "receptive substances" exist, one being excitatory and the other inhibitory. He surmised that when adrenalin reaches a cell, its action is determined by the type of receptive substance there. If the impulse unites with the excitatory substance, contraction results, and if with the inhibitory, relaxation takes place.[3]

In the wake of these stated views, which at the time belonged more in the realm of theory than fact and were not widely accepted, little more of immediate relevance to the story of the development of chemical transmission appeared in print before 1914. And yet, Henry Dale and Cannon himself, as well as other commentators on the evolution of knowledge about chemical mediation, have taken notice of certain occurrences in the period between 1905 and the outbreak of the first World War, which were later seen to be historically significant, though they were not appreciated at the time.

In 1906 and 1907 Walter E. Dixon, a pharmacologist at Cambridge University, had conducted experiments that prefigured later events. Dixon was almost alone in supporting Elliott's views and even argued that parasympathetic nerves might release a chemical transmitter with similar effects. Working with frogs, he found that an extract of the heart prepared after he had stimulated the vagus nerve had the power of inhibiting the frog's heart and, like muscarine (a deadly alkaloid derived mainly from the fungus *Amanita muscaria*), its effect was completely antagonized by atropine. Dixon's observation, meeting with skepticism, went no further; but in 1909 he published a paper with Philip Hamill that drew attention to the similarity between the effects of nerve stimulation and certain drugs, especially muscarine, on the heart.[4]

At or about the same time as Dixon's heart-vagus experiments, two other events took place that were to prove important in the light of later developments. The first was at Johns Hopkins where Reid Hunt, then a young assistant professor of pharmacology, had begun a research on depressor constituents of the suprarenal glands. Noting the presence in aqueous extracts of the adrenal gland of a blood-pressure reducing substance, which on chemical treatment disappeared as the cholin content increased, Hunt inferred as early as 1901 that there might be an ester-like precursor of cholin in the extract. (A chemical compound found in many animal and some vegetable tissues, cholin, or choline, is essential to liver function in the prevention of the deposition of fat in the liver.) In 1906, while testing a number of esters which were made for him by his assistant René Taveau, he found the acetic ester acetylcholine to have an action like choline but one which is a thousand times as strong. Although Hunt's observation was scarcely noticed at the time, its importance was acknowledged when acetylcholine was later found to be a major neurotransmitter.

Also at Johns Hopkins in 1906, the physiologist W.H. Howell began to publish observations that have since been credited with contributing to the knowledge of chemical transmission. Howell not only reported that nerve impulses act indirectly by increasing the amount of diffusible potassium compounds in the heart tissue, he also stated that vagus impulses inhibit the heart by mobilizing potassium ions.[5]

H.H. Dale, who figures prominently in the tale of neurotransmission and neuroconduction, had also been a student of Langley's at Cambridge. In 1904, Dale had taken a post as pharmacologist at the Wellcome Physiological Research Laboratories in London, where he was promoted to director eighteen months later. His first assignment at the Wellcome Laboratories was to work out the pharmacy, pharmacology and therapeutics of the natural fungus, ergot, then considered to be in a state of confusion. Though he admitted to being a tyro pharmacologist and claimed to be unenthusiastic about getting involved in the "ergot mess," he began a long process of testing samples which his able Wellcome colleague, the chemist George Barger, had prepared. From 1904 to 1914, when he left the Wellcome labs, Dale conducted an impressive array of studies on the cardio-accelerator action of the vagus, the action of extracts of the pituitary body, the chemical structure and sympathomimetic (a term he coined) action of amines, in addition to further tests on ergot and investigations of anaphylaxis and histamine.

Dale later recalled that he came across acetylcholine accidentally in 1912 or 1913 as a constituent of a particular sample of ergot, "and therefore as a product of nature." In the course of identifying the compounds of ergot, he found that some specimens contained an unknown substance which inhibited the contractions of the frog heart and stimulated contractions of the mammalian intestinal muscle. Since these actions were abolished by atropine, and since muscarine had the same properties, Dale reasoned that ergot might contain muscarine. This led him to ask another Wellcome chemist, Arthur Ewins, to attempt to isolate muscarine from ergot. In the course of working up a large batch of ergot, Ewins isolated and positively identified the unknown substance as acetylcholine, the intense depressor activity of which was already known through the work of Hunt and Taveau. Moreover, Ewins demonstrated that acetylcholine is a natural product, and not merely a synthetic compound as previously had been thought.[6]

In the meantime, Dale had undertaken a comprehensive study of the esters and ethers of choline, and afterward of acetylcholine which he found to be an extremely active but unstable substance, and summarized his research in a paper published in the *Journal of Pharmacology and Experimental Therapeutics* in 1914. He began by explaining how his investigation had come about and stated that it was desirable to give a comparative account of the action of choline derivatives examined in his search for muscarine.

After describing the action of choline esters of acetic, nitrous and nitric acids, and of choline ethyl ether on various organs and systems, Dale reported two quite distinct types of action on the heartbeat and circulation. One was the depressor, or muscarine, type, and the other, a pressor, or nicotine, type seen only after the first action had been abolished, which was unaffected by atropine but abolished by large doses of nicotine. He pointed out that numerous writers had already noted that the muscarine type action reproduced the effects of stimulating nerves belonging to the cranial and sacral divisions of the autonomic nervous system, and thus, "in the case of organs receiving a double and antagonistic nerve-supply, from this part of the system and from the true sympathetic, the muscarine effect is frequently antagonistic to that of the true sympathetic nerves and adrenine."

The possible physiological significance in the resemblance between the action of choline esters and the effects of certain divisions of the involuntary nervous system was of great importance, Dale concluded, but little evidence was available for discussion. Of all the substances he had examined, acetylcholine was the one, in his view, whose action was

> most suggestive in this direction. The fact that its action surpasses even that of adrenine, both in intensity and evanescence, when considered in conjunction with the fact that each of these two bases reproduces those effects of involuntary nerves which are absent from the action of the other, so that the two actions are in many directions at once complementary and antagonistic, gives plenty of scope for speculation.

Nevertheless, he conceded that "there is no known depôt of choline derivatives, corresponding to the adrenine depôt in the adrenal medulla, nor, indeed, any evidence that a substance resembling acetyl-choline exists in the body at all."[7]

The outbreak of war in August of 1914 shut down further research on the subject by Dale or anyone else for some time, as

Cannon noted twenty years later in his Kober Lecture, which reviewed the development of ideas on the chemical mediation of nerve impulses. Alluding to Dale's 1914 report, he observed: "Here we have definitely the intimation that acetylcholin mimics the action of parasympathetic nerve impulses, much as adrenin mimics the action of sympathetic impulses. It seems probable that this hint would have been promptly seized and worked upon if the World War had not forced the attention of investigators to more urgent matters." Not until seven years had passed was any further progress made; "and then, in 1921, two researches were published, one a contribution of significance which was not appreciated at the time, and the other a recognized forward move in the developing course of events." Whereupon Cannon disclosed that the former research had been conducted by J. E. Uridil and himself, and the latter by Otto Loewi, the Austrian pharmacologist.[8]

By means of a simple demonstration that the vagus nerve produces its effect on the frog's heart by liberating an inhibiting substance, Loewi provided the first proof that chemicals are involved in the transmission of impulses from one nerve to another and from neuron to receptive organ. In the spring of 1921, when following up an idea he later claimed to have conceptualized as early as 1903, and which had only recently re-emerged from his unconscious during a dream, he prepared two beating amphibian hearts by tying off all their vessels except the aorta and then inserting a cannula down the aorta of each through the aortic valve into the ventricle. Afterwards, he pipetted Ringer's saline solution into the cannulas, which were held vertically by clamps, and observed the fluid surging up and down as the ventricles contracted and dilated.

Next Loewi attached levers by a thread to the apex of the hearts to register the rate and extent of contraction on the smoked paper of a kymograph. In one of his preparations he isolated the vagus nerve and placed an electrode on it, stimulating it for a fixed period of time. He then sucked the fluid from the cannula in that heart and substituted it for the fluid in the cannula in the second heart. The perfused fluid from the first heart caused the heart beat in the second to decrease in strength and frequency, leading him to conclude that the vagus-inhibited donor heart contained some substance which also inhibited the action of the recipient heart.

Loewi knew from Howell's previous work that an increase of

potassium could produce the effects he had seen. He found, however, that while the influence of the inhibitory material in the solution was completely blocked by atropine, with an increased concentration of potassium it was not, indicating that potassium is not the inhibiting substance. Unaware of the true nature of the inhibiting substance, he gave it the noncommittal name "Vagusstoff."[9]

Subsequently, Loewi reported in 1926 that during sympathetic nerve stimulation, an accelerating substance is released which, when collected and perfused through a second isolated heart, causes the recipient organ to beat more strongly and rapidly. He named this second undetermined substance "Sympatheticusstoff," or later "Acceleranstoff." But these results were far from convincing. Not only Loewi himself but many others had trouble repeating these early experiments, leading a number of scientists to deny their validity. It took another decade, as Cannon indicated in his Kober Lecture, before Loewi's research would be widely regarded as conclusive.[10]

In the early 1920s, G.N. Stewart and J. M. Rogoff, while claiming that the denervated heart was a poor model upon which to judge adrenal secretion, had reported that even with the adrenals deactivated, afferent stimulation still caused a marked acceleration in the rate of its beat. Holding that they were not called upon to explain their observation, they nonetheless ventured to attribute the increased rate after adrenalectomy chiefly to the rise in arterial pressure accompanying the stimulation. They also referred to variations of venous flow into the heart and increased mobilization of sugar and other changes in the liver as possible accessory factors. The challenge posed by the Cleveland physiologists stimulated Cannon to undertake, in association with his advanced students, a series of studies to validate his use of the denervated heart as an indicator of adrenal function. Although the investigations with David Rapport, J. E. Uridil and Fred Griffith have been described previously, it is well to have another look at them here as a stage in the historical development of chemical transmission.[11]

Cannon's involvement with what came to be known in his laboratory as the mysterious "x-factor" began but did not end with the claims of Stewart and Rogoff. Nor did it originate in his work with Uridil, as might be inferred from his Kober Lecture. Rather, it grew out of the first of two studies he had conducted earlier with David Rapport, both of which were published in the *American Journal of Physiology* in December 1921. In reporting this work, Cannon observed that though the denervated heart rate was usually increased to a relatively slight degree by splanchnic stimulation after removal of the adrenal glands, there were instances of noteworthy acceleration. In a series of experiments, Cannon and Rapport determined that the residual increase of heart rate was neither due to arterial pressure, nor to changes in venous pressure, nor to the improvement of a poor circulation, nor to the ingress into the heart of warm blood from the abdomen, nor to accessory adrenal tissue in the form of outlaying bands or threads of chromaffin tissue. Since their inquiries had provided no explanation, they proposed that some agent other than adrenin, carried to the heart by the blood, was responsible for the faster cardiac rate.[12]

Almost immediately, Cannon, working with Uridil, embarked on another series of experiments in an attempt to identify the unknown accelerating agent. Their findings were also published in the December 1921 issue of the *A.J.P.* Here Cannon reported that in attempting to discover if the phenomenon had its origin in the abdomen, he and Uridil found that the mysterious factor "neither rises from the kidneys nor from the back wall nor from any other abdominal organ—stomach, intestines, spleen or pancreas—except the liver."

Further experiments confirmed this judgment. Because peripheral splanchnic stimulation, in the absence of the adrenals, made the denervated heart beat faster even though the inferior vena cava and portal veins were closed, and because this effect all but disappeared when both the hepatic artery and portal vein were closed or when the hepatic nerves were cut, Cannon and Uridil concluded that "the agent increasing the cardiac rate is a substance given off into the blood stream by the liver."

While there was similarity of action of the hepatic agent and adrenin in that both increased the heart rate and raised the blood pressure, the chemical released from the liver did not resemble adrenin in all respects. It did not, for example, dilate the pupil of the eye, a characteristic action of adrenin. Not only that, but the increments of heart rate varied widely, slight in fasting animals or those in poor condition, much greater when digesting meat or if amino acids were injected into the intestine.[13]

The unknown factor was very much on Cannon's mind when he accepted an invitation to speak to the New York Neurological Society in October 1921. Choosing as his topic "Evidence of Nervous Control of Some Internal Secretions," he reviewed the recent work with Rapport and with Uridil and gave a preview of his then current investigations with Griffith on the nervous control of the liver. Telling his audience that the faster beat of the denervated heart was abolished after severance of the nerves of the liver, he reiterated a point he had previously made in his report with Uridil. "For the present," he ventured to state, "we are inclined to draw the conclusion that probably the effects noted are not due to a true internal secretion produced by the liver, but to a discharge from its cells of amino acids or amines which are sympathomimetic in character."

Since connection between the liver and heart through the inferior vena cava had not been considered previously, Cannon wondered whether it might be possible, by stimulating the cava low in the thorax, to determine if a disturbance induced in the hepatic veins could be carried along the cava to the heart. But when tests made with Fred Griffith produced no effect on the heart, he was led to exclude that possibility. Cannon's next thought was to see if he could collect blood containing the substance produced in the liver and, on injecting that blood, replicate the usual results of hepatic stimulation. Again, after testing, he and Griffith were forced to rule out that possibility as well.

Because his earlier studies had indicated that greater cardio-accelerator action was associated with digestion, Cannon then reasoned that the increase might be due to the influence of material from the intestines—normally removed when the blood passed through the liver—which was not removed if the hepatic nerves were stimulated. This possibility was enhanced by the knowledge that the hepatic nerves constrict and lessen the blood flow in the hepatic artery much more than in the portal vein. Yet after they closed the hepatic artery in a well-fed, digesting cat, the heart rate was not increased and they came up short again.

Nevertheless, Cannon and Griffith kept trying. When testing various foodstuffs on their experimental animals, they found that acceleration of the denervated heart after the nerves of the liver had been stimulated was more marked in an animal digesting protein than when it fasted or subsisted on carbohydrate or fat. The relation between protein feeding and the efficacy of hepatic stimulation, the fairly close chemical relation of adrenin to certain amino acids, and the 1910 report by Barger and Dale that there are amines having sympathomimetic effects—all of these led them to hope that they were at last on the right track. But after a great many further experiments involving the injection of materials ranging from alanine and tryptophane to catalase and liver extracts, they had to admit they were stymied. "After a survey of all the facts," they concluded, "we do not at present see any other assumption to make than that a substance of special and unknown nature is discharged by the liver cells into the blood stream when the hepatic nerves are excited."[14]

In the years that followed, Cannon continued to be absorbed in defending his adrenal research against the attack mounted by Stewart and Rogoff. Locked as he was in mortal combat with the Clevelanders, he recognized that the unexplained acceleration of the denervated heart after adrenalectomy was an incongruous factor that could compromise the work he had done and the conclusions he had formulated. Even so, other events crowded in and other lines of research begged to be explored. When Cannon heard Frederick Banting announce the discovery of insulin at a meeting of the American Physiology Society in December 1921, for instance, he immediately saw its potential and began to make plans to use the new hormone in his research. Although he discontinued his quest for the identity of the accelerating factor for a time, he was not one to leave important questions go unanswered.

That the unsolved problem continued to nag his conscience and pique his curiosity may be seen in a talk he had given to the American Psychiatric Association in June of 1922. Reporting that, from new evidence at hand, sympathetic stimulation evokes from the liver not only a discharge of sugar but also a discharge of some elaborated unknown substance having both cardio-accelerator and pressor effects, Cannon shared his frustration with the audience: "Here we are on the edge of advancing knowledge, and I am sorry to say that I cannot give you further information about the nature of this new substance."[15]

During the next few years, Cannon directed an ever-expanding group of students and fellows in his laboratory in a series of researches on adrenal secretion and sympathetic activity. When Kodama's repudiation of Stewart and Rogoff's findings was followed by a tidal wave of support for the Harvard side of the controversy, Cannon found himself with enough breathing space to probe a variety of intriguing questions. By mid-decade, he was running in high gear.

Late in 1924, in association with J. T. Lewis and Sidney Britton, and with the occasional assistance of other workers in his laboratory, Cannon undertook a series of studies that over the next year and a half resulted in publication of two papers. Their first paper disclosed that they had been mystified by a sudden, immediate increase of heart rate attending emotional excitement, despite the fact that both stellate ganglia had been removed, vagal connections severed, and hepatic and medulliadrenal influences excluded. Only after many experiments did they discover that the mysterious increment was due to accessory accelerator fibers reaching the heart from the upper thoracic sympathetic chain below the stellates.

As Cannon later explained in his Kober Lecture, Lewis, Britton and he had devised a method of separating the heart surgically from the central nervous system while leaving the organism otherwise in a quite normal state. The heart in these circumstances, although not subject to nervous control, could be strikingly influenced by changes in the circulating blood; and it was, as Cannon put it, "exquisitely sensitive to an extremely minute increase in the concentration of adrenin."

As a result, before adrenal inactivation emotional excitement with an attendant struggle might cause a rise in pulse rate of 100 beats or more per minute within a few seconds; but after adrenal inactivation the quick, sharp rise almost wholly disappeared. To the surprise of the investigators, however, animals surviving for some time after operation reacted otherwise. The heart rate of an experimental cat was first tested while it was held quietly in the lap and again after it had been confined on its back within a comfortable holder—the restraint causing the animal sooner or later to begin to struggle and show signs of rage. "Shortly after a brief excited struggle," Cannon recalled, "the rate began to be faster, and at the end of *3 minutes* it might be elevated by 25 or 30 beats per minute."

This display of cardiac acceleration after all known nervous and humoral factors had been eliminated proved to be, as Cannon expressed it, a very disconcerting and mysterious phenomenon. After a great many further experiments under varying circumstances, he had to confess that the situation was indeed baffling. "Why," he asked, "should the phenomenon occur in surviving animals and not during an acute experiment?"[16]

In the winter of 1926–27 Cannon had attempted to answer this question and others by putting Harlan Newton and Raymond Zwemer (a National Research Council fellow) to work with him on the search for the mysterious cardio-accelerator, or "x-factor," as he now called it. With the blood stream the only connection between the heart and the rest of the organism, Newton, Zwemer and Cannon were sure that some chemical change in the blood must account for the phenomenon, and such a change must be caused by an internal secretion. But when, one after the other, the endocrine glands had been excluded from action and yet excitement continued to increase the heart beats, they were almost in despair of finding an explanation.

Since they knew that the curious rise was associated with emotional disturbance and implicated the sympathetic nervous system, they decided to try to learn whether the x-factor arose from some previously unrecognized source in the fore part of the animal, or whether it arose in the hind part. Following denervation of the liver and adrenal glands, the spinal cord in the anesthetized cat was severed at the ninth rib. On recovery from the procedure, struggle involving movement of the forelimbs induced only a slight rise of heart rate (eight beats per minute) compared to an increase of thirty with spinal cord intact. After several more tests, they determined that the source of the x-factor was not anterior to the region of the transection of the cord—"i.e., in that part of the animal where, subject to the cerebrum, excitement and effort occurred."

Next, in order to test whether the sympathomimetic agent came from the hind quarters, needle electrodes were introduced into the lower lumbar cord. Under electrical stimulation, the hind limbs extended and the tail hairs became very bushy. Much to their delight, they observed a slow rise in the heart beat, approximately thirty beats per minute—just as in animals with spinal cord intact. "Obviously," they reasoned, "a substance is given off into the bloodstream by the hind part of the animal that affects the heart in the fore part." What that substance was, however, remained to be seen.

They then decided to extirpate the abdominal sympathetic strands from diaphragm to pelvis to see if the heart rate would remain "flat"—the word put in quotes because it did not mean strict fixity. "This was done," they later reported, "and to our great surprise the slow rise of heart rate still continued!" They also found that animals surviving cardiac denervation for long periods of time had a regrowth of vagal fibers, as indicated by sharp changes in heart rate that disappeared under the influence of atropine.[17]

The impasse thus presented in the fall of 1927 was alleviated by removal of the thoracic chains from the tenth rib to the diaphragm, and they subsequently found that complete removal of the sympathetic chains from stellate ganglia to the pelvis in one piece took them past the hurdle. This work, in conjunction with refinement of the sympathectomy operations, ultimately led to the development of, in Cannon's words, "new organisms in the history of physiology—vertebrates without sympathetic control of the viscera—and they became objects of interest in themselves."

In attempting to find a key to the puzzle of the belated slow rise in the rate of the denervated heart of excited animals, the trio made a good deal of headway with their experiments. But both Newton and Zwemer had commitments that required them to leave the department late in 1927 with many questions unanswered; thus, their findings were not immediately published.[18]

☙ ☙ ☙

In September 1929, Zenon Bacq, a young physician with a fellowship from the Commission for Relief in Belgium (C.R.B. Foundation), arrived in Boston fresh from a fruitful year of postdoctoral study at the Collège de France with André Mayer, who sent him on to Cannon. In later recording some of his impressions of the time he spent at Harvard, Bacq recalled that he "knew practically nothing about the physiology of the autonomic nervous system," but by good fortune "Cannon was looking for a co-worker to tackle a problem which had been growing slowly in his laboratory." Not only was Bacq available, he liked experimental surgery and had been well trained by the surgeons at the University of Brussels. And yet he never could understand why Cannon chose a 25-year old foreign visitor to collaborate with him in an important research—though more than forty years later he would remember it as "a great event, one of the most important of my career."[19]

Cannon's observations with Newton and Zwemer, that the beat of a cat's denervated heart remained perfectly steady, or flat, during excitation if all of the sympathetic chain were removed, had a logical interpretation and working hypothesis. That is, some adrenin-like substance was liberated in the blood stream during activation of the sympathetic nerves. In the fall of 1928, Cannon, with the assistance of Robert Moore, had briefly followed through by proposing to demonstrate the actual production of a cardio-accelerating substance from smooth muscle alone. Cannon and Moore conducted acute experiments on anesthetized cats with denervated hearts; the abdominal sympathetic chains over the sacral promontory were exposed, tied, cut, and the distal ends laid on electrodes. Stimulation of the peripheral ends of the cut chains resulted in an erection of hair over the tail and a slow elevation of the heart rate. The stimulation was also accompanied by a gradual increase in arterial pressure—further evidence of the release from the distal segment of the sympathetic nervous system of a sympathomimetic agent into the blood stream.[20]

In October 1928, Cannon wrote Phil Bard about his experiments with Moore, saying he was hopeful of "at last settling the difficulties which we fell into on previous occasions." He mentioned that Henry Dale, who was visiting, talked about "various chemical agents that are set free at nerve terminals." And he gave voice to the new idea he was mulling over: "It may be that we have been working all along with some substance given off at sympathetic endings quite apart from the true adrenal secretion." When later the results of his investigation with Moore turned out to be inconsistent—some promising, others disappointing—the experiments were discontinued.[21]

A year later, with Zenon Bacq, Cannon made another attempt to investigate the problem with a new series of experiments. In order to exclude all direct nervous interference and to sensitize the heart to adrenin, the technique chosen required the aseptic removal of the sympathetic fibers innervating the cat's heart about a week before the actual experiments took place. Bacq prepared several cats for Cannon, and later reported that the Harvard physiologist was pleased with the condition of the animals as well as with the conduct of the experiments. In all, fourteen experiments were completed in the autumn of 1929, the first six being repetitious of the operations performed previously by Cannon and Moore. In the eight succeeding ones, the heart was partially denervated, the adrenals eliminated or deactivated and, in some cases, the liver nerves were severed in a preliminary surgical operation. The animal was etherized on the day of the experiment; then to complete the cardiac denervation the vagi were cut in the neck, the common carotid arteries tied, the cerebral hemispheres ablated, and the lower abdominal sympathetic strands bared and laid on shielded electrodes. After decerebration, the ether was discontinued.

Shortly thereafter, the animal blew off the ether and Cannon and Bacq were able to stimulate the fibers to the distal portion without causing any pain. In this second set of experiments, two were negative, two displayed a single rise of four beats, and the four others were positive. Of these four, one yielded repeated increase of heart beat—"not longer than six beats per minute, to be sure, but regular and definite." The second had increases of fourteen and six beats. In the third, the heart rate rose slowly after stimulation and in two minutes was at its peak, eighteen beats above the basal, returning gradually to normal six minutes later, and with the hairs standing erect. And in the fourth, the first stimulation was without effect; but in response to a second stimulation, the heart rate rose to eighteen beats two minutes after the stimulus stopped and then gradually returned to the original level after eight minutes. Blood pressure, likewise, slowly rose and fell with the heart rate.[22]

The few positive experiments in the series were not reassuring as to the method used; hence Cannon was left looking for another way to investigate the mysterious cardio-acceleration that would be as reliable as emotional stimulation. The work, however, was interrupted by Cannon's absence in Europe for his exchange professorship, and it was not resumed until September 1930. "When he came back," Bacq remembered, "he was more than ever convinced of the importance of our research. He had seen H.H. Dale in London, who was busy building up the foundation of his work with acetylcholine." Cannon, according to Bacq, had discussed with Dale his ideas about his recent unpublished experiments.

When after ten months' absence Cannon returned to the laboratory, Bacq continued, recovery from bladder surgery forced him to limit his activities and he relied on the younger man to relieve him of much of the heavy physical burden of their investigation. After Bacq had prepared the animals and completed the spinal cord transection under anesthesia, Cannon took over the final preparations and performed the experiment with his assistance—working, in Bacq's words, with wonderful precision and elegance. Cannon also saw to the recording of experiments on the smoked paper of a large kymograph, which measured cardiac frequency and registered arterial pressure. Sometimes, when he was tired and the animal was still in good condition, his young assistant completed the experiment alone. Bacq wrote that the two of them spent hundreds of hours working together in an atmosphere of complete absorption. "The American physiologists on Cannon's staff," he added, "were jealous of me because I spent so much time with their 'boss.' However, I did not realize this until just before my return [to Belgium] in December 1930, so great had been my pleasure to participate at first hand in such an important piece of research."[23]

On resuming their investigation, Cannon and Bacq resolved to produce conditions not only like those under which emotional acceleration of the denervated heart occurred but also which, without anesthesia, would permit local electrical stimulation of sympathetic fibers distributed to smooth muscle. After preparing the animal with the heart denervated, except the left vagal connections, in the usual way, the right adrenal removed and the left denervated, and the liver nerves cut, the investigators allowed ten days for recovery. They then performed an acute experiment in which the cat was anesthetized with ether, its left vagus severed in the neck to complete cardiac denervation, and its spinal cord transected at the base of the seventh thoracic vertebra.

The severance of the cord at this level, just above the sympathetic strands in the lower thorax, effectively disconnected from the brain all remnants of the sympathetic system previously under central control. The result was a heart unaffected by any struggle or disturbance in the fore part of the animal. The whole procedure took so little time that the animal soon eliminated the ether used for anesthesia.

Five or six hours after the first operation, the same procedure was performed on the hind part of the animal, which, separated from the brain, was of course quite without feeling. For excitation of the pilomotor muscles, the sympathetic strand was laid bare at the root of the tail. By use of this method, Cannon and Bacq felt assured, "the effects of electrical stimulation of sympathetic strands became quite as regular and reliable as the effects of emotional stimulation" and the heart rate (recorded by a tambour set against the thoracic wall) was increased in the typical manner. After numerous experiments were performed using this technique, they could confidently state that "we have never failed to obtain the characteristic cardiac acceleration."[24]

Cannon and Bacq then proceeded to study the effect of checking blood flow in the region stimulated. When the blood flow through the hind end of the animal was hindered by closing the aorta and inferior vena cava above the iliac branches before applying the stimulus, the rise in heart rate was much reduced or delayed, despite the fact that the hair of the tail was standing. On the other hand, restoration of the blood flow by release of the ligature caused an increase in heart rate as in the normal animal.

They next injected either ergotoxine or ergotamine, which were known to block the positive action of sympathetic impulses and adrenin on smooth muscle. These agents either partially or totally inhibited the erection of the hair of the tail; nevertheless, the usual slow rise of heart rate was noted, and, in fact, the amount of acceleration increased. Thus, the experiments fully indicated that the sympathomimetic agent was still being released from the tail region, causing acceleration of the heart rate despite the lack of pilomotor activity. Again and again Cannon and Bacq emphasized in their final report that the sympathomimetic agent arose from smooth muscle, i.e., the pilomotor muscles.

To obtain further evidence of the release of a sympathomimetic agent from the tail region upon electrical stimulation, Cannon and Bacq used the salivary gland as a target organ. It had long been known that the denervated salivary gland is susceptible to the action of adrenalin, and any stimulation which would tend to promote an outpouring of adrenin into the blood, whether a reflex one or not, causes a marked increase in the flow of saliva. They repeatedly demonstrated that stimulation of the sympathetic nerves of the tail for about a minute definitely and strikingly augmented salivary secretion. So they concluded that the effects on the denervated heart were due to a chemical substance given off by smooth muscle, based on the following reasoning:

The only connection of the heart with the rest of the body is by way of the blood. The blood is not altered in a physical sense to a degree which influences the heart rate, nor is it altered chemically by the addition of metabolites from active skeletal muscles or the product of an endocrine gland. Yet cardiac acceleration continued after removal of all known sources of internal secretions or after severance of their sympathetic connections. When the sole remnant of the sympathetic supply to pilomotor muscles was abolished, the acceleration disappeared. When the smooth muscle of the tail hairs (or the uterus, bladder or intestines, or blood vessels) was acted upon by sympathetic impulses, the typical result followed. And, finally, they surmised that the effect was not due to the liberation of adrenin from sympathetic ganglia, or from paraganglia in the lower aorta, ruling out possible involvement of the coccygeal gland as well.

In consequence, Cannon and Bacq maintained that electrical excitation, in place of emotional excitement, which discharges sympathetic impulses into smooth muscle, causes a substance to be produced that, upon escaping into and carried by the blood stream, is capable of affecting viscera. They ventured to call the substance produced by smooth muscle a hormone because it fulfilled all the qualifications laid down by E.H. Starling (1914) in suggesting the word. That the action of the substance resembled the action of adrenin Cannon and Bacq readily acknowledged, but in view of the differences encountered during their experiments, they qualified their identification. "If it is adrenin, however," they argued, "it has a different immediate source than that discharged from the adrenals and should be distinguished, therefore, from adrenin."[25]

At the conclusion of his experiments with Bacq, Cannon at once set to work writing up the results for publication. But even before the study was completed, he began to prepare a report of the experiments he had performed with Newton and Zwemer several years previously, and he showed tentative drafts to his former colleagues for their judgment. In a letter to Raymund Zwemer (who had joined the anatomy department at Columbia), written to accompany the manuscript, Cannon described the consistent results that he had recently obtained. "Something is given off from the smooth muscle into the blood stream which accelerates the heart," he explained. "It is possible that it is the same material that is produced by the adrenal medulla. Elliott suggested that possibility twenty-six years ago—i.e., that the nerve impulses act in structures innervated by the sympathetic by local liberation of adrenin in the cells which are brought into action."

Cannon went on to tell Zwemer that the work he and Newton had done was actually the first demonstration that the humoral transmission of nerve impulses, previously tested by Loewi with salt solution on isolated tissues, is a physiological process. He ended on a pensive note, "It is interesting now to look back and consider our efforts in all directions to explain the phenomenon. If we had only considered more seriously what we rather jocularly spoke of as the 'tail hormone,' we should have had this interesting problem settled long ago."[26]

Years later, Bacq would remember that before he left Harvard at the end of 1930, Cannon showed him the report of their work that he had prepared rapidly and with almost no correction. "He had decided to call the substance we had shown to act at a distance from its site of liberation *sympathin*," wrote Bacq, "because we had no idea of its chemical structure although all its properties were those of epinephrine."[27]

The paper by W.B. Cannon and Z.M. Bacq, with the assistance of R.M. Moore, "A Hormone Produced by Sympathetic Action on Smooth Muscle," appeared in the February 1931 *American Journal of Physiology,* immediately after the report by Newton, Zwemer and Cannon on "The Mystery of Emotional Acceleration of the Denervated Heart After Exclusion of Known Humoral Accelerators." It was appropriate that the two articles be published in tandem for, as Cannon contended, they were complementary, many of the facts stated in the first paper being presented again in the second one.

Newton, Zwemer and Cannon began by reviewing the character of the slow acceleration that arises in the heart isolated from the nervous system. The quick rise due to secreted adrenin if the adrenals were left intact might be as great as 100 beats per minute, whereas the slow rise occurring after adrenal inactivation rarely was higher than thirty-six beats—more often between twenty and thirty. "Clearly," the investigators stated, "the effect of adrenin fully covers over the effect of the x-factor, and though it may collaborate with adrenin it can be clearly revealed only after the adrenin effect has been removed."[28]

Although excitement accompanying struggle was the most effective way of bringing about an increased heart rate by the x-factor, the faster pulse caused by repeated struggle was limited.

If they continued, the effect lessened, as though the x-factor was becoming exhausted or the heart was losing its sensitiveness. Why, the authors asked, had not the slow, late rise in the denervated heart been observed before? The answer lay in the fact that the denervated heart as an indicator of change in the organism had not been widely used; equally important, the phenomenon was not clearly manifested if the animal had been fully anesthetized since anesthesia compromised an increase in cardiac rate. Moreover, the acceleration was reduced by the operation itself.

In their continued search for the mysterious x-factor, Newton, Zwemer and Cannon were to enumerate and eliminate all of the possible causes of the acceleration they could think of. They reported that the phenomenon was not due to rise of arterial pressure, increase of temperature, escape of adrenin from the denervated adrenal medulla or cortex, or from the liver; nor due to substances produced by the pancreas, gastro-intestinal mucosa, semilunar ganglia, pituitary body, male gonads, thyroids, parathyroids, or active skeletal muscles. These negative results were as far as their experiments took them—until three years later when Cannon ended their report, "The solution of the mystery of slow acceleration of the denervated heart after excitement and effort, when all known accelerators are excluded, is described in the following paper."[29]

In his paper with Bacq, Cannon reported that in a cat with its heart denervated as described, stimulation of the sympathetic supply to the smooth muscle of the tail hairs and blood vessels causes a gradual increase in the heart rate similar to that reported for the denervated heart in the previous paper. He went on to suggest that the negative results described in the earlier paper, and the positive results in the present one, furnished the first evidence, so far as could be determined, "that cells other than those of endocrine glands, that are innervated by sympathetic neurones, give off into the circulating blood in the course of quite natural behavior a peculiar substance capable of influencing in a characteristic manner other cells having a sympathetic nerve supply."

While the action of this substance, newly named sympathin, was found to be like that of adrenin, it came from a different source and needed to be distinguished from adrenin. After discussing the similarities and dissimilarities between sympathin and adrenin, the authors argued, "Since the evidence indicates that sympathin and adrenin are liberated simultaneously by the same agency (sympathetic impulses), and have similar results on organs under sympathetic government, it is clear that when, in previous reports of sympathetic activity, effects had been ascribed to adrenin alone, sympathin must be understood as being present and coöperating."[30]

Nevertheless, Cannon and Bacq, like others (including Otto Loewi), held to the possibility that the sympathetic substance might actually turn out to be adrenin, developed for local action in smooth muscle cells. The evidence that adrenin and sympathin were similar if not identical had bearing on observations made by T. R. Elliott in 1904 and 1905 concerning sympathomimetic impulses. In their paper, Cannon and Bacq referred at length to Elliott's ideas on the relationship of adrenalin to the sympathetic supply which their present work supported.[31]

Cannon had reason to be pleased with Bacq's collaboration, and stated soon after his departure from Harvard, "I do not think that we have ever had a more energetic, intelligent and enthusiastic young investigator than he." According to Bacq, his paper with Cannon was much discussed in physiological circles in the United States and Europe. At the same time, he reported, Cannon experienced difficulty in attempting to convince reluctant conformist physiologists that chemical transmission was not an unthinkable concept.[32] After returning to Belgium, first in Brussels, then at Liége, Bacq maintained an interest in the work he had begun with Cannon by publishing papers and writing a thesis on sympathin, sympathectomy and the sympathetic nervous system. More than that, he became devoted to championing the cause of chemical transmission of nerve impulses.

Cannon, who repeatedly referred to the release of sympathin from smooth muscle, was well aware that his friend of many years in the zoology department at Harvard, George Parker, had his own notion of the source of the humoral substance in neurotransmission. Parker had long been interested in chromatophores, or integumentary cells, by which a number of animals, such as certain crustaceans, fishes, amphibians and reptiles, may make temporary changes in the color of their skins. When, in June 1931, Cannon delivered a paper on his most recent researches to a meeting of the Association for the Study of Internal Secretions, he acknowledged his friend's stance by explaining,

> Parker suggests that the agent which acts when nerve impulses arrive at the effector cell is a *secretion of the nerve terminals* and not, therefore, a derivative of the cells itself. In other words, the humoral intermediary between the nerve impulse and the responsive terminal organ lies within the cell, according to Elliott, but is produced by the nerve terminal and affects the cell by entering it, according to Parker.

"At present," Cannon was frank to admit, "I do not know of any evidence which will clearly determine which of these two views is correct." The fact that known humoral mediators of nerve impulses come from cells of nervous origin, he continued, favors the production of a substance by the nerves at their terminals—and therefore sympathin might be derived from the sympathetic nerve endings. On the other hand, the absence of response of smooth muscle, lacking sympathetic connections, to adrenin indicates a relation of sympathetic fibres to muscle structure and function. Then, too, it seemed "almost impossible that sufficient substance could be produced by the minute nerve terminals." After weighing Elliott's notion as opposed to Parker's, Cannon ended by leaving the difference unsettled for the time being.[33]

In 1932 Parker published a small book on *Humoral Agents in Nervous Activity, with Special Reference to Chromatophores,* in which he devoted a final chapter to "Neuro-Humoralism." Now it was his turn to acknowledge Cannon's position: "The exact source of the several humoral substances thus associated with muscle has not excited much attention. In the most recent contribution to this subject, that by Cannon and Bacq, the smooth muscle cell is continually referred to as the element from which the humoral substance emanates . . . It seems quite evident from the conditions under which humoral substances of muscle are produced that these substances must originate either from the muscle cells or from the nerve terminals."

Parker, for his part, displayed none of Cannon's hesitation in choosing between the two, for he continued:

> Because of the small size of the terminals, it would be natural to assume that the given substance was produced by the muscle, but

> if the most recent work on chromatophores has any meaning at all, it points most conclusively to the nerve terminals as the source of these materials. Although it is true that the terminals are of small size, it must be remembered that the substances under consideration, such for instance as adrenaline, are enormously powerful, as shown by the degree of dilution that they may suffer and still remain effective, and that therefore a nerve terminal may be amply large enough to secrete them.[34]

Cannon continued to be ambivalent and never addressed the argument again in print. Much as he admired G.H. Parker, he also revered T. R. Elliott. Perhaps he, like Henry Dale, was reluctant to embrace unproved ideas which seemed to fly in the face of the pioneer work on chemical transmission that bore Elliott's name. Years later, Zenon Bacq would recall how happy Cannon was when he showed him Elliott's prophetic note of 1904 while writing up the results of their research for publication in December 1930. Bacq noted that Cannon loved to pay homage to the work of his predecessors or contemporaries with a generosity inherent to his nature.[35]

15 No End of Emergencies

When Cannon journeyed home from England in the summer of 1930 to put himself in the hands of surgeons, he could not possibly foresee that the sudden crisis was but one of several events to pose a threat in the coming years. In a strange twist of fate, the emergency theory that had occupied his thoughts for so long gave way for a time to a number of real-life emergencies that would distract him from his work on chemical mediation and disturb his equilibrium.

Given the economic panic that followed the stock market crash of October 1929, one might think that the first and foremost emergency for Cannon would involve money matters. That, however, was not the case. As far as his work was concerned, just the opposite was true since the generous grant provided by the Rockefeller Foundation put his research in a fiscally secure position that was the envy of all. On top of that, shortly after his return from abroad he learned that his salary would be increased by $2,000 a year, in accordance with maximum salaries in the Harvard Faculty of Arts and Sciences.[1]

Nor is there much evidence that the Cannons were personally discomfited by the crash, for at first there seemed to be little awareness of economic hardship. Marian Cannon Schlesinger, in her family memoir, recalled that she was struck less by any immediate effects of the Depression than she was by the removal of their rented house to a new location across the street on Frisbie Place. Perhaps Mrs. Cannon felt the economic decline more than other family members, because it had an impact on the sale of her second novel, *Heirs,* and, to make matters worse, became the excuse for rejection slips from publishers to whom she subsequently sent manuscripts.

Tenured professors and others who were secure in their posts were affected only gradually, if at all, and even gained an advantage since the dollar's purchasing power rose as prices fell.[2] There came a time, however, when the Depression began to deliver its knocks to Cannon. It did so by pelting him on two flanks, one at work, the other in his home life.

In May of 1932, department heads at Harvard received word that while budgets already sent in for the following year had been approved, business conditions had worsened and it was feared that the University's future income would be drastically reduced. In consequence, they were instructed to exercise care and economy in all expenses, and also to prepare for future cuts in budgets, including salaries. At the Medical School, Dean David Edsall took stock of the situation and in the autumn announced that by using all his resources and cutting costs, he believed the emergency could be met without reducing department budgets. Edsall's warning that no new personnel were to be taken on unless definite commitments had already been made was offset by his expressed hope that, if possible, there would be no reductions of salaries of persons with Corporation appointments and no dismissals of such persons.

Nevertheless, as the Depression deepened, its impact on young people in training—students and fellows alike—became a matter of concern. Cannon, like other department heads, found himself in the unhappy position of having to turn away new candidates for the Ph.D. degree because persons who were already trained had nowhere to go. In one instance early in 1933, on hearing about an application from an N.R.C. Fellow which he felt would have to be rejected, Cannon wrote Edsall, "This is the sort of letter which Dr. Folin and I have repeatedly been receiving for months. There is no position in our departments available. Indeed, men who should leave and give place to others have had to be retained because of no openings elsewhere." This situation would continue to be a problem for Cannon, if not throughout the Depression, at least for a good part of it.[3]

Although financially secure, the Cannons were never wealthy. Nor did they indulge in anything approaching a lavish life-style. Quite the contrary, in fact, for both Walter and Cornelia habitually practiced the frugal ways that they had learned early in life—

regardless of the disapproval of their children who sometimes wished they could take a flyer at champagne tastes. In short, the New England precept of plain living and high thinking suited them just fine. There was, however, one painful lesson that Cannon learned as a consequence of the economic downturn.

During the boom times of the 1920s, he had joined with other medical and professional men who were mesmerized by the silver-tongued promises of the Swedish "match king," Ivar Kreuger, then touring the United States to drum up capital for his business ventures. Cannon, like his friends Harvey Cushing and Jerome Greene among others, couldn't wait to purchase stock in Kreuger & Toll and the International Match Corporation through the banking firm of Lee, Higginson & Company, which recommended the Kreuger securities to its customers. Most of the certificates were held in Mrs. Cannon's name, since it was her earnings from *Red Rust* that made the purchase of more than 200 shares possible.

Cannon had invested in other securities through Lee, Higginson, which were held for his children until they came of age. During the early 1930s, his broker there, Francis W. Sever (who was also a neighbor), while conducting transactions for the family and keeping Cannon apprised of the holdings, commented from time to time on developments in the business and financial world that looked ominous. "Ominous" was hardly the word for the scandal that broke out in the spring of 1932 when the little match king shot himself in Paris. Afterward, it was discovered that Kreuger's multifarious financial dealings had been so strained by the world-wide Depression that the assets and profits recorded for the business were largely fictitious.[4] When a wiser but sadder Cannon subsequently asked Sever what was being done to deal with the crisis, he supposed there was "no telling where the rascality of Kreuger may lead."

We don't know exactly how much money Cannon lost in Kreuger stocks. Surely not as much as Cushing, who could afford to invest far larger sums, or Greene, who lost not only his own money but that of others he had advised (in his capacity as an associate in the Lee, Higginson firm) to buy the certificates. What we do know is that the Kreuger affair completely destroyed the old firm of bankers, and in June Mr. Sever took a position with Kidder, Peabody & Company. Although investigations were proceeding, Sever told Cannon that summer, the Kreuger company had been declared bankrupt. "This is the most disastrous affair and has brought ruin and desolation in many lands," he lamented.

> The firm of Lee, Higginson & Co. has had to go out of business and the partners have lost their private fortunes. There was no security in which Lee, Higginson & Co. had greater confidence and faith, or one which they recommended more highly to their clientele. I feel very sorry that so much money in your various accounts was invested in these securities, but we are all having to start afresh.[5]

When news came of the collapse of the Kreuger empire, Marian remembered, there was "a certain gloom and tension around the house on the part of my father," who was "obviously overcome with guilt." Her mother, who had nothing but scorn for "filthy lucre" as well as for those who saw the pursuit of mere money as an end in itself, only shrugged her shoulders and considered herself more fortunate than most. Besides, her earnings from her "Pueblo" books were safe in a savings bank. Her worst fear was that the nation's hope for better days would be expressed in November with a landslide vote for Franklin Roosevelt.[6]

❧ ❧ ❧

On October 15, 1931, Cannon received a signal honor to mark his twenty-fifth anniversary in the George Higginson Professorship of Physiology. For more than a year, a group of friends and colleagues led by Henry Christian, Cecil Drinker and Alfred Redfield had planned to commemorate the occasion with a celebration, at which a portrait of Cannon would be presented during a day of festivities at the Medical School. Three of Cannon's colleagues outside of Harvard were invited to be the main speakers.

The dean presented the opening remarks at the ceremonies. While noting Cannon's modesty, Dr. Edsall nevertheless proceeded to peel off his sterling attributes—his contributions to knowledge, his scholarship, his teaching ability, his understanding of clinical phenomena, his inspiring intellectual qualities, his strength of character—in seriatim fashion. The first guest speaker was Walter C. Alvarez of the Mayo Clinic. Dr. Alvarez, who as an unknown young physician from the West with no special training had come to Boston in 1913, said he would never forget Cannon's generosity in giving him a place in the laboratory and facilities to work on the problem he had brought with him. He interspersed his talk on

Cannon's contributions to medical thought and progress with personal comments collected from Cannon's students and colleagues.[7]

From Johns Hopkins came William H. Howell to talk about the development of physiology and Cannon's influence on it; and from Cornell came Graham Lusk to speak on the life of a professor. After President Lowell, in behalf of the University, accepted the portrait painted by Marie Page, Cannon addressed the gathering. Whether the portrait was "a good likeness or not, you can judge better than I," he said. "I told the artist if I didn't look like that, it is the way I should like to look." He had insisted he be painted in the brown laboratory coat that he habitually donned at work; in private, he would later express the opinion that the portrait didn't look anything like him but was a great likeness of the lab coat.

After Cannon made a few informal remarks about the satisfactions of his career and thanked the audience for some very happy memories, Mr. Lowell closed the meeting. Coming full circle, he, like Edsall, called attention to Cannon's well-known modesty: "There would seem to be one person here present who doesn't wholly appreciate Dr. Cannon. Dr. Cannon, let me introduce you to the great Professor Cannon. He is a very modest gentleman, but you will find him worth knowing."[8]

Cannon was so touched by the exercises that he confided to friends he would have broken down if they had gone on one minute longer. And yet, there was a harsh paradox to the occasion that few persons at the celebration could appreciate or begin to suspect. Despite the good wishes so affectionately bestowed on Cannon, calamity lay in wait for him. Even as his friends commemorated his long career and envisioned many more years of achievement, he had already begun to display the symptoms of a mortal illness—one that would gravely define his public as well as his private existence until the end of his days.

In the spring of 1931, about six months before the anniversary celebration, Cannon started to experience a bothersome itch and broke out in various parts of his body, arms and legs with what he called an erythema (or urticaria) of the skin. That summer at his farm in New Hampshire, where he was writing his book on homeostasis, he had a miserable time, with only sporadic relief from the condition. Since the itching was frequently accompanied by digestive upsets, he experimented with his diet and for a time came to believe that eggs were at the root of his trouble.

Still, the itching kept coming back. Having ruled out drugs, foecal infection and certain dietary factors, Cannon, in consultation with his friend and physician, Joe Aub, settled on a marked sensitivity to colon bacilli as a possible cause. Dr. Aub made a case note (the first of many) on September 26, 1931, that Cannon "appeared in the office today with intense itching, which felt like little prickles all over his body. His skin was very red and blotchy." Although Aub was alarmed, he could do little but give Cannon a variety of tests and keep on adjusting his diet.[9]

A few weeks later, E. Lawrence Oliver, clinical professor of dermatology at Harvard, wrote Aub in reference to a discussion they had had about Cannon's case: "When I spoke of mycosis fungoides, I used that as an example of the group of malignant lymphomata as well as leukemia." Oliver's letter confirmed Aub's suspicion that Cannon's lesions "remotely resembled some of the early manifestations of leukemia." While Cannon felt well enough to travel with his wife in November to receive an honorary doctorate from the University of Paris and the Baly Medal from the Royal College of Physicians in London, it was a quick trip and they did not tarry.

Aub meanwhile consulted with other doctors—among them, Soma Weiss, George R. Minot and Cannon's dermatologist C. Guy Lane—and it is clear that by early 1932 all of them were terribly worried about Cannon. And yet, other than checking for possible allergies, suggesting changes in his diet, and arranging for exposure to ultra-violet light, they really didn't know what to do for him. Finally, grasping at the straw that fatigue might be exacerbating Cannon's physical state, Aub went to see Edsall. The dean "practically commanded" Cannon to take a restful and relaxing trip in the spring—which he and Cornelia did, going to California and returning by way of the Panama Canal.[10]

Cannon was in such a state that he willingly left his laboratory because he did not want "to have another year so unsatisfactory in actual achievement as this year has been." When he came back from the trip not much improved, his doctors hoped that a summer in Franklin would help. In June, biopsy by Burt Wolbach and other pathologists brought confirmation of the diagnosis, now called lymphoblastoma, probably of the mycosis fungoides type. The prognosis remained dire.

All along, Dr. Aub thought someone in the family ought to

know that the outlook was bleak, and he proposed telling Ida Cannon about Walter when she got back from Europe. "She is a wise person and can advise us whether anyone else ought to be told or not," Aub wrote Lane after the latter had examined the sections Wolbach had sent him. Oddly enough, Cannon, who was already in New Hampshire by this time, appeared to be doing much better and reported that exposure to sun and outdoor exercise had made him more comfortable. In August, Dr. Lane, after seeing Cannon about some lesions on his hands and neck ("probably keratosis"), marveled to Aub, "He looks a different man. The itching is entirely gone and some of the erythema has certainly faded out." A few weeks later, when Aub had seen Cannon's sister and told her the diagnosis, he wrote again to say that he was not planning to go further unless Lane thought it wise to tell someone else.[11]

Cannon continued to do well after returning to Cambridge in the fall of 1932, and Aub reported in October that the itching had entirely stopped, that he was sleeping better and was far more cheerful than he had been. Cornelia watched what he ate and "snatch[ed] indigestibles from under his nose." Against the advice of everyone in the family, he even had "gone off on a tear" to Chicago and Ann Arbor to attend committee meetings and give talks to medical students. Cornelia threw up her hands in resignation and remarked with keen insight, "His motives sound the noblest, but I really think he likes going around and being wined, dined and feted. After all, he reaps in pleasant fashion what he has sown through the years." At year's end, she gave thanks that after examination by Dr. Quinby, who had removed the cancerous growth from his bladder in 1930, Walter was "found perfect." Yet she observed that though his itching was less, he still indulged in a good deal of surreptitious scratching.[12]

Just about this time, Cannon, at the behest of Roy Hoskins, attended a meeting of the directors of the Neuro-Endocrine Foundation and got to talking with Albert W. Rowe, who directed a clinic at the Evans Memorial Hospital in Boston. After Dr. Rowe obtained Cannon's promise to address a gathering of alumni at M.I.T. (this was the occasion for his talk on Biocracy), he further urged him to come to the Evans and have a work-up by his group there. Cannon did go, but not wanting to hurt Joe Aub's feelings, informed him about it first, writing, "This doesn't mean I don't have faith in you, but it seems like a good idea." Rowe duly submitted his report to Aub and told Cannon that the two most outstanding findings of the study were thyroid failure and hepatic dysfunction. His recommendation was that Cannon continue under the care of Dr. Aub.

Not content to stop there, Rowe then called a meeting of all those who had been concerned with Cannon's case. A somewhat wary Aub replied that he hoped only medical people would be included: "I feel this way because I have now told Mrs. Cannon the diagnosis, and she agreed with his sister that it is a great deal better for Dr. Cannon not to know, if that is possible. All through this trouble I have been very discreet as to whom I've told the diagnosis, and I sincerely hope that this conference will not affect the situation."

In addition to Aub, Drs. Wolbach, Quinby, Lane, and Hoskins attended the conference held early in 1933, as well as their own "skin man," Dr. Jacoby. Rowe revealed that the primary diagnosis had been mycosis fungoides, with minor conditions of thyroid failure and hepatic dysfunction. When he expressed his feeling that Cannon should be told about his condition, Aub said it was Mrs. Cannon's desire that her husband not be told. Since Mrs. Cannon's wish was supported by all of the group, Rowe withdrew his recommendation. Before the conference broke up, the doctors discussed the possibility of X-ray therapy.[13]

No sooner had the meeting called by Dr. Rowe ended than Cannon's itching came back with a vengeance and he was started at once on X-ray treatments. The therapy gave him considerable relief, but before long Cornelia wrote her family that he was "having some bad nights lately, a combination of eating the wrong things and getting tired starting up the old trouble." She was particularly anxious to get him away from the laboratory and arranged an automobile trip in April to visit Robert and Ada Yerkes in Orange Park, Florida. That gave Cannon the rest he needed, with the result that X-ray dosage could be halted.

Shortly after his return to Boston, a team headed by John Homans operated on Cannon's hands at the Brigham Hospital. It proved to be the first in a series of operations on the old burns from his early experiments with X-rays that had bothered him for so long and had become almost intolerable. "The miracle is that Father should have been able to go thirty years with these burns," Cornelia wrote the children, "when so many of his contemporaries have lost hands and life."[14]

In a note of thanks to one of the members of the surgical team, Cannon wrote, "The lesions were caused by work which I did when a medical student, using the X-rays for the first time in studying the processes of digestion. That was long before you were born—back in the 'gay nineties.'" Bradford, then in his final year as a medical student, made himself indispensable to his father and took delight in caring for him during recuperation. "How they are going to get along without each other next year," Cornelia wondered, "I do not know."

A second operation followed on the heels of the first, leaving Cannon with both hands bandaged and virtually helpless. At first his spirits were terribly low, but when it appeared that the grafts were taking hold, he was able to leave for Franklin where he cheered up enormously. By mid-June Cornelia reported, "He suns himself, we discuss the news, applaud Roosevelt and damn the American Legion, and are altogether having a little honeymoon, very serene and lovely." For most of the summer, he was able to settle into a daily routine of spending as much time in the sun as possible—the "Puritan father" a sight to behold wearing only a pair of shorts, socks and shoes, Cornelia noted with some amusement. "The result of it is that his skin is getting a beautiful mahogany color and he is almost forgetting what itching is."

Her optimistic outlook was soon shattered. A spell of cold, rainy weather in September brought with it disheartening setbacks, and Cannon's skin, having no sun, began to bother him until, unable to concentrate and feeling nervous and driven, he seemed to revert to the worst stage of his illness. He was so wretched that when they returned to Cambridge, Cornelia resorted to an old ploy she had used before. One day, determined to divert her husband's mind from his agony, she got him to go out and buy a car. That afternoon, he came home "fairly beaming with delight," his itching forgotten with the purchase of a brand new Ford (the Cadillac had been turned in for $49).[15]

All through the long months of suffering, Cannon had borne his affliction stoically, expecting it to be short-term and to run its course any day. Each time the torment subsided, his hopes that the end was in sight were dashed by yet another bout of digestive upsets, sleeplessness accompanied by nervous agitation and, worst of all, an infernal itching. At last, after more than two years, the patience of Job had worn thin. Weary of swinging between the heaven of hope and the hell of despair, Cannon simply wanted to know what was happening to him and why.

Cornelia, in the meantime, was finding the terrible knowledge that Joe Aub had divulged to her a heavy burden to bear, and she began to think of sharing it with her family. By the autumn of 1933, Linda, engaged to marry a young geologist, Harry Burgess, and Marian, beginning her senior year at Radcliffe, were the only Cannon children still living at home. The year before, Wilma had gone to China to marry John Fairbank, who was beginning a study of the Chinese Maritime Customs Service for his doctoral dissertation at Oxford. A few months later, Helen had gone west to spend her freshman year at Stanford University. After graduating from H.M.S. in 1933, Bradford had taken an internship at Barnes Hospital, affiliated with the Washington University of St. Louis, as a first step toward surgical residency. Sometime in October, Bradford received a confidential letter from his mother, who was writing to apprise him of the latest news about his father.

During the summer, she wrote, Father had become so distressed over not knowing what was the matter with him that she decided he had a right to know and informed Joe Aub that she was going to tell him. "Dr. Aub was very reluctant to have me do it," she continued,

> but I insisted. I told him how simply and courageously Father took the bladder trouble and that, though the small trials were hard for him to bear, he was always equal to the big things . . . Well, last night I told him. He took it very cheerfully, of course I tried to be rather casual about it, and asked a few questions, though he soon plumbed to the bottom of my knowledge, said he must read up on the subject and then went quietly to sleep . . . I did not say Hodgkins Disease, which has such definite and painful associations in Father's mind in relation to Professor Bayliss, but lymphomatous infiltration of the skin, very slow in developing, no worse now than two years ago, with the prognosis good.

She said that she planned to tell the girls their father's trouble was serious, so that these last years—and she felt confident they would be many—might be unclouded. She did not, however, intend to tell Wilma since she was so far away and helpless to do anything. Lastly, she assured Bradford that it was a comfort to her to feel his faithful affection, which was the deepest joy to his father.

Meanwhile, Cannon kept Bradford informed about his condition. He wrote him that Dr. Aub, desperate to ease his distress, straightaway had him treated with X-rays and put him on large doses of Viosterol (a combination of vitamins A and D). Aub further advised Cannon to pare down his commitments, and so he was resigning from extra activities like the Sex Committee, the Administrative Board, and the Cambridge Scientific Club. "And that's that. I'm planning to take life more easily and more simply henceforth, with frequent trips to Franklin weekends with Mother, as you have suggested." He also resolved to "strip down to essentials and continue only the work and the relations I enjoy." A postscript to another letter contained a brief comment that gave Bradford no hint as to his father's true state of mind: "I have at last been told the diagnosis of my skin disease. You were told last winter, I understand. Apparently the process is slow in my case and the X-rays appear to hold it in check."[16]

By some mysterious alchemy, a wondrous thing happened when the veil that shielded Cannon from the knowledge of his illness was lifted. He at last knew the truth—if not the whole truth, at least a part of the truth as it was then understood—and his health began to improve dramatically. The X-ray treatments soon stopped and were not deemed necessary for many years after. The vitamin therapy continued, for Cannon was convinced it had a beneficial effect in diminishing the itching. In any event, the combination of X-rays and Viosterol in the fall of 1933 was credited for putting him into a remission and, except for the continued operations on his old X-ray burns, he was granted a period of relatively good health that lasted for some time.[17]

All through his ordeal, there was one person to whom Cannon wrote candidly about his physical problems. After confiding in Walter Alvarez when he participated in the twenty-fifth anniversary celebration held two years before, Cannon had kept him apprised of his condition. Not only was Alvarez a skillful, caring and sympathetic physician, he was also an expert on nervous indigestion and mind/body relationships. In a letter of November 2, 1933, Cannon poured out his feelings to Alvarez:

> It appears that my medical colleagues told Mrs. Cannon last winter their diagnosis of my skin condition and did not tell me because they thought it would worry me! This fall I was so puzzled by the continuance of the redness and itching that Mrs. C. felt that I ought to know the facts. She *knew* I was a matter-of-fact person, and not so much disturbed by reality as by uncertainty. And so I was informed that the condition is lymphoderma. Apparently it is a rather slowly developing case, for it has been going for about 3–1/2 [sic] years without much progress. X-rays mitigate the itching considerably (about 1/4 an erythema dose, occasionally), but some opposition has been expressed to their use, because it has been suggested that the condition itself may be the consequence of my early exposure to X-rays.

Such a consequence was for Cannon "not a convincing idea, for the relation seems remote." For others, however, it appeared that prolonged exposure to radiation might well have been a predisposing factor in what, twenty-three years after he had stopped using X-rays, was ultimately called a case of mycosis fungoides with development of lymphatic leukemia.

Once Cannon's prognosis improved, one of the first things he did was to ask Aub if he could rejoin his clubs and committees, for he had missed them and was quite lonely. "Already I had come to terms with my future," he wrote in acknowledging one of Alvarez's wise and friendly letters, "in the conviction that really nothing was changed—at best life is uncertain, and, in my case, it remained so!" He ended his letter on an upbeat note. If he could only carry through the plans he had made to extend his studies on smooth muscle into the central nervous system, it would be his best work yet.[18]

Though Cannon found that "reducing the irritability of the nervous system" could lessen the intensity of his skin pathology, at no time did he refer to self-regulating or homeostatic mechanisms when discussing the state of his own health. But he *had* noted a similarity between the steady state of an individual and that of society at large, elaborating these ideas in an imaginative analogy that formed the epilogue to his book *The Wisdom of the Body*. Cannon's epilogue, as well as his speech delivered at M.I.T. on the subject of "Biocracy," captured the attention of businessmen and social scientists, some of whom attempted to incorporate his ideas in their own constructs of social, economic and political theory.[19]

For the most part, Cannon was content to leave social and political theory to others. It was the threat that fascism increasingly posed to scientific research and to the lives of his foreign students and fellow scientists that would involve him in the politics of the following years. Although conscious of the political changes occurring in Europe in the early 1930s, it should be noted that initially Cannon, like many other enlightened academics and physicians, did not see very clearly the dangers of fascism.

In May 1932, when Anton Carlson at the University of Chicago wrote a letter to *Science* suggesting that physiologists boycott the coming International Congress of Physiology in Rome because of Mussolini's dictatorship, Cannon and A.V. Hill, both of whom were members of the International Committee, were furious. "I do not see much point in refusing to associate with our Italian colleagues because we do not approve of certain acts of their government," Cannon wrote Hill in England. "Certainly Carlson would kick strenuously if we refused to have anything to do with him because we didn't like the way Chicago was governed." "What an uncivilized old ass Carlson is!" Hill replied, adding, "I must say too that I think Cattell is almost as stupid to have published it. It can't do any good and it may do much harm. The best thing, I suppose, is to hush it up and to say nothing more about it."[20]

There were others who thought like Hill. Early in 1933, when the anthropologist Franz Boas petitioned the Council of the National Academy of Sciences to protest the increased tendency in Europe to control scientific work by interfering with freedom of thought and opportunities for research for political and racial reasons, the Council, in its wisdom, decided it "did not feel it could take the action requested." A few months later, however, after the first dismissal of Jewish scholars and scientists from German universities, the response of scientists to Boas' pleas changed.

When Boas wrote Cannon and asked if he would join an emergency committee organized to raise funds to help ousted scientists continue their research, Cannon promptly agreed. As chairman of the board of the Ella Sachs Plotz Foundation at Harvard, Cannon told Boas, he had already made provision to send the biochemist Richard Willstätter the $400 he had requested, and the Plotz Foundation would be willing to help others.[21]

Scientists on both sides of the Atlantic were anxiously watching the German situation. That June, Cannon wrote a cordial note to Otto Loewi in Graz, Austria, to say that Loewi's Dunham Lectures had "made a lasting impression on the Harvard group," and appended an expression of concern to his letter:

> It is a great satisfaction to all of us who are watching events in Germany and Austria that thus far Austria has fought effectively Nazi influence. Anything that will reduce the arrogance and the assumed power of the present German leaders seems to me valuable if it has no other effect than to make them more cautious. You will be interested to know that a prominent professor from the University of Berlin took a special trip from New York to Cambridge last week to tell one of his colleagues that many of the university men of Germany were deeply ashamed of the way in which the Government had acted and that they were exercising all the influence they could to modify the repressive policies which had been announced.

Despite the manifest outpouring of concern, it had already become apparent that the problem of helping refugee medical scientists and physicians was so complex that an overall strategy would be needed. When one of Franz Boas' close associates, Dr. Benjamin Harrow, sent a request to Cannon to sign a petition urging the government to lift the bars against immigration, he refused. "It seems to me," Cannon replied, "that with approximately twelve million men out of work in this country and with universities shutting down or refusing to increase their personnel . . . it would be unwise to modify the existing immigration laws. I believe that a move in that direction at the present juncture might be likely to result in a development of race feeling in this country which, though not comparable with that which has already developed in Germany, might be seriously disturbing."[22]

Cannon's fear of a possible explosion of anti-Semitism in the United States was not a measure of his political timidity. During the 1930s, anti-Jewish feeling was part of the American social environment. Constantly stirred by the strident anti-Semitic remarks of radio preachers like Father Coughlin and Gerald Winrod, as well as by political mountebanks like Gerald L.K. Smith and William Dudley Pelley, such rhetoric became part of the ritual message mouthed by nationalist groups. It was even given a kind of public respectability in speeches by congressmen and senators such as

John Rankin of Mississippi, Martin Dies of Texas, and Robert Rice Reynolds of North Carolina.

Nor did Cannon need instruction as to the presence of anti-Semitism at Harvard. He was well aware of President Lowell's strenuous efforts during the 1920s to establish a quota system to restrict the entrance of Jewish students to Harvard. He also knew that certain hospital staff in Boston had a hand in driving Oscar Schloss, Harvard's Jewish professor of pediatrics, back to Cornell from whence he came because they would not refer patients—which severely limited his income.

Harvard was not alone in such practices. The leaders of many universities across the country, in proper and gentlemanly voices, adopted exclusionary quotas against Jewish students and teachers. Indeed, eminent medical school deans like George Hoyt Whipple at the University of Rochester or Milton Winternitz at Yale—who often lectured their faculties and students on the importance and necessity in the medical profession of that fragile element "character"—made little effort to conceal the operation of their own anti-Semitism in the administration of their schools.[23]

While Cannon was unwilling to sign the petition, in his letter to Harrow of June 2 he related that he was helping to raise funds in aid of the ousted Jewish professors through use of his name as well as his influence on the board of the Plotz Foundation. "You will not think for a moment, therefore," he ended, "that my interests and sympathies are not engaged."

Committees continued to spring up almost overnight to deal with the emergency. In October 1933, Cannon received a letter from Bernard Sachs, a neurologist in New York, about the organization of a new committee—this time a Committee in Aid of Displaced Foreign Physicians. Sachs wrote Cannon that the committee not only was anxious to secure posts for the refugees to enable them to carry on their work but would also endeavor to "avoid assignments that might work injury, or introduce unfair competition with, American physicians." The members of the executive committee, with George H. Baehr as secretary, included such notables as Alexis Carrel, Emanuel Libman, and William H. Park in New York; David Riesman in Philadelphia; and J.M.T. Finney in Baltimore. Sachs invited Cannon to join the executive committee, urging, "The need is great; we must begin activities soon; so let me have your 'yes' as soon as possible."

The "yes" was not forthcoming. Cannon begged off on this occasion, saying that he couldn't undertake any work because he had been ill and, at the urging of his physician, was reducing the number of tasks consuming his time and energies. As a matter of fact, he had resigned from a national committee to which he had belonged for over a decade, and lessened other obligations as well. He was tempted to join the new committee and would like to give his time freely and generously, he told Sachs, but at this particular moment he couldn't do it.[24]

Along with the inception of American emergency committees in 1933, there was similar activity abroad. By the autumn an Academic Assistance Council headed by Lord Rutherford and A.V. Hill, among others, had been launched specifically to aid displaced German scholars, physicists and medical scientists to carry on their research in Great Britain. Henry Dale wrote Cannon to explain the urgency of the situation and to assure him that funds from the Ella Sachs Plotz Foundation would be immensely welcome. Already the Council was literally overwhelmed with requests for help from the initial flood of refugee medical scientists in England.

Cannon, in turn, wrote Hill that he was prepared to place a grant of $3,000 from the Plotz Foundation at the disposal of the Academic Assistance Council. "It should be understood by the Council that the grant is not to be regarded as a continuing grant," he pointed out. "The Trustees feel that the change of policy which we make by voting money for maintenance and not primarily for the promotion of scientific work is strictly an emergency move, and we prefer to await developments before deciding to continue in that manner."[25]

Later that year, Cannon received several letters from the Committee in Aid of Displaced Foreign Physicians in New York, this time from its secretary, Dr. Baehr. Asking Cannon for the benefit of his help and counsel, he wrote that they were making a determined effort to place a limited number of specially qualified foreign medical men in temporary posts in medical schools, hospitals and research laboratories. Furthermore, Baehr reported that the committee would maintain a registry of refugee foreign physicians, from which institutions in this and other countries might obtain information concerning persons qualified for positions which could not be filled locally.

This time, Cannon not only accepted the invitation to be a member of the general committee, he also agreed to receive a worker into his laboratory who would be on the same footing as fellows of the National Research Council or the Rockefeller Foundation—that is, provided for by outside funding. Almost at once, he was sent the names of several displaced physiologists, and just as quickly he rejected them. "Unfortunately," Cannon wrote Baehr, "I do not see how any of the men named on the list would fit into the work which is being done here. As I have explained in earlier correspondence, my experience has shown me that it is highly undesirable to introduce into an organized group of workers someone who brings his own problems and who works in a field very different from that studied by the group."[26]

But when he received the name of Ernst Fischer, Cannon was more receptive and wrote to his friends in England and the United States for their views of Fischer as a physiologist. W.O. Fenn in Rochester sent a letter of recommendation, which was immediately followed by another from A.V. Hill in England. "I know Ernst Fischer of Frankfurt very well indeed and thoroughly approve of him," Hill wrote, calling him "a thoroughly capable worker and an intelligent person." Although "not quite in the front rank as a research worker," Hill believed that Fischer would make an excellent member of a team. "He is fond of teaching; he is a person with a broad culture and intelligent outlook and I should be very happy myself to have him about my laboratory." In fact, Hill was arranging a summer post for him at the marine biological laboratories in Plymouth, England.[27]

Intrigued by the letters of recommendation, Cannon decided that he would ask Harvard to appoint Fischer temporarily to his department. When Elliott Cutler interceded to request approval of the appointment, Worth Hale, the assistant dean at the Medical School, refused. When Cutler asked for an explanation, Hale told him he had been instructed by Dean Edsall (who was then away) that President Conant (who had succeeded A. Lawrence Lowell) and the Harvard Corporation had a new policy. To wit, the University would henceforth distinguish between the applications of foreign scholars coming as refugees through the auspices of the Emergency Committees and those with recommendations and support from one of the large foundations, on the grounds that the latter were nominated strictly in accordance with their intellectual qualities and training, whereas the former were nominated for "sentimental reasons" and needed more careful selection.

Cannon, who had been careful about vetting Fischer's application, was outraged and when Edsall returned, asked the dean whether he (Cannon) had been acting contrary to Harvard policy. Edsall made a special effort to mollify Cannon, explaining that while Conant and the Corporation had expressed willingness to take on suitable persons, they were unwilling to have individuals make decisions to bring men to the Medical School and then leave the University with the obligation to care for them when emergency funds ran out. At any rate, that was his understanding of their action, rather than any stated official policy. As for himself, the dean had been confused by the great haste that had been pressed upon him; he had not intended to obstruct taking Fischer on, but "merely to avoid assuming moral responsibilities that we might not be able to carry out later." He said little about the "sentimental reasons" advanced by Hale, save to deny that he had ever used such words in his instructions. To point out the good will of the University toward the German refugees, Edsall called Cannon's attention to two new immigrants, Kimmelstiehl in pathology and Schatzki in radiology, who had recently been appointed to posts.

Cannon had to accept the decision, for Fischer had already been assigned to the University of Rochester (which he had previously visited as a research associate in physiology) for a period of two years. In 1935, he received a permanent post at the University of Virginia School of Medicine.[28]

Prior to the elections of 1933 in Germany, few in Europe or the United States anticipated the rapid changes that would accompany the Nazi victory. A.V. Hill used the occasion of his Huxley Memorial Lecture in November of that year to call attention to these changes. "It seemed impossible, in a great and highly civilized country," he declared, "that reasons of race, creed, or opinion, any more than the color of a man's hair, could lead to the dramatic elimination of a large number of the most eminent scientists and scholars, many of them men of the highest standing, good citizens, good human beings. This, nevertheless, has happened; the rest of the world of learning is gasping and wondering what to do about it."

Hill followed his prologue with a catalogue of some of the changes taking place in Germany—among them that thousands of professional men, lawyers, doctors, teachers, were prevented from following their profession; that tens of thousands of tradesmen and workers lost their means of livelihood; that 100,000 were imprisoned in concentration camps for no cause beyond independence of thought or speech; that over a thousand scholars and scientists had been dismissed, among them some of the most eminent in Germany. Many of these men were patriotic citizens who had fought in the German armies in the war, some were from families that had been in Germany for centuries, not all were even Jewish. Hill confessed that he found it "difficult to believe in progress, at least in decency and common sense, when this can happen almost in a night in a previously civilized state."[29]

Hill's charges were promptly answered by Johannes Stark, a Nobel laureate in physics writing from the Physikalisch-Technische Reichsanstalt, Berlin-Charlottenburg, and published in *Nature* early in 1934. The Nazi government had not introduced any measures against scientific teaching and research, Stark maintained. Rather, they wished to restore freedom of research that had been restricted by the preceding government. The measures which the National Socialist government had instituted that affected Jewish scientists and scholars, he assured his readers, were only an effort to correct the great influence and monopoly exercised by Jews in hospitals and scientific institutes—indeed, in all academic posts.

According to Stark, only a very small part of the 600,000 Jews who earned their living in Germany had been affected by the new National Socialist legislation. As a matter of fact, no Jewish civil servant was affected who had been in office before August 1914 and had served at the front or whose father or son had fallen in the war. He made little of Hill's charges that a thousand scientific workers and scholars had been dismissed, but said that in reality many Jewish scientists had voluntarily given up their posts. The truth concerning those who were in concentration camps had little to do with their desire for freedom of thought; rather they were guilty of high treason or of actions against the community.

Hill found Stark's political anti-Semitism distasteful and made a half-hearted effort to correct a few of his facts. When Stark again returned to the fray in the pages of *Nature,* Hill did not trouble to dignify his remarks with a reply and instead made an appeal in behalf of the Academic Assistance Council.[30] A few years later, however, there would be new ramifications to the debate when Cannon joined forces with a group of physicians and scientists to mount an attack on Nazi treatment of Jews and to refute Stark's defense of their policies.

16 *Chemical Mediation of Nerve Impulses:*

II. Sympathin E and Sympathin I

LATE IN SEPTEMBER of 1930, about three months before Zenon Bacq returned to Belgium, Arturo Rosenblueth, a young physician from Mexico, arrived at Harvard as the holder of a John Simon Guggenheim fellowship. The new fellow arrived at an opportune time—a time when Cannon's health and energy had begun to decline and he needed someone to assist him on a regular basis. Rosenblueth captured Cannon's attention with his brilliance. For despite the fact that he had limited research experience before joining the physiology department, he was a man of striking, even dazzling, intellectual ability. In relatively short order, he would become an indispensable fixture in Cannon's laboratory as well as his favored collaborator.

Arturo was the eighth child of a Jewish immigrant father from Hungary and a Mexican-American mother. While growing up in Mexico, the boy displayed a love of learning and precocious accomplishment, adding a mastery of English and French to his native Spanish and playing the piano so well that he seriously considered becoming a concert artist. As a teenager, however, he was inspired by the works of the French mathematician Henri Poincaré to set his career goals in another direction. Thereafter, he began what would be a lifelong preoccupation with the relation of mind to matter, as it was manifested in the tension between the subjectivity of theory and the objectivity of fact in science. Moreover, he placed a premium on proper scientific method in order to find the right balance between theory and fact, putting emphasis on mathematics as the purest and most precise form of scientific theory.

Rosenblueth matriculated in 1918 at the National School of Medicine in Mexico City and spent six months in 1923 studying medicine at the University of Berlin. But it was from the Ècole de Médecine at the University of Paris (where he enrolled in a course taught by Charles Richet on the experimental method of Claude Bernard, which reinforced his interest in scientific methodology) that he finally earned his M.D. degree in 1927. Returning to Mexico City, he tried to establish a practice in psychiatry but soon became bored. For the next few years, he held the post of Chief of Investigation at the Institute of Hygiene; at the same time, he taught physiology at the National School of Medicine and saw patients at both the general and psychiatric hospitals in order to supplement his income.[1]

In search of an opportunity to fulfill his scientific aspirations, Rosenblueth once more looked beyond Mexico. In 1929/30, he applied for and received one of the first of the newly-established Guggenheim Latin American Fellowships, which had been instituted to achieve "a more intimate understanding between the republics of North and South America through the closer relationships of scholars and professional men in those communities." Assured of sufficient funds for travel and twelve months of study, he boarded a steamer for Boston, where he proposed undertaking courses in physiology, biochemistry and physical chemistry at the Harvard Medical School.

At Harvard, Rosenblueth was introduced to the rich intellectual environment of the research laboratories. He soon came into contact with Cannon, then at the age of fifty-nine, who, with his ready smile and open countenance, impressed him as one of the most amicable men he had ever met. And when Cannon invited him to conduct some experiments under his direction, the invitation came as an exciting opportunity for an outsider. Scarcely a month after his arrival, Rosenblueth reported to the Guggenheim Foundation that the physiologist "of legitimate world-wide reputation" had treated him "with singular attention and kindness, such as I have never encountered in any of the other professors at the other foreign Universities that I have attended." For his part, Cannon immediately saw that his new fellow was "extraordinarily gifted" for the pursuit of physiological investigation; he was impressed by the speed with which the young Mexican grasped complicated laboratory techniques in record time.[2]

The pair discovered that they were a good match, each making up for the other's shortcomings. Rosenblueth provided the older man with a spurt of enthusiasm and inspiration at a time when most scientists his age were expected to have run out of steam. Cannon, a seasoned mentor, provided a fount of invaluable ideas and experience such as the young novice could only wonder at. Although Rosenblueth consulted Cannon on all scientific matters and readily absorbed his notions on the philosophy and methodology of science, the two did not immediately venture to undertake an investigation together.

In the autumn of 1930 when Rosenblueth came into the laboratory, Cannon had resumed his experiments with Zenon Bacq, their search for the mysterious "x-factor" ending in the discovery of "sympathin." Rosenblueth was an observer during the last phase of Cannon's work with Bacq and quickly became involved with other young investigators in the laboratory who were studying various aspects of the autonomic nervous system.[3] Not only did he embrace life in the laboratory with a guileless passion, he proceeded to launch some exhilaratingly high-flying balloons.

For his initial research, Rosenblueth teamed with Teodoro Schlossberg, a holder of the Harvard Buenos Aires Fellowship who communicated more readily in Spanish than in English. They proposed to try to eliminate the delay in experiments following denervation by sensitizing the organ with various drugs. Employing blood pressure as an indicator in decerebrated cats, they found only cocaine to be an effective sensitizer. They soon observed a similarity between the role of cocaine as a sensitizer of smooth muscle to the influence of adrenin and its role as a sensitizer to the action of sympathin released by stimulation of the sympathetic fibers. Hence they concluded that their evidence favored the view that adrenin and sympathin were identical. It was not a view that Rosenblueth would hold onto.[4]

In the months following his experiments with Zenon Bacq, Cannon broadcast the accomplishments of his laboratory by reviewing the previous work on medulliadrenal secretion and reporting more recent studies that resulted in the discovery of sympathin. In June 1931 when addressing the annual meeting of the Association for the Study of Internal Secretions, Cannon told the audience of observations, both his own and those of other investigators, that provided evidence of a chemical mediation of nerve impulses. Substances given off from or near effector organs innervated by sympathetic or parasympathetic fibers of the autonomic nervous system, he stated, are capable of influencing other organs similarly innervated, just as nerve impulses do. "If we consider that these substances are set free by the nerve impulses as they come to the effector cells," he argued, "we may regard the substances as chemical mediators of the impulses." His own work had resulted in the discovery of a circulating mediator of widely distributed sympathetic impulses, which he called sympathin. "Quite possibly sympathin is the same substance as adrenin," he noted, "but as it has a different source, it may be justifiably given a different name."[5]

Cannon undertook two new investigations in 1931, one successful, the other not. The observation of Rosenblueth and Schlossberg that cocaine sensitizes smooth muscle to both adrenin and sympathin suggested that the drug might also be a simple indicator of the action of sympathin. He reasoned that devising an anesthetic which would not interfere with the action of sympathin might allow for less difficult and complicated surgical techniques. It would also be interesting to know whether gland cells (such as in the liver) and smooth muscle would give off sympathin when subjected to sympathetic stimulation, and whether sympathin and adrenin, which under normal conditions would be produced simultaneously, were synergistic.

An attempt to address these issues was made in Cannon's first collaboration with Arturo Rosenblueth, with Rosenblueth taking the lead. They performed two series of experiments, a short preliminary one in animals with brains pithed transversely, and the second under "Dial" anesthesia, which did not interfere with all the effects of sympathin they wished to study. The nictitating membrane was first denervated, either by severance of the cervical strand or, more often, by removal of the superior cervical ganglion. Thus sensitized by this previous denervation and the addition of cocaine injected intravenously, it provided them with a quantitative measure of sympathetic activity either by action of adrenin or sympathin. By means of minimal or weak stimulation of the splanchnic nerves and the lower abdominal sympathetic nerves, they were able to call forth a small amount of adrenin and of sympathin respectively. On further stimulating the two sets of nerves, they demonstrated that adrenin and sympathin are synergistic—i.e., they collaborate in affecting structures innervated by sympathetic nerves.

The experiments of Rosenblueth and Cannon confirmed the observations of Cannon with Uridil and Griffith in 1921–22 that stimulation of the hepatic nerves causes a *prompt* discharge of a substance into the circulation, which makes the denervated heart beat faster and elevates blood pressure, much as does adrenin secreted from the adrenals. This reaffirmation provoked the interesting question of whether the cardio-accelerator agent, given off by the liver when stimulated through its sympathetic nerves, was not itself sympathin. There was, however, a subtle difference, and in their published report, the authors were unable to state from the evidence they had gathered that the two substances were in fact identical.[6]

Prior to his investigation with Rosenblueth, Cannon, in association with two of his N.R.C. fellows, Norman E. Freeman and Robert A. Phillips, attempted to demonstrate the humoral action of "vagus substance" in the circulating blood. Just as it had been found that when the sympathetic division of the autonomic nervous system is stimulated, a humoral substance is liberated into the blood stream which has an adrenin-like action on distant organs subject to sympathetic control, so had it been argued (Otto Loewi being the first) that there is likewise a parasympathomimetic substance for the cranial division of the autonomic system. Cannon posed the question, "Might not the failure to demonstrate [the existence of such a substance] in circulating blood be due to the use of too small an area of vagus distribution?" He and his colleagues thought it worthwhile to try to demonstrate the presence of a vagus substance in the blood stream by excluding the liver and applying methods of sensitizing response organs. But when they afterward described their attempts to detect parasympathetic activity on the rate of the denervated hearts of cats, using a variety of test organs (iris, submaxillary gland, tongue, stomach and intestines), they were able to report only negative results.[7]

Arturo Rosenblueth, on becoming ever more facile and energetic in his role as scientific investigator, applied for and received an extension of his Guggenheim fellowship. Having got his feet wet in the business of physiological research, he plunged in deeper with a zest typical of the man and was soon churning out papers in considerable numbers—some allegedly written in record time. Indeed, Cannon, who was meticulous in vetting his students' papers, had a hard time keeping up with his output. In 1931 and 1932, Rosenblueth published, alone or in collaboration with others, no less than seven papers in addition to those written with Schlossberg and with Cannon. Two further papers submitted in 1932 were published in 1933, with five others following. And he kept up the pace for a number of years.

One of the first papers that he published during this time was a report of research conducted with Norman Freeman on the reciprocal innervation in reflex changes of heart rate. This was followed by a second paper reporting research undertaken by Freeman with Rosenblueth's help on reflex stimulation and inhibition of vasodilators in sympathectomized dogs. Two very brief papers by Rosenblueth—one with René Gayet, a Rockefeller fellow from France, on the influence of diet on the adrenin content of the adrenals, the other with Robert Phillips on sympathin and the hepatic sympathomimetic hormone in the dog—appeared in 1932. Yet another paper by Rosenblueth, with Philip Bard as second author, concerned their study of the innervation and function of the nictitating membrane in the cat.[8]

Rosenblueth followed up the foregoing lines of research with two investigations that he undertook on his own. They differed somewhat from the standard approaches of the day in that he attempted to apply mathematical reasoning to the solution of physiological problems. Years later Horace Davenport remarked of Rosenblueth, "I can't say how profound was his mathematical skill, but it was certainly far beyond that of most of the mathematically illiterate physiologists of the day."[9] Still later Cannon's son, Bradford, who as a Harvard medical student did some physiological experiments with Rosenblueth, noted that he filled a gap in his father's background because of his mathematical skills—something that Cannon felt was a great shortcoming in his own career.[10]

In the first of these studies, published as "The Mode of Action of Adrenin and the Quantitation of Adrenin by Biological Methods," Rosenblueth observed the responses of the nictitating membrane, blood pressure, rate of the denervated heart, and inhibition of the non-pregnant uterus of the cat to varying doses of adrenalin. He stated that his mathematical treatment of the results corresponded to that employed by A.V. Hill in 1910 to analyze the mode of action of nicotine and curare on the striated muscle of the frog. After demonstrating that the sympathomimetic action of adrenin is brought about through a chemical process rather than a physical one, Rosenblueth proposed that adrenin combines chemically with some hypothetical substance in the muscle and that the contraction is proportional to the amount thus formed—in line with the same postulation of Hill for the action of nicotine and curare. This was the simplest explanation, in his opinion, which would account for all of the experimental data. The concept he expressed also bore a similarity to that advocated by J.N. Langley: i.e., that the agent (adrenin) combines with differentiating receptive substances, which are present in the responsive cells.

Rosenblueth found that all biological methods for the quantitative determination of adrenin could be simplified by determining the constants of the hyperbola of the preparation with only two injections of the standard, and in this way the amounts of adrenin contained in the unknown solution could be readily calculated. Reflecting his predilection for mathematical reasoning and expression, his paper was replete with logarithms, formulae, curves, plots, ratios, equations and other supporting paraphernalia—hence prefiguring his later involvement with the M.I.T. mathematician Norbert Weiner in helping to bring about the theory of cybernetics.

In a companion study, "The Chemical Mediation of Autonomic Nervous Impulses as Evidenced by Summation of Responses," Rosenblueth reported quantitatively the responses of different structures innervated by the autonomic nervous system to varying frequencies of stimuli. In every case, several readings were taken and the data obtained were tested in semi-logarithmic and logarithmic plottings. From analyses of the plottings, he deduced that the relations between the frequency of the nervous stimuli and the magnitude of the response obeyed a general hyperbolic law and, furthermore, this law applied to widely differing types of responses, such as contraction, inhibition, secretion, and the rate of heart beat.

Rosenblueth theorized that each quantal (all-or-none) nervous impulse liberates a quantal (constant) amount of a chemical mediator, symbolized as M, which combines with a hypothetical substance, symbolized as H, in the effector and produces a response that is proportional (not all-or-none) to the amount of MH formed.

He believed that the secretion of the adrenal medulla offered an interesting exception to the hyperbolic law, suggesting that this atypical behavior was possibly correlated with the atypical innervation of the gland by preganglionic fibers (according to Elliott), and with the embryological identity of the adrenal medulla and the postganglionic neurones. "In accord with this suggestion," he argued, "the production of the mediator might depend upon the postganglionic neurone." In a discussion of spatial and temporal summation in the responses of organs supplied by the autonomic nervous system, he concluded that the effects of spatial summation are of scant value compared to those of temporal summation.[11]

In another study on electrical potentials in smooth muscle, Rosenblueth in collaboration with Chester Leese and Elizabeth Lambert conducted experiments by stimulating electrically the nictitating membrane, pilomotors, bladder and uterus of the cat, afterward comparing the responses obtained by stimulation of the sympathetic with those evoked by adrenin. They also investigated the influence of ergotoxine, which, as Cannon and Bacq had previously shown, does not inhibit the production of sympathin. The trio came to several conclusions: that the electrical disturbances attending contraction of smooth muscle, as previously reported, are not comparable to those exhibited in skeletal muscle; that two distinct electrical phenomena of muscular origin were observed, an initial potential that precedes contraction and a delayed potential that is simultaneous with the mechanical response; that the initial potential is simultaneous with the production of sympathin while the delayed potential is a function of contraction; and that the initial potential and contraction are separable phenomena.[12]

When the paper by Rosenblueth, Leese and Lambert appeared in the *American Journal of Physiology* in 1933, it was followed by another contribution of Rosenblueth's—this time written in collaboration with David McKay Rioch from the anatomy department at H.M.S. Rosenblueth and Rioch reported experiments on the nictitating membrane of the cat and their results: that the smooth muscle under influence of adrenin is not refractory, that the responses to adrenin are contractions and not contractures (defined by H.S. Gasser in 1930 as an instance of contraction without conduction), and, finally, that cocaine increases the contractions obtained by nervous stimulation but does not affect the electrical responses.[13]

These studies, and others like them, prepared Rosenblueth for a major investigation that was also published in 1933, his name appearing as second author and Cannon's as the primary one. Briefly titled "Sympathin E and Sympathin I," their report was a novel view of chemical mediation that would be argued forcibly and subjected to a great deal of questioning. It would also arouse strong opposition that, for many, cast a shadow of doubt on the worth of Cannon's contribution to the chemical transmission of nerve impulses.

Before looking at the particulars of Cannon and Rosenblueth's report, it would be beneficial to examine the progress made by English and European investigators in elucidating the role that acetylcholine plays in the action of the parasympathetic division of the autonomic nervous system. While Cannon focused his attention almost exclusively on the sympathetic division of the autonomic, he kept abreast of studies by Otto Loewi, Henry Dale and others which attempted to unravel the mysteries of the inner workings of the parasympathetic and, more especially, the role of chemical substances in effecting the transmission of their impulses.

Loewi's classic experiments with the frog's heart in 1921, providing the first solid evidence that neurochemical mediation is a viable concept, were not without their ambiguities. Despite the doubts of some who considered his interpretation too simplistic and others who were unable to consistently reproduce his findings, Loewi forged ahead with a substantial series of papers on humoral transmission that he and his coworkers published over a period of fifteen years.

After finding that the action of the vagus substance is abolished by atropine, the search for its chemical equivalent was greatly simplified, and as early as 1922 Loewi argued that "Vagusstoff" might be a choline ester. But it was not until the publication of several further papers that he and his collaborator E. Navratil ventured the statement (1926) that the vagus substance had to be acetylcholine. What was especially significant for future developments was that Loewi discovered that the vagus transmitter, when liberated, is extremely susceptible to destruction by an enzyme that is not merely an esterase, as Dale had earlier presumed, but is specifically a cholinesterase, with most of its action readily inhibited by the alkaloid eserine (or physostigmine).[14]

In the mid-1920s, Loewi's identification of acetylcholine as a chemical mediator was still speculative, for there was no known evidence that it is present in the bodies of animals. It had never been isolated from an animal organ and was in fact known only as a synthetic drug, or a by-product of plant metabolism. Adrenaline was the sole chemical substance that mimics nerves known to be present in the body—that is, until 1929 when Henry Dale and Harold Dudley, while surveying the histamine content of various tissues, isolated and identified acetylcholine in the spleen of the horse and the ox.

After the war, Dale had increasingly performed the functions of the executive head of the Medical Research Council (or National Institute for Medical Research, as it later was called), being appointed its first director in 1928 while remaining in charge of his department. Although he tried to keep time free for experiments and discussions with colleagues, his days were filled with administrative duties that took him away from the laboratory. Pulled in two directions, he nevertheless was able to transcend the situation and for a time remained productive as a laboratory investigator.

In addition to his paper with Dudley, Dale published in 1929 a review of the chemical factors responsible for the control of circulation, reporting that acetylcholine mimics normal nerve stimulation to produce vasodilation and slow muscle contractions. Proposing that both Loewi's "Vagusstoff" and Dixon's muscarine-like substance were in fact acetylcholine, he suggested that its role within the parasympathetic system might be analogous to Elliott's postulate that adrenalin is a sympathetic chemical stimulator. But there were still pieces of the puzzle that did not quite fit together.[15] On the one hand, as reported by Dale and John Gaddum in 1930, the chemical action of eserine on denervated muscle greatly enhances the effect of acetylcholine and nerve stimulation, thus corroborating the earlier finding of Loewi. On the other hand, they also found that atropine paralyzes the vasodilator actions of acetylcholine while leaving the actions of the nerve intact—a result that was difficult to explain.

Aside from the puzzling action of atropine, one of the holes in Dale's view of neurotransmission and the role of acetylcholine involved the unclear relationship between acetylcholine and adrenaline. Another was the so-called "Vulpian phenomenon," i.e., the anomalous response of striated muscle to stimulation of nerves distributed to blood vessels. What's more, since acetylcholine is an evanescent body compound that is present at nerve endings only briefly and in minute amounts, Dale could not obtain enough of the chemical to provide conclusive evidence for his ideas.[16] And that was as far as matters had gone when Cannon reviewed the status of the work thus far accomplished on the chemical mediation of nerve impulses and set out on a new line of research on sympathin with Arturo Rosenblueth.

Cannon and Rosenblueth took their point of departure from the prior experience of the Harvard group, as well as the experience of others like Zenon Bacq after his return to Belgium, which indicated that the effects of sympathin on various test structures resembled those of adrenin. Instead, they proposed to present evidence that sympathin was not only different from adrenin, but that there were two sorts of sympathin. Working together in the autumn of 1932, they conducted a series of experiments on cats under Dial anesthesia whose reacting organs were usually sensitized by cocaine. After ergotoxine was administered, an injection of adrenalin caused the expected fall in blood pressure; but stimulation of the lower abdominal sympathetic chain or the hepatic nerves caused a rise in blood pressure, accompanied by a faster heart beat.

Since both sympathin and adrenin influence the heart rate, they decided to do a further experiment, in which the heart was denervated, the adrenals tied off, the liver nerves placed on shielded electrodes, the abdomen closed and ergotoxine injected. Whereas stimulation increased the heart rate from 104 to 148 beats per minute and raised the blood pressure, the injection of adrenalin increased the heart rate from 104 to 146 beats per minute but unexpectedly lowered the blood pressure. The difference between the actions of the two agents led them to determine to carry out tests on other parts of the body.[17]

At this point, Cannon put some ideas down on paper and sent them to Rosenblueth in the form of a letter. Asking what Rosenblueth thought of the following suggestions, he listed four hypotheses and probed their significance:

1. That sympathin is the immediate agent to both contraction and inhibition of smooth muscle;
2. that adrenin is a stimulating agent capable of duplicating the nerve impulses;

> 3. that ergotamine blocks both the motor and inhibitory action of sympathin, but blocks only the motor action of adrenin;
>
> 4. that ergotamine does not block the action of either sympathin or adrenin on cardiac muscle.
>
> If the foregoing are correct, it seems to me that we could explain the phenomena that we have been obtaining. The rise of blood pressure after ergotamine produced by sympathin would not be due to the faster heart rate because adrenin produced that effect, but would be due to the fast rate in the presence of high peripheral resistance. The reversal of adrenin would be due to the uninterrupted stimulation of vasodilators.
>
> If the suggestions given above are correct, the ideal structure for us to study is the leg, because it has in it both vasoconstrictor and vasodilator elements associated with faster reactions. We ought to be able to decide definitively the first three points mentioned above by use of a plethysmograph on the leg. I should like to work on this next Tuesday morning.[18]

In their next experiment, Cannon and Rosenblueth found that stimulation of the hepatic nerves led to contraction of both the leg and the spleen (also selected as a test site), with a concurrent rise in blood pressure. Similar results were observed when adrenalin was injected. After administration of ergotoxine, hepatic nerve stimulation induced a rise of blood pressure, with minor changes in leg and spleen volume. When adrenalin was administered again, the blood pressure fell, as now expected, with variable changes in volume of spleen and leg. These results led them to conclude that the ergotoxine induced and increased paralysis of the contractive action of adrenin, leaving its inhibitory action unimpaired. The fall of blood pressure was due, therefore, to lessened peripheral resistance because of vasodilation—just as Cannon had hypothesized.

The results obtained with the changes in leg and spleen volume, while highly suggestive, were by no means clear-cut, and so the investigators sought other means of establishing the difference between the actions of sympathin and adrenin. Next Cannon and Rosenblueth determined to use as indicators smooth muscle organs, which are exclusively contracted or relaxed by adrenin and do not require the use of ergotoxine. They settled on the nictitating membrane and the non-pregnant uterus. The former had already proved to have an admirable sheet of smooth muscle which, when denervated, registered quite conveniently its contractions under the influence of adrenin or sympathin; the latter, which in the cat relaxed when adrenalin was injected into the blood stream, had been found to be a satisfactory and delicate means of registering inhibition.

Using cocaine to sensitize the reacting organs, they observed that when the liver region is the source of sympathin, the nictitating membrane contracts under stimulation and the uterus does not relax. But when the gastro-intestinal tract is the main source, both organs respond characteristically—i.e., contraction of the membrane without relaxation of the uterus when the hepatic nerves were stimulated, and the typical reaction of both organs when the current was applied to the splanchnics.[19]

While acknowledging striking ambiguities in their recordings that challenged interpretation, Cannon and Rosenblueth nevertheless proceeded to draw inferences from their accumulation of data, which they numbered for the sake of clarity. The last three of the eight items they enumerated derived from Rosenblueth's hypothesis that a substance, A from the outside (adrenin, for example) or M (a local product dependent upon a number of nerve impulses in the stimulus), united in the cell with a hypothetical substance, H, thus making the combination AH or MH, which evoked a response proportional to the amount formed. At this point, new evidence was added from Cannon and Rosenblueth's experiments to show that the adrenin-like substance (produced when sympathetic impulses affect smooth muscle) became differentiated for positive and negative action. It followed that H must be regarded either I (inhibitory) or E (excitatory). Hence the combination, after nerve stimulation, would be ME in a contracting muscle and MI in a relaxing muscle.

Sympathin therefore was defined as the chemical mediator of sympathetic nerve impulses, which in the cell induced a typical response, contraction or relaxation, and on escaping from the cell into the blood stream, induced effects in organs innervated by the sympathetic. These effects were excitatory when sympathin came from the structures which were only excited to action by nerve impulses; they were excitatory and inhibitory when it came from both contracted blood vessels and inhibited muscular walls of the intestine. When both effects were present, as sometimes in the stimulation of the duodeno-hepatic nerves, severance of the duodenal branches, which excluded the source of the inhibitory sympathin, resulted in a change from the double effect caused by the excitatory sympathin alone. From all of the facts and recordings they had examined, Cannon and Rosenblueth felt justified in maintaining that there are two kinds of sympathin: sympathin E (excitatory) and sympathin I (inhibitory).

Though they admitted that several investigators had reported results that did not fit their interpretation, e.g., Brinkman and Van Dam in 1922 and Lanz in 1928, they explained that these prior experiments were performed by perfusing salt solution through the organs, "and that may have altered the sympathin—perhaps by removing the differential feature, E, and leaving an undifferentiated agent resembling adrenin." Langley's views expressed in 1921 were more to their liking. "He assumed a chemical combination between adrenin, for example, and receptive substances in the responding cells" that were supposed to belong to two classes—those which cause contraction and those which cause relaxation.

Thus, they reasoned, "Langley's receptive substances would correspond to the substances E and I here postulated," and, in accordance with their accumulation of evidence, they believed that the further step they had taken was justified. Anticipating the question why ME, which induced contraction in the cell where it originated, did not induce contraction when it entered a cell inhibited by sympathetic nerve impulses, and, similarly, why MI, entering a contractile cell, did not cause inhibition, they responded, "One can only speculate on answers to these questions."[20]

Cannon's paper with Rosenblueth was published in the June 1933 *A.J.P.* Another paper with the same title, this one by Cannon alone, announced the discovery of the two sympathins to the Association of American Physicians. Still a third paper, comprising a more general overview of the latest developments in chemical mediation, was delivered by Cannon at the FASEB meetings in Cincinnati. Cannon's papers are models of clarity and condensation, typical of his straightforward and unassuming style, while the paper with Rosenblueth is complicated, less organized in its presentation of facts and much more difficult to comprehend. Since the latter's ponderous tone and intricate, somewhat confused, structure are in keeping with Rosenblueth's less accomplished but more grandiose style, not a few readers have wondered whether it may have been written largely by the younger investigator.[21]

Because of its speculative nature, and because its conclusions did not accord with the theories and findings of other investigators, Cannon and Rosenblueth's paper on sympathin E and I met with a good deal of skepticism. Ironically, Zenon Bacq, who had assisted Cannon in the discovery and naming of sympathin and who held his teacher in high esteem, was the harshest of all its critics. Indeed, at the very time that the Harvard investigators were conducting their experiments and well before their paper came into print, Bacq had already begun to express contrary views.[22]

Near the end of 1932, Bacq informed Cannon that he was "trying to get the absorption spectrum of sympathin in the ultraviolet," and if he succeeded, "the structure of sympathin will be definitely stated; as far as I know nothing has been recorded against the identity of sympathin and adrenalin." Cannon expressed interest in Bacq's observations and requested more information about the absorption spectrum studies. "We have some very interesting new evidence that in its biological effect sympathin is different from adrenin—indeed, that there may be two kinds of sympathin," he wrote, adding, "we are eagerly pursuing this lead at the present time and hope to have the matter settled before long."

A few weeks later, Bacq dashed off a note to Cannon after "one of the most exciting weeks in my life," during which he had "obtained most beautifully the inhibition of the denervated uterus by sympathin and also the retraction of the retractor penis. This contraction is absolutely the same as that evoked by adrenalin." Bacq went on to publish a report that stimulating peripheral ends of the severed sciatic nerves resulted, after about two minutes, in a cessation of the movements of the denervated intestines. He concluded that the actions of sympathin and adrenalin are identical—a view that was in conflict with the one promulgated by Cannon and Rosenblueth in their paper on sympathin E and I.[23]

Cannon and Bacq continued to debate the issue in their letters, the latter maintaining that sympathin and adrenalin are identical, the former holding fast to his belief that they differ. "Why should we think that the chemical work of any gland cell is perfect, and gives 100% of a definite compound?" Bacq asked early in 1933. "I rather think that from the adrenals and from the sympathetic nerve endings there is given off a mixture of diphenols having sympathomimetic properties, these phenols being adrenalin, noradrenalin (Tiffeneau), noradrenin (Annau etc.), all substances in close relation but different only by small variations of the chemical structure of the lateral chain of adrenalin."

Although Cannon wrote "no" in the margin when he read Bacq's comment, the disagreement did not close any doors. Neither one moved the other from his stance, but they continued to exchange information and found the liberal expression of ideas stimulating. Bacq looked forward to a free and open discussion of all the issues when he returned to Boston for a visit in the fall.[24]

Bacq's skepticism regarding sympathin E and sympathin I was not based on intuition alone. When discussing adrenergic transmission nearly forty years later, he recalled that in 1932 he had become acquainted with "the highly significant paper" published in 1910 by Barger and Dale on the relations between chemical structure and sympathomimetic action. "One fact stayed in my mind," Bacq remembered, "DL-noradrenaline is more hypertensive than DL-adrenaline. One day while I was visiting Forneau at the Institut Pasteur in Paris together with Tiffeneau, this fact was mentioned, and I can still hear Tiffeneau saying that he could not understand why 'nature' could not use a more active and simpler molecule than adrenaline."[25] When Bacq completed his thesis on the biological and physico-chemical properties of sympathin compared to those of adrenin, he stated among its conclusions that he was unable to separate two closely related substances, adrenaline and noradrenaline—both highly likely candidates for sympathin.

Neither he nor Cannon realized at this time how close to the truth the young Belgian had come. Many years would pass before it became generally known that the Parke, Davis Adrenalin® Cannon often used to inject his laboratory animals was actually a mixture of epinephrine and norepinephrine—and that the substance liberated from the adrenal medulla is likewise a mixture.[26]

Soon after arriving at Woods Hole in the summer of 1933, Bacq wrote Cannon that there are "scientists with whom it is painful and dangerous to disagree," and others with whom it is a pleasure not to agree. He assured Cannon that he counted the Harvard scientists among the second group and expected "to enjoy with you and Dr. Rosenblueth a warm and fruitful discussion."

We know comparatively little about Bacq's return to Harvard

in September and October. An ailing and weary Cannon, then in the midst of getting his department going for the year and preparing lectures for the celebration of the Beaumont Centenary, had to squeeze his young friend into his busy schedule as best he could. It appears that the visitor spent much of the time with Rosenblueth, the two men completing a study of calcium and potassium ions and attempting to isolate sympathin I from sympathin E. Unfortunately, it simply was not possible to obtain positive evidence of the properties of sympathin because, at the time, the chemical techniques were too primitive and paper chromatography did not yet exist. Though Bacq failed to convince Rosenblueth and Cannon that their hypothesis was faulty, all parted on friendly terms.

The good faith continued, and in April 1934, after receiving two reprints from Bacq with touching inscriptions, Cannon wrote him, "As one gets older, the affection and respect received from former students becomes more precious." Meanwhile, Bacq was fast becoming one of the chief spokesmen on the European continent for the theory of chemical transmission. In May, he attended a congress at Nancy of the Association des Physiologistes where he presented a review of the pharmacology of the autonomic nervous system, especially its sympathetic division. He wrote Cannon the meeting had been "a great triumph for your ideas. Nobody has contested the reality of chemical mediation, and the pharmacological implications of this theory." Promising to send on the reports and discussions of the congress, Bacq expressed the hope that "Rosenblueth and yourself will make a severe criticism of my paper. It will be more interesting to me than congratulations."[27]

Even though Cannon was well aware of Bacq's open opposition to the two sympathins, when he got around to examining the Belgian's review, he must have been taken aback. In essence, Bacq explained that while Cannon and Rosenblueth's hypothesis was grounded on the idea that there existed a fundamental qualitative difference between sympathin's ability to excite or inhibit, he himself was inclined to think that excitation and inhibition were no more than varations of the same phenomenon, which remained to be elucidated. He conjectured that two adrenergic agents existed: where the sympathetic response was inhibitory, the agent was adrenaline; where the sympathetic response was excitatory, substance M was noradrenaline, which, as with all the primary amines, does not have very weak inhibitory effects, and whose action is not reversed by ergotamine. Stated otherwise, according to Bacq, Cannon's sympathin I was equivalent to adrenaline and sympathin E equaled noradrenaline.

Cannon responded to Bacq's "interesting discussion" with two pages of criticism from Rosenblueth and himself, comments he hoped Bacq "would take in the spirit in which they are sent" because they were "directed toward clarifying opinions and suggestions." He listed and rebutted a number of ambiguities they had detected in Bacq's argument, objecting to his attribution of the mediator going from the nerve outside the reacting cell, or that it arose from minute nerve terminals, as well as his statement that the mother substance of sympathin is inactive, oxidized adrenin, to name but a few.[28]

The debate continued for the remainder of 1934, with Bacq in August sending Cannon seven pages of explanations, diagrams and citations to the opinions of other scientists in defense of his report. Letters flew back and forth in October. Nor was the discussion confined to establishing the identity of sympathin. In November, Cannon sent a blunt criticism of a report by Bacq, Brouha and Heymans on vasomotor reflexes in sympathectomized dogs. Although he called their work a clear exposition of the facts, their observations did not accord with the facts observed by investigators at Harvard, who were now forced to conduct further experiments in order to clarify "the phenomena which both in this Laboratory and in the Belgian Laboratories have been found difficult to explain."

Their final exchange of the year, for the most part, concerned an essay that Bacq had been invited to write (at Cannon's suggestion) for Asher's *Ergebnisse der Physiologie* on the humoral mechanism of autonomic nerves. When the review was published a year later, Cannon expressed his delight at having it at just that time. He told Bacq that he had agreed to prepare, in collaboration with Dr. Rosenblueth, a little book on the autonomic nervous system and its effectors, in the preparation of which Bacq's essay would prove useful.[29]

It was not until early 1935, just before Cannon departed on a sabbatical which would take him away from the laboratory for the better part of the year, that he and Rosenblueth submitted their next major paper for publication, "A Comparative Study of Sympathin and Adrenin." It was an attempt to reply to Bacq's criticisms and clarify their differing views. In conducting further experiments on cats, they had compared the effects of sympathin and adrenin on a number of indicators—blood pressure, heart rate, leg vessels, nictitating membrane and pupil of the eye, and the pregnant and non-pregnant uterus, taking into account the influence of acetylcholine, eserine, ergotamine and atropine on the various organs. The observations they made and the data they collected led them to conclude with a strong defense of their concept of sympathin E and sympathin I.

Addressing Bacq's alternative suggestions, they rejected his ideas that sympathin was equivalent to adrenin; that the two mediating substances were adrenin (inhibitory) and noradrenalin, or non-methylated adrenin (excitatory); and that potassium ions, liberated when the sympathetic was stimulated, accentuated the excitatory and lessened the inhibitory influence of sympathin. Since Bacq had so far not been able to provide hard evidence to back up his theories, which went counter to the physiological facts that confronted the Harvard team, he was, for all intents and purposes, out on a limb. Yet he was only one of many who found it hard to swallow the sympathin E and sympathin I theory of Cannon and Rosenblueth.

For instance, Henry Dale, responding to a request from Bacq, had written in February 1934 that he would try to send him a small sample of noradrenaline:

> It used to be manufactured by the Hoechst Farbenfabrik, under the name "Arterenol." It is, as regards pressor actions, the nearest to adrenaline of all the substances we examined, but, like all the primary amines, it is very weak in the inhibitor axis. If Cannon's "sympathin E" were really a separate entity, "Arterenol" would be the best candidate for this function of any of the substances that I have ever tested.[30]

☙ ☙ ☙

All the while that Cannon was proposing sympathin as the chemical mediator of the sympathetic nervous system, Dale had maintained his interest in acetylcholine as a neurotransmitter for the parasympathetic. But by the 1930s he had become a senior spokesman for science, a renowned figure and administrator who, for a time, stopped doing research altogether. Although he identified ways in which acetylcholine simulates autonomic nerve stimulation, he personally had not been responsible for other laboratory findings pertaining to neurochemical research. He did not, for example, discover that acetylcholine is a natural body product (Dudley did), nor did he develop critical experiments (like Loewi) that reawakened scientific interest in the notion of neurochemical transmission. Nevertheless, luck stepped in to make Dale a major player in the verification and acceptance of the chemical transmission of nerve impulses.

The first stroke of luck for Dale occurred in 1932 when he attended a meeting of the German Pharmacological Society in Weisbaden and heard two members of the Physiological Institute of the University of Berlin report on new techniques. One was Wilhelm Feldberg, who explained how the release of an acetylcholine-like substance had been detected in the blood during nerve stimulation after the injection of eserine. The other was Feldberg's colleague Bruno Minz, who described the way in which the eserinized leech muscle preparation afforded new and far more sensitive means of identifying minute amounts of acetylcholine in the organism.

The second stroke of luck came about a year later when, after the Nazis seized power in Germany, Jewish and non-conforming academicians were thrown out of their posts, Feldberg among them. Dale lost no time in rescuing the young physiologist (Alex Forbes would later call him "Dale's little present from Hitler"). By the autumn of 1933, Feldberg was safely ensconced in the Hampstead laboratory of the National Institute for Medical Research. As Bacq, who worked for a short time in Dale's laboratory, observed, Dale knew well how to choose his associates. "Without Feldberg," Bacq mused in retrospect, "the research on chemical transmission of nerve impulses might well have taken a different course."[31]

With Feldberg's arrival, Dale again became active in laboratory research and before long the Hampstead group turned out several important papers, two of which appeared under the authorship of Dale and Feldberg in 1934. They found evidence that atropine-resistant nervous effects, as well as atropine-sensitive effects, were transmitted to the effector cells by the liberation of acetylcholine. Another report, by Feldberg et al., completed the work he had started in Berlin on the appearance, after splanchnic nerve stimulation, of an acetylcholine-like substance that involved preganglionic sympathetic neurons. Still another set of experiments, published by Feldberg and Gaddum, showed that by stimulating the cervical sympathetic ganglion and using Feldberg's techniques, they had been able to collect and identify acetylcholine—revealing in the process that release of acetylcholine is responsible for transmission of preganglionic impulses past the synapses.[32]

The results of these investigations, and others like them, gave Dale the evidence he needed to support his theory of acetylcholine transmission, extending it beyond the parasympathetic system to the sympathetic nervous system and including synaptic transmission. This contradicted earlier research that assumed a clear distinction between the parasympathetic and sympathetic divisions, with adrenaline, or an adrenaline-like substance, indicated for the function of sympathetic nerves and acetylcholine for mediation of the parasympathetic system. Hence, Dale suggested that in lieu of the concepts now rendered obsolete, new terms were needed to indicate the two kinds of chemical transmission.

Accordingly, he proposed the word "adrenergic" for use where adrenaline and the sympathetic are concerned, and "cholinergic" for acetylcholine and the parasympathetic. In light of the fact that many organs are served by both sympathetic and parasympathetic fibers (for control in the proper maintenance of bodily function—an exquisite example of Cannon's theory of homeostasis), and because their respective actions and interplay could better be understood, these new definitions were judged to be more realistic and were seen as a step forward.[33]

At a meeting of the Physiological Society in London in 1934, Dale and Feldberg reported the first experimental evidence for transmission of acetylcholine at motor nerve endings. A full-length paper explaining chemical neuromuscular transmission was postponed awaiting additional experimental work, for they knew full well that opposition on the part of those championing the theory of electrical transmission of nerve impulses, led in England by John Eccles, would be strong. Throughout 1935 and 1936 the work went on at Hampstead. In the latter year Dale's laboratory group discovered that by injecting acetylcholine into the muscle's artery, it would be delivered rapidly to the motor end plate, thus demonstrating that contractions were elicited purely on the basis of acetylcholine injection.[34]

While Dale was directly involved in only part of this research, his role in the validation of neurochemical theory was crucial in that he conceptualized and directed the work in his laboratory. As far as the successful accomplishment of this work was concerned, there can be little doubt that Feldberg was pivotal. It was Feldberg who took part in each of the experiments dealing with the transmission of acetylcholine. His expertise was applied continually in resolving issues of ganglionic and neuromuscular transmission and in the establishment of the function of acetylcholine within the autonomic system. Feldberg did not, however, share in the prize awarded by the Nobel Committee to Henry Dale and Otto Loewi in 1936 for "their discoveries relating to the chemical transmission of nerve impulses."[35]

Nor, for that matter, did Cannon. The research of Dale and Loewi and their coworkers resolved only part of the mystery of chemical transmission. There still remained the explication of the mediation of the sympathetic system and the identification of the substance involved in the transmission of its impulses—which would prove to be a far more complicated matter. As Dale himself pointed out in his 1934 review of chemical transmission, progress in the knowledge of Loewi's accelerating substance and its role as a mediator had come largely from Cannon's laboratory at Harvard and from the researches of visitors from foreign countries who had worked there. Briefly touching on Cannon's discoveries with Bacq and Rosenblueth before focusing on his own work, he recommended, "We may safely leave the details of the chemical transmission of peripheral sympathetic effects to the further investigations of Cannon and his school."

Notwithstanding Dale's apparent seal of approval, he really did not place much confidence in the two sympathins theory. But he was too much of a politician to say so out loud. When some time later he received a manuscript of a forthcoming paper from Otto Loewi, he replied: "A piece of direct evidence like this is just what we wanted to brush away the complicated tangle of indirect deductions based on a half-understanding of physiological detail, which had led Cannon and his coworkers to the sympathin E and sympathin I story. I never believed it; but it was difficult to find exact conditions for obtaining direct evidence."[36]

Time and again Bacq would point out that neither Cannon nor Rosenblueth had much grounding in basic pharmacology, nor did they acknowledge that their theory required a reorganization of research with the significant participation of biochemists (a strategy that Dale had successfully adopted). Lacking the ability or the willingness to go off in another direction, Cannon gained little ground in elucidating this riddle, no matter how hard he tried. He knew he was in a bind, for he wrote his son in October 1935, "I want, more than anything else, to know what difference there is between sympathin E and I, and how to explain increased sensitivity after denervation. But I can't see how to tackle the question."[37]

Further research with Rosenblueth over the next several years could not explain the discrepancy between the opposite actions of adrenin and sympathin other than through the two sympathins hypothesis. When presenting a report on the evidence for chemical transmission at the 1935 International Congress of Physiology, along with another paper comparing the effects of sympathin and adrenin on the iris (which was presented at a session in Moscow presided over by Bacq and G. Lilejstrand of Stockholm), Cannon adhered to this view. And he continued to hold firm to sympathin E and sympathin I when putting together with Rosenblueth the volume on *Autonomic Neuro-Effector Systems* that was to be published by Macmillan.

Coincidental with the publication of Cannon and Rosenblueth's book in 1937, the first pieces were fit into the puzzle that the Harvard investigators had been unable to solve. But it must be said that they were put in place by others and followed along the lines that Zenon Bacq had tried to lay down. Two papers, one by R.L. Stehle and H.C. Ellsworth, the other by K. I. Melville, published in the *Journal of Experimental Pharmacology and Therapeutics*, reported on the action of arterenol and suggested that it might be the substance that was liberated when the hepatic sympathetic nerves were stimulated. A more significant report, appearing the following year in the same journal, was published by C.M. Greer, J. O. Pinkston, J. H. Baxter and E.S. Brannon, working out of the department of pharmacology at Vanderbilt. While evidence was still sketchy, the accumulated effect of these papers contributed to a consensus that had begun to build.[38]

It appeared to Greer, Pinkston, and their colleagues that "although the catechol amine, nor-epinephrine, may not be identical with 'sympathin E,' the close similarity between the responses which are induced by this compound and those induced by 'sympathin E' seems to warrant serious consideration of it" for the role of sympathomimetic agent. They believed that their results, as well as those of others, were best explained by the tentative assumption that there are at least two adrenergic mediators involved in the transmission of impulses from sympathetic nerves to effector cells.

Whereas Cannon and Rosenblueth proposed that the nerve impulses liberate a single, *undifferentiated* mediator (presumably epinephrine), which must combine with a cellular constituent (E or I, depending upon the effector cell) in order to become differentiated into the respective sympathins, the Vanderbilt group believed that the data were more simply and completely explained on the basis of at least two *already differentiated* mediators, one inducing responses that resemble very closely those evoked by norepinephrine and the other inducing responses similar to those evoked by epinephrine. In essence, they were suggesting that the effects of the two adrenergic mediators differ only in degree and not in kind.[39]

One of the Vanderbilt investigators was well known at Harvard. Jim Pinkston had earned his doctorate in Cannon's department and spent several years there. Although he had never published with Cannon, Pinkston worked alongside of Rosenblueth on some of his projects and had been present in the laboratory when Bacq made his return visit in 1933. Cannon wrote Bacq in April 1937 about a paper Pinkston had presented before the Pharmacological Society, "in which he urged that noradrenaline is the equivalent of sympathin E. I asked him about his knowledge of your suggestions, and it was evident that he had them in mind."[40]

Cannon went on to tell Bacq that in the preface to his book on autonomic neurotransmitters, he had minimized as much as possible the importance of theory, saying, "It is too bad to place theories at so high a valuation. They should be regarded as useful tools and nothing more." What he actually wrote in his preface was much more elegantly stated; it was a fundamental truth that could not be gainsaid:

> We wish to be explicit in declaring that we look upon the theories which we have projected as quite tentative modes of conceiving facts now available. New facts may appear which may effectively destroy these theoretical concepts. If fresh evidence should require us to

> change our views, we shall part with the old theories with no more regret than we should feel in relinquishing a once helpful tool for a newer and more efficient instrument.

But unhappily, letting go was easier said than done and both Cannon and Rosenblueth held onto the theory of the two sympathins far longer than many believed was justified. Some even blamed Rosenblueth for having led Cannon down the wrong path.[41] In the long run, however, most of Cannon's critics praised his efforts in behalf of chemical transmission, for his work not only stimulated the discovery of novel methods and important ideas, it also provided an important advance toward the solution of old problems and opened up new ones.

In the fullness of time, later commentators have come to put Cannon's work in proper perspective. In his 1956 monograph, *Noradrenaline: Chemistry, Physiology, Pharmacology and Clinical Aspects,* Ulf von Euler, for one, made a point of giving him his place in the history of chemical transmission, observing that "not before the classical series of experiments of Cannon and Rosenblueth (1933) did it become evident that sympathetic nerve stimulation led to the appearance of some factor other than adrenaline." And in 1975, Hermann Blaschko and A. David Smith in *The Handbook of Physiology* referred to Cannon as an outstanding pioneer in the field whose work "is in no way overshadowed by the discoveries of Otto Loewi or Ulf Von Euler, with whose names the identification of the adrenergic transmitter is mainly associated . . . He gave a unifying and comprehensive interpretation to the meaning of adrenosympathetic stimulation. This work has retained its validity ever since."[42]

When Henry Dale, after winning the Nobel Prize in the fall of 1936, replied to Cannon's message of congratulations, he indicated that he was well aware of those who might have shared in the award, but did not. In what was perhaps a silent salute to Cannon's contributions to the discovery of chemical transmission, he alluded to the influence of luck, "which cannot be excluded from these awards. So often it must be the right impression, made at the right moment, which brings it to a man no more deserving than many others. If all my friends who deserve it could receive the award, we should need more fortunes of many explosive millionaires to meet the demand."[43]

17 *Around the World in 1935*

IN THE AUTUMN of 1934, Cannon opened his mail to find two letters that pleased him enormously. The first one invited him to go to China the following spring and serve as a visiting professor in physiology at the Peiping Union Medical College (P.U.M.C.) from mid-April until the first of June. He was to lecture as much or as little as he wished, to spend time in the laboratory, and generally to bring wisdom and light as he saw fit. There was even a generous honorarium of $2500 to go with the invitation.

To his wife's amazement, he promptly wrote a radiant letter of acceptance, conditional on Harvard's letting him take the time as part of a sabbatical (in all his years at the Medical School Cannon had never taken a sabbatical). Cornelia, recognizing that Cannon's request for leave could hardly be denied, immediately began to plan for Helen and Marian to share in the trip, so they could all have a happy reunion with their "Chinese children," Wilma and John Fairbank.

No sooner had Cannon accepted the invitation to go to China than he received the second letter. This time it was from Professor Pavlov, who paid him the great compliment of asking him to give the opening address at the fifteenth International Congress of Physiology when it convened in Leningrad in August of 1935. An added bonus was that while in Russia all of his expenses were to be paid by the Soviet government. Cannon, of course, accepted Pavlov's invitation with pleasure. The only problem that remained—where to spend the time between June and August—melted away when he decided to visit Japan before going to the

Congress. Not only did Walter have friends there, but it would, as Cornelia put it, "be happier for him, cheaper and less depressing than spending all that time in Russia."[1]

For some months previously, the Cannons had been toying with the idea of attending the Physiological Congress in Russia and had hoped to find a way to combine it with a visit to the Fairbanks in China. Cornelia even went to see Dr. Aub about the effects that such a trip might have on her husband's health. Although her visit was exploratory, she discovered that Aub was all for it, thought he ought to go, and would take it up with the dean to see what pressure could be brought to bear to persuade Cannon to leave his laboratory.[2]

During the summer of 1934, they began to read books about life in the Soviet Union. One called *Escape from Soviet Russia* they thought particularly revealing of the plight of the bourgeoisie in Russia at that time. Some weeks later, they received a visit from the wife of a Russian professor and hung on her words as she related her experience with the O.G.P.U. (forerunner of the K.G.B.) in Russia. The terror, although she tried to make light of it, was all too real. Then she told them a sad story about one of her friends, a peaceful professor of biology who was arrested, disappeared in jail for months and finally committed suicide—thus bringing home to them the inescapable facts of *Escape from Soviet Russia.*

Once Cannon had accepted Pavlov's invitation, everything fell into place and he got right down to work on his address. Although the speech in essence would constitute an up-to-the-minute report of the state of knowledge about chemical transmission of nerve impulses, it would be much more. As a preface to his scientific exposition Cannon planned a digression into conditions favorable and unfavorable for scientific research, a topic he had considered many times before. During the remainder of 1934, he spent a good deal of time quietly working in his study at home on the address.[3]

From the outset Cannon saw his address as an opportunity to speak frankly to an international audience about the assaults on academic freedom then cropping up all over the world. He knew, of course, about the effects of Nazism but was less sure of the impact that Soviet policy was having on scientific inquiry. Throughout the late 1920s and early 1930s, a spate of laudatory articles and books had appeared in the United States on the rapid development of medicine and science in Russia under the Soviet government.

For instance, following a trip to Russia during the summer of 1932, Lewellys Barker, professor of medicine at Johns Hopkins, informed readers of the *Scientific Monthly* that he was surprised not only by the difficult living conditions he had observed, but also by the progress of the nationalization of medicine that the Minister of Public Health had instituted. He especially noted that scientific institutes engaged in medical research were cropping up in a number of urban centers. "I am impressed," Dr. Barker wrote, "with the fact that Soviet authorities desire to foster and encourage medical research . . . I could not help but have real admiration for the scientific workers in Russia who are endeavoring to hold the torch aloft despite the difficulties encountered."

Perhaps the most enthusiastic report on scientific development in the Soviet Union was given late in 1934 by Karl Compton, President of M.I.T., who had gone to Russia as a member of the National Academy of Sciences. "Russia," he stated in a public address, "is basing her entire hope of economic betterment and improvement in the standard of living on the success of her scientific and technical program, into which she is putting a larger amount of money than into any other aspect of her national program, not even excluding the military and defense item."[4]

Despite such rosy accounts, Cannon had doubts about the freedom of scientific inquiry in Russia. Some of these doubts were fostered in part by his memory of Pavlov's early opposition to the Soviet government, and in part by more recent reports of the attitudes and beliefs that appeared to guide the repressive acts of Soviet officials. The British social historian Sidney Webb warned in *Current History* in 1933 that the Soviet government "is a dominant and intolerant autocrat in intellectual matters." It exercised a ruthless repression "on great personages and humble folk alike," often without open trial, which led to "severe sentences of imprisonment or punitive relegation to places where existence is but prolonged agony, or even to secret execution."

Concerned lest his remarks to the Congress inadvertently cause distress for Pavlov and other Russian physiologists, Cannon turned to two of Pavlov's former students, Boris Babkin and Vasily Boldyreff, for the latest information about the conditions under which medical scientists lived and worked. Desiring to know the attitude of the Soviet government toward scientific research, he asked whether physiologists were free to follow their own interests, or were their investigations undertaken for the special service of the state?

In their replies, Babkin and Boldyreff acknowledged that the Soviets had in recent years appropriated large sums of money for medical research, but they maintained that much larger appropriations were devoted to spurring industrialization and increasing the strength of the Red Army. As for the future of pure physiological research in the Soviet Union, Cannon's informants took a decidedly gloomy view. Babkin (seconded by Boldyreff) wrote, "The pure research workers are scientists belonging to the pre-revolutionary period. Their training, their scientific ideas and ideology belong to the 'past' . . . The great majority of the new generation of scientists . . . are unable to undertake and are not interested in purely scientific problems." In fact, he did not know "of any promising young physiologists with originality of thought who could replace 'the old guard'."[5]

Cannon also wrote to Horsley Gantt, who had been to the Soviet Union in 1933. His reply gave a picture of Pavlov at work; although Pavlov was eighty-five years old, he was in excellent mental form. Gantt believed that scientists were relatively better cared for than they had been five or six years ago, but added that the government in general was becoming more interested in the accomplishments of applied and industrial science than in pure science.[6]

Nor did Cannon confine his inquiries to physiologists and workers in medical science. He also wrote (at the suggestion of Alan Gregg, who had succeeded Richard Pearce as head of the Division of Medical Sciences at the Rockefeller Foundation) to A.A. Horvath at the U.S. Bureau of Mines, as well as to Solomon Lefschetz in the department of mathematics at Princeton. For Horvath, "Pavlov's Institute is an exception, and its research and independence are a stage show for the foreign countries as well as for the home consumption." Lefschetz, who had attended a mathematical congress in the Soviet Union the previous summer, knew of no restrictions for research workers in one direction rather than another; but he deemed it likely that some had received preferential treatment. This was true in any country, he said, and had little to do with the attitude of the government toward scientific endeavor and much to do with political cleverness of some of the scientists.[7]

Of all the people whom Cannon consulted, Babkin was the one he depended on most. Cannon was well aware of the political fallout that might follow the introductory section of his address at the Congress and upon completing it, he asked Babkin to read it critically lest some observations had crept in that might offend his hosts. He was delighted when Babkin found fault with only a few of his remarks, which were easily altered. By the end of February 1935 Cannon had no more worries about his address and prepared with a light heart for his trip abroad.[8]

☙ ☙ ☙

After a relaxing voyage that began in San Francisco and was broken by a splendid week in Hawaii, Dr. and Mrs. Cannon, accompanied by their daughter Helen (Marian had decided to go on ahead in October), arrived in Kobe, Japan, early in April. Barely there long enough for a preview of the country which they would visit later, they went on to Shanghai where a welcoming committee led by Dr. J.H. Liu, who had been a student in Cannon's physiology class at Harvard in 1915, greeted them.

With the three leading medical men of China as their hosts, they embarked on a dizzying round of sightseeing and social events followed by a quick trip to Nanking via Hangchow. Cannon, registering his first impressions, was prompted to quote his old friend, Rev. Samuel Crothers, "Their look had turned from day before yesterday to day after tomorrow." In both Shanghai and Nanking, his public lectures received enthusiastic applause. At the National Medical College in Shanghai he spoke on the process of scientific research and on chance and the prepared mind; to students at the Army Medical College at Nanking he presented an introduction to the nature of emotional excitement.[9]

Mrs. Cannon was invited to speak, too. At a tea in their honor in Shanghai, she talked about her recent volunteer work for the Birth Control League of Massachusetts, in which she had become a leading member and was taking an active role. Impressed by the critical need for reform in such an over-populated and impoverished country as China, she learned, to her surprise, that there already was a birth control movement under way, which had been initiated by the medical profession and was now seeking public support.[10]

On reaching Peiping, the Cannons were met at the station not only by Wilma and John and Marian, but also by the entire P.U.M.C.

physiology department. For the next three months, they lived with the Fairbanks at 10 Ta Yang Yi Pin Hutung—a compound of houses set round a courtyard—where the senior Cannons had their own quarters. In addition to a "number one" houseboy, a cook, and an assortment of other domestic help, two rickshaw boys had been added to the household staff of servants for their personal use. No sooner had they settled in than Cannon presented himself to the physiological laboratory at P.U.M.C., where he was welcomed by Dr. Lim and given every facility for conducting his researches.

Since the early 1930s, Robert K.S. Lim, the professor of physiology at the medical college, had been sending Cannon papers from his laboratory. So impressed was Cannon by the quality of the research reported that when Lim asked him to accept one of his young associates, A.C. Liu, as a postgraduate student for a year, Cannon readily agreed. It was a happy judgment. When Dr. Liu arrived in 1934 on a Rockefeller fellowship, he proved to have all of the qualities Lim had claimed for him.[11]

Lim was equally pleased when Cannon wrote a glowing report of Liu, since it also stood as testament of the growing scientific stature of his department—a development that had not escaped the attention of Roger Greene, the director of the Peiping Union Medical College. After Greene learned from friends at Harvard that Cannon wished to visit his daughter in China before going to Russia for the meetings of the Physiological Congress, he had conspired with Alan Gregg of the Rockefeller Foundation to invite him to serve as visiting professor of physiology at the P.U.M.C.[12]

Cannon prepared in his usual manner for the lectures he was planning to give at P.U.M.C.; most of them revolved around his work on the emotions and sham rage, homeostasis and sympathectomy, and the chemical mediation of nerve impulses. When the family went to hear him speak, they found the hall so crowded that they couldn't get in. "Until May 8 he is to lecture three times a week," Cornelia noted in a letter home, "and meanwhile he has started in a research with one of the young assistants and things are going busily. We find we shall have to release him from social engagements for he takes his P.U.M.C. duties very seriously and wants to give them every ounce of his strength."

There followed for Cannon some weeks of work with congenial people, both in the laboratory and outside. In the meantime, his wife ("as energetic and eager-minded as ever," Cannon wrote his son) and daughters indulged in a whirl of social and cultural activities: they visited fairs, took weekend excursions, had lessons in brush painting, went to the Chinese theatre, and marveled at the beautiful buildings of the Forbidden City. As if that were not enough, Cornelia on her own made rounds with the nurse of the local family planning association and indulged her growing interest in birth control—a practice she would follow throughout the trip.

It seemed to Cornelia "too bad" that Walter was missing the enjoyment of what was "surely one of the loveliest cities in the world." Once, she even complained to her family that he "might as well be in Cambridge, for, except for his ricksha ride to the P.U.M.C. every day and the weekends in the country, he sees only reostats and smoked drums and the insides of cats." Still, Cornelia had to admit that he was in the best of health, free from worry and enjoying himself to the utmost.[13]

On rare occasions, he did agree to take time out for visits with his family—for instance, to the mission hospital at Tung Chow run by Dr. Canby Robinson or to the Ming Tombs. Not all of the sights they saw were lovely. On one weekend trip to a sacred mountain in the western hills, Cannon was so upset by the pathetic beggars they met that he left the mountain trail to avoid them. He also had an opportunity to see primitive medicine at work when "a filthy old man" gave a demonstration of acupuncture by driving a needle into the arm of a coolie who complained of stomach pains. Though the coolie looked better and felt better afterward, Cannon was alarmed and remarked that the possible sepsis from the procedure absolutely disturbed him.[14]

Whatever his distress at the "appalling readiness" of the Chinese to tolerate unsanitary conditions, there is no denying that Cannon was having a wonderful time. When he posted a note to Phil Bard in mid-May, he could hardly restrain his enthusiasm:

> After dining about five times a week on bird's-nest soup, shark's fins and bamboo sprouts—and liking them! I still have a vigorous g.i. tract and an optimistic outlook. Friends here have kept me busy with lectures, demonstrations, conferences and visits—besides dinners. It is all very pleasant because they are so kind and appreciative . . . What a marvelous old place—beauty and squalor, rickshaws, camels, wheel barrows, trains and automobiles, women in pants and men in dresses, etc., etc. If you ever have a chance, come by all means.

Given a free rein, Cannon did not limit his lectures to the P.U.M.C. but crafted some for the general public as well. He found his mailbox full of invitations from any number of medical schools, colleges and organizations including, among others, Yenching and Tsinghera Universities, the Peiping Society of Natural History, the National Normal University, and Sigmi Xi of Peiping. He even had a request for a lecture from the Rotary Club of Peiping. All told, Cannon gave almost a score of popular lectures on a variety of subjects stemming from the physiological investigations he had conducted throughout his career.

Cannon was charmed by the Chinese. He felt attracted to the common people, so friendly and with such a delightful sense of humor. Moreover, he was continually astonished by the widespread familiarity of his Chinese audiences with English; so much so, that during one of his public lectures he decided to test his audience. In a room festooned with the flags of various universities and colleges, Cannon, speaking in English, asked his auditors if they had ever heard the cheer of a college whose banner he did not see there: "Down with Harvard, down with Yale, all I know I learned by mail." The laughter from the audience assured him that they not only could appreciate a joke but that they knew their English well.[15]

In May he began a new series of special lectures to physiologists and research workers at the P.U.M.C. At the end of the month he capped these lectures with demonstrations of the technique used for denervating the heart and the method of doing a sympathectomy.

For Cannon, who frequently referred to himself as "a laboratory hermit," the high point of his stay in Peiping was the opportunity it gave him to engage in research with Dr. Lim and his young associates, among them H.C. Chang, T. P. Feng, and S.C. Shen. All were well-trained investigators, and Cannon took pleasure in introducing them to problems that stemmed from his work on neurotransmitters. With Dr. Shen he undertook a study to determine whether sensitization by ganglion-removal would occur in smooth muscle innervated by parasympathetic fibers, which eventuated in publication. A special bond developed with the group that Cannon later reinforced with suggestions on how to advance the research begun during his visit.[16]

During the course of these investigations, Cannon was delighted to learn that Chang had given him a Chinese name—"Yuan Shi Kai." Chang explained that the Chinese name must not only have the same sound as "Cannon," it must also embody the two aspects of his character: kindliness and devotion to work. Cornelia, writing home that Walter was thrilled over his research and that the department was equally thrilled over him, described a special moment: "They have had a Chinese robe of beautiful imperial blue tribute silk, a cap and shoes, made for him and he is a thing of beauty in it, so enthralled with it that he now wears it to Chinese dinners and looks most dignified and Chinese; even his eyes tilt a little, to match."

Although Cannon's stay at the P.U.M.C. was a short one, Roger Greene was well satisfied with the results of his visit. "I feel that we have received full value for the expenditure which was necessary to bring Dr. Cannon to Peiping," he wrote Alan Gregg. "He has worked early and late, and has started some of the younger members of the department of physiology on problems of some interest."

A year after Cannon's visit to China, Lim excitedly wrote him of the new developments in physiology by a few of his former students:

> In the laboratory, we have been carrying on very much on the same lines as when you were here. Dr. Chang is still hunting acetylcholine in the placenta. Dr. Feng has made some interesting studies on the behavior of the neuromuscular junction. Dr. Shen is still analyzing the changes which take place during fattening of the liver in chicks. Mr. Shen is working with Feng, but continues to make observations along the lines he started off with you . . . I have three good assistants, who are working with me on the medullary "sympathetic" centre . . . The direction in which this work is going is a direct effect of your visit here—a good example of serendipity![17]

Cannon had told Lim that he desired to pay his respects to Dr. Sataké in Sendai, Japan, after finishing his duties in Peiping. He had long wished to meet and know Sataké, one of Japan's most distinguished physiologists. Indeed, he felt indebted to him. More than a decade before, Sataké's laboratory had found evidence that independently supported Cannon in his debate with Stewart and Rogoff on the emergency function of the adrenal medulla. Given the political tensions existing between China and Japan after the

Japanese invasion and conquest of Manchuria in 1932, arranging Cannon's trip required delicate planning, which was ultimately supplied by Lim with the collaboration of Yas Kuno, the director of the physiological laboratories at the Japanese University of Mukden. Throughout the existing political turbulence, Lim and Kuno had maintained friendly professional relations and it was through their joint efforts that Cannon was invited to lecture at the Imperial Universities of Sendai, Tokyo, and Kyoto.[18]

On the way to Japan, Cannon, accompanied by his wife and some of their children, visited Mukden and Korea. In Mukden he spent much of his time with Dr. Kuno, who worked on problems of temperature, and who, Cornelia later noted, "was a very charming gentleman, and was well liked by the Chinese of the region for his consideration and the respect he paid them. He had none of the air of the swaggering conqueror which the Japanese military displayed. In fact he got along so well with the Chinese and was so much liked by them that soon after we saw him he was recalled to Japan."

When Cannon later went on to Russia, he arranged with A.V. Hill, a fellow member of the International Physiological Committee, to put Yas Kuno on the governing board of the International Physiological Society as an indication of Kuno's good repute with scientists in the West.[19]

In contrast to the visit to Mukden, the trip to Korea was a disaster. Through a misunderstanding, some professors went to meet the train on which the Cannons were expected and "lost face" when the visitors did not appear. They refused to try again. Cannon made haste to present himself at the Japanese Medical School and express regret at the mix-up (for which he was not in the least responsible), but he was rebuffed and not admitted to the presence of the injured officials. It was "a revelation of the sensitive Oriental dignity," Cornelia noted with exasperation, "but the main impression it left on us was its absurdity."

Putting the unpleasantness behind them, they went on to Japan, which they found "marvelously beautiful," and (a welcome change from China) clean and neat, orderly and efficient. Cannon's reception could not have been warmer. Delegations always met him at the train stations, bowing deeply from the waist and bestowing huge bouquets of flowers on Mrs. Cannon. Both were overwhelmed by the attention they received. Cannon was particularly moved by his meeting with Sataké. "No more kind or considerate reception have I had anywhere than that which I received from Professor Y. Sataké in Sendai, Japan, in 1935," he reminisced almost ten years later, but added in light of later events: "When I look back upon the evidences of good will shown by this gentle Japanese scholar I find it impossible to believe that he would approve the brutalities and sneaking deceits of the military group which has latterly dominated his government."[20]

Arriving in Tokyo in a blaze of welcome, they were greeted by a large group led by Dr. Kato. Cannon's first talk at Keio University received a great deal of attention. "It might have been a Hollywood idol or Babe Ruth or Miss America," Cornelia reported, "instead of the interpreter of emotion, and the exponent of the movements of the digestive tract." On the basis of their experience, she urged that no one "should ever dream of traveling without a distinguished husband or father." Aside from the usual round of sightseeing and lectures, Walter saw Dr. Goto, a biologist who was an old friend from college days, in hospital just before he died, while Cornelia met with two activists of the birth control movement, Dr. Majima and Baroness Ishimoto (the Margaret Sanger of Japan).[21]

Bidding farewell to their children in Kyoto, the Cannons went on to Vladivostock, and were met there on July 9 by an In-Tourist agent who made all the preparations for them as guests of the Soviet government. They were to travel "soft" on the Trans-Siberian Express with a two-berth compartment to themselves. After boarding, their porter took good care of them and supplied them with hot tea at any time of day. The food though monotonous was not bad, but the flies were almost more than Cannon could bear. At one point during their journey, they got off the train to try to buy wild strawberries from a peasant woman at one of the stations. Not only was the transaction unsuccessful, but Cannon had his watch snatched from him on the platform. Fortunately, a guard was on hand to seize the thief and rescue the watch, but he refused the reward that Cannon tried to give him.

Cornelia, who kept a record of the trip, condemned the false advertising of "deluxe accommodations" on the Siberian express, which was incredibly dirty, hot and uncomfortable. At the same time, she praised the Russian people, whom they liked very much, as friendly and pleasant, ready to laugh—a most attractive type of human being. As for her early impression of the Communist state, she commented wryly, "Plain living and high thinking never seemed to me drab but I have seen nothing suggesting anything else in Russia so far."

The trip across Siberia lasted eight days and took the Cannons through undeveloped country where settlements were few and far between. It reminded them somewhat of the American west until, increasingly, they became aware of a large number of "convict" camps—low barracks surrounded by high barbed-wire fences and guarded over by towers with armed men in them standing watch. At times, they even saw groups of men from the camps, "some whose faces bore unmistakable marks of culture," at work on double-tracking the trans-Siberian railroad. Even more ominous were the trains of wooden freight-cars they passed, loaded with men and women being shipped to the far East. "The face of one woman I shall never forget," Cornelia lamented long afterward, "so utterly despairing was the expression. It haunts me to this day."[22]

Finally they reached Sverdlovsk, where they were met by Dr. Andreyev, a former pupil of Pavlov's who had been sent to meet them. With him were a delegation of the city's officials. They were taken to a room in a newly-built apartment house and were fed well, without the accompaniment of flies. Then they were shown factories in the region, to one of which they were taken by airplane. The plane was small with just enough room for the three of them and the top of the passenger section had been removed. "The result," as Cornelia described it, "was that we sat on seats without a projection of any kind to hold onto if the plane should desert the horizontal by so much as a few inches. Not only were we defenseless against the possibility of falling out of the plane, but we were nearly frozen by the cold air from which we had no protection." Nevertheless, they made it back to Sverdlovsk without incident.[23]

On the last day of their visit, Cornelia had a misadventure. While Dr. Andreyev took Cannon to visit a local hospital, she went off with her camera to take pictures of the many log cabins in town. When she photographed the compound where the Imperial family had lived out the last weeks of their lives, she got into trouble with Communist officials, was arrested and taken to what she later learned was the headquarters of the G.P.U. for Siberia. She had neglected to carry proper identification, but after she pointed to a picture of herself with her husband in a newspaper, her guards contented themselves with confiscating her film before permitting her to return to her hotel.

Andreyev, who with Cannon had been anxiously awaiting her arrival, was horrified to have her under guard of the O.G.P.U. Although Cornelia tried to pass the matter off, she spent the next few hours living over in her imagination "the sufferings of the thousands of innocent persecuted persons who have lived in this country." When Andreyev later reported the incident in Moscow, she received a handsome apology from the government for her arrest. Still later, friends assured Cornelia that if instead of log cabins she had taken pictures of new factories, all would have been well, for there were, of course, under the new and enlightened regime, no log cabins in Siberia.[24]

On leaving Sverdlovsk the Cannons were taken by airplane to Kazan. Flying over the Ural Mountains and the great farming region beyond, they were greatly impressed with the villages and countryside below them. Even so, Cornelia was impelled to remark:

> Walter is feeling the queer atmosphere of oppression and throttling that seems to underlie the simplest doings. It began with my arrest and he said later, "Don't you feel as if you'd like to get out of this country?" There is a queer kind of horror that seizes you now and then, in these outlying regions where the veneer is thinner than I imagine it is in Moscow. The bear shows his claws in Siberia.

On the boat to Gorky, their equilibrium somewhat restored, they compared river life on the Volga to what the Mississippi must have been like in Mark Twain's day.[25]

Toward the end of July, the Cannons arrived in Moscow and had nearly two weeks to themselves before it was time to go to Leningrad for the meetings of the Congress. Their days were full of interest as they saw for the first time the various enticements of the new regime—the parks, the theatres, state and collective farms, the beautiful subway, and the interesting museums crowded with visitors. "There was obviously a passion for improvement among the people who had been for centuries inadequately provided with opportunities for education," Cornelia later recalled. "They were all seizing it now, to the point of exhaustion, I was told." While in Moscow, she had an opportunity to visit an abortorium, one of many then functioning; some years later she was informed that they had been closed by order of the government.

In one of her letters home, Cornelia recorded another side to their visit in Moscow: "We are quite comfortable in our hotel and have nothing of which to complain, but our hearts are heavy with the tales our medical friends tell us of what is happening to them and theirs in this land of suspicion and menace. They talk to us as they probably have not talked to a soul since the Revolution. It seems to be a relief to them to talk to someone whom they can trust not to betray them even though what they say is not critical of the regime."[26]

❧ ❧ ❧

When it was time to leave Moscow, they took a train for the 350-mile trip to Leningrad and were met there by Pavlov who, limping along beside Cannon, told him that he had almost died from pneumonia the previous spring. Taken to the Hotel Metropole, they were given a splendid suite with a room for Bradford (who had recently joined them), as well as a chauffeur and automobile (which had served Kirov before his assassination). They found a striking difference between Moscow and Leningrad: the former proletarian with a jerry-built appearance, while the latter gave an impression of aristocracy, lavishly planned with parks and boulevards, magnificent buildings and everything on a sumptuous scale. The people, however, were no different at all in appearance; in both places everyone was shabby.

Nothing had been spared when the fifteenth International Congress of Physiology convened on the morning of August 9. It was, after all, the first large scientific gathering to be held in the new Russia, the U.S.S.R. When the plenary sessions opened in the beautiful Uritsky Palace, there were even earphones provided for simultaneous translation in several languages. "In the superb entrance hall," Cornelia recorded in her diary, "beautifully decorated with flowering plants and palms, was a huge statue of Lenin, but except for that we might have been back in the days of the *former* tyranny. The great auditorium was packed, about 2,000 people. Pavlov, looking like a beautiful old saint, presided, flanked by Prof. Hill of England and Prof. Lapicque of Paris."[27]

First Pavlov greeted the delegates with a brief but stirring speech. As seekers of truth, he told them, they all must help to insure that the strictest justice would prevail in international relations. Next came the Commissar for Health Kaminski, who was followed by another Communist official and the aged president of the All-Union Academy of Science before Pavlov, without further ado, turned the meeting over to Cannon.

Cannon stood on a lower platform directly in front of Pavlov, surrounded by cameras and spotlights. He spoke in a firm clear voice, and was frequently interrupted with applause. His words, painstakingly wrought in the quiet of his Cambridge study, took on a momentous quality in that assembly of people who came from countries in turmoil and on the verge of war.

The title of his address, "Some Implications for Chemical Transmission of Nerve Impulses," gave not a hint of what was to come. After a heartfelt tribute to Pavlov for his pathbreaking research and steadfast support of the importance of scientific inquiry, Cannon focused the introduction to his scientific report on the perilous political state of the world and its devastating impact on science. "During the past few years," he began,

> how profoundly and unexpectedly the world has changed! Nationalism has become violently intensified until it is tainted with bitter feeling. Governments whose strengths seemed deeply rooted in fixed tradition have vanished like phantoms, only to be replaced by the strange new forms and new agents . . . Creative investigators of high international repute have been degraded and subjected to privations. Some universities have been closed; others have been deprived of their ideal social function of providing a sanctuary for scholars where search for truth is free and untrammeled and where novel ideas are welcomed and evaluated.[28]

Speaking of the worldwide economic depression that shackled scientific research, Cannon went on to compare conditions in the United States, where money made available for research had been greatly reduced in recent years (a policy he believed was incredibly short-sighted), with the manner in which science was encouraged and furthered in the Soviet Union, where the availability of funding "for the development and prosecution of scientific studies is relatively greater than in any other country of the world."

Cannon did not content himself with a litany of the woes that had recently been levied against scientists. Instead, he reminded those present about their obligation to preserve freedom of inquiry. "The attention of the investigator," he told the assembly,

> must not be confined to immediately useful prospects, nor to a so-called "nationalistic science," nor to any group of political ideas. Such limitations are fatal because they confine the free spirit of inquiry . . . Every nation needs the advantages which flow from a free interchange of students and ideas, and since we have all received benefits from the past, we should recognize that we all owe equally a service to the future.

At no time during his address did Cannon mention Nazi Germany by name. The message, however, was not lost on the German delegation, who for a time considered registering a protest with the Congress but then thought better of it. John Fulton, the Yale physiologist who was in the audience, later characterized Cannon's words as "bold and forceful." Within a few days, both the *New York Times* and the *Moscow News* bestowed their own accolades on the speech. The *New York Times*, in a story filed by its foreign correspondent Walter Duranty, noted that the Russians had been pleased by the speech and proclaimed in a headline that Soviet prime minister Molotov had specially congratulated Cannon afterward. Others who heard the speech later thanked Cannon for having the courage to say "what some of us ought to have said long ago," and called it the "*pièce de résistance*" of the Congress.[29]

In addition to the scientific sessions that followed the opening of the Congress, delegates were treated to an orgy of sightseeing excursions to such gorgeous pre-Revolutionary edifices as the Hermitage, the Winter Palace, and the Fortress of St. Peter and St. Paul, as well as a tour making the obligatory round of orphan asylums, state farms and people's parks. For Cannon, there were opportunities to see old friends like Juan Negrin from Madrid and Zenon Bacq from Liège. A special treat was an extended visit with Pavlov.

"Mrs. Cannon and I lunched with him and his family at his apartment and spent a very delightful day at the great new laboratory for experimental neurology outside Leningrad," Cannon later wrote in his diary. "It was obvious that he was not well. I think he was rather appalled by the responsibilities of the huge development of his work and anxious that the great expenditures involved in the new establishment should be justified by important results. He greatly wished it might have come twenty or twenty-five years earlier."[30]

Cornelia also made a memorandum of the delightful day spent

with the Pavlovs, "walking and talking, sitting in the warm sun on the porch speaking German most of the time, having lunch and dinner." She described the various foods served "in true Russian family fashion; luncheon with fish, caviar, baked biscuits filled with meat or cabbage, cake and coffee for dessert. At dinner we had soup, fish, chicken, carrots with sour cream, potatoes, cucumbers, and then delicious raspberries and Mrs. Pavlov's name day cake for dessert."

Pavlov's wife, a devout woman, worshipped at the village church whose steeple could be seen from Pavlov's laboratory. His daughter, Vera, however, told Cornelia that her father was not a religious man but said he wanted to do something for religion before he died. Although Pavlov did not go to church himself, he believed that every man should be free to go or not go as he chooses. Before they left, Cornelia was given a tablecloth embroidered with Mrs. Pavlov's initials on it, and vowed she would "treasure it always in memory of a sweet woman, a great man, and a very happy day."

The huge laboratory at Koltusi, eighteen miles from Leningrad, interested Cannon greatly. He felt deep sympathy for Pavlov, burdened at his age with the heavy responsibility of directing such a place. Cornelia, too, was interested in seeing Pavlov's laboratory and thought the dog kennels in which he kept his experimental animals beautiful—in fact, "better than the accommodations of most of the children that we have seen."[31]

The final festivity extended to the Congress delegates was a great banquet at the Palace of Catherine II, a place of utmost splendor and luxury. It was a very lavish affair—lavish to the point of opulence—with an abundance of food and wine "appropriate to an Oriental potentate," Cornelia observed, "but which was inconceivable in a proletarian country. I never saw anything like it." It had been a long time since the Russian physiologists had seen so much wine offered so freely, and they indulged to an excessive degree. At the end of the evening, a great block of ice cream in the shape of a dog was carried in and, amid the applause of the diners, placed before Professor Pavlov.[32]

The Russian physiologists and medical men, for the most part, were charming and friendly hosts who made a pleasant and lasting impression on the Congress members. Some of the visitors from the United States thought they detected a decided difference between Russian Communists and American Communists. Others observed that the Soviet government had provided the Russian delegates with new suits for the occasion, each one the exact replica of the other (as though cut from the same bolt of cloth). Of all the Russians, it was Pavlov, the grand old man of physiology, who once again made the deepest impression. Pavlov's death a year later severed Cannon's strongest tie with Russia, but for the rest of his life he thought of the people he had come to know in Moscow and Leningrad with nostalgia.[33]

Before departing Russia, Cannon and other members of the International Committee had an interview in Moscow with Molotov, who discussed with them the proposed establishment by the Soviet government of prizes for scientific achievement somewhat like the Nobel Prize. Cannon did not think much of the idea; he suggested instead that access to scientific journals from the rest of the world would be a fundamental service to Soviet scientists. Molotov seemed to think Cannon's suggestion trifling and unimportant and brushed it aside.[34]

When the Cannons with Bradford left the Soviet Union (Cornelia with her precious diary and family letters strapped securely inside her corset), their train crossed a bridge where the boundary between Russia and Finland was marked. The three of them took deep breaths of relief and sighed at being in a free land once more. Later at Helsingfors, one of their physiological friends asked them, "Well, did you sigh when you crossed the boundary bridge?" They looked at him in astonishment and replied, "How did you know?" "Everyone does," he said.[35]

From mid-August to early September, the trio traveled through Finland, Sweden and Norway, resting and seeing the sights. They couldn't help comparing Russia and the Russians unfavorably to neighboring Finland—clean, free, and infinitely more civilized. They found even more to admire in Sweden, where the Liljestrands and the Berglunds took them in hand and Cannon had a busman's holiday visiting laboratories. A friend of Ida Cannon's took Cornelia to visit birth control clinics; in Stockholm she admired Swedish idealism harnessed to common sense and efficiency, not just "talk and a dream of the future."

In Norway, they were captivated by the natural beauty of the fjords. When they visited Oslo, Cannon renewed yet another friendship, this time with Otto Mohr, whose Dunham Lectures late in 1933 on breaking developments in the science of genetics as related to the practice of medicine had stimulated extraordinary interest at Harvard. Dr. Mohr's mother-in-law, Mrs. Anker-Moller, Cornelia was delighted to learn, was one of the world's pioneers in birth control work and a champion of women.

The last leg of their journey took them from Bergen to Scotland, where they visited Edinburgh and Glasgow. At their hotel in Edinburgh, Cannon's old friend from the war, John Fraser, was waiting to greet him and gave a dinner in his honor. In Glasgow, they caught up with the latest international news from Geneva and glimpsed the inevitability of another war looming in the future. Sailing from Liverpool, they arrived back home in late September, just in time for the beginning of a new academic year.[36]

Cannon's speech at the Physiological Congress catapulted him into the spotlight as an international spokesman for science. After his return to Boston, he was frequently asked to speak or write about what he had seen in the Soviet Union. In a letter written to Boris Babkin, for example, Cannon reported that everywhere he went in Russia, there was "evidence of great activity in the building of factories, houses, hotels, railroads, bridges and other needed structures." He was truly impressed by the care given to the laboring classes; but he found himself "oscillating between admiration for such signs of new activity and humane interests in the Soviet Union and a very deep horror of cruelties and injustices inflicted upon certain members of the community without explanation or legal process." He could not understand why the new government was repeating the injustices of the old regime—injustices that had led to the Revolution itself.

Babkin had given little indication of his personal feelings when he originally helped Cannon with his address. But he now confirmed that what Cannon had seen with his own eyes coincided with what he knew to be true:

> The Russia of today is a slave state, as many, or probably all, states were once. The Egyptian pyramids and Roman aqueducts were built by the hands of slaves. In the same way now in Soviet Russia they build the canals, factories, palaces, etc. The only difference is that in the old days the kings and the aristocrats ruled the state, and in

> Soviet Russia a few doubtful characters, who declared themselves "proletarians" are ruling by terrorization of the blinded masses.[37]

Some time later, Cannon received a letter from L. Cherniansky, of the U.S.S.R. Society for Cultural Relations with Foreign Countries, who asked him about the situation of Soviet scientists. In reply, Cannon again gave credit to the governing body of the Soviet Union for supporting scientific studies. He was concerned, however, that the Soviet scientists had difficulty in obtaining information regarding work done elsewhere because of the lack of foreign scientific periodicals. "It is well to recognize," Cannon told Cherniansky, "that scientific effort crosses national boundaries and that those engaged in scientific research profit greatly by knowing one another's methods and results. Full exchange of information would be highly valuable."[38]

Cannon's report card of his strongest impressions about the Soviet Union gave both high and low marks. On the one hand was "construction, construction everywhere," on the other was the prevalence of flies. The only fly swatter he had seen "from Vladivostok to Leningrad was brought from China. Making these simple instruments could be a new industry, fly killing could become a major sport! And the Commissar of Health would rejoice." He gave high marks to Soviet care of working people as shown by fewer hours of labor in the harder jobs, by provision of parks, playgrounds and outdoor entertainment, by opportunities for mothers to rest and recuperate from childbirth, and by maintenance of vacation resorts. Most of all, he lauded the selection of promising young boys and girls with unusual talents to receive special training and treatment for the development of those talents.[39]

Of the number of people who approached him to tell them what he had seen in Russia, there were some who were strong supporters of the Soviet Union and some who may well have been Communists. Cannon, though courteous in his rejoinders, refused to tailor his views to accommodate their agendas. When Herbert Goldfrank, a Friend of the Soviet Union, asked him to sign a manifesto of the International Committee of the Friends of the Soviet Union, Cannon replied that he could not in good conscience sign it, "in spite of my enthusiasm for all of a positively beneficial and admirable character that the U.S.S.R. is doing." In a subsequent letter to Goldfrank he wrote:

> Just as I regard the cruelties and injustice of the United States as something which should be recognized along with the favorable conditions, so likewise, in judging the Soviet régime, I think that in the long run it is for the best interests of everybody that we should keep clearly in mind the fact that there are evils still existent which subtract considerably from the great achievements which the Soviet régime has to its credit. Perhaps, as a man of science devoted to truth, I may be excused for insisting on the whole truth being respected.[40]

Just as Cannon could not wholly endorse the Soviet Union, neither could he wholly condemn it. Indeed, his insistence "on the whole truth" made him spring to Russia's defense if he felt that the country was being unfairly attacked. When the Friends of the Soviet Union requested a statement from him about a piece on Russia that appeared in the *Saturday Evening Post* near the end of 1935, he replied that what he most objected to was that the article was one-sided. Not only did it present a false picture of the total situation, it also left out a great many positive things. Anyone who wanted to cast slurs on the American scene, he maintained, could focus on such things as gangsters and graft and false advertising and lynchings, and thus make quite as strong a case against us as was made in the *Saturday Evening Post* against the Soviet situation. Cannon ended his letter with a moment of reflection:

> Persons who have had the advantage of historical perspective, however, realize that affairs are better now than they were a few years ago. Anyone who knows the past experience of the United States is aware that our development has been similar and that we have gone through stages of slackness, inefficiency and filth. I commend, to anyone who wants to read about our past, Dickens' "American Notes." We weren't pleased by his characterization of our manners and customs, but I have no doubt that what he saw was true at the time of his visit.[41]

18 *The Spanish Medical Bureau*

PERCEIVED FAR and wide as a friend to Russia, Cannon in the spring of 1937 was invited to accept the presidency of the American Russian Institute. The offer intrigued him. Since the time of his visit to the Soviet Union, he had wanted to establish closer relations with scientists there and thought of his laboratory as a place that would welcome bright, young Russian students. Initially, Cannon considered accepting the presidency; but caution prevailed and he decided to write to William A. Neilson, the president of Smith College and the outgoing president of the Institute, before coming to decision. Neilson wrote back his candid judgment that if the new president were to be more than an ornament, "he ought to be within easy reach of New York or in New York itself." He told Cannon:

> I think it would strengthen the Institute greatly to have you accept and for this reason I dislike advising against it, but my frank opinion is that the political problems that keep arising in connection with the Institute are such that it is not safe for people like you and me to appear at the head of it and not have an important part in the framing of the policies. I suspect that you do not find it any easier to get to New York than I do and can as little spare the time.[1]

Cannon took Neilson's advice to heart and promptly refused the presidency of the American Russian Institute. There were other reasons for his refusal, and not the least were reports of stepped-up government repression in the U.S.S.R. which troubled him and added to his growing doubts. In June, when he was invited to participate in the twentieth-anniversary celebration of the Russian

Revolution, he replied that his enthusiasm for the Soviet experiment had been waning and "that elements of suspicion and fear which existed in the Soviet Union when I was there nearly two years ago have been enormously accentuated by the events of the past six months." In addition, he could not help feeling that matters seemed to be going "in the direction of disloyalty, violence and terror." Under the circumstances, he was inclined to withhold his participation in the proposed celebration.[2]

Perhaps if Neilson's advice to avoid becoming involved in organizations with headquarters in New York had arrived sooner, Cannon might also have turned down the national chairmanship of the Medical Bureau to Aid Spanish Democracy—a post he had agreed to fill just a short time before. But if he balked at endorsing government policy in Soviet Russia, he had no such qualms when it came to backing the cause of the Spanish Republic.

Spain was by no means a new interest for Cannon; his visit there in the summer of 1930 had had a profound effect on him. His former fellows, Rosendo Carrasco-Formiguera and Jayme Pi-Suñer, made a point of introducing him to young medical scientists who not only discussed their research but also informed him with great passion of the hardships they had suffered under the recently deposed military dictator Primo de Rivera, and of their eagerness to establish a democratic form of government. He particularly enjoyed meeting Dr. Juan Negrin, the professor of physiology at the University of Madrid, who showed him plans for construction of a university city on the outskirts of Madrid, which in Negrin's description were "a part of my intention to advance freedom and science in Spain." Cannon later remembered that everyone he had met and every place he had gone during his visit added to the hopes for a brighter future for Spain.[3]

In the years that followed his visit, Cannon kept abreast of events in Spain and corresponded not only with Carrasco and Negrin but also with Georges Marañon and other friends there. After the establishment of the Second Republic in 1931, he had written Marañon, "Ever since we returned from Spain I have told my friends that in my opinion Spain offered the most interesting spectacle to watch in Europe—even more interesting than that provided by Russia." In 1933, after Negrin had accepted an official post in the new government, Cannon had written him to suggest that the Jewish scholars being driven out of Germany might be invited to Spain. "My sole interest," he explained, "is that of doing two things at once: relieving the distress of persecuted leaders of intellectual activities, and promptly putting into action at Madrid the investigative work which I know you dreamed of."[4]

Following the electoral victory of the Popular Front in February 1936, conservative factions rallied behind the right-wing National Front, which openly appealed to the military to save Spain from Marxism. The military rising had started in Morocco on July 17, and within a matter of days, spread to the garrisons of metropolitan Spain. In October, General Francisco Franco, one of the leaders of the elite army corps, declared himself "head of state" and took full command of the insurgent forces. Cutting a swath through the hastily assembled Republican militia, he arrived at Madrid in November. Against all odds, the Republican security forces managed to defeat the better trained and better armed military in Madrid. While the Republicans held the industrial zones, the Nationalists held the agricultural areas. One consequence was to be an increasingly acute shortage of food in the cities of Spain throughout the war.

Right from the start, both sides had appealed for help from other countries. Italy and Germany (even though they appeared to favor the non-intervention pact promoted by the Western democracies) responded by sending transport and troops, planes, tanks and munitions to Franco's Nationalist rebels. Save for slight support from France and Mexico, only the Soviet Union had come to the aid of the Republicans. The successful defense of the capital city in the fall of 1936 was helped enormously by the arrival of military supplies from the U.S.S.R. and by the International Brigades—volunteers who came from a score of different countries.[5]

With a full-blown civil war raging in Spain, liberal-minded Americans saw the rebellion against a democratic form of government by the military insurgents, aided by Mussolini and Hitler, as an extension of the threat of fascism. In October, Cannon had joined with others to sponsor the medical division of a newly-established aid organization known as the American Friends of Spanish Democracy. He later lent his name to petition the State Department to issue a permit for sending medical units and supplies to Spain, as an exception to the embargo that had been instituted with passage of a new Neutrality Act.[6]

"It has been a great satisfaction to learn," Carrasco wrote Cannon early in 1937, that "you have carried your sympathy for us up to actively joining the doctors' committee of the Friends of the Spanish Democracy. Some ambulances with their personnel and materials have already arrived and have been heartily welcomed. I am very glad to know that you have personally contributed to this friendly and generous help from dear America." Yet, when Roger Chase, the secretary of the medical division of the American Friends of Spanish Democracy, asked if he would head a national fundraising campaign, Cannon turned him down.

It soon became apparent that the medical division would function better as a separate entity and it was reorganized as the "Medical Bureau." Throughout its life-span, the Spanish Medical Bureau was a loose coalition of liberals, socialists, and communists—all unified by a common aim of providing humanitarian aid for the Spanish Republic. In March 1937, Chase again approached Cannon and asked him to be national chairman. This time, not only did Cannon say yes, some months later he also agreed to chair a chapter of the Medical Bureau in Boston.[7]

Cannon devoted himself to his duties with a will, especially buoyed by the knowledge that the new premier chosen in May to head the Loyalist government of the young republic was his friend Juan Negrin. Almost at once, Cannon was drawn into speaking engagements and letter-writing campaigns aimed at gaining support and raising money for medical aid to Spain. In his first major speech before a huge rally held at the New York Hippodrome on April 17, he explained to the audience how he had come "to support the 'Reds' in the war which has been waged in Spain during the past nine months."

First, he recalled his experiences in Barcelona and Madrid where in the spring of 1930 he had visited his students, met Juan Negrin, and listened to their dreams for a democratic form of government. Then he invoked a memory of an even earlier time—his experiences in World War I when he had become acquainted with conditions confronting those struck down in battle and learned that chances of recovery from serious wounds were directly related to prompt medical and surgical treatment. Although the Spanish Medical Bureau endeavored to bring aid to the victims of the civil war by establishing hospitals and mobile units, more was needed. He ended by earnestly imploring the rally to do "all that we can to mitigate the sufferings of those who are gallantly fighting for those ideals, for *our* ideals, in a world where dictatorship and tyranny have boldly set up their challenge."[8]

Scarcely more than a week later, market day on April 26 in the Basque town of Guernica, the firebombing of defenseless civilian populations by German Condor Legion fliers in the service of the Nationalists caused an international furor. Cannon joined with others crying out in protest when he wrote Carrasco in early May, "The destruction of Guernica sent a wave of horror through many American breasts. The whole procedure was so utterly wanton and uncalled for in so far as military advantage was concerned that it seems incredible that anyone could have ordered the act to be done." Despite the outcry, Picasso's mural depiction of Guernica—according to one art critic, "the most celebrated, widely reproduced and universally recognizable political painting of the twentieth century"—did not change Franco's regime one whit or shorten his forty-year reign by so much as a single day. Moreover, the wholesale slaughter, however incredible it seemed to Cannon and others in 1937, set a pattern that would be followed in World War II by air forces on both sides of the conflict.[9]

Cannon's speech at the Hippodrome not only signalled to the public at large that he had taken up the Spanish cause, it also marked the start of the wholesale use of his name in association with the Medical Bureau. Almost at once, he became known as a leading supporter of the Spanish Loyalists and began to receive letters and calls from all over the country. One letter that he prized highly came from Marian Craig Wentworth, who had been a schoolmate of his in the St. Paul Central High School. When Cannon's name was read out in support of the Loyalists at a public meeting in New York, she wrote him that it gave her a profound thrill "to feel you were there helping [and] that you care." She was glad to have a man like Cannon espouse their cause, and she thanked him for what he was doing.[10]

Cannon's service as national chairman of the Medical Bureau lasted until it was disbanded in 1939. Throughout the entire time of his association with the organization, the major thrust of his involvement—sending humanitarian relief to a war-torn country—never wavered. Over the course of its existence, the Bureau raised hundreds of thousands of dollars, sent numerous shipments of mobile ambulance units, medical supplies and relief to the Loyalist side of the conflict, and coordinated over 100 doctors, nurses and ambulance drivers to serve in Spain.[11]

☙ ☙ ☙

The satisfaction gained in developing a successful medical relief program nevertheless brought in its wake problems Cannon had not anticipated. Early on, the Medical Bureau found that raising money for mobile ambulance units was the most efficient means of providing speedy medical relief and surgical aid to the wounded. At Harvard, a number of undergraduate students, with Cannon's encouragement, canvassed their dormitories for funds to buy and outfit an ambulance to send to the Spanish Republic. The campaign fulfilled its goal, and that summer an ambulance proudly identified as a gift from Harvard students was sent to New York to be shipped to Spain. Prior to being loaded at the dock, the ambulance took part in a peace parade organized by the American League against War and Fascism.

Two Harvard sophomores, Merwyn K. Hart, Jr., and Sidney Q. Curtiss, who read about the parade in the newspapers, were outraged and wrote an angry letter of protest to the *Harvard Crimson*. Charging that the funds gathered in behalf of the Medical Bureau had been used for partisan propaganda purposes—"to help Moscow in Harlem," they called it—they demanded an investigation by the student council. The students' complaint was immediately picked up by the *New York Times* and other newspapers, including one in Portland, Maine, which advanced a new charge: that the Medical Bureau was an outpost of communism in the United States.

Cannon, for his part, wrote a calm letter of rebuttal to the *Crimson*, pointing out that the funds for the ambulance had been collected for humanitarian purposes, not propaganda, and that the ambulance was to be shipped to Spain under a special permit issued by the State Department. "The legitimate Government in Spain," he concluded,

> is fighting not only a group of Spanish fascists under Franco but also invading fascists from Germany and Italy. In these circumstances can one question the desirability of letting an ambulance appear in a parade which is destined to aid the struggle against fascism in a foreign land where a Government elected by the ordinary democratic process is struggling to maintain a democratic spirit against powerful fascist foes?[12]

Cannon's letter brought support from a number of his fellow faculty members, as well as strong praise from Felix Frankfurter, the Harvard law professor. "I cannot withhold a word, however inadequate, of admiration and gratitude for the lesson in intellectual responsibility—which through these two sophomores—you have been conveying in the last few days to a much wider audience," Frankfurter wrote.

The student council, however, was less impressed by Cannon's letter and launched an investigation of the charges. Before its members could report their findings, an editorial written by Vinton Freedley appeared in the *Crimson* of October 1, 1937, which summarized the allegations and ended,

> Where the libellous letter dissolves into sound and fury is that the equipment was not used "to aid Moscow," but as part of a parade against War and Fascism. There are more than several people in the United States who are against War and Fascism and who are not Communists . . . Mssrs. Hart and Curtiss would do well to retain a lawyer if they intend to spend their days making criminal accusations after carefully misreading the morning paper. Statements of this sort, whether verbal or written, do great harm to the accused, and in their impact bring shame to the University, and in their disproval [sic], ridicule.

Several weeks later, the student council exonerated both Cannon and the Medical Bureau but apparently could not forbear sending a rebuke in Cannon's direction. "As Harvard men," their report concluded, "we strongly deplore the appearance of the ambulance in the parade"; they also thought the Medical Bureau should never have been granted the right to collect funds "for a cause so avowedly partisan." Still later, Cannon was pleased to receive a kindly letter from the president of the *Crimson,* Caspar Weinberger, thanking him for his courteous cooperation. It helped to take the edge off an unpleasant episode.[13]

As chairman of the Medical Bureau, Cannon frequently received requests from headquarters in New York and from local chapters for advice, written statements, speeches and attendance at meetings. At the end of July 1937, Cannon was invited to speak at a dinner in honor of Edward K. Barsky, the chief surgeon of the American hospitals then on leave from Spain, about the program of medical aid to the Spanish Republic. Unfortunately Cannon was taken ill and had to send a message instead, but he did get to meet Dr. Barsky at a luncheon in Boston several weeks later. In October, Cannon was invited to speak at a dinner in honor of Herman F. Reissig, the executive secretary of the North American Committee, just back from an extended visit to Spain. Reissig, an ordained minister who had served several Congregational churches, had a reputation for being not only a polished lecturer but also a contributor to various journals, such as the *Christian Century* and the *Presbyterian Tribune*. He was in the years that followed a person who worked very well with Cannon.[14]

For a speaking tour to the Midwest that autumn, Cannon met with Norman Bethune and Jay Allen to discuss the material each would cover in the lectures they were scheduled to deliver in Chicago. Bethune, a Canadian, was a distinguished thoracic surgeon who some months before had given up a large practice in order to establish a blood transfusion program in Spain to assist those wounded in battle as well as civilian war victims. Allen, a foreign correspondent in Europe for almost a decade, had conducted well-crafted interviews with political leaders in various countries. One of a band of journalists who were allowed to accompany the advancing insurgent troops in Spain during the early days of the war, he electrified world opinion with his reports of mass executions in the bull ring after the city of Badajoz had been taken. From the outset, Allen became a supporter of the Spanish Republic and later served as a canny adviser to Cannon about the various political changes in the Medical Bureau during the last days of the Loyalist government.[15]

Cannon also included on this tour visits to Detroit and Rochester, Minnesota, where he found an increasing reluctance on the part of audiences to speak out in support of the Spanish Republic. Some time later, when Jay Allen wrote to ask him if the Mayo brothers would lend their names as sponsors of a relief ship carrying medical supplies to Spain, Cannon in reply gave him an account of his experiences during his visit to Rochester to address a public meeting on the Spanish situation. "I agreed to do this," he wrote,

> on condition that the Mayos, whose guest I was, would have no objection. "Will" Mayo gave authorization to proceed. No sooner was the meeting announced than a hullabaloo broke loose. I was

> accused of being a communist and a dangerous "Red". . . As you probably know, the Mayo Clinic was based originally on a Catholic hospital and Catholic interests are still very strong. Neither of the Mayo brothers attended the meeting and it was rumored that indirectly they put pressure on members of the staff to say nothing whatever about the whole incident after it had ended. Neither Charlie nor "Will" Mayo spoke to me regarding the meeting during the four or five days after the meeting while I was in Rochester. From this experience I feel quite sure that they will not stick their necks out.[16]

Much as conservatism within the medical establishment was a fact of life that Cannon had to deal with, this incident should not suggest that the Medical Bureau had little support from physicians. On the contrary, there was a strong network of medical scientists and clinicians from coast to coast who rallied to the cause of the Spanish Loyalists, including, among others, Thomas Addis, Leo Eloesser, and Howard Naffziger in California; Anton Carlson and Percival Bailey in Chicago; Evarts Graham, Carl Cori, and Joseph Erlanger in Saint Louis; Adolf Meyer and Henry Sigerist in Baltimore; Haven Emerson, Ernst Boas, and Florence Sabin in New York; C.E.A. Winslow and John Peters in New Haven; and a veritable phalanx of friends at Harvard headed by Alex Forbes, Stanley Cobb, and George Wislocki.[17]

Early in 1938, the Medical Bureau merged with the North American Committee in Aid of Spanish Democracy to become the leading American organization engaged in helping Spain. Bishop Francis McConnell was appointed to share the national chairmanship of the new organization with Cannon, who fully expected to get some relief from the heavy demands made on him. However, the Bishop was seriously ill and unavailable for most of the year. As the civil war dragged on, the calls on Cannon's time and energies only increased—even at times pulling him away from his teaching and research at the Medical School. Frequent requests calling for his travel to New York and elsewhere for meetings and public appearances annoyed him—especially when local people or others would do just as well. Furthermore, he was rankled by wasteful spending and poor management of affairs at the headquarters office. That April, for example, he was asked to join a trainload of delegates going to Washington, D.C., to lobby for the lifting of the arms embargo. He did so but found it a frustrating experience, for things were not as he had been told. "Altogether the affair was poorly managed because there was no leader," he grumbled.[18]

Such incidents dismayed Cannon and left him feeling helpless; but he continued to drive himself for the sake of providing relief to the suffering people in Spain. As Franco's Nationalist forces mounted offensive after offensive and the bombing and bombardment of cities increased, a large part of the civilian population seeking food and shelter became displaced. In the spring of 1938, an emergency appeal to aid them was launched, followed by a campaign to send a relief ship to Spain, which was to contain food, clothing and medical supplies for civilian victims of the civil war. After attending a mass rally in Madison Square Garden in June, Cannon noted optimistically, "It was one of the most enthusiastic and effective meetings that I have ever attended," and he wrote Dr. Negrin that $12,000 had been raised.

Nevertheless, the news from abroad was hardly encouraging. Along with reports that the Loyalist forces in Spain were rapidly losing ground, Hitler's bold move into eastern Europe was further cause for alarm. "We watch from day to day the changes in Spain and in the world at large," Cannon wrote Carrasco in August. "Always we hope that something may happen which will break the morale on the side of the insurgents, or bring such pressure to bear on the supporters of Franco that he will have to stop his aggression. In a world so unstable as the present world almost anything may happen at any moment."[19]

The critical situation in Spain, exacerbated by a severe food shortage in the industrial sections, raised fears about malnutrition among the civilian population, particularly in children. As early as February 1938, Thomas Addis of Stanford University had sent Cannon an outline of a program for supplying vitamins that he had also sent to Henry Borsook at the California Institute of Technology. Borsook believed that it was necessary to first obtain accurate information about the condition of Spanish children and then, if justified, he would urge the Medical Bureau to sponsor the project. Cannon, for his part, wondered why Dr. Barsky or someone else in Spain hadn't informed the Medical Bureau what foods were lacking and what vitamin supplements were needed.[20]

A related concern came to the attention of the Medical Bureau at this time through alleged reports from Ernest Hemingway and others, who claimed that there were 40,000 cases of pellagra in Spain, and a call came in for supplies of nicotinic acid to treat the disease. When the Medical Bureau's relief ship, the *Erica Reed,* sailed for Spain at the end of October, it contained not only food, vitamins and medical supplies, but also a small amount of nicotinic acid to fight the pellagra epidemic, along with a portrait of Cannon and messages of good will for Negrin and the Republican government.

Cannon believed that supplying essential food elements to prevent diseases like pellagra was every bit as important (if not more so) than supplying nicotinic acid to cure them, but in his message, he called attention to the "magic chemical" that would bring relief for an age-old disease. "We have great pleasure, therefore, in being able to present this new and beneficial discovery to harassed people in Spain," he wrote. "The advances of science can bring the greatest health and wealth to man, but only where there is freedom of research and action. Scientific men will be ever grateful to the Spanish people for their heroic stand against the invaders of their country and against the enemies of free democracy."[21]

After the *Erica Reed* reached Spain, Cannon received an acknowledgment from his Spanish friends for his efforts in their behalf. A letter signed by Negrin, Carrasco-Formiguera, J.M. Bellido, Rafael Mendez and a dozen others testified to the courage of Cannon and those in the Medical Bureau and North American Committee who were helping the cause of the Spanish Republic. "By this time your name has become popular in Spain as that of the main promoter of the assistance, both material and spiritual, we are receiving from North America," they wrote him. "We feel ourselves joined by all democratic Spaniards and we pray you, our most admired master and beloved friend, to accept the cordial and sincerest expression of our affection and of our gratitude." Later, Carrasco wrote Cannon that Negrin had taken his portrait and put it in a prominent place in his private quarters.[22]

The trials of chairing the Boston chapter as well as the national organization added to Cannon's sense of being overburdened. At one point in 1938, he complained, "I should not continue to be Chairman of the local chapter because I am too old for that task, and also because I have very heavy duties connected with my

University position and the outside obligations which it implies." True, the Boston Bureau had great success with raising money for Spain by sponsoring a series of dinners, concerts and other well-attended social events, but Dorothy Hickie, the local secretary on whom Cannon depended, filled his ear with tales of incompetence and dissension on the part of its officers. Not that he needed to be told. He found it impossible to get the executive committee to pull together, and one of their meetings in October 1938 was so dominated by radicals who had their own agenda that both Cannon and Miss Hickie walked out in a state of agitation. Another blowup with leftists at a December meeting presaged a showdown in the near future.[23]

Because the Spanish ecclesiastical hierarchy and the parish priests were staunchly aligned with the right-wing National Front, there were some who tried to define the civil war in Spain as a struggle between those who believed in God and those who did not. Cannon saw this definition as "a mischievous falsehood," but there is no denying that the Spanish aid cause was decidedly unpopular in the Boston community at large because of heavy Roman Catholic opposition. He often came under personal attack from the official organ of the Catholic Church, *The Pilot,* for his support of the Loyalists. Indeed, the attacks on the Spanish Republic and the Medical Bureau by the Catholic Church and newspapers in Boston became so fierce at one point that Cannon asked Crane Brinton, the historian who was then president of the Harvard chapter of the American Association of University Professors, if it might not be worthwhile for the A.A.U.P. to undertake an examination of the relation of a liberal university to an illiberal, reactionary community. "German professors have informed me," he told Brinton, "that it was because of failure to take a definite stand against limitation of thought and opinion that the German universities have gone into eclipse after the Nazi régime."[24]

Perhaps most galling and time-consuming for Cannon were the accusations that the Spanish aid organizations were merely Communist fronts. These accelerated in 1938 and were abetted in the autumn with the appointment in the U.S. House of Representatives of the Dies Committee investigating un-American activities. This was not the first time that a legislative body had seen fit to examine the activities of the Spanish Medical Bureau. In 1937, Cannon had been drawn into internal politics in Massachusetts when Senator Thomas M. Burke of Dorchester spearheaded a drive to investigate the activities of groups engaged in subversive propaganda in the Commonwealth, particularly focusing on the raising of funds for the Spanish Loyalists. Cannon wrote Frederick W. Cook, the Secretary of the Commonwealth, and warned that Burke's claim that "the Loyalist party is a communist goverment along the same lines as the Soviet" was a grotesque perversion of the truth that could well have serious consequences. "It was by raising the bugaboo of communism that Mussolini was able to grasp power in Italy and Hitler was able to dominate Germany," he pointed out. "If we, as Americans, are not sufficiently skeptical to demand facts and are stupid enough to be swept off our feet by hysterical threats, the chance may arise that 'it can happen here'."

However much Cannon railed against it, the "bugaboo of communism" would not go away. In September of 1937, still fuming over the Burke investigation, he wrote Negrin that the attitude of the American people had been "deeply affected by the common charge that the Loyalist Government is red, bolshevistic, and communistic." Not only that, but a New England newspaper had declared the Medical Bureau to be "a Communist organization with headquarters in Russia!"[25]

Early in 1939, action to continue the Dies Committee was taken up by the House Rules Committee. On January 22, Cannon, whose name had been listed in the Committee's partially printed hearings, received a telegram from Kent Keller, a congressional staff member. Noting that Cannon was one of the "distinguished citizens whose Americanism has been called into question," Keller urged him to wire Chairman Sabath of the Rules Committee, if he agreed that persons attacked by the Dies Committee were entitled to defend themselves before a more unbiased Congressional body. Keller also suggested that he write to majority leader Sam Rayburn, as well as to his own congressman, in protest of the Committee's conduct.

Cannon promptly complied with Keller's suggestion. "There is nothing reprehensible or illegal in anything that the Medical Bureau has been engaged in since its initial organization," Cannon informed Rep. Adolph Sabath. In protesting the activities of the Dies Committee, he stated that it was particularly abhorrent to him to see his efforts in a humanitarian enterprise "subjected to inquisitorial attention." And he urged that the committee, whose members were "strongly biased by absurd fears" and "highly prejudiced," discontinue its unfair treatment of American citizens.

Cannon's letters to Sabath and others in protest of the activities of the Dies Committee, added to the sentiments of many others, helped to make the remainder of its existence relatively harmless. While his name remained on the roster of the Dies Committee, he was neither investigated nor called up before it. In 1941, Martin Dies failed in his bid for a Senate seat and his committee lapsed into desuetude. Although the Dies Committee was gone, after the war the House Un-American Activities Committee would come back stronger than ever.[26]

Some of the criticism that Cannon received throughout the course of his work with the Medical Bureau was not always easy for him to brush off. Nor, for that matter, was he always successful in defending himself from that criticism. One such attack came from one of his former students, Richard H. Miller, a surgeon at the Massachusetts General Hospital and secretary to the Harvard medical faculty. Miller, who had received a letter asking him to make a contribution to purchase medical supplies for the Spanish Republic as a special tribute to Cannon, saw the request as part of a Red plot. "I wish to protest against this," he wrote the editor of the *Harvard Alumni Bulletin* in October of 1938, "and to say that every well-informed person knows that this so-called Spanish Republic is not a democracy, but a communist state aided both materially and spiritually by the present régime in Russia."

Cannon responded with his own letter to the editor, in an effort to persuade the readers of the Harvard alumni magazine that the Loyalist government was not communistic, and more especially to explain how an appeal for a contribution in his honor came to be sent out. To Miller he also sent a private note. "Apparently," Cannon wrote,

> there were persons who thought that it would be appropriate to use my friendship for Dr. Negrin to collect a fund which would help to buy food and clothing and medical supplies for the stricken people of Spain. I had nothing to do with this effort, which was intended as a pleasant surprise to me. When I learned of it, I did not feel justified in checking it because the purpose was strictly humanitar-

> ian. It seemed to me quite in accord with the ideals of the profession to which you and I are devoted.

Miller was not persuaded and replied that he still believed the Spanish government was "in spirit a soviet, communistic and atheistic government." Then he added a sentence that cut Cannon to the quick: "The fact that so many Harvard men have sincerely congratulated me, in the past forty-eight hours, is clear proof . . . that very many others agree with me." Cannon could not let Miller's boast of congratulatory support go unanswered. "It happens that I am a sensitive person," he wrote back, "and like to be well regarded by my former students . . . The thought that many of them approved of your statements in criticism of the action which was taken in my name will make me feel that I am subjected to a censure which I have no means of answering."[27]

If Cannon felt unable to respond to such censure, his colleague in the physiology department, Alexander Forbes, did not hesitate to challenge Miller, with whom he was on friendly terms. "A man of your intelligence and usual fair-mindedness really should ascertain the facts before appearing in print as you have done," Forbes wrote Miller reprovingly. He went on to inform the surgeon that he was personally acquainted with Carrasco-Formiguera and other Spanish physiologists who backed the Republican government and its premier, Juan Negrin—all of them "professional men of the finest type, whose word can be thoroughly trusted."

Negrin himself did not want a communist Spain, Forbes argued. He did not doubt that Russia had lent aid to Negrin's cause, but he was certain it was much less than the aid Hitler and Mussolini lent to Franco. "I dislike bolshevik Russia as much as you do," he declared. "But the fact that Russia has backed Negrin, I am convinced, represents far more the fight against Hitler and Mussolini, who I believe are today the most menacing enemies of world peace, than anything else. And certainly the opposition to them in which we share, and which, incidentally, affords support to Negrin, does not make republican Spain a communist state."

Miller, loath to end a friendship of many years, was evenhanded in his reply to Forbes:

> You have good reasons for your opinion, and I have good reasons for mine. You have your evidence, which you believe, and I have mine, which I believe, and they don't agree. That, as I said to Dr.

> Cannon, is bound to happen when men are on opposite sides of such a vital struggle. I've had my say, and unless I change my mind, and something new comes to my attention, I shall not continue the controversy further.[28]

Now this was not the first time that Alex Forbes had tried to answer critics of Cannon and the Spanish Republic. Earlier in 1938, he learned that his eldest brother, W. Cameron Forbes, who had traveled to Spain with Ellery Sedgwick, the editor of the *Atlantic Monthly,* in response to Franco's invitation to look at the Nationalist side of the civil war, had implied in public that "all supporters of the Loyalist cause are villains and criminals of a low order." Feeling that what "Cam" was bruiting about was "too grossly partisan to be plausible," Alex sent him part of a letter Dr. Carrasco had written to Cannon explaining the situation in Spain from the Republicans' point of view and urged him to examine both sides of the war impartially.[29]

Cameron Forbes was no more convinced by Alex's arguments than Richard Miller would later be. He contended that the government Franco had established was "orderly and friendly, the people contented and the situation comparatively normal." In the bulk of Franco's forces, he said, "90% are Spaniards and are the flower of what is best in the young manhood of White Spain. It is in fact a civil war. Both sides are helped from outside by things they do not produce within, such as airplanes, tanks and certain types of weapons and munitions; the Reds drawing their supply from France and Russia and the Whites from Italy and Germany." As for the embargo, he believed that shipping arms from America into Red Spain would merely mean prolonging the struggle. "It can't change the ultimate issue. We had better keep our hands off."

Alex realized that further discussion with his brother would be futile. And indeed, Cameron Forbes and Ellery Sedgwick continued to stick to their party line, downplaying the importance of the aid that Franco was getting from Italy and Germany and harping instead on the support given to Negrin and the Spanish Republic by the French, which was next to nothing, and by the Russians, which had dwindled to very little indeed.[30]

Not long after Cannon answered Miller in the pages of the *Harvard Alumni Bulletin,* two further letters to the editor followed on the subject of Spain. One was from Walter Muir Whitehill, the director of the Boston Athenaeum, who angrily reported that a friend of his, an elected deputy of the Right, had been gunned down and killed in Madrid soon after the outbreak of the civil war. The other was from an alumnus, Herbert M. Warren, who had lived in Spain for eight years until September 1936, "when I decided three months of witnessing anarchy, murder, and wholesale robbery, in so-called 'Democratic Spain,' enough for me."

Cannon, responding to both charges with a letter to the editor, said the dates were important. Recalling testimony in Carrasco's letter, he acknowledged that when the Spanish government was initially faced with rebellion by its army, "it placed weapons in the hands of any who were willing to fight on its side. Thus, irresponsible persons, many nourishing hatreds and seeking revenge, were given power without being controlled."

While conceding that during the latter half of 1936, "most lamentable consequences ensued," he also claimed that atrocities like those which aroused the bitter feelings expressed by Whitehill and Warren had long since stopped. Nevertheless, in spite of efforts to broadcast the truth, the government of Dr. Negrin was still being blamed for the reign of terror that had occurred more than two years previously—months before he even became premier.[31]

A number of Cannon's academic colleagues rallied to his side after reading the series of letters in the *Harvard Alumni Bulletin*. Arthur Holcombe, a professor of government, for example, called his response to Miller's criticism "a masterpiece of dispassionate and rational argument." Holcombe, venturing to speak for the Harvard community, went on to tell Cannon, "Your friends will all be proud of your behavior under fire, and friends of all good causes will be eager to secure your leadership in the troubled times which seem to be ahead."[32]

Ever since passage of a Neutrality Act in 1935, supported by a consensus of the American people who did not want to be drawn into another war, the U.S. government had maintained political neutrality with regard to acts of aggression against other countries. Early in 1937, Congress passed a new Neutrality Act, which added acts of aggression *within* other countries to the hands-off policy—in effect, blocking the civil war in Spain from any official government action. An embargo resolution that was part of the Act made it unlawful to export arms, ammunition or implements of war from any place in the United States. When it became evident, as the war in Spain lengthened, that the Loyalists had little chance of triumphing against a Nationalist army fortified with the most modern instruments of destruction from Germany and Italy, friends of the Spanish Republic endeavored to find a way to lift the embargo. Unfortunately, they did not meet with success.[33]

Late in 1938, the Medical Bureau and North American Committee again turned its efforts to having the embargo lifted, especially with an eye toward preventing a second Munich agreement with Spain as the new victim. When Herman Reissig asked Cannon to send out a form letter in an attempt to rally support from the medical community, to his surprise, he found Cannon reluctant to accede to the request. To be sure, Cannon replied, it was important to lift the embargo, but he believed there were more important things the Medical Bureau ought to be doing. He especially balked at changing the direction of his efforts from medical relief to political action. "I have had enough experience with myself to know my limitations," Cannon protested. "My life has been spent in the quiet of a laboratory and not in the world of affairs, and I know nothing or next to nothing of strategy in a public arena. In short, I am so certain of my incapacity to lead in a political attack that I cannot agree to take even the first step in that direction."

On the other hand, Cannon had always felt that the refusal of the U.S. government to supply munitions to the Spanish Republicans was a dreadful mistake. And so he asked that his name be included in an appeal to President Roosevelt by several leaders of the Spanish aid movement, and then drafted a statement that Reissig might use in a general letter to be sent out from his office. "Our action during the past two years has made the world not only unsafe but actually dangerous to democracy," Cannon wrote.

> If Hitler and Mussolini are to have their way with democratic Spain, the march of fascism will gain enormous impetus. Whether we shall allow the Spanish Republic to be sacrificed to fascism rests upon us. By lifting the embargo against shipments of munitions to Spain we would inspire new courage in the upholders of democracy everywhere and aid the Spanish people in their gallant struggle for human liberty.

Cannon at year's end expressed the hope that when Congress met in January, it would take action to "reverse the deplorable policy" enacted two years previously. But time was running out for the Spanish Republic.[34]

With the fall of Barcelona and the collapse of the Catalonian army early in 1939, it became apparent that the end of the Loyalist government was at hand. In March, fighting broke out in the streets of Madrid between communist and anti-communist factions and Franco's forces easily took command of the starving capital city. The victory of the Nationalists saddened Cannon. "The debacle in Spain is a bitter disappointment to me," he wrote Harvey Cushing. "I had earnestly hoped that in the struggle between the totalitarian states and the Spanish effort to establish a democracy, success would rest with the Republic. The United States played an ignoble rôle in the disaster, I am convinced, by refusing to sell munitions . . . The formal struggle in Spain is over, I fear, but perhaps there may be something yet in Wendell Phillips' dictum, 'Liberty knows no defeat.'"[35]

On April 1, the long and brutal civil war ended with the complete and unconditional victory of General Franco. The war in Spain was hardly over when Cannon received a letter from Norman Thomas, head of the Socialist Party in the United States, that upset him even further. Thomas informed Cannon that he was going to withdraw his party's support from the Medical Bureau and the North American Committee, on the grounds that he couldn't trust Communists and their sympathizers within the organization to equitably direct the relief of the Spaniards who sought refuge in southern France. "We want to throw all our influence on the side of organizations like American Friends, which do not subordinate relief to partisan politics," he continued. "In getting out, we feel compelled regretfully to make a public statement." Cannon was stunned; members of the Socialist Party were one of the mainstays of the Medical Bureau and North American Committee. What was most alarming to Cannon was the threat of a public statement charging that the Medical Bureau was distributing relief solely to Communist refugees.

Cannon replied at once, asking Thomas for evidence of the charges of partisan distribution of relief. The publication of his claims, he pointed out, could only elate the Red baiters. Even more important, it would create a political uproar that would disrupt the Medical Bureau and the North American Committee as an effective fund-raising organization at a time when money was desperately needed for relief. Furthermore, it would provide an opportunity to bring charges against officers of the Medical Bureau for raising money under false pretenses. "I most sincerely hope," Cannon ended his letter, "that you will not make public that statement which you have prepared until you have definite facts in hand. As a man of science I want the facts and when I have them in hand I shall be ready for action."[36]

At the same time that Cannon replied to Thomas, he telegraphed Roger Baldwin, chairman of the executive committee of the North American Committee, and asked for further information. Within a day, he received letters from both men. Baldwin, believing that "the time has come for us to take stock of our relation to the changed situation in Spain," focused on the future of Spanish aid rather than its past. He suggested that the best thing to do in the circumstances was to dissolve the Medical Bureau and the North American Committee as such and consolidate some of its staff and facilities with the remnants of other Spanish relief agencies. The new consolidated committee was to be wholly free of any political connections and thoroughly representative of American interests. In consequence, it no longer would direct its efforts to aid Spanish refugees in France through the politically connected International Coordinating Committee in Paris but, preferably, turn over all money collected to the Quakers (American Friends Service Committee) for distribution. "The Socialists," he added in a postscript, "agree to this solution and will make no statement if this is done."

Thomas, in his letter, wrote Cannon his approval of Baldwin's plan and then assured him that he had leaned over backwards to avoid Red baiting and did not want to weaken the legitimate liberties of Communists. "We should expect the Communists to claim the same rights we claim for ourselves; what we object to is their use of the name and reputation of prominent American liberals to advance covertly their own political program and above all, to build their own political organization, with its tight discipline over all its members."[37]

Subsequently Cannon confessed to Dorothy Hickie that he did not understand the attitude of the Socialist leader, and believed that Thomas regarded the prestige of the Socialist Party to be of more importance than the suffering of the Spanish refugees. He found the partisan nature of the bickering especially distasteful, for he had long held that party politics had no place in the Medical Bureau. Early in 1938, for example, when some members of the North American Committee in Boston and New York attempted to place representatives of the Communist Party on the executive committee of the Boston chapter of the Medical Bureau, Cannon had blocked them. Members of all political parties were welcome in the chapter *as individuals,* he wrote John Sherman, the national organizer of the North American Committee, but he objected strenuously to introducing politics into the organization. "We have thus far been running our affairs without reference to any demands from the outside except those which come from Spain, and I am sure that if we begin to play politics we shall get into trouble."[38]

In mid-April of 1939, convinced that there was no longer a need for the existence of the Medical Bureau and eager to be freed from his exacting duties, Cannon resigned his chairmanship. He did not take the step of resigning without careful consideration. "I have never felt so harried and oppressed as I have felt for the past months," he confessed to Reissig a short time later. "I am sure that any doctor knowing my condition would insist on my getting rid of a good deal of the load which I am carrying." The very next day, at the urging of his physician, Dr. Joseph Aub, Cannon wrote Dorothy Hickie that he wished to resign as chairman of the Boston chapter as well.[39]

In responding to Cannon's resignation from the national organization, both Norman Thomas and Roger Baldwin expressed their regret as well as their appreciation of his service in the cause of Republican Spain. To Baldwin Cannon wrote at the end of May, "Though I am no longer Co-Chairman of the Medical Bureau and the North American Committee, I have not ceased my efforts to aid the Spanish Loyalists." Reporting that he not only had former Spanish students to look after but also others appealing for aid, he continued, "I am hoping that I can still be useful in the larger campaign for the refugees and also in the care of individuals."[40]

Dorothy Hickie's message of appreciation touched Cannon deeply. "You have been a tower of strength to the friends of Spanish Democracy all over this country—and wherever there are Spaniards." She went on to speak in behalf of the Spanish people in Boston, many of whom had never even seen Cannon: "To them you represent all that they have ever expected an American to symbolize. You were their friend when everyone else seemed [to be] turning against them. Every time the newspapers denounced you they were cut to the heart . . . when they speak of you it is almost with awe."

In acknowledging Miss Hickie's tribute, Cannon described his recent meeting in New York with Dr. Negrin, who "threw his arms around me and gave me a bear hug!" Perhaps it was the former Spanish premier's show of affection that impelled Cannon to say that he would be willing to serve as honorary chairman of the Boston Bureau—so long as he was not required to perform any duties.

In July, when Negrin was once again in New York, he wanted to visit Cannon at the family farm in New Hampshire, but was unable to do so. He made it a point, however, to add his message of heartfelt thanks to Cannon for all he had done for Spain. It was a brief note but it must have meant a great deal to Cannon when he read what Negrin wrote: "I know that you saw our struggle as I did, as one in the common interests of all democratically minded peoples. So few saw it as such in time. There were many dark moments when the thought that you had understood gave me comfort."[41]

To Cannon's way of thinking, he and Negrin were indeed of the same persuasion, having been motivated by their democratic ideals. Others might accuse him of being naive, of not knowing how deep the waters were into which he had waded, and even of being a dupe of the Communists—if not a "Red" himself or a fellow traveler. But to the end of his days Cannon never regretted his support of Spain. Nor did he look back on the aid given by the Medical Bureau to the people of a country torn asunder by civil war with anything but pride and a sense of honor. Sadly, the commitment he had made to the brief and shining moment of the Spanish Republic was swept away in the defeat of the Loyalist government by a wave of pathos and frustration. His angle of vision, however, was soon to be broadened by a series of events that would plunge the continent of Europe into all-out war.

19 *Refugees from War-torn Europe*

CANNON'S ALARM at the spread of totalitarianism, and especially the increasing threat to the international community of scientists and scholars posed by the rise of fascism and nationalism, impelled him, despite his reluctance, to devote a good deal of his time and energy to political activism. His personal commitment to safeguard the unrestricted pursuit of knowledge and scientific inquiry took a variety of forms—not only in joining with others to deliver medical aid to the Spanish Republic or in continuing efforts to bring foreign students and fellows into his laboratory, but also in his willingness to do whatever he could to help individual medical scientists and practitioners escape from an all-consuming European conflagration. Whether motivated by democratic ideals or humanitarian spirit, Cannon's involvement cannot be separated from his sense of holding an honored place in a long line of men of science who served society. It was a lineage that stretched from those who came before him, like Bowditch and Ludwig and Claude Bernard, to those who would come after—his students and their students and on into the future.

While there was an outpouring of sympathy for the refugees who fled Germany after Hitler's rise to power, few anticipated the extent to which conditions in that country would worsen. Repressive actions against unwanted inhabitants steadily intensified until they reached a new height with enactment of new legislation announced in the autumn of 1935 by the National Socialist Party Congress at Nuremberg. These so-called Nuremberg Laws initiated a fresh wave of persecution of a character not previously seen. The disenfranchisement of Jews, Christian non-Aryans, and political or religious dissenters brought the threat of pauperization and exile to untold numbers of Germans. "Apart from all questions of principle and of religious persecution," declared James MacDonald, a high official of the League of Nations Commission for German Refugees,

> one portentous fact confronts the community of states. More than half a million persons, against whom no charge can be made except that they are not what the National Socialists choose to regard as "Nordic," are being crushed . . . Tens of thousands are today anxiously seeking ways to flee abroad, but except for those prepared to sacrifice the whole or greater part of their savings, the official restrictions on export of capital effectively bar the road to escape, and the doors of most countries are closed against impoverished fugitives.[1]

Despite such dire warnings, variations on the theme of Aryan or Nordic superiority continued to be played out in Nazi Germany. Late in April 1938, little more than a month after the Anschluss of Austria, the Nobel laureate Johannes Stark of Physikalisch-Technichen Reichenstalt Berlin-Charlottenberg published a long article in *Nature* which again gave expression to his political views. In contrast to his article in 1934 that defended Germany against A.V. Hill's charges of uncivilized behavior by the Nazis, this time Stark was on the offensive. He argued that there are two principal types of mentalities in physics. Those evolved from the founders of research in physics, "the great discoverers from Galileo and Newton to the physical pioneers of our time," he called pragmatists; they were almost exclusively Aryan and predominantly from the Nordic race. The others he called dogmatists, saying, "If we examine the originators, representatives, and propagandists of modern dogmatic theories, we find amongst them a preponderance of men of Jewish descent." Not only did the Jews play a decisive part in the foundation of theological dogmatism, they were also, for the most part, the authors and propagandists of Marxian and communistic dogmas.[2]

Following the appearance of Stark's article, a group of scientists—not only from the physical sciences but from biology and medicine as well, Cannon among them—prepared a statement that mounted a counter-attack:

> In essence, Professor Stark has attacked all theoretical physics and, by obvious implication, scientific theory in general. He cloaks his attack by playing upon the words "pragmatism" and "dogmatism" and by introducing the official racialism of the Nazis to divide physicists into good, i.e., non-theoretical and "Aryan," and bad, i.e., theoretical and Jewish. He names five men who fall under the second category: Einstein, Born, Jordan, Heisenberg, and Sommerfeld, only one of whom is in reality a Jew.

The very men, the scientists continued, whom Stark pretended to venerate—Kepler, Galileo, Newton, Maxwell—had all insisted that separating experimental research from theory was futile. Moreover, his charge that theoretical physicists were dishonest publicity seekers had implications "in its sneering reference to textbooks and lecture halls, that strike at the very roots of science. If science is not to be disseminated as widely as possible, so that it may spread its benefits to all mankind, what purpose is it to serve?" The statement concluded with ideas that Cannon had often expressed:

> We are firm believers in democracy. We feel that any attack on freedom of thought and belief in one sphere, even if it is as non-political a sphere as theoretical physics, is simultaneously an attack on democracy itself. When men like Albert Einstein may no longer pursue their investigations, whether the reason is their creed, race, or beliefs, all mankind suffers the loss. They must be defended in their right to speak the truth as they understand it. If we American scientists wish to avoid a similar fate, if we wish to see the world continue to progress and prosper, we must bend our forces to that end now.[3]

Cannon and the others then circulated the statement to their friends and colleagues to sign. One of the people Cannon initially approached was Harvey Cushing, who at the time was quite ill and didn't like the notion of adding his signature. As a matter of fact, he indicated that replying to Stark's article in *Nature* would only serve to call attention to it, and at the same time dignify it and lead to futile polemic. "Stark's letter riled me when I first read it," Cushing told Cannon,

> but I have reached an age [or my blood vessels have] when I can't afford to get riled to the extent of stirring up my adrenals lest I blow out a fuse. You should understand this on pragmatic grounds, if anyone does.

> I consequently don't let myself get unduly agitated over Franco or Mussolini or Hitler or even the Japs though I sometimes have the feeling that it would be better for the world if some of them were towed out to sea and *spurlos versenkt.*

Cannon did not make a fuss about the rejection. He understood that Cushing was not well and confessed to only slight disappointment. Replying that he had "a lurking admiration" for Cushing's "somewhat rigid backbone," he recalled William James's advice about truncating one's interests in the later years of life—an ideal which he himself seemed to have reversed. "Although I once belonged to thirteen committees—and a committee, as you doubtless know, is 'an organization which keeps minutes and wastes hours!'—and now belong to only one . . . I seem always to be encumbered with something new to take the place of the old. There is the recurring hope, however, that freedom from obligations is just around the corner."[4]

During the late spring of 1938, with a mounting tide of refugees desperate to get out of Europe, it became apparent that the various organizations concerned with aid to displaced medical scientists and physicians would have to be reorganized. One of the strongest groups supporting the relocation of medical refugees by the Emergency Committee to Aid Displaced Foreign Physicians was based in Boston. In May Jacob Fine, a member of a committee made up of Cannon, David Edsall, and Mrs. William Herman of the Emergency Committee, asked Cannon, among others, to approve proposals that aimed to strengthen the national organization. The group wanted to establish a coordinated national program, which would feature a central clearing house in New York (the major port of entry) for registration and placement of the émigrés along with a network of affiliated branches in a number of key cities.

Cannon thought the statement admirable, especially the part suggesting a unified effort through the creation of a nationally coordinated organization with local branches. The crisis brought about by Nazi cruelty and injustice was indeed serious, but properly managed, he believed, "the process of absorbing excellent men from abroad is sure to enrich our country."

Over the summer months, meetings were held to lay out plans for the formation of a branch office in Boston, to be known as the Boston Committee on Medical Émigrés. Its executive committee included Dr. Edsall, who was chairman, Stanley Cobb, who was vice chairman, Tracy Putnam, Oliver Cope, Mrs. Herman, and Jacob Fine. Later, when Cobb asked Cannon if he would fill a vacancy on the executive committee, he refused. Cannon said he would be glad to help in any way he could, but he was already pulled in too many directions and must not increase the number of meetings and attention to detail that a place on the executive committee would require.[5]

In 1939, after a National Committee for the Resettlement of Foreign Physicians was established under the aegis of a National Coordinating Committee for Aid to Refugees, the Emergency Committee in New York changed its name and devoted itself mainly to aiding displaced foreign medical scientists. The first order of business for both the reorganized committee in New York and the new Boston Committee was to seek financial support for the purpose of hiring traveling agents to make contact with interested groups of physicians and institutions at the state and local levels. It was the hope of the committees that members of the medical profession would cooperate in this effort "to the end that the refugee physicians, who have no desire to compete with American physicians in congested areas," might be given opportunity to use their skills and their training in locales where their services were most needed.[6]

From the start, the Boston Committee's efforts were guided by certain events taking place in Massachusetts. In particular, some members of the Massachusetts Medical Society had proposed a resolution that made full citizenship a requirement for licensing in Massachusetts. Any number of states already had this requirement for licensing, and without a license no one could practice. There was a difference of opinion between practicing physicians and academic physicians over the proposed resolution, but it was the practicing physicians who held the upper hand. In November 1938, the executive committee of the Boston Committee asked Cannon and others to contact Dr. C.C. Lund or members of the legislative committee of the Massachusetts Medical Society to see if they could get the resolution tabled.

There was a problem, and it was not a new one. On the one hand, a fear existed among practitioners that the deluge of refugee physicians seeking asylum in America would exacerbate the overcrowded condition of the profession. On the other hand, the executive committee believed that to put the resolution into effect would virtually close the doors to all medical émigrés and thus, they said, "oblige us to disavow our heritage as a democracy, offering asylum for political refugees. It would ill become the birthplace of American freedom to permit this to happen."

Though the council of the Medical Society appeared to be supportive of the views expressed by Cannon and the executive committee, by some strange happenstance the resolution went to the state legislature, and the bill (No. 1407) to prevent refugees from practicing medicine in Massachusetts was put forth to the House of Representatives Committee on Public Health. This turn of events required a new action; in February 1939, Cannon was delegated to contact Dr. Lund again and to appear at the hearing of the bill. The news soon came, however, that the bill had in fact died in the legislature.[7]

With that out of the way, the Boston Committee proceeded to carry out its program. Stanley Cobb reported that grants had been obtained from the Rockefeller Foundation and others to employ field agents to ascertain what was needed in different sections of the country in the way of medical personnel. And here again Cannon was called on—this time to supply the names of sympathetic friends and colleagues whose aid might be enlisted to facilitate the work of these agents. In February and March of 1939, Cannon sent out a raft of letters to acquaintances in medical schools and medical societies throughout the land.[8]

During this same time, various study groups were established around the country to prepare refugee physicians for examination and practice. There was, for instance, a study center operating in New Haven under the direction of Dr. H.M. Zimmerman, where fourteen refugee physicians were being trained for examination by instructors at the Yale Medical School, who volunteered their services and offered lecture courses in biochemistry, pathology, pharmacology, and internal medicine. In addition, opportunities for clinical observation in various out-patient departments and for attendance at hospital rounds and teaching conferences became available.

Many hospitals accepted foreign physicians as interns for the year beginning on July 1, 1939. This was looked upon as an arrangement that benefited both the hospital and the refugee physicians whom they employed. The internships not only granted the newcomers the opportunity to acquaint themselves with American medical methods and American patients before they began to practice, they also gave them the hospital experience that was required in some states (but not all) in order to qualify for licensing examinations.[9]

From the beginning of the effort to rescue German medical scientists who had been ousted from their posts, some of the younger men fell through the safety net that their American and English colleagues were attempting to construct. True, members of the scientific community in increasing numbers were appalled by the fascist onslaught against scholarship and intellectual freedom. At the same time, it was difficult to make room for foreign exiles when budgets were already constrained by the Depression and when the academic ladder was clogged with unemployed graduates with advanced degrees.

One of the unfortunates was Bruno Minz, who had received his medical degree in 1930 from the University of Berlin. He remained in Berlin as a research assistant at the Physiological Institute in collaboration with Wilhelm Feldberg, who was working on the physiology and pharmacology of histamine and acetylcholine and other so-called tissular hormones. Minz's study of the acetylcholine-eserine reaction on the leech muscle led in 1932 to the elaboration of a specific and quantitative biological test for measuring acetylcholine over a wide range of concentrations. In so doing, he was able to demonstrate the importance of acetylcholine for adrenal discharge and for the humoral transmission of sympathetic impulses.

Both Minz and Feldberg were Jews. While Henry Dale invited Feldberg to join his laboratory after Hitler came into power in 1933, nobody offered a place to Minz. Fleeing Germany, he made his way to Paris and presented himself for a post at the Collège de France. French physiologists who were acquainted with Minz's research arranged a fellowship for him so he could continue his studies. Unfortunately, given the rules and regulations of the government, they could not offer him a regular post. Still, Minz's investigations employing the eserine leech body wall preparation proved extraordinarily useful and he was able by this technique to isolate a substance which seemed to strengthen the action of acetylcholine liberated by the nerve during electrical stimulation.[10]

In 1935, Minz wrote Cannon and asked if he might have a post in his department. But Cannon's department was already filled to capacity with competent young physiologists who were looking for posts. Regretfully, he told Minz that he was unable to invite him to work in his laboratory. Early in 1938, the French physiologist André Mayer wrote Cannon in Minz's behalf, and added his high regard for Minz and his ability to that already stated by Louis Lapique, René Gayet, and Leon Binet. Cannon, in turn, wrote to several individuals who might have a place for Minz, like David Marine at the Montefiore Hospital in New York, but was told that none was available. Minz struggled along in Paris as best he could, but in April of 1939 he wrote Cannon about the harsh conditions under which he was living and working. Bewailing his bleak situation, he asked for advice. Cannon promptly got in touch with the Boston Committee on Medical Émigrés and sent Minz's impressive list of publications to see what they could do for him. Apparently the committee could do nothing.[11]

In 1940 Minz's brother-in-law in New York sent Cannon word that Minz, who had fled to North Africa, was looking for a fellowship that might allow him to obtain a visitor's visa from the U.S. Consulate in Marseilles. Some months later Dr. Fine of the Boston Committee told Cannon that Minz was working in a military hospital in Saida, Algeria, and that the committee was ready to supply funds to anyone who could offer a teaching position for him to be eligible for a non-quota visa. While Cannon had previously indicated that he would like to have Minz in his department because of mutual research interests, he could not accept him without the promise of continued support after the initial grant had expired. Instead, he suggested that one of his colleagues, David Rapport at Tufts or Fred Pratt at Boston University, might be willing to receive Minz, but without result. All roads were blocked and, as far as Cannon was concerned, the matter ended there.

Despite the loss of some of the most promising years of his youthful career, when World War II ended Minz returned to Paris and eventually became Research Professor at the National Center of Scientific Research, Department of General Physiology, the Sorbonne. By the 1950s he had published many research papers and a book on neurohumoral transmission, as well as a book that resulted from a seminar series delivered at the University of Illinois in Chicago.[12]

If Cannon was unable to help Bruno Minz gain refuge from the European debacle, he was more successful when he acted in behalf of another young medical scientist who came to his laboratory from France. Henry Haimovici, a Romanian émigré at the University of Marseilles, was a young surgeon who had been awarded a prize for his study of the diseases of peripheral blood vessels. The prize attracted the attention of the Rockefeller Foundation and in 1938, when he expressed interest in acquiring a physiological basis for further work in surgery, they awarded him a fellowship to conduct research in Cannon's laboratory at Harvard. Within a year Haimovici concluded several studies, including an investigation on the specific dynamic action of animals wholly deprived of the sympathetic nervous system, as well as a collaborative study with Cannon on the sensitization of neurons of the central nervous system when deprived of their neural connections.[13]

Despite the promise of his various investigations, when France and England declared war in the fall of 1939, Haimovici, out of a sense of duty to his adopted country, gave up his fellowship and returned to France for military service. He survived the battle of France; but when he was demobilized and returned to Marseilles, he discovered that the Vichy government barred him from his post at the medical school and even deprived him of the right to practice medicine in France or any of its colonies. Early in 1941, a desperate Haimovici wrote Cannon about his difficulties in the hope that he might be able to continue his research in the United States.

Cannon would have liked to take Haimovici back into his department, but Harvard had already received more than forty displaced scholars and passed a rule not to accept any more save in rare and special circumstances. Instead, he turned to Jacob Fine at the Boston Committee for support of Haimovici and began a letter-writing campaign to the surgical departments at the Mayo Clinic, Johns Hopkins, and other places to see if Haimovici's surgical talents and his skill in conducting physiological research could be utilized elsewhere. All to no avail.[14]

In a chance conversation, however, Cannon discovered that John Homans (the surgeon who had operated on his hands at the Brigham Hospital) was very interested in peripheral vascular disease and had a small fund to support an investigator to study the influence of tobacco and nicotine in contracting blood vessels. Encouraged by the news, Cannon inquired if the Rockefeller Foundation would consider giving an institutional grant to help provide a temporary salary for Haimovici, either through Tufts or Boston

University. After further assistance was obtained from the Emergency Rescue Committee in New York and from the Hebrew Immigrant Aid Society, Louis Zetzel at the Boston Committee informed Cannon that they had secured an affidavit for Haimovici and asked if he would make out a "moral affidavit," which assured the government of the applicant's character. "I think such a form filled out by you," he told Cannon, "would carry more weight [with] any Consul than a merely routine one filled out by the one giving the affidavit of support."

In the end, Haimovici was able to conduct his research for Homans at the Pratt Clinic of Tufts and later at the Beth Israel Hospital in Boston. The work went well and after two years, he received a fellowship from Dr. Emanual Libman to carry on the study he was doing on Buerger's disease at the Mt. Sinai Hospital in New York.[15]

The stories of Haimovici and Minz illustrate how Cannon exerted himself and exhorted his colleagues in attempting (not always successfully) to provide safe haven in the United States for European exiles. They were, however, only two of a steady stream of young scientists and physicians seeking to escape the spread of war in Europe who appealed to him for help—among them, Jean Reboul of France, Wilhelm Raab of Austria, Aldo Luisada of Italy, David Nachmansohn of Germany and Julien Maes of Belgium. (And the numbers increased to a considerable degree after the fall of the Spanish Republic early in 1939.) Cannon's sympathy for these young men knew no bounds. Nevertheless, his greatest investment of time and energy was in helping an older man: a longtime friend and fellow worker on the chemical mediation of nerve impulses, Otto Loewi.[16]

During the years following Loewi's Dunham Lectureship at the Medical School in 1933, he and Cannon had written to one another about their mutual interest in chemical transmission as well as of the dangers posed to European countries by German aggression. In March 1934, for example, Cannon wrote Loewi that he was watching "the course of events in Austria, anxious, of course, lest there should be a spread of the dark clouds from Germany into your country. I most certainly hope that you will be spared the degradation of humane spirit which has manifested itself in your native land." For a time, despite the unsettled Austrian situation,

Loewi was able to keep on with his scientific work at the University of Graz. As late as January 1938, when newspapers reported a projected Anschluss of Austria by Germany, Cannon wrote Loewi that he still had hope that "Austria will not be subjected to the cruelties and barbarities which have afflicted Germany."[17]

Whatever hope Cannon had for Loewi's well-being evaporated when the Germans invaded Austria and marched into Graz. The Nobel laureate was arrested and thrown into prison, together with a large number of other Jewish citizens. The day before, on March 11, he later recalled, he had been working in the laboratory with one of his associates on an experiment, the last of a series showing that afferent nerves in contrast to efferent nerves contained not a trace of acetylcholine. When someone came in and announced that the Nazis had taken over Austria, his thoughts were so occupied with his recent finding that he didn't comprehend the full impact of the news. Its significance was soon brought home to him with brutal force. "I awoke from a deep sleep at three o'clock in the morning," Loewi remembered,

> when a dozen young stormtroopers, armed with guns, broke into my bedroom, took me downstairs, and pushed me without any explanation into a waiting prison van that took me and others to the city jail. Later that night I was joined there by our two youngest sons, Victor and Guido. Our two elder children, Hans and Anna, were not in Austria at that time, and, of course, did not return.[18]

At the end of March, Loewi's son Hans wrote Cannon and Henry Dale, as two of his father's dear friends, to inform them of the distressing event, and asked them to take action. Before the letters from Hans could reach Cannon and Dale, they had already learned about Loewi's arrest from newspaper reports. Part of Dale was incredulous at first and he wrote Frau Loewi that he didn't believe there was any truth to the reports of arrest, that it must be a terrible mistake. Nor did he credit the trumped-up charge that Loewi was a Communist or a Communist sympathizer, observing that he had an intimate acquaintance with her husband for over thirty-five years, and had "known him as a man wholly devoted to his science with results known to the whole world, and completely aloof from any kind of political activities." Indeed, Dale said that he was prepared to tell any authority that Loewi was not a

Communist, and he was willing to do anything, in any form, in any way, in his behalf.

But another part of Dale realized the awful truth. Even before he wrote to Loewi's wife, he had alerted Professor Beznak at the Institutum Physiologicum in Budapest of Loewi's predicament, impressing upon him the notion that anything that had an air of organized public or official protest would likely do more harm than good. Telling Beznak that it would not be necessary to call upon colleagues for monetary contributions, Dale pledged, "If Loewi could manage to get out of Austria, and come to England, I believe he would be assured of a welcome. I would certainly make some arrangement to help him to renew his work."[19]

Cannon also lost no time in trying to help his old friend. Soon after receiving the letter from Hans, he addressed members of the American Physiological Society about the sad fate of Otto Loewi. All agreed that an inquiry through the U.S. State Department would be helpful. Cannon thereupon consulted Felix Frankfurter, who suggested that a strong letter regarding Loewi's contributions to humanity might make the State Department bring pressure for his release from custody. Cannon wrote such a letter, had it co-signed by distinguished colleagues, and sent it to George Messersmith, then an assistant secretary of state. Messersmith's rejoinder offered little help. "It is our primary duty," he wrote early in May 1938, "to offer full and effective protection to American citizens and American interests, and hence any deviation from this policy could not be other than prejudicial. Accordingly, while fully understanding your solicitude, it is felt that we cannot take the step you recommend on behalf of Professor Loewi."[20]

Loewi was released from prison after two months, and his sons a month later. Ordered by Nazi authorities to leave Austria by the end of the year, he was forced to have the Swedish bank in Stockholm transfer the money he had been awarded with his Nobel Prize in 1936 to a prescribed Nazi-controlled bank. His wife was made to remain in Austria while the Nazis dickered with the Italian government to dispossess her of some real estate in Italy that had been in her family for over a hundred years. The couple would not be reunited until 1941.

Loewi had not been allowed to attend the International Congress of Physiology in Zurich, and it was September before he was able to secure permission to leave Austria. That autumn an opportunity arose for him to work in a laboratory in Brussels, but before taking a position under the Belgian Foundation Franqui, he stayed with Dale in London for a few weeks. Cannon, who had cancelled his plans to attend the Congress because of trouble with the old X-ray burns on his hands, had heard only rumors about his old friend since his release from prison. When he asked Dale for the latest news, Dale replied that Professor Lacqueur of Amsterdam had managed to obtain financial support for Loewi from Dutch and Swiss pharmaceutical houses.

In addition to the money Lacquer had raised, which amounted to about 280 English pounds per annum for three years, the Society for Protection of Science and Learning in England (formerly the Academic Assistance Council) added 100 pounds per annum for three years and there was a grant of 50 pounds for one year from the American Joint Distribution Committee in Paris. Applications for additional aid had been made to the Emergency Committee in New York and the Josiah Macy Foundation as well. In sum, Loewi and his wife were financially protected but, as Dale pointed out, he had reached the age of sixty-five and because of his ejection from his country, lost his pension, forfeited all of his property, and came away a pauper.[21]

Some weeks later, Loewi wrote Cannon, too. He knew it was difficult to get an appointment at his age and he was grateful for his good fortune. "Thank God," he exclaimed, "I feel absolutely healthy, lively, and eager, keen on work." For all his gratitude, Loewi before long began to look elsewhere for opportunities. Perhaps he was motivated by reports that a national drive, spearheaded by Harlow Shapley of Harvard and backed by James B. Conant and Herbert Hoover, among others, was being launched to provide and distribute among cooperating universities in America support for some of the foremost academic exiles.

In May of 1939, Loewi wrote Cannon that although he appreciated everything that had been done for him, he didn't feel comfortable being a guest in someone else's laboratory. He suggested that there might be a place for him in the United States where he could have the opportunity, beside working privately, to help others in their scientific work and to inspire young people. "Beside this," he freely admitted, "you can easily imagine that I am rather tired of Europe, not knowing today what will happen tomorrow," but at the same time he assured Cannon that there was no need for him to feel pressured by this inquiry "as it is not an S.O.S."[22]

In mid-June Ernst Boas informed Cannon that the Emergency Committee in New York was considering Loewi's case. It was reported that officials at the University of Buffalo and the University of Rochester had expressed sympathy and interest in receiving him, but he would have to be financed from outside funds. Boas was optimistic and anxious to help Loewi. Cannon, however, worried that even if they could get the money, what would happen after temporary support for Loewi ceased? Who would then take on the obligation of supporting him? Later, upon inquiry Loewi was told that some of the funds he had obtained in Europe might be transferred to any institution for his use in America, provided that it afforded him the opportunity to continue his scientific work. But first he needed to have a definite offer, a statement of the scientific conditions under which he would be working, and an estimate of the sum that would be necessary to insure his support.[23]

While visiting friends in England in September, Loewi found himself trapped by the outbreak of the second World War. It seemed to him impossible to go on with scientific work when his money was tied up in a conflict that could well take away the last limit of his resources. Within weeks, however, Cannon got word from Dr. Emanual Libman at Mt. Sinai Hospital in New York that the Dazian Foundation, of which he was chairman, had decided to recommend a grant of $2,000 a year for at least two years. This was followed by Dr. Boas's assurance that further funds for Loewi would be available from the Emergency Committee and private individuals. "The important thing at present," Boas told Cannon, "is to get him some kind of a university appointment, which will enable him to enter the United States on a non-quota visa. Can you obtain this?"

Straightaway Cannon approached Dean Burwell and President Conant at Harvard on Loewi's behalf, but without success. Writing to Boas that November, he took pains to explain the University's position. After taking in more than forty men who were excluded from one country or another because of political or racial reasons, the Corporation declared that they would not assume further responsibility except in extraordinary cases. The placement of Loewi was especially difficult because he had already passed the customary age of retirement. Even though he was assured of support for three or four years, he might live for many more years thereafter and be unable to obtain a post elsewhere. Cannon told Boas, "I hope that I have made the attitude of the Harvard authorities clear to you. There is no lack of sympathy and consideration for scholars in distress, but the University has done so much that there is a conviction that henceforth, caution must be exercised."[24]

Cannon was still trying to find a way to help his friend when he learned there was every possibility that Loewi would be offered a research professorship at the New York University College of Medicine. The formal offer had not yet been made, but surely would be. In mid-December, when Cannon was certain that the position at N.Y.U. was assured, he wrote a candid letter to Loewi, essentially covering the same ground about the status of his chances at Harvard as he had in his letter to Ernst Boas. Knowing that Loewi had been disappointed many times and not wishing to add to his discouragement, Cannon first begged his forgiveness for not telling him before. Then he continued,

> This morning I received from my former associate here, Dr. Homer W. Smith, who is Professor of Physiology at New York University, a statement that the Faculty of the College of Medicine had voted to invite you as research professor of Pharmacology. This news brought great joy to me. I feel sure that the financial support for the present will be forthcoming and that we shall soon have the deep satisfaction of knowing that you are back in research again.

For Loewi, however, the offer from N.Y.U. was problematic, and his first impulse was to reject it. Cannon's letter had not yet arrived when Loewi wrote him to explain his circumstances. He had heard from the Emergency Committee that enough money (how much was not revealed to him) had been collected to add to what he already had so that he could exist in the United States for four years. He had also received a letter from the dean of New York University College of Medicine from which he learned "that the College will accept me as a kind of Research Professor." But at the same time the dean drew his attention to the fact that the accommodations for his work ought not to be expected as good, and perhaps even worse than what he then had in England. "Furthermore, he informs me that they could not give me any financial support. I write you this in order that you [might] know and understand from the foregoing the reasons which obviously obliged me to refuse the offer."[25]

After Loewi received Cannon's letter, he detailed again his

reasons for not accepting the offer from N.Y.U., adding that Henry Dale was "quite of my opinion." If Cannon had been informed about the conditions of the offer, Loewi ruefully commented, "you would not have been, I think, so happy about my appointment." Nor did he try to hide his disappointment that efforts to secure a place for him at Harvard had met with failure because, as Cannon wrote, the Corporation would not take in any more refugees except in exceptional cases. "I am sorry that I was wrong in believing I would represent such an exceptional case and—in consideration of what I have done for the progress of our science—I would in my plight earn a little more than only sympathy expressed to me by so many universities in the U.S.A."

Cannon deeply regretted Loewi's reaction and felt there had been misunderstandings about both the offer from N.Y.U. and the financial support. He nevertheless accepted his decision and trusted that one day, when the war was over, Loewi would be able to return to his post in Graz. Actually, Loewi's rejection of the offer from N.Y.U. is understandable, given that he had been made cynical by previous offers of help that hadn't come through—or when they did, were fraught with conditions that were not easy for him to accept. Almost at once, however, he began having second thoughts and wrote Cannon that, of course, it was not a strict refusal and there might still be other possibilities.[26]

During the winter of 1940, many of Loewi's friends on both sides of the Atlantic were concerned about his future. "As soon as Loewi's letter of December 29 came to me," Cannon wrote Ross Harrison at Yale,

> I showed it to Professor von Brücke, who is spending two years with us and who is a staunch friend of Loewi's. Immediately he sat down and wrote to Loewi a letter in which he urged him to reverse his decision and to cable acceptance of the New York offer. Whether there is any possibility of doing that at this time I do not know. I feel that Loewi has made a mistake in refusing the offer, and I fear that this refusal precludes any further effort to aid him in this country.

In February, Dale wrote Cannon a lengthy letter of explanation. Both of the letters that Loewi had received were open to misinterpretation. "He actually showed me his letter of reply to Boas, and I should not have advised him to send it if it had been, as he appears to have suggested to you, a direct refusal. What he actually said was that he found himself unable to accept the invitation on the information thus far given to him."

After Dr. Boas obligingly provided Loewi with more explicit details—a sum of $10,000 would be available to maintain Loewi over a period of time—Dale advised his friend to accept the offer on the Emergency Committee's terms, for he was certain that it was the best opportunity still open to him for renewing his activity in science. In due course, Loewi agreed to the conditions of his appointment at N.Y.U.—much to the relief of Cannon, who reassured him that he had made the right decision. "I really believe that you will find conditions favorable to your research at New York University and that there will be an atmosphere of friendliness and a freedom from the distractions of warfare which also will be to your advantage as a scientific investigator."[27]

Later, Cannon revealed to Dale that he had smoothed the way for a successful close to the negotiations by informing Boas of Dale's testimony that Loewi's desire to have the money applied for three years instead of four "was not because Loewi would live extravagantly but because he has had such harsh experiences and knows so little about living conditions in New York that he is fearful of being caught again in a state of actual need." Cannon confessed to Dale that "probably Loewi should not know about the impression he has made in New York—it might be disturbing. I hope that he will come with generous tolerance for changed conditions and . . . that if he will be patient, matters will be arranged in a way favorable for his continued activity." After all of the difficulties and misunderstandings in connection with the arrangements, Boas was only too glad to put the Loewi case behind him.[28]

The tale of another distinguished Austrian scientist well along in years provides an interesting comparison and contrast to that of Otto Loewi; it also expressly illustrates the cruel and disastrous effects of the Nuremberg Laws that were passed in 1935. Ernst Theodor von Brücke, the grandson of Ernst Wilhelm von Brücke, one of the founders of modern physiology, was trained in medicine and physiology in Leipzig and Vienna. In his late fifties at the time of the Anschluss, he had held the chair in physiology at the University of Innsbruck for over twenty years. Shared scientific interests in problems of neurophysiology had led him into a friendship with Alexander Forbes, who ran the laboratory of electrophysiology in Cannon's department at Harvard. When Forbes had paid a visit to Brücke in the summer of 1934, Austria showed little effect of the political changes in the country next door.

In the spring of 1938, however, Forbes learned from A.V. Hill in London that the Austrian physiologist had been abruptly dismissed from his professorship at Innsbruck. Unlike Loewi, he was not thrown into prison; also unlike Loewi, he was not Jewish. The reason given by the Nazi authorities for Brücke's expulsion from his post was that both his wife and his mother, although of Protestant faith, were of Jewish descent. Forbidden to teach or do research, he became "discouraged, despondent, downcast, and extremely doubtful" as to what the future had in store for him. He realized that he would have to leave Austria and find a position abroad—the sooner, the better—and had begun to reach out to friends in England and America to help him.[29]

Other reports of Brücke's altered circumstances soon came to Forbes. One of Brücke's American students asked Forbes to help, ending his plea that "every year Prof. Brücke has to spend here means ten years from his life." Selig Hecht at Columbia heard from a Mr. Sidgwick in England, who stated that Brücke was safe at the moment, but emphasized "the danger of worse things to come" and respectfully asked him "to exert everything in your power to make his coming to America possible." Hecht left no doubt of his feelings when he relayed the news to Forbes: "All the expletives have already been used up on one occasion or another against the Nazis, so there are almost none left for this particular piece of idiocy. I hope that something can be done for Brücke."

Forbes wasted no time springing into action. First, he informed Cannon about Brücke's situation. He also got in touch with Herbert Gasser at the Rockefeller Institute, vowing that he himself would guarantee the bond required by the U.S. Departments of State and Labor so Brücke would not become a public charge. Advising Sidgwick of his actions, he added, "I hope it will be possible to raise enough money to assure Brücke a research fellowship for at least a year and possibly more. I am most anxious to do whatever can be done for him, both because of my high regard for his standing as a physiologist and because of my personal feelings toward him as perhaps my best friend in continental Europe."[30]

In June, Forbes went off to Europe and stopped in London to

confer with Sidgwick and Dr. Barbara Travers, who pressed him to undertake the task of bringing Brücke to the United States. Before he knew it, Forbes had agreed to provide or find funding for the Austrian for two years and to get him an appointment at Harvard.

Cannon, after learning of Forbes's commitment, went to Dean Burwell and proposed that Dr. von Brücke be appointed a research fellow in the physiology department with the prospect of outside financial support for two years. Concerned about what might be a problem for the University at the end of the two years, the dean promised he would see President Conant about the matter. After a time, Cannon received an answer that Burwell was willing to approve a recommendation of Brücke as visiting lecturer for one year "if the matter is really urgent and the man is in positive danger." Cannon sent word on to Forbes, telling him that he thought it was as near a definite promise as the dean could give under the circumstances, and might be used to secure a visa for Brücke.

That summer, after attending the International Congress of Physiology in Zurich, Alex Forbes met Brücke in Munich and discovered that he had considerable property, some in Germany, some in Switzerland and New York—and might be able to keep part of it from the Reichsbank. His wife had escaped from the country, Forbes later wrote Cannon, and had "about $10,000 in Switzerland and America, which is safe. Also, she is a fairly eminent gynecologist, being a corresponding secretary for Austria of the International Women's Medical Association. She can probably support herself at her profession."[31]

Forbes had found his friend in seemingly good health, but the shock of his sudden reversal of fortune made him speak seriously of suicide. Alarmed, Forbes gave Brücke a letter to a young acquaintance of his in the American Consulate in Vienna, Brewster Morris, who was in a position to help. After meeting with Consul Morris, Brücke wrote Forbes he could get the visa for the States at once if he had an official invitation from a university or hospital. With such an invitation from Harvard, he believed he would have no trouble immigrating to America; he begged Forbes to do what he could to procure one for him.

Brücke's letter was confirmed by Morris, who instructed Forbes about the procedure to be followed. To begin with, Brücke needed a contract from a recognized institution of learning in the U.S. indicating the nature of the position and the remuneration to be received. Then, to expedite the case Morris suggested that the State Department be asked for an advance ruling on whether Brücke qualifed as a non-quota applicant and that the Department advise his office of their findings. Morris ended his letter with a hasty note, "I arrived here a few weeks ago, but have not had time to see much of Vienna or its surroundings thus far, as our office has been simply swamped with work—all Jewish refugees—and I have had to work almost every evening."[32]

In September, Forbes went over the entire situation with Dean Burwell and obtained his promise to recommend the appointment of Dr. von Brücke for one year, with the probability of renewal for a second year, but no longer. Later that month, after President Conant had given his approval, Forbes sent the full details to Brewster Morris in Vienna via the diplomatic pouch. Some weeks later, Brücke's appointment at Harvard as research associate in physiology was passed by the Administrative Board; the next day it was passed by the Faculty of Medicine and sent to the Corporation for ratification.

In November Consul Morris notified Forbes that the papers appeared to be in order and he hoped Brücke would have his visa in a few days. "It will be just as well," he added ominously, "for the attack on the Jews is being renewed with increased bitterness, as a result of the shooting of the German diplomat in Paris the other day by a Polish Jew." (Kristallnacht was November 9–10, 1938.) Brücke, too, hoped he would soon be able to get out of Austria.[33]

Nevertheless, it was February 1939 before Forbes, after a long silence, heard from Brücke again. From month to month the Austrian had hoped to be able to give his friends some definite news, "but, alas, this is not possible even now." He could only report that he was working incessantly from morning till night to settle the thousands of details necessary before even having a prospect of being able to leave. In spite of his frequent inquiries, and in spite of the fact that the officials appeared to favor him as far as they could, he had not yet received the permission without which he dared not go out of the country.

What made life so difficult for those who were being ousted by the Nazis were not only the restrictions placed on them by local officials, but also the paper walls that had been erected at consular offices to discourage persons who desired to immigrate to the

United States for permanent residence. After describing a host of problems that beset him, such as paying taxes, disposal of property, etc., Brücke ended his letter to Forbes:

> Four lawyers are working for me, and it takes a hundred petitions and months and months to get all these permissions, if you can get them at all . . . It is absolutely impossible to imagine all these exhausting difficulties for one who has not [yet had] this experience himself. I must say that I hardly find leisure or quietness of mind to read a book. After all that, I need not tell you how ardently I am longing for a moment when I shall be allowed to do my work, scientific work, again and how thankful I am for the prospect to be able to do so. I hope it will be soon!

Brücke subsequently wrote that he expected to leave in a few weeks, but, again, it was months before there was further news from him. At long last, in June a letter came from Switzerland, informing Forbes that after visiting Basel and Liège, London and Cambridge, he would be arriving in New York on the 15th of August. One month later, his wife, Dora Teleky-Brücke, also arrived in New York. Intersecting their separate arrivals was the outbreak in Europe of the second world war.[34]

Alex Forbes was delighted to have Brücke with him in Boston. As he explained in a statement not long after his friend's arrival, he had for some years considered him "one of the best experimental physiologists in the field of the nervous system in continental Europe." Resuming his studies on reflex inhibition and fatigue in nerve, the Austrian exile went on to collaborate with Forbes on a series of experiments that examined both sensory and motor nerves in frogs and cats; they focused on the temporal parameters of electrical nerve activity during prolonged stimulation, with special reference to inhibitory mechanisms.[35]

Cannon was pleased with Dr. von Brücke's arrival as well, for the war was rapidly making inroads on his department. In the spring of 1941, after Forbes had entered his country's service as a Lt. Commander in the U.S. Navy, Cannon looked into the matter of continuing financial support for Brücke, who was then in the second year of his appointment. Harlow Shapley at the Harvard Observatory, one of the strongest supporters of émigré scholars at the University, offered advice—and then added a personal observation: "The serious difficulty is that Mr. Conant, the bravest man in the world in most respects, at least from the standpoint of international manslaughter, goes extremely conservative every time one of these refugee matters comes up."

Despite the good intentions, Brücke's tenure at Harvard was cut short. One evening in June 1941, after performing an experiment and appearing to be in excellent condition, he went home with a slight toothache. In the middle of the night he died in his sleep.[36] Though the problem of Dr. von Brücke ended there, it was immediately replaced by another problem: finding a job for his wife, Dora. That proved even more difficult for a variety of reasons—among others, that she was not an easy woman to deal with. Cannon and Forbes wrote to everyone in her behalf, yet even the Kentucky Frontier Nursing Service, which was in need of personnel, failed to offer her a position. Forbes supplied funds to help support her, but by 1942 letters ceased to come to him or to Cannon from Dora Teleky-Brücke. Eventually she opened her own practice in Boston and retired to Switzerland in 1950.[37]

In the meantime Otto Loewi, despite seeming to be the more unfortunate of the two Austrian physiologists, had settled in comfortably within a year of arriving in New York. His wife was released by the Nazis in Austria and they lived together on 93rd Street near the East River. For twenty-one years, as Henry Dale later noted, American generosity and respect for his scientific accomplishments would provide a happy evening to Otto Loewi's career. He spent much of his time writing articles and giving lectures, and passed each summer at the Woods Hole center for marine biology on Cape Cod until his death at the age of eighty-nine, in December 1961.[38]

20 *From the Body Physiologic . . .*

WHILE CANNON's interest in neurotransmission would have a prominent place in his research for the remainder of his career, this is not to say that it was his only concern. Far from it, for as chief of a large department and as teacher and mentor of young men and women, he was busy all the day—either directing, and sometimes collaborating in, the investigations of others or heralding the accomplishments of his research group. In his later years, it is true, he had less and less contact with the first-year medical students, who held him in awe, aware that he was one of the great men of American science, and flocked to the few lectures he gave each spring in the physiology course. Yet he remained as keen as ever on training a younger generation of workers in the physiological methods that could, and often did, lead to the frontiers of medical knowledge.

With the generous grant from the Rockefeller Foundation providing a measure of stability that others lacked during the years of the Depression, Cannon's department was relatively well off. Indeed, when in 1931 Ludwig Kast of the recently-founded Macy Foundation approached him with an offer of support for his research, Cannon replied that his needs were already taken care of but left an opening in case he should require aid in future. (He later obtained a grant for an investigation conducted by Hallowell Davis and his collaborators on the course of nerve impulses passing into and through the central nervous system.)[1]

If money for his research was not a major problem for Cannon, the comings and goings of staff members could, and often did, upset his routine. Early in the decade, he learned that Alfred

Redfield was leaving the department to accept a joint professorship at the new Biological Institution in Cambridge and the Oceanographic Institution at Woods Hole. Cannon, sincerely sorry to lose Redfield, nevertheless saw the loss as an opportunity to inquire if Philip Bard would consider leaving Princeton and return to Harvard as an assistant professor of physiology. "I feel sure," he wrote Bard, "that your researches would add an important element to the apparently spontaneous interest in the nervous system and its functions and disorders that has sprung up in this region."

Bard wasted no time or words in replying that he would accept an invitation to take Redfield's place "with alacrity, joy and enthusiasm." Cannon was delighted and promptly made a formal offer, exclaiming, "There is great rejoicing over the prospect of having you back here. We are all congratulating ourselves." Bard fulfilled Cannon's expectations in every way and spent the next few years at Harvard in fruitful research, conducting some experiments with other workers in the department, such as Arturo Rosenblueth and James Pinkston. He also began a productive partnership with David Rioch, newly arrived to assist George Wislocki of the anatomy department, thus forging a fresh spirit of collaboration between the two departments.[2]

Much to Cannon's satisfaction, the interdisciplinary approach to studies of the nervous system he was then fostering was coming into its own. "We have an unusual gathering of persons who are interested in the functioning of the nervous system," he noted with pride at the end of 1931. "Stanley Cobb, Tracy Putnam and Frank Fremont-Smith at the City Hospital make one group; Alexander Forbes, Hallowell Davis, Philip Bard and myself constitute a group here in the Medical School; Cushing and his collaborators are at the Brigham Hospital; and Crothers, Ayer and Campbell represent more strictly the clinical interests." This was a trend that Cannon would encourage and enlarge on for years to come, and he was fond of quoting Bowditch's remark that "the study of the nervous system is the true field of battle for physiologists, all other investigations, however interesting and important, being of the nature of skirmishes."[3]

When Phil Bard returned to Boston in the autumn of 1931, he brought with him his graduate student, Chandler McC. Brooks, who never forgot the magic of the time he spent at Harvard. On arrival, Brooks found "an atmosphere of refinement and academically oriented culture" at both the University and the Medical School, with admirable students and teachers and, in the physiology laboratory, a group of cooperative but self-reliant men. Under the direction of Cannon, the students and associates who were engaged in the study of physiology "formed one of America's most congenial and accomplished scientific families," and some formed friendships that endured for a lifetime.

"It was Cannon," Brooks later remembered, "who suggested the problem I initially attempted and it was Cannon who told me gently to try something else when it became apparent that I would never solve it." Once every week or so, he continued, Cannon would make a little tour of the laboratories. "He looked in, and one had the opportunity to ask him something or to tell him what you were doing. He showed interest but never remained for long." Brooks was aware of only one restriction: that Cannon should not be disturbed by petty matters, and not at all when he was in his laboratory.

Brooks recalled that Arturo Rosenblueth, who was "too facile, brilliant, well-informed and certain in opinion for most of us to handle," had an interesting association with Rioch, who was capable of dealing with him. ("Arturo published about two papers a month and still had plenty of time to argue with David Rioch and others.") And that Magnus Gregersen, "with much energy, enthusiasm and happy noise," conducted studies of thirst, salivary gland activity, blood volume determinations and related matters that correlated well with Cannon's earlier studies of thirst and traumatic shock. Then there was Percy Stiles who taught the dental students; Roy Hoskins and Milton O. Lee, who edited the journal, *Endocrinology;* Chester Hampel and J. O. Pinkston, both of whom would go on to the American University at Beirut; René Gayet and Mayliss Guillaumie from France, Teodoro Schlossberg and Oscar Orias from Argentina. Along with the score or more of workers who lived in Brooks's memory were the support staff: secretaries like Minna Weiner and Agnes McLeod, assistants like Elizabeth Bright and Tom Barnett, Pat the Irish janitor, Irving the animal man, and many others who were solidly behind the research and teaching of the department and who contributed to its sense of well-being.[4]

Bard, assisted by Brooks, conducted studies on the functioning of the central nervous system. Working with cats, he studied their behavior, especially the emotions of anger, fear and sexual excitement, after various intracranial operations. He had obtained a rage reaction, Cannon reported, never before seen in animals from which only one cerebral cortex had been removed or in animals without the hypothalamus. Further study, it was hoped, would yield much information about the neural organization for emotional expression. Bard's other research was directed toward securing added insight into the nature of cortical localization, as well as work with Pinkston on the problem of temperature regulation.

Although Cannon informed the dean that the organization of his department was likely to remain a "going concern" for some time to come, it was not to be. In the spring of 1933, the department was thrown into crisis when Bard was offered the professorship of physiology at Johns Hopkins. Alfred Redfield rushed to Cannon's side to express his concern over the possibility of Bard's departure. "He feels as I do that if Dr. Bard should go," Cannon wrote Edsall in March, "there would not only be a great weakening of the neurological group but also a serious loss to physiology in Boston."[5]

The dean did everything he could to mount an attractive counter-offer that would keep Bard at Harvard, but within a week the young physiologist made up his mind to go. "You probably have noted that I have lost Bard," Cannon wrote F. T. Lewis of the anatomy department in May. "His cooperation with Dr. Rioch made them a working pair unequalled anywhere else. I looked forward to the issue of great things from their combined interests and energies, but . . . I regard the production of men as part of our function, and they must go elsewhere to take on larger responsibilities than we can give them here." Lewis replied that Bard's appointment at Hopkins was cause for congratulation, then added a wry comment: "Without him Harvard was rated by Cattell as having the strongest department of physiology in the country, whereas anatomy held lower place than tenth . . . So that it would seem that Dr. Bard could be spared for his high promotion."[6]

Years later, Bard disclosed that he nearly turned down the offer because of his deep attachment to Cannon and the pleas of Dr. Aub that he stay at Harvard in consideration of Cannon's poor health. At the time, however, he wrote Cannon that there were many factors involved in his decision—the most difficult one he had ever had to make:

> Let me say perfectly frankly and very sincerely that leaving you is to me a tragedy. This can be tempered only by the feeling that in any event I can be your representative and perhaps I will be able to spread your influence to a certain portion of American physiology . . . I have no doubt that behind this invitation to come to Johns Hopkins is a feeling that the next best thing to Dr. Cannon himself is Dr. Cannon's influence.

Cannon, in turn, told Bard that he had such faith in his ability "that I could not help wishing for you a position of great strategic advantage. That you will have at Johns Hopkins. You will go to it not only with my best wishes, but with my full confidence that you will have an illustrious career." Touched by the personal comments in Bard's letter, he said he would preserve and read them "when the blue devils have me on the run."[7]

Apart from his reassuring words, Cannon was left with the daunting task of replacing Bard as assistant professor of physiology. That summer, when Zenon Bacq was at Woods Hole, Cannon sounded him out about any possible interest he might have in such a post. But Bacq declined the overture because, as he later wrote, he did not enjoy the American way of life and decided he would prefer to stay in Belgium since he had a strong attachment to Wallonia and to Europe.

Eventually, it was Arturo Rosenblueth who took Bard's place. Loath to leave off his absorbing researches and return to Mexico when his Guggenheim fellowship was due to run out, Rosenblueth had requested renewal of the grant for 1931–32. After some dickering, and with a glowing recommendation from Cannon and added support from Hans Zinsser, the committee acquiesced. In September 1931, following a whirlwind courtship, he married Virginia Thompson, a beautiful and intelligent Radcliffe student from Oregon, thus cementing his attachment to America. Appointed an instructor in physiology in 1932–33, Rosenblueth the next year took over the tutorial reponsibilities after Bard left. In the fall of 1934, armed with an offer of a job in Mexico, he received a three-year appointment as assistant professor, which was later renewed.[8]

Actually, the problem in Cannon's department during the depths of the Depression was not with people leaving but rather with people staying because they were unable to find employment elsewhere. By 1935, however, it looked like the log jam was about to give way. When Cannon gave instructions to Hal Davis for running the department during his impending sabbatical, he told him to be on the lookout for personnel for next year, especially if Gregersen, and possibly Pinkston, received a call to go elsewhere.

In the spring Davis wrote Cannon, then in China, about anticipated personnel changes. "It is quite a different situation from the last few years," he remarked, "to have three of our younger men find positions all at once, and it is most encouraging as a sign for the future." Davis continued that Dr. Hampel had accepted an offer from Beirut, Dr. Pinkston had unofficially been offered a post in the department of pharmacology at Vanderbilt, and Dr. Derbyshire also had an unofficial offer from Frank Hartman, who had moved to Ohio State. Donald Lindsley, nearing the end of his N.R.C. fellowship, had an offer to join the Brush Foundation in Cleveland, while Margaret McKay Sawyer, depending on her husband's situation, might also be leaving next year. "All in all, it looks like quite an exodus," he concluded. A few weeks later, Cannon heard that Magnus Gregersen had a tentative offer of a professorship at the University of Maryland, which he would in all likelihood accept.[9]

After Cannon returned to the Medical School in the fall of 1935, he informed C. Sidney Burwell, the new dean who had replaced David Edsall when he retired, that "the department had been cleaned out by acceptance of positions elsewhere by persons trained here." If the physiology department appeared to be "cleaned out" at midpoint in the decade, the cupboard did not remain bare for long. "The news has reached me that you have an unusually large number of students this year," Pinkston wrote Cannon late in 1936. "According to Miss Bright, the Laboratories of Physiology are almost as much of a beehive of activity as they were during 1930–32. It would appear that physiology is now in the post-depression renaissance." A short time later, Cannon confirmed Pinkston's observation when he answered an inquiry from Louis Gross, director of laboratories at Mt. Sinai Hospital, about taking one of his men for further training:

> The situation is one of such crowding that I really do not feel that we can place another man at this time. Every single room that we have is taken, and all the places in the combined rooms as well. I was holding a place for a possible unforseen arrival, and that was taken about six weeks ago by Dr. Luco, of Chile. I not only have

> graduate students, candidates for the Ph.D. degree and advanced medical students electing work in physiology, but representatives from Belgium and France also. In addition to the crowded state of the Laboratory there is the necessity that I keep the supply of animals going and also direct the work of the investigators and see that they get out of their difficulties. It is altogether a rather heavy burden.[10]

And in fact, for the remainder of the 1930s the physiology department had a full house. There was an unusually large group of foreign fellows that included Jean Reboul and Daniel Bargeton from France, Julien Maes from Belgium, J. V. Luco from Chile and J. Aste-Salazar from Peru, K. Lissák from Hungary, U.U. Uotila from Sweden and P. O. Therman from Finland. It was Luco who later remembered with some amusement that when he presented his first records of a study on the defatiguing effect of adrenaline to Cannon, "his face expressed some doubt and with great tact he said, 'Two Nobel Prize winners have found different results.' I replied, 'I will call on you again after a series of twenty experiments is completed.' 'That is the right answer,' he told me, adding, 'The Nobel Prizes may be wrong, the cat is always right.'"[11]

There was also Harry Friedgood, who had a stipend from the Sex Research Committee to study sexual behavior in various laboratory animals, as modified through controlled physiological disturbances in such endocrine glands as the anterior hypophysis, ovaries and thyroid. For a time Friedgood received additional support from the Macy Foundation, so he could show students at the Beth Israel Hospital the practical medical importance of their physiological studies. Fiorindo Simeone (who had published a research while still a Harvard medical student and was now a National Research fellow) worked with Cannon and Rosenblueth to complete a study on the sensitization of sympathetic ganglia by partial denervation, though his interest in surgery ultimately overshadowed his interest in physiology. Among others were Carlyle Jacobsen, a fellow of the General Education Board, and Harold Wiggers, a Porter Fellow of the American Physiology Society, whose father, the chairman of the physiology department at Western Reserve, had sent him to Cannon for further training.[12]

Late in 1935, Cannon once again had cause for concern about his teaching staff. A rumor reached him in November that Hal Davis was being considered for the chair in physiology at Cornell, vacated when Herbert Gasser took over as director of the Rockefeller Institute. Alerting the dean to the situation, Cannon said that he realized advancing Davis to a permanent appointment at Harvard "would in fact be selecting him as my successor," since the University had the right to request his (Cannon's) retirement when he reached his 66th birthday in 1937. Nevertheless, he took pains to express his high regard for Davis's ability as a physiologist as well as his skill in experimentation and in stimulating others.

In the spring, when the time for faculty appointments rolled around, Cannon recommended promotions for Drs. Forbes and Davis in the physiology department. There was no problem gaining approval for Alex Forbes's rise to a full professorship. Not only did his distinguished contributions to electrophysiology well deserve recognition, he had been an associate professor for a full fifteen years. Besides, no salary was involved—not to mention Forbes's financial support of the enterprises with which he was concerned. Promotion for Forbes was easy enough; advancing Davis was something else again.[13]

After a meeting of the medical faculty in April, several professors wrote President Conant to oppose the idea of placing Hal Davis in a position where he could be considered a successor to Cannon. Henry Christian, nearing the end of his own career, was one of the first to express his doubts:

> There is one phase of the situation in Physiology which with Dr. Cannon present I did not feel that it was proper to bring up and that concerns Dr. Cannon's age, health and inevitable retirement before a great length of time. Dr. Cannon becomes sixty-five next October and he is not a well man, as you most likely know. I just wonder whether it is advisable to create Associate Professor titles in a department so near the time of the retirement of the chief.

George Wislocki wrote in a similar vein, warning of the danger in promoting Davis because it might embarrass the committee in their choice of Cannon's successor if he should take early retirement. Cecil Drinker, for another, hastened to tell Conant that, in his view, Davis had become too highly specialized to teach medical students. "I believe that were we faced with the problem of filling the Professorship in Physiology at the present time," he ended, "we would be better served if we appointed a man such as Landis in Philadelphia or Best in Toronto."

Edwin Cohn, on the other hand, wrote enthusiastically in support of Davis, and so did Herbert Gasser, in response to a confidential inquiry from Conant. On May 1, 1936, Cannon sent a long, and very strong, recommendation for the promotion of Hallowell Davis to President Conant, who had told Dean Burwell he was not in favor of it. It had been claimed that Davis was narrowly limited in his research interests, but the main argument against his advancement (as far as Conant was concerned) was that he had not been recognized by a firm offer of a position elsewhere. Cannon hotly denied both allegations, and pleaded so persuasively in behalf of Davis that his promotion later went through, along with that of Alexander Forbes.[14]

Irrespective of Davis's promotion, the question of Cannon's retirement had already come under discussion. Early in 1936, Burwell told the president that he had conferred with Cannon about whether renewal of the Rockefeller Gift for Research in Physiology was contingent on his staying at Harvard (it was). The dean, who gathered that Conant would approve of keeping Cannon on past the age of sixty-six, quoted from a letter he had recently received from Cannon. In it, Cannon stated his willingness to continue his services to the University; indeed, it would be of some satisfaction to him to complete forty years of teaching.

We have Cannon's description of how matters developed, for he wrote to his son Bradford in February:

> A few weeks ago the Dean told me that Dr. Gregg of the Rockefeller Foundation would visit the Medical School soon and that the question of renewing the grant to the Laboratory would be raised. He suggested that I be prepared to make a statement. I told him that the term of the present grant would end in September 1937, and that the University would have the right to ask for my resignation in October of that year, for then I shall reach my 66th birthday. I could not inquire about renewal of the grant unless I could know what my future would be. Well, in short, the question was put up to the President, and he said that he would be glad to see me in the service of the University as long as my health would permit! So that was that.
>
> And then Dr. Gregg came and said that he would advocate a renewal of the grant from the Rockefeller Foundation for a five-year period beginning in 1937. If that should go through, it would provide for support of my researches until the end of the academic year when I am 70 years old. I guess I'll be ready to quit then anyhow.

Cannon ended his letter by telling Bradford, "I wanted you to know of it because it means that, if all goes well, and I keep my health, I shall be able to hold my job until you are well settled in your career. Hurrah!"[15]

In November, the Rockefeller Board voted to appropriate $75,000 "for the researches in neurophysiology of Professor Walter B. Cannon" for a five-year period from September 1, 1937, to August 31, 1942—a renewal of the gift made current since 1930. When Alan Gregg informed Cannon of the vote, he pointed out that the word neurophysiology, as distinguished from physiology, should be interpreted broadly. Cannon, in reply, expressed his gratitude for the trust the Foundation had placed in him, and agreed with the change from physiology to neurophysiology. "We shall doubtless be concerned with the relations of the nervous system to muscles in a continuation of our study of chemical mediation of nerve impulses and also to glands of internal secretion," he assured Gregg. "After all, the system is of no avail without its instruments."

Little more than a month after the renewal went into effect, however, Cannon dutifully followed convention and offered his resignation to the president. Mr. Conant replied:

> May I congratulate you on your sixty-sixth birthday? I can assure you that the President and Fellows have no desire whatsoever to see you give up your valuable work in the Harvard Medical School. As a matter of fact, I mentioned your situation many months ago, and therefore I feel quite certain without consulting the Corporation further that we wish you to continue your important activity as long as possible.[16]

◆ ◆ ◆

When it came to producing a synthesis or overview of the various lines of research that came out of his laboratory, Cannon was a master builder. His speeches and review articles, painstakingly wrought, were models of seamless fusion forged out of separate facts and fragments, bits and pieces from this or that observation—all coming together in a perfectly logical development of his thesis. The total effect such fusions conveyed, to scientists, to clinicians and to a larger general public, was that his investigations, taken as a whole, had a cumulative impact which transcended their worth as individual contributions to the stockpile of medical knowledge.

We have put the focus on Cannon's research during the 1930s almost exclusively into his study of chemical mediators. But what about those research interests, it might well be asked, that no longer absorbed the main thrust of his time and energies? Was there a place in his laboratory for further work on homeostasis and sympathectomy, for example, or on the inter-relations of endocrine glands and bodily effects of the emotions? Did Cannon ever concern himself with recent investigations on traumatic shock or later contributions to understanding of the digestive process and gastrointestinal motility? The answer in part is that he sometimes returned to old research interests or, more likely, passed them along to his students to explore further. And in part the answer is that former interests were never abandoned, nor did others who appreciated Cannon's contributions to their fields of interest allow his role to be forgotten.

In 1932 alone, the year of publication of *The Wisdom of the Body,* Cannon delivered a nation-wide radio address on the effects of strong emotions, prepared with Cecil Drinker a revision of "The Dog's Gift to the Relief of Suffering" (a defense of animal experimentation), and published an article on thirst with Magnus Gregersen. In addition, a report on the physiology department noted that Margaret McKay Sawyer and Theodoro Schlossberg had continued studies on the importance of the sympathetic system for the maintenance of homeostasis and Robert A. Phillips, in cooperation with the Harvard Fatigue Laboratory, was working on animals deprived of a sympathetic system.

The following year, 1933, marked the 100th anniversary of William Beaumont's first publication, in book form, of his remarkable and fundamental series of researches on the physiology of digestion. "To signalize this centenary in Michigan," it was announced, "it is peculiarly fitting that the annual Beaumont Lectures of the Wayne County Medical Society, Detroit, should be given this year by Dr. Walter Bradford Cannon . . . under the general title of 'Some Modern Extensions of Beaumont's Studies on Alexis St. Martin.'" Cannon's lectures were subsequently printed in the Michigan State Medical Society's *Journal* and in monograph form. No less a master of exposition than Harvey Cushing took his hat off to Cannon for producing lectures that he judged "absolutely

A-1" and "full of meat, even if it may be old things newly expressed."[17]

Cannon followed up his Beaumont Lectures by contributing a chapter on "Hunger and Thirst" to the *Handbook of General Experimental Psychology,* published in 1934. He also gave three prestigious lectures that year, each providing guidelines for clinical workers in distinctly separate fields. His Kober Lecture to the Association of American Physicians, as noted previously, told the story of the development of chemical mediation. That lecture was preceded by the Hodgen Lecture, which Cannon delivered on the physiology of the emotions before the St. Louis Medical Society in January; it was followed by his Caldwell Lecture to the American Roentgen Ray Society meeting in Pittsburgh in September. The Caldwell Lecture, a review of the work Cannon did some thirty years earlier on the digestive tract, attempted to map out a territory for future studies. Not long after, an editorial in the *British Medical Journal* noted that his "summary of the facts and of the lacunae in our knowledge of the physiology of the stomach" had influenced the trend of a discussion at a recent meeting of the Royal Society of Medicine.[18]

Cannon's Hodgen Lecture on the significance of the emotional level received its share of attention as well. As the leading authority on the physiology of the emotions, his views from first to last were listened to with rapt interest. An address to the American College of Physicians on "The Role of Emotion in Disease," published in the May 1936 issue of the *Annals of Internal Medicine,* was in such demand that Cannon claimed he had never prepared a paper which aroused a greater degree of interest. "Furthermore," he wrote the president of the College, "for the last five months requests have come from all parts of the United States for reprints of the article. I hope that it has had a useful influence in medicine."[19]

Judging from the number of invitations he received, Cannon was on everybody's list of speakers, or so it seemed. In February 1934, Elliott Cutler apprised him of arrangements for the American Surgical Association meeting to be held in Toronto, at which there would be a symposium on traumatic shock. "We are most anxious that you give us at least a fifteen-minute paper on this topic," wrote Cutler. "I know I cannot overemphasize the desire of the members of the American Surgical Association to have you present." Cannon, after agreeing to open the symposium, not only asked his son to advise him about the paper he was preparing ("my interests have strayed widely from shock in the past dozen years"), but also suggested that they attend the meeting together. His talk was a triumph, reviewing nervous as well as toxic factors in the clinical picture of shock, and also reporting new findings based on the research of Norman E. Freeman in his laboratory.[20]

Clinical interest in Cannon's findings was stimulated as well by his work on the sympathico-adrenal system and sympathectomy. Norman Freeman collaborated with two other surgeons, R.H. Smithwick and J. C. White, at the Massachusetts General Hospital in a study of adrenal secretion in man. White and Smithwick's volume on *The Autonomic Nervous System: Anatomy, Physiology, and Surgical Application* came out in 1935—a time when surgeons were just beginning to understand the effect of interrupting the cardiovascular and other visceral nerves. "Here was a chance for pioneering, and it was taken," Cannon later stated, observing that White and Smithwick, among others, had made "impressive advances in applying surgical technique to the treatment of pathological states resulting from abnormal autonomic activity."

White drew heavily on the work of Cannon and his collaborators in summarizing both the scientific information then coming to light and its practical significance. Cannon, well aware of White's efforts, wrote him after publication that John Fulton, in his review in *Science,* had praised the chapter on anatomy but called the one on physiology sound but dull. "The dullness, I am happy to note," Cannon continued, "is not attributed to you but to the physiologists whom you consulted. Since you have mentioned as physiological consultants only Drinker, Rosenblueth and myself, the onus for the dullness is definitely settled. Please write something else which involves physiology and the three of us will get together and either set the chapter to dance music or put it into poetry for Dr. Fulton's pleasure."

White was entertained by Cannon's letter and added a tidbit of his own:

> Strangely enough, on reading through Fulton's review, I find that he has put in a good deal of information about Raynaud's disease that was never written in the book. The practical application of our findings on the importance of not cutting the post-ganglionic fibres was never printed, and was barely worked out in the laboratory at

> the time the book appeared. I must have told him about it subsequently because his statements are quite correct, and I wish they could have been included in the chapter on Raynaud's disease. It amused me, nevertheless, to know how liberally he had called on his own imagination.[21]

Public interest in homeostasis continued far beyond initial publication of *The Wisdom of the Body,* requiring a second edition to be issued in 1939. According to one reviewer, it was "already one of the most influential books in the development of fundamental theory of modern physiology." In the years between, Cannon spoke several times on the subject—for example, when he delivered the Mary Scott Newbold Lecture in Philadelphia late in 1934 on "Stresses and Strains of Homeostasis," and again at the New York Academy of Medicine a year later, when he took part in a lecture series on the Art and Romance of Medicine. A *New York Times* editorial of October 5, 1935 reported, "The opening lecture by Professor Cannon of the Harvard Medical School on 'The Wisdom of the Body,' before an audience that overflowed the lecture hall of the Academy, gave illustrations not only of how 'fearfully and wonderfully' we are made, but by what an intricate and seemingly miraculous mechanism the stability of our bodies is maintained."[22]

In 1938, Cannon again returned to the subject of steady states—this time when he accepted an invitation to give the William Welch Lectures in New York, one on new aspects of homeostasis, and another on homeostasis in senescence. Two years before, he had conducted a study of the effects of sympathectomy on the heart rate and muscular performance in the dog with Lucien Brouha, a C.R.B. fellow, and D. Bruce Dill from the Harvard Fatigue Laboratory. This work, which responded to observations by Heymans, Bacq and Brouha in Belgium that sympathectomized dogs differed in their behavior from sympathectomized cats, became the subject of Cannon's first Welch Lecture.[23]

The second of the Welch Lectures was similar to a chapter on ageing of homeostatic mechanisms that Cannon contributed to a volume edited by E.V. Cowdry, which in turn grew out of a conference called by the Macy Foundation to consider a scientific approach to the problems of the ageing process. Cannon's daughter Wilma had come to the second lecture and, he wrote Bradford, cheered him by her presence in the audience that evening. He also told him about a New York businessman, Elisha M. Friedman, "a personage of considerable achievement," who was impressed by the epilogue of *The Wisdom of the Body* and had sent copies to his friends, including Secretary of Agriculture Henry A. Wallace. Friedman wrote articles on the relation of biology to social organization in which he used illustrations from Cannon's writings. Cannon found him alert, well-informed and highly stimulating; the two men entered into an association in which they exchanged ideas that would prove to be mutually beneficial.[24]

Toward the end of 1935, Cannon had plans under way to begin work on two new books. "Dr. Rosenblueth and I are working up 'sympathin' for a volume to appear in a series of biological monographs, under the title, 'The Autonomic Nervous System and its Effectors,'" he wrote his son. "And besides, a few weeks ago Mr. Norton asked me if I had anything I'd like to publish, and I told him about the little volume of Beaumont Lectures which was never used . . . I am thinking that the title 'And Inwardly Digest' or 'Let the Good Digestion Wait' might be good." Six months later, he was reading proof on the popularized advice book for Norton, now called *Digestion and Health,* but had slow going with the other book over the summer months of 1936.

In late fall, however, he reported triumphantly: "I've finished preparing the first nine chapters of the book on 'Neuro-effector Systems.' Dr. R. has four chapters to top them, and then I enter again with a final fourteenth chapter. It all ought to be ready by the middle of November." The final chapter that Cannon wrote traced the development of knowledge about sensitization of autonomic effectors to chemical mediators, beginning with two unsolved problems handed down by nineteenth-century physiologists: the paradoxical pupillary reaction and the Vulpian phenomenon. Over time, both of these problems had been found to be related to chemical mediation of nerve impulses, their explanations laying in the increased sensitiveness of denervated structures to circulating chemical agents.

The following spring, Cannon wrote Alan Gregg that his monograph was approaching publication. He pointed out that practically all of the work done in his laboratory, as described in the book, had been supported by the grant from the Rockefeller Foundation, and that indeed, it was Dr. Pearce's interest in the early observations that drove him to support the researches in the first place. Since Cannon was hampered by Rockefeller policy from publicly expressing his gratitude for the aid of the Foundation, he had decided to dedicate the book to the memory of Richard M. Pearce.[25]

Autonomic Neuro-effector Systems, published in 1937 by Macmillan in their Experimental Biology Series, was reviewed with comments that ranged from glowing admiration to wary respect. Three reviews that appeared in English journals are of particular interest. One in the *British Medical Journal* began warmly, "No laboratory has been more in the forefront of these investigations, particularly as regards the adrenaline-like mediators, than that of Professor Cannon, and it is a pleasure to have available in the form of a monograph an account of the most important investigations in the whole field, and written by workers who have themselves made history in this branch of research." The reviewer went on to issue a caveat that most British authorities would probably not be prepared to follow the authors in their postulation of sympathin E and I, but it would seem wise to postpone judgment pending further work. "That the outcome will be of great importance to medical science, as to general biological concepts, cannot be doubted."

Another reviewer writing in *Lancet* found the book to be "as exciting as a detective story." The discovery of the role of chemical substances in the conduction of nerve impulses, in his eyes, was one of the most fascinating and important advances in neurophysiology. Citing the work of Elliott, Cannon and Dale as milestones in the history of this discovery, he noted with relish that the subtlety of the investigations and the inferences drawn from them "make one feel that Dr. Watson has at last scored off Sherlock Holmes."

Neither of these reviews was signed. A third review in *Nature* was by "J.H.G."—probably John Gaddum. "A monograph has now been published which gives, for the first time, an authoritative exposition of the views of the Harvard School," he stated. "The authors have both written many papers describing investigations of the properties of adrenergic nerves. They have shown that stimulation of these nerves causes the liberation of at least two different substances. The clear presentation of the evidence for this conclusion forms the most valuable and impressive feature of the book." But this reviewer found the interpretation of this evidence to be problematic. "The theory that drugs combine with a special receptive substance in the motor organ provides the only reasonable explanation of the specificity of drug action, but Cannon and Rosenblueth's theory that this receptive substance is liberated with the drug into the blood stream seems unnecessarily complicated, and has met with opposition." Calling the book argumentative and dismissive of the theories of other investigators, "J.H.G." in the final analysis noted that it was, in fact, likely to lead to much discussion.[26]

Given the extent of Cannon's increasing involvement at this time in organizations and causes that pulled him away from the laboratory, it might come as a surprise that his research continued to be productive. Indeed, it is rare, in the assessments of Cannon's scientific contributions, for more than passing mention to be given to the final phase of his physiological research—perhaps because his work on supersensitivity of denervated structures remained unfinished at the time of his death. Many would attribute the authorship of the monograph that was finally published in 1949 more to Rosenblueth than to Cannon.

Cannon's facility for seeing the myriad facets of a physiological problem led him back to an observation he had made years before: the development of heightened sensitiveness of the heart, after sympathetic denervation, to circulating adrenin (and later sympathin). Extensions of this observation were to provide him with many examples of sensitization to chemical substances by denervation, peripherally in effector organs and in the ganglia of both divisions of the autonomic nervous system, and centrally in the spinal cord. As Philip Bard has written, Cannon was on the track of something fundamental; it was physiology's loss that he was unable to follow through.[27]

Starting in the mid-1930s and continuing for the next several years, Cannon's research group embarked on a series of studies dealing with the sensitization of denervated structures and the formulation of a general law of denervation. The understanding that he and his associates sought to achieve (less than perfect prior to knowledge about the mechanism of uptake and reuptake) was to provide penetrating insights for what later developed into a major field of physiological inquiry.

In 1935, a step toward understanding the sensitizing effect of sympathetic denervation was taken by Chester Hampel, a postdoctoral fellow in Cannon's department, who traced the course of increased sensitivity of smooth muscle after severance of its sympathetic fibers and demonstrated that postganglionic denervation led to greater sensitization than preganglionic denervation. That same year, in a research conducted in China with S.C. Shen, Cannon found that after removal of the ciliary ganglion, the constrictor muscle of the iris became highly sensitized to acetylcholine, the chemical mediator of nerve impulses in sympathetic ganglia.[28]

Since observations on both smooth and skeletal muscles suggested that denervation augments their response to the natural chemical mediators of the nerve impulses affecting them, Cannon reasoned that there might be a general law that denervation increases sensitiveness to stimulating agents. The interesting question, for him, was whether this would prove true also for nerve cells, and he aimed his research over the next several years toward that end.

After performing experiments that utilized the nictitating membrane of the cat as an indicator, Cannon, in collaboration with Rosenblueth, concluded in 1936 that acetylcholine, either injected or applied locally, has a much greater effect in stimulating the superior cervical ganglion when chronically deprived of its preganglionic innervation than when it is acutely denervated. (Though this finding was later disputed by a worker in Henry Dale's laboratory, Cannon and Rosenblueth were able to successfully replicate their results.) The authors attributed the enhanced sensitiveness to a greater permeability of denervated structures to various stimulating agents.[29]

Near the end of 1936, Cannon wrote his son, "Dr. R[osenblueth] and I are working on the physiology and pharmacology of the sympathetic ganglion, with the hope that it may be preparatory to going into the spinal cord." But not until more than a year later was there a breakthrough. "Suddenly yesterday experiments which had been dragging sadly turned to a revelation that was really exciting," he exulted. "It looks as if we might have the beginnings of evidence for chemical mediation in the cerebral processes. Time will tell, of course; but just now I can hardly wait for Monday to come with the opportunity to take the next step in the quest."

When Fiorindo Simeone's experiments suggested that the denervated nictitating membrane of the cat is hypersensitive to the first nerve impulses reaching it after regeneration of the nerve fibers, Cannon set out to determine whether the partially denervated ganglion would similarly become sensitized to nerve impulses. In the spring of 1938, an article by Simeone, Cannon and Rosenblueth in the *American Journal of Physiology* reported that it does. Moreover, the authors believed that their demonstration of sensitization within neuronal systems had important bearing on the theories of synaptic transmission, inasmuch as the electrical theory did not appear to account satisfactorily for the increased sensitivity of the partially denervated ganglion to nerve impulses, whereas the chemical theory did.[30]

Work on sensitization was not the only research that appeared to go forward in fits and starts. Nowhere was this more evident than in the investigation that Cannon had undertaken with Harry Friedgood on his old nemesis, the thyroid. In the late autumn of 1935, Cannon reported that Friedgood had been able "to produce ovulation in the rabbit by cervical sympathetic stimulation—an anterior pituitary effect." Though it looked as if they had the thyroid under control in the same way, the work was still in progress.

Soon after, Cannon became excited over the appearance of a cat with phrenic-cervical sympathetic anastomosis that showed the same signs he had described many years ago. "The animal eats voraciously and is losing weight," he observed. "She is very restless, and furtive, just as the animals were in 1914. The probability is that the effect is produced via the anterior pituitary acting on the thyroid, and not directly on that gland." Before long, however, expectations were dashed when, to their great disappointment, "the cat with high metabolism (phrenic sympathetic anastomosis) died in the metabolism chamber for some reason not discovered." Undeterred, Friedgood and Cannon began preparing other animals in the conviction that the results they had obtained were too significant to neglect.

One year later, in December 1936, Cannon wrote Bradford: "It looks very much as if we had had success again in joining the phrenic with the cervical sympathetic. Two animals with the metabolism raised far beyond the extreme variations of normal metabolic rates. I think I've learned the trick of uniting the nerves—much greater care in approximation than we have given hitherto." Still later, Cannon reported that the wide range of influence of the anterior lobe of the pituitary in actions on the thyroid gland, the adrenal cortex, the ovary, and possibly the pancreas rendered the recent work done in his laboratory of utmost importance.[31]

In the last years of the 1930s, scientific controversy was much on Cannon's mind—though he had little taste for entering into conflict himself. He chose to present his views on the matter to a young audience, members of the medical honor society Alpha Omega Alpha—once in an issue of *Pharos* published in May 1938, and again, to a chapter at the University of Minnesota in 1939. His article, "Scientific Controversy: Its Possible Harm and Value," has as its centerpiece a reprise of the debate, primarily with Stewart and Rogoff, over adrenal secretion. Although, he contended, "one now hears relatively little" of the conflict of evidence, and the conclusions of the Harvard group had received support from various quarters, curiously enough two false impressions continued to persist. First, because conditional control of adrenal secretion had once been in question it was still in question, and second, because the adrenal medulla may be excluded from action without causing death, its usefulness to the organism was still denied.

"The harm which may result from controversy," Cannon warned neophyte investigators, "includes not only neglect of unfavorable evidence, confusion and misjudgment; it may cause a certain blindness to the significance of facts which are quite evident and which invite further study." Nevertheless, he was hopeful: "When differences between investigators result in fresh efforts to learn the truth, in refinements of technical processes, in more discriminating experiments, in more alert attention to questioned phenomena, in more cautious inferences—instead of what Harvey called 'pretty disputations'—the chances are favorable, I believe, that a polemic may do more good than harm."[32]

In 1939, when Cannon was invited to speak at the fiftieth anniversary celebration of the founding of the University of Minnesota Medical School, he used the occasion to address a controversial matter he had long shied away from. In "The Argument for Chemical Mediation of Nerve Impulses," he discussed and compared the electrical theory and the newer chemical theory of synaptic transmission of nerve impulses. "Each theory has its ardent advocates, and a vigorous controversy prevails between them," he told his audience. "For purposes of the present discussion these advocates may be called the 'electragonists' and the 'chemagonists'—'agonist' meaning contestant or combatant!"

Briefly skimming over the evidence for electrical transmission, Cannon focused on the arguments of John Eccles, a student and then collaborator of Sherrington's at Oxford, and others. The better part of his talk, however, he devoted to evidence for the chemical theory, relying almost exclusively on the observations of Arturo Rosenblueth, not only with himself but also with J.V. Luco, K. Lissák and A. Lanari, among others. What Cannon did not convey in his talk was the heated nature of the debate.[33]

Eccles was at this time the leading proponent in Britain of the electrical nature of synaptic transmission and a major opponent of Dale and others who championed chemical mediation—the controversy sometimes being referred to whimsically as "soup versus sparks." (In America, Alex Forbes some years before had playfully pinned the tag "axonologists" on physiologists who worked on nerve action potential; the possibility that synaptic transmission might be chemical in nature he dubbed "soup at the synapse.") As early as 1934, Eccles had taken exception to what he perceived to be discrepancies in an article by Rosenblueth and David Rioch on "Temporal and Spatial Summation in Autonomic Systems." The paper was rich in mathematical analyses and Eccles, finding disagreement between the figures and curves, called Rosenblueth to account.

Forbes, who knew Eccles well, had rushed to Rosenblueth's defense and asked for details of the criticism. "I do not of course doubt the existence of chemical mediators—at least for postganglionic endings," Eccles replied to Forbes, "but I do question the experimental evidence which supports the hyperbolic relationship between the responses of autonomic systems and the frequencies of stimulation of the corresponding nerves." He then proceeded to rake Rosenblueth and a number of statements in several of his articles "over the coals." Adding that other British physiologists felt as he did about Rosenblueth's unreliability, he recommended "a course in Langley to Rosenblueth before he does any more theorizing."[34]

Despite what appeared to be a lighthearted touch in a note to Rosenblueth headed "Eccles Heckles," Forbes had a high regard for his formidable opponent and took great care in drafting a lengthy letter, which countered each and every complaint that had been put forward. Eccles backed off from further debate, but he continued to consider Rosenblueth a seriously misguided antagonist and the affair did not help the latter's scientific reputation—or Cannon's, for that matter.

A few years later, Eccles published an article in which he was sharply critical of the work on sympathin. So ferocious was his attack that Henry Dale (though critical of the work himself) felt compelled to apologize for his "sparring partner" in a letter to Cannon of February 1937. But by the mid-1940s Eccles had become converted to the idea of chemical mediation, though he held fast to his belief in the electrical nature of neurotransmission. He went on to share a Nobel Prize in 1963 with Hodgkin and Huxley for "their discoveries concerning the ionic mechanisms involved in excitation and inhibition in the peripheral and central portions of the nerve cell membrane."[35]

When the academic year began in the fall of 1938, Cannon was hopeful that several of the investigations he had under way were about to deliver their promise. In particular, he intimated that an experiment he had undertaken with Koloman Lissák showing the presence of adrenaline in sympathetic nerve trunks would uncover important new possibilities. Early in 1939, he wrote optimistically to J. T. Lewis that many of his projects had at last come together:

> Work here in the Laboratory has gone on quite favorably. A young Hungarian investigator (an International Rockefeller Fellow) and I are just publishing a paper showing that adrenaline is found in all parts of the ultimate sympathetic neurons. The term "adrenergic fibers," therefore, is quite appropriate. Also, I have just finished another research with a young French Rockefeller Fellow [Haimovici] showing that neurones in the central nervous system are sensitized to chemical agents when they are partially deprived of their connections with other neurones, i.e., partially denervated. We now have enough evidence, I think, to say that there is a law of denervation. At least I am proposing to use that as a title in giving the Hughlings Jackson Lecture in Montreal in April, when I shall summarize all our evidence.

The lecture was, in fact, a sequel to Cannon's final chapter in his 1937 monograph on autonomic neuro-effectors, and presented an overview of work from his laboratory on the increased sensitiveness to chemical stimuli appearing in structures that were deprived of their proper nervous connections. In it, he proposed "a law of denervation" and defined it thus: "When in a series of efferent neurones a unit is destroyed, an increased irritability to chemical agents develops in the isolated structure or structures, the effect being maximal in the part directly denervated."

To illustrate his law, Cannon undertook a discussion of the responses of smooth muscle, of glands freed from nervous control, of skeletal muscle, and of nerve cells to chemical stimuli. He ended by invoking Jackson's interest in epilepsy and suggested there was work to be done on cortical cells that might further illustrate the law of sensitization of denervated structures. "You asked about the reception of the Hughlings Jackson Lecture in Montreal," Cannon wrote his son in May. "Well, there seemed to be a willingness to listen to it, and later [Wilder] Penfield told me that it opened a new approach to problems of epilepsy on which he and his collaborators have been working. I'm going into that realm, myself, next fall."[36]

Cannon ended the decade brilliantly, with 1939 a banner year for his scientific publications. His "Law of Denervation" and the revised edition of *The Wisdom of the Body* provided integrating concepts that contributed in a fruitful way to scientific thought. He also completed reports of his research with Rosenblueth on the effects of preganglionic denervation on the superior cervical ganglion; with H. Haimovici on the denervation of motoneurones in the spinal cord; with K. Lissák on the presence of adrenaline in adrenergic neurones; and with L. Brouha and D.B. Dill on the regulatory mechanism affecting blood sugar in sympathectomized dogs (a follow-up of their 1936 report).

Cannon argued in his article with Lissák, "The results are consistent with the view that adrenaline is liberated at the endings of adrenergic fibers, and that sympathin, which escapes from the stimulated region into the blood steam, is this adrenaline modified in the affected cells." In an address delivered at the New York Academy of Medicine that featured the findings of Lissák and himself, he was quoted by the *New York Times* as stating, "The evidence strongly indicates therefore, that Elliott's suggestion, made in 1904, is correct, that the ultimate sympathetic neur[ons] influence glands and cardiac and smooth muscle by the adrenaline which they set free at their terminals, and that adrenaline secreted from the adrenal medulla and circulating in the blood stream has on these structures the same influence as the nerve impulses because it is the same substance as that which they produce."[37]

Cannon never passed up an opportunity to give credit to Thomas Renton Elliott for his ground-breaking research on chemical mediation. In the sporadic but cordial correspondence they engaged in throughout the decade, Cannon, an enthusiastic admirer of Teddy Roosevelt, appeared to take great pleasure in addressing his letters to "My dear T. R." Of all the friendships that Cannon held dear over the years of his career, Elliott occupied a unique place. Surely he deserves a few more lines in this biography.

In the summer of 1930, when Cannon was in London faced with a medical emergency, the British clinician had given sound advice and, equally important, a boost to his morale. The two men in their letters (and one further meeting) regularly compared notes on their respective health problems, with Elliott, who suffered from lumbago and sciatica, often writing while confined to his bed. When he attributed his troubles to old age, Cannon observed that investigators in their active years paid little attention to the changes involved in growing old. He looked forward, however, "to a time when we shall have 'professors of geriatrics' just as we now have professors of pediatrics. Then there will be experts to take care of such as we."[38]

Elliott, firmly committed to the practice of medicine, nevertheless kept up with the developing story of chemical mediation—particularly through reports emanating from the laboratories of Henry Dale and Walter Cannon—and always had an incisive comment or two at the ready. At the end of 1930, when Cannon wrote him excitedly about the discovery of sympathin, Elliott responded with ideas for further experiments or suggested reading. However "marvellously exciting" he found Cannon's physiological findings, he was even more impressed by work that had direct clinical applications. His imagination was fired when Harvey Cushing wrote him that the tiny pituitary was proving to be the "absolute monarch of so many and diverse functions in the kingdom of the body."

Subsequently, in response to news of the two sympathins, Elliott told Cannon how pleased he was to hear "of vivid progress by you, a friend of 30 years, in that field of physiology where we started together." Having said that, he again expressed his fascination with Cushing's work, this time a recent study of the clinical picture associated with basophile pituitary tumour. "Observations in medicine have given us wilder fairy tales than physiology in her reasoning mood could ever have constructed," he declared. "Medicine is the more romantic field." He ended his letter by straddling both worlds: "Still, my affection remains with physiology and I like your work best of all."

Later in the decade, Cannon informed Elliott of two recent papers coming from his Laboratory that he wished he could dedicate to him:

> They are concerned with the evidence of chemical events at the motor plate and skeletal muscle and at the synapses in sympathetic ganglia. A very close resemblance between the occurrences in these two places is completely in accord with the suggestion which you made about thirty years ago—when we were both young! In telling the story of the transmission of impulses through a sympathetic ganglion I have mentioned to my audience your brilliant early work in physiology and have said that it was a tragedy that you were not persuaded to remain a humble physiologist instead of going onward and upward into clinical medicine. Anyhow, it must be a real satisfaction in these later years to see your ideas so completely confirmed.[39]

Top, Cannon and Otto Loewi, Dunham Lecturer, 1933; bottom, Bruce Dill and Lucien Brouha, colleagues at the Harvard Fatigue Laboratory, 1936.

Cannon in the 1930s: This page, portrait for 25th anniversary celebration as George Higginson Professor at H.M.S., 1931, and with Chim and Panzee, two of Robert Yerkes' chimps; facing page, in Chinese costume on visit to Peiping Union Medical College, 1935, and pencil drawing by his daughter Wilma.

International Congress of Physiology, Leningrad, 1935: This page, Cannon delivering speech at opening session, and group photo with Erna and Fred Gibbs, A. C. Liu, C. M. Brooks, Zenon Bacq; facing page, group photo with W. Feldberg, Otto Loewi, Zenon Bacq, C. Bykov, and Dr. and Mrs. Cannon at banquet.

Top, Dr. Joseph C. Aub in his office at the Huntington Hospital, Boston. Bottom, Dr. and Mrs. Cannon with Bradford, St. Louis, 1937.

Some international figures in physiology: From the top, Otto Loewi, Ernst von Brücke, Robert K. S. Lim, Juan Negrin, B. A. Houssay.

Cannon and grandchildren at the family farm in Franklin, N.H., summer of 1944.

Portrait of Cannon, showing ravages of illness and aging, in the last year of his life.

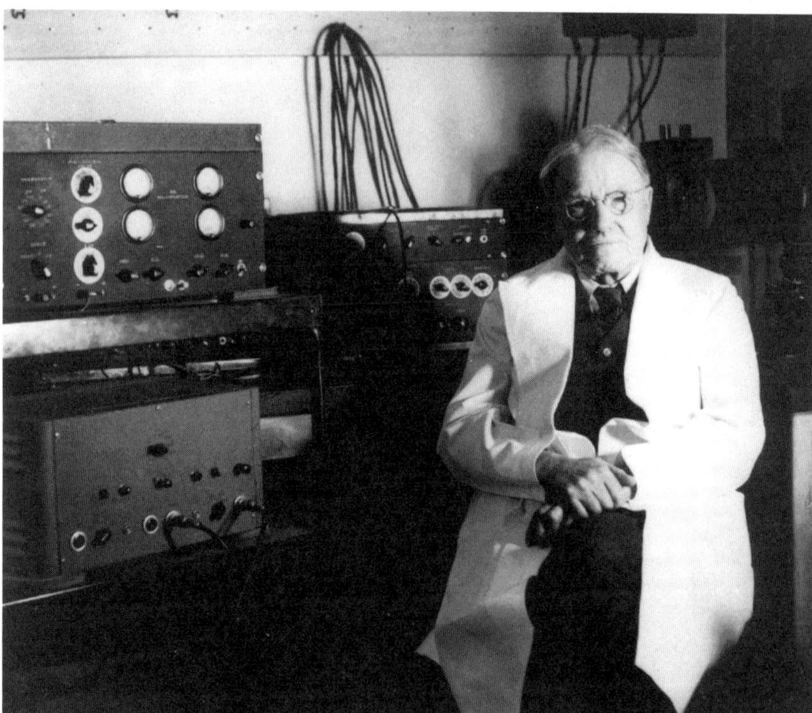

Cannon in Mexico City, 1945: Top, with Arturo Rosenblueth and Juan Garcia Ramos; bottom, seated in front of a kymograph in Dr. Rosenblueth's physiology laboratory.

21 . . . to the Body Politic

BY THE EARLY 1930s, Cannon had become the most senior member of the Harvard medical faculty by virtue of his years of tenure. As such, he was called on not only to act as spokesman for the faculty but also for service to the Medical School and the University at large. The dean's office came to regard him as one of its most trusted advisers, a man whose counsel on knotty problems brought both wisdom and experience to the table. Whether for matters of curriculum or admission, personnel or budget, students or teachers, it was Cannon to whom the dean most often turned when he was faced with setting or changing policy at the School. And Cannon rarely disappointed; he was for David Edsall and others a rudder that held steady through good times and bad.

An issue that Cannon played a key role in resolving surfaced early in the decade. It was a complicated matter, and it involved not only the Medical School but also the Division of Medical Sciences, the Faculty of Arts and Sciences, and Radcliffe College. As one of the initiators of the Division of Medical Sciences in 1908, Cannon had maintained an interest in the affairs of the program, not least because of its close relations with the Faculty of Arts and Sciences, which awarded the Ph.D. degree. His long-standing allegiance to Radcliffe went back even further. As a student, he had tutored and lectured in zoology there; not only was his wife a graduate of the college but three of his daughters became Radcliffe students. Moreover, Cannon sustained his department's tradition of being open to women—the precedent established in the 1890s when Ida Hyde, the first female member of the American Physiological Society, was given a place to work in the physiology laboratory. It is not surprising, therefore, to find him engaged in a heated debate that raged over the granting of Ph.D. degrees to Radcliffe graduate students during 1932–1933—the final year in the presidency of A. Lawrence Lowell.[1]

Early in December 1932, Dean Edsall wrote Cannon and others that President Lowell had raised questions about Radcliffe women taking courses in the medical sciences at the Medical School. It was his understanding and the Harvard Corporation's, Lowell pointed out, that it was desirable to train these women to be of direct value to the School as assistants in research there. Their acceptance, however, should not be taken as a right on the part of Radcliffe, or because they might be of use elsewhere after their training. Lowell also expressed the opinion that female students should not be admitted to the same courses as the male medical students but should be instructed in a distinct and separate way.

Cannon replied almost immediately to Edsall's letter and sent a copy to Ada Comstock, the president of Radcliffe, to whom he was both friend and adviser. He began by going over the ground rules whereby professors in the laboratory departments of the Medical School (and of the School of Public Health), who trained students for higher academic degrees, acted in accordance with their colleagues in the Graduate School of Arts and Sciences. And he testified that in his experience in the physiology department, men and women students had worked in the same room and had uniformly behaved in a thoroughly respectable manner. As for Lowell's idea of training women in the medical sciences solely for positions as research assistants at the Medical School, Cannon called it "an impossible situation." If anyone received an offer for more favorable opportunity elsewhere, he argued, "I should not wish, nor know how, to prevent her from leaving. It seems to me quite as unwise in a University as in the State to make laws which cannot be enforced."[2]

When Edsall at a subsequent meeting of the Administrative Board (of which Cannon was a long-standing member) brought the matter to open discussion, the Board declared that their understanding was very different from that of Mr. Lowell. They were unanimous in their desire to continue to train superior women. No department had ever recommended that female students be taught separately from the medical students; indeed, they all agreed that such a thing would be both expensive and impossible. When the dean revealed that Lowell had suggested that a Doctor of Science degree under the Faculty of Medicine might replace the Doctor of Philosophy in Medical Sciences awarded by the Faculty of Arts and Sciences, the suggestion encountered vigorous opposition. Hans Zinsser, for one, voiced the prevailing view that it would be a serious detriment to the departments of the Medical School if they were deprived of the right to accept candidates for the Ph.D. degree in their respective disciplines.[3]

The Cannon family followed the tug-of-war with a great deal of interest. Cornelia Cannon, an ardent alumna of Radcliffe, never forgot her gratitude to her alma mater. Not only did she send her daughters there, she also raised funds for the college and donated some of her earnings as lecturer, monologist and writer. Throughout the course of the battle, she gave a blow-by-blow account in her family letters, and especially to her daughter Wilma in China.

The first inkling of trouble was when she and Cannon had paid a holiday call on Miss Comstock "and heard about Mr. Lowell's latest efforts to put Radcliffe in its place before his resignation takes effect. This is all sub-rosa, but he would certainly like to shake off Radcliffe . . . Mr. Lowell is evidently afraid of the weakening and demoralizing effect of the female sex." Early in 1933, Cornelia indignantly reported the president's "latest effort to knife Radcliffe." She was scandalized that he wanted to "throw her out, divorce her from Harvard, and try to get the Medical School to refuse to allow women to come up for the Ph.D. degree in physiology, etc." When the professors flatly refused, she continued, he tried to "abolish the Ph.D. at the Medical School! It is the most astounding idea, like taking a sledgehammer to kill a flea." In March, she related that the teachers of Harvard were "fighting tooth and nail in defense of the education of women" and that Cannon was "on the battle line, I promise you."[4]

In the weeks that followed, members of the Division of Medical Sciences committee from both the Medical School and the Graduate School of Arts and Sciences met with President Lowell to address some of the issues he had raised. The core of their concern, as Cannon delicately put it, was to establish closer and more helpful relations between the scientific departments of the Medical School and similar departments in Cambridge than had previously existed.

Addressing his colleagues' criticism that the group in Cambridge knew little of conditions in Boston, he suggested two actions. One was that someone from the Medical School (like Hallowell Davis of the physiology department, who had served for some time as secretary of the Division of Medical Sciences) become a member of the Administrative Board of the Graduate School of Arts and Sciences. The other was to invite George Chase, dean of the Graduate School, to come and see for himself what was going on at the Medical School in the way of training graduate students in the sciences.

Cannon's proposal helped to resolve whatever sticking points remained. After receiving a corroborative statement from Edsall and then speaking with Lowell, Dean Chase prepared to go ahead under existing conditions as before. On March 18, Cannon sent Ada Comstock a transcript of Edsall's statement, so that she might complete her record of the actions that led to what was likely to be the final surrender of the attempt to deprive the Division of Medical Sciences of the right to recommend the Ph.D. degree.[5]

Not long after the close of this incident, Cannon was asked to speak in behalf of the professors at President Lowell's last faculty meeting at the Medical School. His remarks at the meeting held on June 16, 1933, do not appear to be on record. But we have Mrs. Cannon's word for it that her husband "planned a charming, witty speech, full of words about the fine things [Lowell] has done and soft-pedalled the less admirable things which he has tried to do but failed in!" If Cannon indeed carried out his plan, he would have followed in the footsteps of Dr. F. C. Shattuck, who played a similar role in 1909 at Charles W. Eliot's last official meeting with the medical faculty. While Shattuck said it was a privilege to pay homage to the president and proceeded to praise him to the rafters, he did not let Eliot off scot-free. One thing is certain: however Cannon structured his remarks in 1933, they would have been delivered with as much grace and courtesy as Fred Shattuck had summoned up almost twenty-five years earlier.[6]

That summer Ada Comstock, in recognition of Cannon's staunch support during the recent battle, wrote him that he had been appointed to the Board of Trustees at Radcliffe. By the time Cannon read her letter, all eyes were on the new president of Harvard, James B. Conant.

Years later, Conant would recall, "I tell my friends that 1933 was quite a year for Germany, America, and me—Hitler rose to power, Franklin Roosevelt took office, and I became president of Harvard." In the autumn, at ceremonies held in Sanders Theater celebrating the 150th anniversary of the founding of the Medical School, Cannon for the first time heard Conant speak. The president, Cannon wrote his son Bradford, spoke clearly and impressively of the old and the new medicine, and of his hope that the Cambridge and Boston departments of the University would draw more closely together.[7]

When it came to calls for service to Harvard, whether marking occasions, both happy and sad, or committee work, Cannon was an odds-on favorite. The testimony of his peers bears witness to the value of his presence. Dr. Edsall, in appointing Cannon head of a committee in 1932 to select members of a new Administrative Board, wrote him, "There can be no question in regard to yourself. Everybody would insist that you continue as a member of the Board, so that you will certainly have to recommend yourself." In the spring of 1933 when Cannon was worn down by continuing health problems, he asked to be relieved of some of his committee assignments. Edsall hastened to reply, "It is difficult to get a really suitable and effective person to take your place, but that is always the case when a good man leaves any work, and I believe that the relief of your multiple duties is important enough to demand that we get somebody else."[8]

No matter what the circumstances, Cannon rarely turned down an opportunity to pay tribute to old friends and colleagues. In May of 1933, for instance, he spoke at a dinner marking Harvey Cushing's retirement at which portraits of both Cushing and Henry Christian were presented. Some weeks later, he wrote Cushing, "I have just seen in the *New York Times* the news that you are going back to New Haven. Although we have lived in the same neighborhood for many years, I have not seen you nearly so often as I have wished. Nevertheless, it has been a satisfaction to know that you were close by. The thought of your leaving, therefore, comes with a wrench."[9]

Cushing's retirement was but one in a series of losses at the Medical School. The following year came the shock of Otto Folin's death. "I know that you will be eager to learn about the circumstances of Dr. Folin's unexpected death," Cannon wrote a former student of Folin's in October 1934. It all began with an infection picked up during the summer, he related. Cannon had gone to Folin's home to tell him about plans for a celebration in his honor at the Medical School and the arrangements for the presentation of his portrait.

Cannon continued that Folin had decided to go to the hospital for a minor operation, and did quite well until his blood pressure began to drop suddenly. He became unconscious and passed away quietly the next day. Folin's untimely death was mourned by the entire medical faculty, none more than Cannon; he was especially touched to learn from Mrs. Folin that her husband had regarded him as his dearest friend. Again, Cannon presided over the unveiling of a portrait, but this time it was an occasion that had sadly turned into a memorial service.[10]

Still other losses followed with the retirements of Milton Rosenau as professor of preventive medicine and David Edsall from the deanship. Edsall's retirement might have given Cannon an opportunity to ease up on his committee work and advisory duties, but apparently such a thought never entered his head. "I have just heard the news that you have been appointed the new Dean of the Harvard Medical School," he wrote C. Sidney Burwell early in 1935, wishing him success in his new venture. "I have already served under four deans during my experience as a teacher here, and if my knowledge of the past and my judgment can be of any use to you I shall be very glad indeed to help."[11]

By the time Cannon returned from his trip around the world, Burwell was on the job. The incoming medical dean was to work closely with the young president during his first years in office, for a new and novel spirit of cooperation had taken hold among the various faculties at Harvard. That new spirit was a motivating force when it came to planning one of the most sumptuous university-wide events ever held—the celebration of Harvard's 300th anniversary. And, much to Cannon's satisfaction, his old friend Jerome Greene was called back to Cambridge to direct the entire operation.

The main event of the Harvard Tercentenary held in September 1936 was a two-week Conference of Arts and Sciences, with a gathering of about 2700 expected to attend. According to the program, "Seventy-two eminent scholars from fifteen different countries are to present the principal papers and lectures of the Conference." Three of the five symposia scheduled for discussion had been selected "as important and timely problems which invite an approach from the standpoint of different disciplines." To Cannon went the signal honor of chairing the first session in a symposium on "Factors Determining Human Behavior" held in Sanders Theater, which featured talks by E.D. Adrian, J. B. Collip, Jean Piaget, and C.G. Jung.[12]

One week later, at a Tercentenary session held at the Medical School, Cannon chaired another symposium, this time on "The Nervous System; Central and Sympathetic." After Cannon's introductory address (later published as "Experimental Neurology in the Harvard Medical School"), an impressive array of speakers showcased the interdisciplinary spirit at the School. James C. White spoke on surgery of the sympathetic nervous system, Arturo Rosenblueth on chemical mediation of nervous effects, James B. Ayer and Henry R. Viets on use of prostigmin and other drugs in the diagnosis and treatment of myasthenia gravis, Hallowell Davis on electrical activity of the human brain, Frederic Gibbs and William Lennox on electrical activity of the brain in epilepsy, Stanley Cobb on cerebral circulation, Soma Weiss on syncope and collapse, and Tracy Putnam on pathogenesis of multiple sclerosis.[13]

Cannon scarcely had time to recover from the exhausting round of Tercentenary exercises before learning that he was expected to participate in another of Harvard's ceremonials, one of her grand old traditions. Near the end of 1936, President Conant sent Roger I. Lee, a member of the Committee on Honorary Degrees, the names of seven Harvard professors who had never received honorary degrees from the University, and whose distinction was such that he felt they deserved recognition. From this list Cannon was the one chosen to be honored in 1937.

Early in the new year, when notified of his selection, Cannon wrote Conant of his pleasure, remarking, "After all a Mother's commendation is best—she knows the defects as well as the merits! I shall place the note among my most private papers, but I shall not forget it." When the honors were handed out at the Commencement held on June 24, Cannon was the recipient of a Doctor of Science degree. Among other honorees, Robert Frost was awarded the Doctor of Letters and Jerome Greene, together with Cardinal O'Connell, the Catholic Archbishop of Boston, received the Doctor of Law degrees.

For Cannon, one of the happiest aspects of the affair were the warm wishes of his colleagues at the Medical School—among them,

George Minot, Henry Christian, Hans Zinsser and Arlie Bock. "Your degree was fitting and highly merited and adds dignity to the whole institution of honorary degrees," observed Zinsser, while Bock fondly noted, "The example you have set for all young men is now topped off . . . I admire your enthusiasm and spirit and hope it continues to guide you through many another year."[14]

Whenever the Medical School lost a member of the faculty—whether by death or retirement or transfer to another institution—it meant, of course, that a replacement had to be found. During the thirties, there was an unusually large turnover at the School, with many of Cannon's old friends on the faculty fading away to be replaced by a new generation of younger men. Time after time, he was asked to sit on ad hoc committees convened to consider candidates for this or that professorship. Convinced that a school is only as good as its teachers, he accepted such calls as a sacred duty and did his utmost to insure that Harvard stayed among the premier medical institutions in the world.

For example, in surgery early in the 1930s, Cannon took part in the selection of both Edward D. Churchill to replace the ailing E.D. Richardson at the Massachusetts General Hospital and Elliott Cutler to succeed Harvey Cushing when he retired from the Moseley Professorship at the Peter Bent Brigham. Each of these men more than fulfilled the search committee's expectations of keeping Harvard in the top rank. A few years later, President Conant named Cannon chairman of a Committee on Biochemistry, not only to consider a replacement for Otto Folin but also to chart the future direction of biochemistry at the Medical School. Although Cannon relinquished his place on the committee when he went abroad in 1935, Joe Aub kept him informed of the deliberations, which resulted in bringing A. Baird Hastings to Harvard as head of the department and promoting Edwin Cohn to full professor.[15] It was not long before Cannon had a hand in another key appointment that would give added lustre to the medical faculty—that is, naming Otto Krayer to direct the School's program in pharmacology.

One of Sidney Burwell's first actions as dean of the Medical School was to recommend that the president appoint a committee to consider the future of the pharmacology department. The professor, Reid Hunt, a very distinguished pharmacologist since the beginning of the century, had not been productive of late and was ready to retire. The associate professor in the department was Worth Hale, who for twenty years had not published much, if any, and also served as assistant dean at the Medical School. The committee would be asked to consider a change in the department's goals as well, so that henceforth in considering the action of drugs on tissues, the emphasis would be not on the drug but on the tissue.

In October 1935, Cannon was named chairman of the committee, with Baird Hastings, Fred Hisaw, Soma Weiss and George Wislocki as members. Wasting little time, the new committee, propelled by Cannon, produced a plan for the development of pharmacology and experimental therapeutics, which called for a university-wide approach involving biology and chemistry at Harvard College, pharmacology at the Medical School, and therapeutics at the affiliated hospitals. Cannon was well pleased with his handiwork and sent a copy of the proposal to Henry Dale for comment, assuring Burwell that Dale was a very wise man and that he would be much influenced by Dale's judgment.

Dale, in reply, wished Cannon and his colleagues every success in promoting their program, writing in part that what the memorandum had to say about the present position of pharmacology was true of the world at large, and not merely of America. "It hardly needs to be said that therapeutics, as it is now developing, on the basis of an attempt to deal with the *causes* rather than the symptoms of disease, will have an ever diminishing relation to pharmacological studies of the Schmiedeberg type. There is still, I think, important and interesting work to be done along those lines; but it is likely to have at least as much bearing on physiology as on therapeutics."

This was not the first time that Cannon, in his capacity as chairman of the pharmacology committee, consulted Dale. Some months previously, he had asked for information about the German pharmacologist Otto Krayer, whom he had met at the International Congress of Physiology that summer. Cannon had been impressed by Krayer and was beginning to make discreet inquiries about his qualifications for the post at Harvard; moreover, he knew that Krayer was well-known in England and that Dale especially had followed his career.[16]

Krayer was, when Hitler came into power in 1933, professor

extraordinarius at the University of Berlin. At the time Dale, concerned with finding aid for scientists who had been ousted from their posts by the Nazis, wrote Cannon:

> Men are beginning to try to find opportunity elsewhere, who are neither Jews nor politically suspect, and whose positions in Germany are not threatened, but who simply cannot tolerate the present conditions in their own country . . . Krayer, the best of the younger pharmacologists in Germany, was ordered to go to Dusseldorf to replace Ellinger [who had been dismissed from his post on racial grounds], and, because he politely asked to be excused, was excluded from Laboratories and Libraries in Berlin, though not discharged from his position.

Determined to get out of Germany, Krayer obtained a Rockefeller Fellowship and was given leave to go to England, where he worked for a time with E.B. Verney at University College, London. There he and Verney demonstrated that a vagal reflex would release acetylcholine into the coronary circulation—further proof of work Krayer had done earlier with Wilhelm Feldberg in Berlin showing that direct stimulation of the vagus released acetylcholine. At the end of his fellowship, he refused to return to his homeland and instead accepted a position as professor of pharmacology at the American University of Beirut.[17]

Clearly, Krayer was a person of high moral character and steadfast opinion. Equally important, as a student of Trendelenberg trained in cardiovascular physiology and pharmacology, he had made a specialty of studying the effects of drugs on tissues. In the spring of 1936, largely due to Cannon's efforts, he received an invitation to come to Harvard as a visiting lecturer in pharmacology for the coming academic year. Although Krayer was unable to accept the invitation, Dean Burwell agreed to his proposal that he remain in Boston for eight weeks following attendance at the Tercentenary celebration in September.[18] Krayer's lectures were a stunning success. As a matter of fact, for the first time in years students applauded at the end of a class in pharmacology.

Krayer was invited to spend some time in Cannon's laboratory (at his own request) and was given every opportunity to make the most of his visit. Arrangements were made for him to give additional lectures outside the classroom, which were attended by outstanding clinicians like Soma Weiss, assistant to George Minot at the City Hospital. Cannon himself vetted those lectures, as did Elliott Cutler and Oliver Cope of the department of surgery. All were favorably impressed with Krayer, who spoke English very well. Yet there remained some doubt as to his suitability as head of the pharmacology department.[19]

Dean Burwell wondered whether Krayer's record thus far justified the conclusion that he was a productive scholar of the first order. Even Cannon really didn't know how far he would develop as a investigator. But after reviewing the letters of support previously received from eminent pharmacologists like Dale and Verney—letters which particularly stressed the important research Krayer had been involved in—Cannon and his committee were ready to come to decision.

Early in 1937, the Committee on Pharmacology had its final meeting and recommended that Otto Krayer be invited to come to the Medical School as an associate professor for a term of five years. In addition, the committee produced a "Proposal for a Diversified Study of the Treatment of Human Illness." In his letter to President Conant, Cannon emphasized that the future of pharmacology would reside in the study of disease and that it would also come "in no small measure as the direct and inevitable result of advancements in physiology." He continued,

> In my conversations with Dr. Krayer, I learned that his mode of teaching was definitely based on physiological systems and tissues instead of being based primarily on drug classifications—the traditional method of presentation in this country. I believe that one important reason for the enthusiasm which he aroused in the students was due to his mode of approach to the subject that he teaches.

When he later apprised Dale of the committee's decision, Cannon explained why the offer to Krayer to head the department did not carry with it a full professorship: "We had to make that recommendation because of the long service of Dr. Hale, who is also Associate Professor. Placing Krayer above Hale in the circumstances might have made difficulties." But he added that there could be no doubt of his advancement if during his term of appointment, Krayer demonstrated what Cannon believed were his excellent qualifications as a productive scholar.[20]

To Krayer Cannon wrote, "The Committee had no question

regarding the quality of the investigations which you have achieved, and they also felt that in favorable circumstances you would not only carry on researches yourself but would stimulate others to do likewise." Krayer, after some hesitation, accepted the offer, expressing the notion that his decision was based on selfish reasons. "I think if anywhere, then at the Harvard Medical School will it be possible for me to find a place for further study," he wrote Cannon.

> In collaboration with you and other members of the physiological group, and in close contact with your clinicians, of whom so many have sprung from physiology and pharmacology, it should be possible for me to fill in the gaps of my knowledge. How could I justify my teaching, and how could I satisfy your excellent student body, if I did not see there a possibility of improving on the scientific and investigative side of my knowledge as well as an opportunity of enlarging its scope into the clinical field?[21]

Krayer was an instant success at Harvard and his work went well. Scarcely a year later, however, for some obscure reason, he decided he was not the person to develop pharmacology at Harvard after all and was on the verge of accepting an offer of a professorship at the P.U.M.C. In September 1938, following Krayer's visit to Beirut, Jim Pinkston observed in a letter to Cannon, "I think it is extremely unfortunate that he has decided to abandon Harvard for Peiping Union Medical College." When Krayer made his position known to the students, there was a great uproar. The students objected strenuously to his leaving, and even held meetings to persuade him to change his mind.[22]

Subsequently, many of his colleagues at the Medical School—including, among others, Elliott Cutler and Burt Wolbach, as well as Ernest Tyzzer and Stanley Cobb—wrote letters in Krayer's behalf to President Conant. Cobb felt they had not done enough for him; Wolbach fretted that he might be an associate professor for life; Cutler and Tyzzer were worried about his staying at all. Burwell mollified the critics and offered Krayer an appointment without limit of time, which he duly accepted.

In March 1939, Burwell wrote Conant that Krayer was on his way to Europe and would be back in time for the appointment of a new University Committee on Pharmacotherapy. Some time later, a considerable sum of money was granted to the Medical School for a five-year period to test the possibility of extending pharmacology toward clinics and amplifying its resources in contacts with physiology, biochemistry, organic chemistry, and endocrinology. "The experiment will be an interesting one," Cannon remarked in a letter to Pinkston early in 1940, "and I am hopeful of such good results that the financial support will be continued."

Eventually, it was as he had predicted. Krayer became a full professor, and in the years until his retirement in the early 1960s, he made the Harvard department a major force in the world of pharmacology.[23]

Increasingly in the 1930s, Cannon, like his friend Henry Dale, was perceived by the scientific community as one of its senior statesmen. Not just at home but abroad, not just locally but nationally, not just at Harvard but in the world beyond its ivy-covered walls, his presence was desired at functions that signalled landmark occasions—dinners and celebrations that paid homage to the shining moments in medicine, science and higher education. To top it off, a goodly number of awards, medals and degrees were bestowed upon him with undisguised affection.

In 1931, a signal honor was conferred on Cannon with the award of the Baly Medal by the Royal College of Physicians. Lord Dawson, in presenting the medal, called Cannon's contributions to knowledge "a monument of industry and achievement. Hardly a year has passed, since Professor Cannon took his M.A. degree in 1897, but has shown a harvest from his labours." The medal carried with it, he told Cannon, "the gratitude and affection of your London confreres, and is an emblem of the understanding and friendship which unite the medical profession, in your country and ours."

Other tributes followed in rapid succession. In the early part of the decade, Cannon was elected to membership in the Academy at Halle, Germany, and to honorary membership in the Physiological Society (England) and in the Academia Nacional de Medicina of Spain. In 1934, he was awarded a Gold Medal by the National Institute of Social Sciences. Honors continued to come his way in the latter part of the decade. Washington University in St. Louis, the University of Barcelona, and the University of Madrid awarded Cannon honorary degrees. Medical and scientific academies and societies as far-flung as Budapest and Argentina, New York City and Sweden, elected him to honorary membership.[24]

In the spring of 1939, Cannon received word that he had been made a foreign member of the Fellows of the Royal Society. "I find that he was much pleased," Cornelia Cannon wrote her family, "more so than he usually is over his honors, for so many of his friends in England are members that it makes the recognition precious to him." Notices about Cannon's election to the Royal Society promptly appeared in *Nature*, the *Times* of London and the *Daily Telegraph*, and he received a letter of congratulation from Henry Dale. "The number never exceeds fifty, and they represent the best scientific work in all subjects and in all countries outside the British Empire," Dale reported. "Cushing, Flexner, Morgan and E.B. Wilson are the other American representatives of the medical and biological sciences, and it is very welcome to all your friends to see your name take its proper place with these."[25]

Meanwhile, Cannon found himself riding circuit on the Eastern seaboard as he added to a growing list of organizational affiliations, some more obligatory than others. In the second half of the 1930s, not only was he traveling in behalf of medical aid to the Spanish Republic or to assist refugees from war-torn countries, he also had a regular calendar of meetings of boards and committees, foundations and government agencies, in which he participated. After he had accepted membership on the Board of Scientific Directors of the Rockefeller Institute, his wife noted in her family letter of April 27, 1936, that Cannon had gone "to Philadelphia, Washington and way stations, to confer on medical and research policies. He is becoming the American medical Nestor and his life promises to become increasingly interesting as he is more and more in demand for the loan of his accumulated wisdom."[26]

A year later, almost exactly to the day, she exulted over a new job Cannon had taken on—chairmanship of an Endocrinology Committee to dispense $100,000 (later increased to $150,000) of foundation money over a three-year period. The Endocrinology Committee had come into being through a report prepared for the Markle Foundation by a special committee of the National Research Council. As head of that committee, Cannon was largely responsible for the 83-page monograph, *An Appraisal of Endocrinology*, that was published late in 1936. It set forth an account of research activities, past, present and future, on the ductless glands, and proposed the establishment of a fund to provide financial support and development of a research program in endocrinology. Last, but not least, it suggested specific plans to establish a committee like that for sex research, either under the N.R.C. or under the Markle Foundation, or jointly under both.

In order to devote himself to his new duties, Cannon resigned from the Dunham Lectureship Committee at the Medical School and some of his other long-term committee assignments. Although his new post on the Endocrinology Committee came without salary, he would have all his expenses paid and all the secretarial help he needed. It was the kind of inspirational work he was particularly qualified to do, in Cornelia's view, with an important opportunity to guide and stimulate research all over the country.

Cannon remained as chairman of the committee for the next several years and played an important role in evaluating applications for grants to support work in endocrinology. Among the many projects he approved was one by Frederic Gibbs, a member of Hal Davis' group at Harvard, who was using a new method developed by Albert Grass in a study of the electrical manifestations of brain activity. Right from the outset, it became apparent that there was considerable overlap of interests between the two National Research Council committees—the one on endocrinology and the other on sex research. Robert Yerkes and Cannon, between them, worked out a *modus operandi* that allowed for transfer of applications from one group to the other, depending on the nature of the project, which over time benefited a large number of investigators.[27]

One of Cannon's stopovers on his regular round of meetings outside Boston was the headquarters building of the National Academy of Sciences and the National Research Council in Washington, D.C. Not only did he attend gatherings there of the Sex Research Committee and, after 1937, the Endocrinology Committee, he also became a member of the Council of the National Academy and of the Executive Board of the N.R.C. As such, he frequently became engaged in matters of national policy concerning medicine and science.

As far as Cannon was concerned, it was important to have a frank discussion of the issues involved, without resort to polemics or personalities. He did not hesitate to speak his mind, and he expected others to do the same; he also expected a courteous hearing for diverse and disparate opinions. In 1937, he accepted appointment to the Medical Advisory Committee of the American Foundation Studies in Government. It was a post that would draw him into a fierce debate on the future of health care in America.

That year, the foundation published a two-volume report, *American Medicine: Expert Testimony Out of Court*, which presented the views of the medical profession on the state of American medicine. It was based on a survey of 2100 doctors, who found that in many communities, given the harsh economic climate, adequate medical care, research work and medical education could no longer be kept up to standard without support from local, state and, possibly, federal government. The foundation made no recommendations of its own, but let the doctors speak for themselves.[28]

Following publication of the American Foundation report, a letter was circulated by Hugh Cabot of the Mayo Clinic and Robert B. Osgood of Boston, two physicians of the highest standing, in which they called upon their profession to "assent to certain general propositions." This in turn led to the formation of a Committee of Physicians, with John P. Peters of Yale acting as secretary, to formulate a set of "Principles and Proposals" for the improvement of medical care. Cannon was one of the first to join with the signers of the document.

From the thesis that "the health of the people is a direct concern of the government" right through to the proposal that all public support of medical care be coordinated by a federal department of health, these propositions were certain to be attacked as a surrender to socialized medicine. Which is exactly what happened. After describing the course of events, an editorial writer in the *New York Herald Tribune* of November 8, 1937, concluded:

> But when all the excitement has subsided, we believe it will be appreciated that these proposals are not radical but realistic, and that, if the profession lines up squarely behind them, "socialized medicine" can be forestalled by a conservative system of subsidies to adequate medical care, controlled and directed by doctors.

The same day as the *Herald Tribune*'s editorial, Cannon, after reading over the "Proposals," wrote John Peters that some of them, "which have been attacked as being radical and subversive of the independence of medicine," had actually been in effect for many years and were recognized as appropriate in American practice.

Peters was grateful for Cannon's comforting words and observed, "By some strange stroke of fate, I find myself continually being asked to serve with you and [Henry] Sigerist these days. I am honored to be in such company. [C.E.A.] Winslow, who is often with us, calls it our 'red cell.'"[29]

The most obdurate opposition to the proposals was mounted by Morris Fishbein of the American Medical Association, who wrote a hostile editorial in *JAMA*, and by Eugene S. Kilgore, a San Francisco physician, who urged the signers to withdraw their support. Cannon was so pestered by letters from Dr. Kilgore that he asked Osgood what could be done about him. "It seems to me that he has developed a good deal of heat and botheration about some *proposals*," he remarked. "I had conceived these proposals as a basis for discussion and not as final dicta, nor as something to be dealt with in terms of personalities or appeals to authority."

To Kilgore, Cannon expressed his dismay at the manner in which the "Principles and Proposals" had been received. "This does not seem to me at all the way to solve the problems confronting the medical profession and the public in need of medical attention," he wrote him. "If the medical profession takes the attitude of doing nothing, it is likely to be treated as the profession was treated in England and in being subjected to the regulations laid down by politicians, instead of itself taking the lead in the orderly evolution of medical practice."[30]

Cannon continued to urge that the medical profession "must clean house itself or have the Government step in and do the job." His cautionary message would have relevance for events that happened long past his lifetime. After publication of the report from the Committee on Costs of Medical Care, the American Foundation turned its attention to undertaking a study of medical research. Once again, in a letter to Esther Lape of the foundation of February 1939, Cannon was emboldened to gaze into the future. While anticipating the trend of the modern debate between holistic medicine, on the one hand, and cellular and molecular biology, on the other, he held fast to the notion that purpose and use were central to understanding the workings of the body as a system:

> . . . there is another direction in which medical research is likely to proceed aside from diving into the secrets of the processes of the individual cell. That direction is toward the understanding of the way in which the more complex organisms operate as a whole. It is

> perhaps significant that much of the most important work in physiology in recent years has been concerned with the collaboration of the parts of the body. For medical purposes this interest is quite as important as the interest in the reactions and internal events of the bodily units of the cells.[31]

On the last day of 1938, Cannon received a telegram that astonished him. It contained the totally unexpected news that he had been elected to the presidency of the American Association for the Advancement of Science. In some part chagrined at the thought of taking on a new job (even a temporary one) on top of all his others, he was nonetheless pleased with the tribute from his peers. Not only that, but the Medical Bureau and North American Committee for Spanish Democracy, as Cornelia waggishly remarked, was "lyrical with joy for, of course, anything that distinguishes their chairman reduces the communistic taint on his orthodoxy."

"The past few days have been somewhat exciting," Cannon wrote his son in St. Louis two days later, "with election to the presidency of the A.A.A.S., congratulatory telegrams, and a cable message from Dr. Negrin announcing the award of a degree of Doctor Honoris Causa by the University of Madrid. Also a cable from Moscow expressing best wishes for the new year in work for 'science, democracy and progress of humanity,' and signed by Orbeli (Pavlov's successor), Propper and Koshtoianz. How sadly my reputation for respectability would suffer if *that* became public!"[32]

Following a press release from Science Service bearing the headline, "Socially conscious scientists have authoritative spokesman in Prof. Cannon, new A.A.A.S. president," a number of newspapers and magazines carried items about him. Esmond R. Long, writing in *Science,* reported that Cannon had been unanimously elected president of the Association and was the first physiologist to fill the year-long position. Congratulatory messages followed on the heels of the announcement. "I think that they showed themselves an organization of very good judgment," Sidney Burwell wrote. "I congratulate them and the Medical School as well as yourself."[33]

Cannon was called into service almost at once and soon had a new round of meetings to add to his schedule. Early in January,

F. R. Moulton, the permanent secretary of the Association, informed him that his major duties for the coming year would be to attend Executive Committee and Council meetings and to give to others the benefit of his judgment and advice. He would also be expected to preside at the semi-annual meeting in Milwaukee in June and the annual meeting in Columbus in December. Other than that, Moulton looked forward to drawing on Cannon's experience and expertise for the Research Council on Problems of Alcohol, whose work the A.A.A.S. but lately had undertaken to sponsor.[34]

As head of such a prestigious association, Cannon's views were soon made public. An editorial writer in the *Boston Globe*, after interviewing two notable men of science, Walter Cannon and Albert Einstein, on the relationship between scientists and politics, reported on February 9, 1939: "Their replies are in vigorous agreement. The hour has passed, Dr. Cannon points out, when men of science may risk indifference to trends in public affairs in the belief that their proper role is one of seclusion." Continuing that the most reputable Association of Scientists in this nation had lately begun to explore the realities of contemporary political life and thought, in contrast to the spectacular decline of science under the rule of fascism in Italy, Japan, and most of all, Germany, the editorial ended, "Science ignores politics at its peril."

Soon after, the *New York Times* reported that Cannon was one of four distinguished professors who had broadcast a warning to the public to safeguard democracy while condemning totalitarian governments, especially in Germany and Italy, as forces tending to stifle science. He had joined with Columbia University professors Robert Lynd, sociologist, Harold Urey, Nobelist in chemistry, and Franz Boas, anthropologist, in the radio broadcast under the auspices of the Lincoln's Birthday Committee for Democracy and Intellectual Freedom. It was part of a drive for support of democracy as a prerequisite to intellectual freedom that had been initiated by an A.A.A.S. resolution passed two years earlier.

Cannon's year as president of the Association passed quickly and after it was over, he looked back on it "in many respects as a very pleasant experience." His main complaint, he wrote Dr. Moulton early in 1940, was that the many business meetings interfered with the real interest of the Association, "i.e., the presentation and discussion of the results of scientific study." It seemed to him "grotesque that a group of scientific men who have had large experience should coop themselves up in a room and spend the period of a great scientific meeting talking about affairs" which could just as easily be relegated to some other time slot.[35]

But Cannon, as retiring president, was not quite finished with his duties for the Association. He still had to make a presidential address; even though the time scheduled for it was a year away, he had already begun thinking about it. "I have not yet decided in what direction to turn for a subject for my Presidential address next December," he wrote Moulton in January 1940.

> Your suggestion of a new turn to what might be called "social" or "mental" homeostasis is stimulating. I may go in that direction or I may take up the role of the investigator in society and consider the conditions which are most favorable for his services. But even that may prove less interesting than something else!

Another person with whom Cannon corresponded about his address was Elisha Friedman, the New York businessman who had used illustrations from *The Wisdom of the Body* and Cannon's other writings for his articles on the relation of biology to social organization. Ever since their meeting in 1938, when Cannon gave the Welch Lectures on homeostasis, they had entered into a lively exchange of ideas, as expressed in their own writings and those of others, on the application of biological principles to stabilizing economic conditions as well as to maintaining equilibrium in human society.

Though Friedman exhorted Cannon to take as the subject of his address the suggestiveness of physiology for social organization, Cannon by the autumn of 1940 had not yet made up his mind to do so. "In view of the fact that there will be two sessions of the meeting of the American Association for the Advancement of Science devoted to the social homeostasis of which you have been a leading exponent," Friedman wrote Cannon early in October, "it would be regrettable indeed if you did not use the prestige of your office as president to support those two programs." He invited Cannon to call on him if he needed help with the social and economic aspects of his paper.[36]

Some weeks later, Cannon replied that he had finally solved his dilemma after a chance meeting at the American Academy of

Arts and Sciences with an old friend, Henry Dennison. Dennison, a successful businessman, had recently spent an evening discussing some of the points made in the Epilogue of *The Wisdom of the Body* with Edwin Gay, a professor of economics at Harvard, and he convinced Cannon that a further discussion of social homeostasis would be of timely importance.

Not until Cannon was asked in mid-November to wire the title of his address for a preliminary announcement, did he disclose that his subject would be "The Body Physiologic and the Body Politic." The preparation of his talk amid a pile of obligations atop all the incidental tasks associated with the headship of a department, Cannon confessed to Dr. Moulton, had threatened to overwhelm him. Begging off from an added assignment that would have required him to prepare a fifteen-minute talk for broadcast, he underscored for Moulton that he was, after all, retiring—"indeed, a very retiring President."[37]

Cannon gave a draft of his paper to several members of his family and friends for close reading. Henry Dennison, for one, made a number of suggestions and returned it to Cannon with his warm commendation. His son-in-law John Fairbank, for another, urged taking care to distinguish between terms and concepts that had specific biological meaning and their application to the social context lest false analogies creep in. Fairbank called the paper "a contribution of the greatest value" and offered several observations—all of which Cannon gladly incorporated into his text.

On December 27, 1940, Cannon performed his final duty as outgoing president to speak before the A.A.A.S. annual meeting in Philadelphia. Cornelia described the occasion: "Terrified by his responsibility, dressed in white waistcoat with particular attention to white tie and all the fixings, [he] stood up in Convention Hall where [Wendell] Willkie was nominated . . . and gave a superb paper . . . It offered, most modestly and tentatively, the homeostatic workings of the body as a possible model for some kind of homeostasis in society." Afterward, Dr. Moulton called the address "magnificent" and told her that her husband was "the foremost scientist in the country."[38]

Judging from the comments after the meeting, "The Body Physiologic and the Body Politic" enjoyed a fine reception by the assembly. The *New York Times* had a lengthy dispatch from William L. Laurence, which began, "The most efficient and stable human society would be a body politic modeled after the organization of the human body," and proceeded to quote a goodly portion of the speech. "Dr. Cannon's society would be a 'biocracy' in which the myriads of differentiated cells would be organized into functional organs all cooperating in a dynamic democracy in which any form of dictatorship by any individual organ would lead to degeneration and death," ran another article in the *Times*. The *Philadelpha Inquirer* reported that Cannon replied to dictatorships by lashing out "at those who resist social change and called on them to look at the human body and learn from it how a society can make itself strong and free."

An editorial in the *New Republic* describing the A.A.A.S. meeting noted that

> Dr. Walter Cannon of the Harvard Medical School drew a savage portrait of our society under the guise of comparing the superior efficiency of the human body to the frailties of the social organism. One speaker after another indicated a belief that control of much of the world's area has got into the hands of "ape men" and expressed the conviction that science must help to restore control to civilized human beings.

Elisha Friedman, who had obliged Cannon with a thorough critique of his manuscript, was delighted with the finished product. "I think your speech influenced President Roosevelt's thinking as revealed in his inaugural address, when he compared man and society," he noted with glee near the end of January. Cannon felt happy about the talk himself—happy that it was all over so he and Cornelia could look forward to a vacation in Florida with their friends, George and Florence Pierce.[39]

PART IV

World War II

22 *A World At War*

ALL THROUGH 1939, the Cannons watched in horror as the world appeared to totter on the verge of ruin. "I am in such a state of mind that I do not feel as if I ever wanted to see Europe again," Cornelia declared in the spring after the takeover of Czechoslovakia. As one aggressive act followed another, their days were spent glued to the radio and reading newspaper reports of the Nazi-Soviet pact, the move of Russia into Poland, and, finally, the entrance of England and France into war against Germany.

"I feel at times as if the world was shaking under my feet," Cannon early in 1940 wrote Dr. Lissák, who had returned to Hungary. "No one knows what the outcome of this ominous war is going to be. The only faith which I cling to is that human society must persist and that there are forces at work which must establish a sort of social homeostasis in order that human society may persist."[1]

Despite the threat of world war coming ever closer, Cannon nonetheless found much to be cheerful about at home. Late in 1939, he received the joyous news that his son Bradford would be returning to Boston. Not only that, but he and his wife Ellen were expecting their first child. "You cannot imagine the fun I look forward to," Cannon told the happy pair, "making pictures, telling stories, and having 'the time of my life' playing with my grandchildren. I shall seem to be entering my second childhood, but what more innocent joy could one have."[2]

The end of Bradford's surgical training in St. Louis was a high water mark in the changes that had taken place in the Cannon family during the decade. Near the end of 1932, when Bradford was a fourth-year Harvard medical student, his father, on a train heading west, happened to meet with an old friend, the suregon Jack Yates, and told him that Bradford was interested in a career in surgery. Their conversation led to Yates writing to Evarts Graham, the eminent surgeon at Washington University Medical School with whom Cannon had expressed a wish to have Bradford train. A few days later a telegram came from Dr. Graham, all but offering the young man an internship at Barnes Hospital. Bradford promptly wired back that he would like to come but wished to be considered on his own record rather than as the son of a distinguished father, and that he would send his transcript the next day. George Crile's son, Barney, the top man in Bradford's class, had also applied for a place in Graham's department. Within a week, both young men were set to begin their training in St. Louis after graduation.[3]

Though it was a wrench to have Bradford go so far away, his parents believed that it was essential for him to "see different points of view, meet different people, different medical men, and cease being a pure Bostonian." They were not wrong. Neither was Bradford disappointed with his years in St. Louis—despite occasional bouts of homesickness. Almost five years later, he neared the completion of his training in general surgery and was contemplating his professional future when he became engaged to be married.

While home for the holidays late in 1937, Brad Cannon met Ellen DeNormandie at a Christmas Eve party. The two were instantly smitten with one another and by the time he went back to St. Louis, their future together had been sealed. When Ellen came to dinner with the Cannon family, they all fell in love with her themselves. "She is a fresh lovely girl," Cornelia reported in her New Year's letter, "lovely to look at and so intelligent and sweet, with no lipstick or finger coloring. You can imagine how that went over with Father!" Everyone was in raptures over Ellen DeNormandie, whose father was an obstetrician/gynecologist and whose sister Alice was married to the young M.G.H. surgeon, Oliver Cope. The match between Ellen and Bradford appeared to be made in heaven and evoked memories for Walter and Cornelia of their own courtship. Indeed, the marriage ceremony occurred on June 25, the date that the senior Cannons had wed, with the father of the groom serving as best man.[4]

When Bradford was home at Christmas, he had looked into job possibilities in Boston to help him decide about his future. There were several options open to him, and although Cannon, as always, took an interest in his son's career plans and would have liked nothing better than to have him return to Boston, the choice was Brad's to make. "If Dr. Graham thinks you would do well to stay in St. Louis until an opening appears," Cannon wrote him early in 1938, "you had better pay careful attention to his advice. He is very influential and a wise counselor. Do let us know as soon as you can your final decision."

A short time later, Bradford decided to remain at Barnes Hospital for a year of intensive practice in reconstructive surgery with the top-notch team of Drs. Vilray Blair and Barrett Brown. He was so well pleased with the program they arranged for him that he stayed on for a further year in 1939 as their associate. At long last, ready for more independence and increased responsibility as a plastic surgeon, Bradford believed that the time was right for his return to Boston. Once everything was settled in St. Louis, he and Ellen looked forward to contributing to the summer filled with grandchildren in New Hampshire.[5]

In August of 1940, Marian Cannon married Arthur Schlesinger, Jr., the brilliant young son of a Harvard history professor. With war clouds hovering over the United States, the family was fully aware that the wedding, held at the Franklin farm, might well be one of the last celebrations for which they all would be together. Throughout the 1930s, the Cannons had seen the break-up of their happy (albeit chaotic) home life as the children married and left the nest. Despite the dispersal, the family remained as close as ever, with the Franklin farm serving as the magnet that drew them together.

Though Wilma had started the exodus when she went to China in 1932 to marry John Fairbank, they had since returned to Cambridge where John taught Chinese history as a member of the Harvard faculty. Linda, who had a degree from Simmons College in social work, had married Harry Burgess, a young geologist; in 1936, the same year that the Fairbanks came back to Massachusetts, the Burgesses moved to Montana. Much to the delight of her family, Linda came back East whenever she could, and brought an increasing brood of children with her. Helen, following a long engagement, married Douglas Bond directly after her graduation from Radcliffe in 1937. The son of Earl Bond and May Newson's sister, Grace (old friends from Walter's and Cornelia's childhood days in St. Paul), Douglas was pursuing a career in psychiatry, like his father. The young Bonds, like the young Cannons, were expecting a baby in 1940.

Nor were the children the only ones to leave the nest. In 1938, Cannon's sister Bernice bought one of the oldest houses in Cambridge and left Frisbie Place for Winthrop Street. The following year, Cornelia—by then deeply immersed in volunteer work for the Mothers Health Council (the former Birth Control League) of Massachusetts and serving as its president—rented an apartment on Prescott Street near Harvard Yard, having decided that rattling around in the big University house was too depressing. "The house at 6 Frisbie Place now seems desolate," she lamented in May 1939. "Only three remnants of a household of nine wander about the lonely and reverberating rooms. I shall be glad to move to the apartment where there are no ghosts." When moving day came, Ida Cannon relocated to an apartment of her own just down the street.[6]

In February 1940, Henry Dale (who had joined in the World War I study of shock and allied conditions as a member of the British Medical Research Committee) wrote Cannon that he was back doing war work on problems of shock and blood replacement. At the same time, in reference to America's neutrality, he expressed the wish "that your great country could find it possible to take a more decided attitude, so that the chance might be increased of saving the essentials of civilization." Cannon could not let Dale's comment go by. "If the United States had not had such unfortunate experiences during and after the last war, I do not think we should be so shy of giving support to the side which the great mass of population favors at the present time," he retorted. The lack of influence of the American government in condemning international aggression had come as a shock to many, he continued, "and since then one evil action after another has been dealt with in a way which has been deeply disturbing." In the final analysis, this failure to act had led many thoughtful Americans to feel that their country should not become directly involved in the present struggle.

Following the German invasion of western Europe in May, Cannon wrote again to Dale to say that the United States was now awake and alert to the dangers to its liberty. He warned, however, that although many would rejoice to see American resources put at the disposal of the allies, the country was vast and would move slowly toward that end. Dale was happy to note the movement of American sentiment, and only hoped it would not be too slow. What was needed, he bluntly told Cannon, was "not argument, or criticism, or even theoretical sympathy, but just armaments and particularly airplanes."[7]

In the midst of his exchanges with Dale, Cannon was faced with having to re-examine his own stance on American neutrality. As one of 500 signers of a peace resolution drafted by the American Association of Scientific Workers and forwarded to President Roosevelt by the A.A.S.W.'s president, A.J. Carlson, he came under heavy criticism. Many of his friends were dismayed to see Cannon's name among the signatories—and none more than his former student, Magnus Gregersen, who sent him a letter that was, to Philip Bard's way of thinking, "perhaps a bit intemperate." Cannon, replying at once, told Gregersen that his horror of war was based on his experiences in 1917–18. Furthermore, he was disturbed—and had been for a long time—by the actions of the dominant group in England that included Neville Chamberlain. They had refused to support action against Japan's seizure of Manchuria and against Italy's attack on another member of the League of Nations. They had "supported a futile and spurious Neutrality Committee" in its relations to Spain and had "sold Czechoslovakia down the river." What enthusiasm, he asked, could anyone possibly have for such a group? He had approved the resolution because it recommended support of "all reasonable programs" to preserve peace for the United States and bring peace to the world. He had to confess, however, that recent events had given new meaning to the phrase "reasonable programs."[8]

Recent events, of course, referred to the Nazi move into western Europe. Within days of the invasion, the *Evening Standard* of London reported, "Horrors now being perpetrated by Hitler's troops in France and Belgium have brought sweeping changes in American opinion." According to an article in the *New York Times*, Dr. W.B. Cannon, president of the American Association for the Advancement of Science, and others had undergone a "change of heart" since they had signed the peace resolution sent to the President some weeks ago.

> Dr. Cannon, explaining why he no longer subscribes to the peace resolution, says: "Never for a moment did I understand that the resolution meant peace at any price. We know now that there is a powerful wild beast on the rampage in the world. No one is safe. I am willing to go as far as anyone in protecting the United States from danger. This goes so far as supporting France and England in their conflict with Germany short of sending men."[9]

A.V. Hill, the British physiologist then in the United States on official government business, could not help comparing the attitude he encountered in the spring of 1940 with the offers from Americans only too eager to "come over and help" during the first World War. "I shall be glad to get away from all the ineffective sympathy, pessimism and pacificism one finds here and be back with some people who have some fight left in them," he wrote Cannon near the end of May. He had seen the *New York Times* that morning and was very glad to read that Cannon had "left the 500 who think science is more important than civilization." He went on, "We have all made mistakes, but America has made just as many and as bad ones as the other democracies. Now we are in the soup, though I still think we shall pull out in the end." If such were the case, he was sure that America would be able to continue its way of life without sacrifice or discomfort; if not, he feared that the light of civilization would go out all over the world. He ended his tirade somewhat sheepishly, "Sorry! but the worm has turned."

"I have had both bouquets and brickbats as a consequence of the publication of my attitude in the New York Times," Cannon wrote W.W. Norton some weeks later. "I am glad that I let my change of attitude be known, and, as I wrote to A.V. Hill, in so far as my name carries any weight at all, the paragraphs quoted from me must have been more important than inclusion of my name with 500 others."[10]

Cannon had indeed changed his attitude, but he did not labor the point. For the remainder of 1940, whenever he communicated with Hill or Dale or with his other British friends, he confined his remarks to expressions of concern for their safety or empathized with them for the hardships they were enduring, especially as the bombardment of London intensified. At one point in the Battle of Britain, he wrote Dale, "The staunchness of the English people, their splendid morale, and their high-spirited resistance to the attacks by the Germans have awakened in the American people, from one end of the country to the other, a deep and persistent admiration and sympathy." In another letter, Cannon lauded the heroic deeds of the R.A.F. and the British Navy, which "rouse our greatest enthusiasm."[11]

In July, Cannon had discouraging news to report to Dale about the Zander children. The daughter of Rudolf Magnus, the Utrecht physiologist, Gretl Zander (who with Alison Dale had been a house guest of the Cannons' during the 1929 International Congress in Boston) had asked her father's influential friends for help to insure the safety of her children for the duration of the war. Gretl and her family had fled Europe for England, where her husband, a German, was interned as an enemy alien. It was her hope that she could send her four children to America, which was then providing a haven for a limited number of British children. "The outlook is not good," Cannon told Dale:

> Although theoretically anyone may come here on a temporary visa, an American consul . . . would probably refuse to grant such a visa to a person of German birth, because it is granted on the assumption that the person in question will return to his native country after a year. Since Gretl's children cannot do so, the securing of visas for them is extremely doubtful. I am very sorry that they cannot share the benefits which are being enjoyed by English children.[12]

Nevertheless, obstacles were somehow overcome and everything was made ready for the reception of the Zander children in America. But by the time arrangements had been completed, the journey across submarine-infested seas was judged to be too dangerous. In October 1941, Gretl wrote a lengthy letter to the Cannons in which she described her family's life in England. She was glad she had not sent the children away after all, and Cannon was glad too to learn they were safe and her husband was with them again. When he replied, he had news of his own that he proudly reported: "Mrs. Cannon and I now have seven grandchildren—among them identical twin boys, born to Bradford and his wife only a little over a month ago." He wrote that Wilma and Linda were in Washington, where their husbands were working on defense projects, Bradford was a surgeon in Boston, and Helen and Marian were living in Cambridge, one married to a young psychiatrist, the other to a young historian.[13]

Another refugee whom Cannon tried to help at this time was André Mayer. Cannon never forgot his gratitude to Mayer, a professor at the Collège de France, who had offered him laboratory space in Paris when he was working on shock in World War I. He had since become, according to Cannon, the foremost French physiologist and a successor to Claude Bernard. While on a mission for the Vichy government to get vitamins for the children of unoccupied France, Dr. Mayer had turned up in Cambridge with his family. "They will probably stay here until the war is over," Cornelia indicated in her family letters in March 1941, "as they are Jews and must fear for their children." Their son Jean, who was in the French Army, had escaped from German imprisonment and was now in command of 150 men in unoccupied France—"quite a feat for a 21-year old boy."

Cannon did his best to help André Mayer, both by approaching L. J. Henderson about a Lowell Lectureship for him and submitting his name to Hal Davis for consideration as a Dunham Lecturer. He even offered to pay research expenses if Mayer should care to enter his laboratory and work there. Later Cannon helped with arrangements for André Mayer to do consulting work in Washington with the French delegation to the United Nations in matters of nutrition, public health and welfare.[14]

In November 1941 Cannon wrote A. V. Hill, "There is an anxious effort to speed production of war machines on a huge scale so that not only the United States but also other countries opposed to the tripartite totalitarian powers may be fully armed. Already your great leader, Churchill, has recognized handsomely the service which my country is performing. At the moment we are on the perilous edge of a break with Japan. The next week or two may determine whether we plunge into the war on your side." Cannon made an effort to explain where matters stood, for he was certain Hill had not really grasped "how almost impossible it is to have a unified attitude of people in this vast land."

Hill's answer, written on Christmas Day—the very day he received Cannon's letter—alluded to Pearl Harbor as he rejoiced, "At last we are open allies." When it came to Cannon's explanation of America's (and his own) change of heart, Hill was diplomatic: "We have all gone through these phases of pacifism, appeasement, complacency, 'it never can happen to us': I (like you) deplored our attitude in the past—we have committed many follies . . . under a cloak of hypocritical idealism, of confused thinking, of other-worldliness. We have learned at last what the real world can be like, if good men are not prepared to fight and sacrifice, if needs be, for decent human relationships." Hill ended his letter, "Nothing can give those who love Americans greater satisfaction than to feel that you are in this job now with us, not only for the war but for the building of a new world after it."[15]

With the escalation of hostilities, Cannon, like Dale, once again was drawn into problems of shock and transfusion, an interest that he had laid aside some twenty years previously. But this time, in contrast to his active participation in seeking both the cause and treatment of shock during World War I, Cannon's role would be played out almost entirely in committee meetings. In the spring of 1940, Lewis H. Weed, Chairman of the Division of Medical Sciences of the National Research Council, asked Cannon to come to Washington in order to organize and head a Committee on Shock, Transfusion and Blood Substitutes. Although Cannon had just recently written Robert Yerkes that he was overwhelmed with responsibilities and tried (again) to resign from the Sex Research Committee, he did not hesitate to answer the call for service to his country. A group meeting at the N.R.C. on May 31 decided there should be two subcommittees; one on shock, to which Cannon appointed Alfred Blalock as chairman, the other on transfusion and blood substitutes, to which he appointed Cyrus Sturgis as chairman. Cannon's role in connection with the two sub-committees became that of adviser and coordinator.[16]

A Subcommittee on Blood Procurement was soon added and then turned over to the Red Cross, with Canby Robinson as director. In July, Cannon suggested that his colleague at Harvard, the physical chemist Edwin J. Cohn, apply his knowledge of protein chemistry to the refinement of elements in bovine plasma which might be a means of replacement of bovine blood for human blood in the treatment of shock and hemorrhage. "His development of methods of processing human albumin and the recent studies of bovine albumin have been the outcome of that suggestion," it was later reported. "The present Red Cross drive for human blood grew out of our activities." Some time later, Dr. Weed, in praising the work of the Shock Committee, singled out the development of blood substitutes by Dr. Cohn and others "as one of the major contributions of medicine to the present war effort."[17]

In the summer of 1941, Phil Bard, as a member of the Shock Committee, began conducting experiments on nervous factors in the shock complex, in response to a call for information regarding the possibility that the depressed state is "brought about reflexly by nervous impulses set up in the injured tissues." Early in September, Cannon, intrigued by a report from Bard, wrote him, "Your experiments appear to be quite definitely favorable to the existence of an important nervous element in the production of the shock state." He asked to be kept informed about progress, especially with regard to the *modus operandi* of the investigation into the role of the nervous system.

Despite the promise of Bard's work, Cannon took care that the nervous element in shock should not be confused in the public's mind with emotional stress. The following spring, when asked to talk to a group of newspapermen about shock, Cannon issued a word of warning, "Wound shock is to be distinguished sharply from so-called 'shell shock'—a very bad term. That is an hysterical breakdown that occurs when a man has been subjected to an emotional stress which is beyond his limits of endurance."[18]

As a member of the Shock Committee, Bard tried to relieve Cannon of some of the administrative duties that his aging friend found so burdensome. For instance, Cannon told A.N. Richards, who had been appointed head of the Committee on Medical Research under the Office of Scientific Research and Development (O.S.R.D.), that he would be glad to serve as a consultant to his newly-formed committee but did not wish to be involved with travel. Thereafter, Bard agreed to head a survey committee that Dr. Richards appointed to travel around the country and report what was being done in work on shock. Early in 1942, after Cannon had read the first report of the survey committee, he wrote Bard he was pleased with his recommendation that future concentration on practical matters "should be conducted by a committee and not by *me*." Bard's advocacy of an administrative committee, Cannon confessed, gave him a sense of freedom he had not enjoyed for many months and he looked forward to getting down to work on a book summarizing his laboratory research on "Hypersensitivity of Denervated Structures"—something he'd been aching to do for a long time.[19]

During the final years of his life, Cannon, while cognizant of his flagging energies and reluctant to take on time-consuming committee work, nevertheless felt duty-bound by a sense of service to society and an unwillingness to desert old friends. Among all the activities that took him away from his laboratory, none touched his heart to the extent that Spain did.

In the early months of 1939, an estimated 500,000 Spaniards—soldiers and civilians, men, women and children—fled from Catalonia before the advance of the "liberating army" of Franco's Nationalist forces and crossed over into Southern France. Reports of conditions along the French frontier from correspondents like Gênet, who wrote for *The New Yorker*, Eric Muggeridge, an agent for the Foster Parents' Plan for Children in Spain sponsored by the Duchess of Athol, and Edgar Mowra, an American newspaperman, portrayed the misery of the Spanish refugees in heartbreaking terms.

Muggeridge, for one, informed the American public that the Deladier government in France, faced with the problem of such a tremendous influx of refugees, decided that, of course, the Spaniards must be allowed to cross the border, but then herded them into concentration camps instead of providing decent convalescent centers. Mowra, writing from Perpignan on January 27, described the refugees as "the most pathetic people I think I have ever seen . . . The world must have a heart as hard as stone to allow such things to happen." Still others complained that what the French government was doing for the wretched refugees did not even represent a decent minimum. Food, shelter, sanitation, medical care for the sick and wounded—all were said to be non-existent or in short supply. Many died before they could reach the end of their long march.[20]

Cannon, anxious about the safety of his friends, cabled Camille Soula, a physiologist at the University of Toulouse, for news of the Spanish refugee scientists. Soula immediately showed Cannon's cable to Drs. Bellido and Carrasco, who had sought sanctuary with him, and they read it with deep feeling. For his part, Soula wrote

Cannon, he was devoted to helping the refugees and had opened a center of welcome where the university people of Spain and Catalonia could gather. In this endeavor he was aided by French colleagues, the Socialist government of Toulouse, and the University, but was opposed in an "underhanded manner by the ignoble government from which I hope France may soon be freed."[21]

Late in March, Cannon received news from Bishop McConnell of the North American Committee that heartened him. Harold L. Ickes, the Secretary of the Interior, had agreed to be honorary chairman of a new Emergency Campaign to raise $1,000,000 for the relief of the Spanish Republicans who had fled to France and North Africa. Soon after, John Reich, the secretary of the Committee on Spain of the American Friends Service Committee (A.F.S.C.), who regularly supplied up-to-date information about the situation in France, asked Cannon if he would help raise money to finance the resettlement of Spanish refugees in Mexico and Latin America. Though appreciative of what the Quakers were doing in behalf of the Spanish refugees, Cannon replied that he did not feel he could take part in any new enterprise directed toward fund raising.

But Cannon, unable to turn his back on his Spanish friends, still hoped to be useful in the larger campaign, which in June was renamed the Spanish Refugee Relief Campaign (S.R.R.C.). Secretary Ickes continued to serve as honorary chairman of the new organization, which had Bishop McConnell as national chairman and Herman Reissig as executive secretary. At the same time, the Physicians and Scientists Committee headed by Dr. Edward Barsky became the Medical Aid Committee for Spanish Refugees, a division of the S.R.R.C. In the fall of 1939, the outbreak of war in Europe made the situation of the Spanish refugees even more desperate. The S.R.R.C. redoubled its efforts and Cannon allowed his name to be used on an appeal letter in behalf of Spanish physicians and surgeons in France.[22]

In Cannon's view, there were only two ways that the humanitarian efforts the Medical Bureau had promoted could continue. One was to provide relief to the hundreds of thousands of Spanish refugees in France, the other would be to bring pressure to bear on Franco's government to safeguard those who had opposed him. There was widespread fear that should the Republicans return to Franco territory, they would be treated as traitors, thrown into prison, put on trial and perhaps executed. On March 16, 1939,

Cannon wrote a letter to President Roosevelt, urging him not to recognize Franco's new government until the United States had received assurance that humanitarian conduct would prevail in Spain.

Meanwhile, in a special effort to help some of the exiled physicians and scientists who sought new lives for themselves after the demise of the Spanish Republic, Cannon enlisted the aid of the Boston Committee on Medical Émigrés and other groups that were dealing with the emergency of rescuing displaced persons from Europe. His attempts earned him some measure of success but also a good deal of frustration. Given restrictive immigration policies as well as the lingering effects of a harsh economy, it soon became clear that previous resettlement of more than 300 Germans and Austrians and Italians had left few opportunities for the Spanish physicians and scientists seeking employment in the United States at the turn of the decade. When Cannon saw Juan Negrin in New York in May 1939, the former premier reported that he was looking for places that would take in the Spanish refugees. So far, only Mexico, which had been sympathetic to the Republican side in the civil war, displayed any real enthusiasm for welcoming them.[23]

After the collapse of his government, Dr. Negrin had gone to Paris where he established an organization known as SERE (Servicio de Emigración para Republicanos Españoles, or the Spanish Refugee Evacuation Service) to aid the Spanish Republicans in France and North Africa. In December 1939, French officials raided the offices of SERE and carted away their files of correspondence. Within a matter of days, all of the material taken was returned to the SERE office, but the act of seizure rankled. The officials remained hostile to the Spanish exiles, especially Negrin and Alvarez del Vayo (a foreign minister in the Republican government), who were made to take up quarters outside of Paris.[24]

In the aftermath of the seizure of SERE records and at the instigation of the Spanish Refugee Relief Campaign, a Continental Conference to aid Spanish refugees convened in Mexico City on February 15–18, 1940. The purpose of the conference was to further the mission of Negrin's Spanish Refugee Evacuation Service, first by supplying relief for those who were still in the French concentration camps, and then by arranging their migration to Latin American countries. The Conference more than proved its worth in eliciting fresh assurances of cooperation from Mexico, Chile, Argentina, and Cuba for resettlement of Spanish refugees in Central and South America.

Two members of the American delegation, Samuel Guy Inman, who was treasurer of the S.R.R.C., and Herman Reissig, its executive secretary, introduced a resolution evenhandedly protesting the treatment given the Spanish refugees in France, while recognizing that the French government had spent millions of francs for their support. This resolution was voted down and another put in its place that flatly condemned the French government for repeated instances of ill-treatment, neglect, and abuse of the refugees, who had spent a year of misery and despair in the concentration camps. The leadership in support of the counter-resolution was assumed by a Communist who had been a member of the Spanish Cortes, and he was backed by Douglas Jacobs, the campaign director of the S.R.R.C.[25]

Later in February, Douglas Jacobs was relieved of representing the Spanish Refugee Relief Campaign in any way, shape or form. Nevertheless, there continued to be dissension within the S.R.R.C. and it ran along a political fault-line. Not long after the Conference, a mimeographed copy of an order attributed to General Menard, who was in charge of the refugee camps in France, was received from the International Coordinating Committee at Paris, calling for repatriation of the Spanish refugees by March 15. This order, if authentic, was freighted with such an overwhelming indictment of official French policy that it was up to the S.R.R.C., as a responsible organization, to check its reliability before raising a hue and cry.

When a confusing statement was published in the *New York Times,* Reissig issued his own statement for the S.R.R.C.: "Our concern is simply and wholly that the full right of asylum be granted each individual refugee who is unwilling to return to Spain . . . a forced return of a refugee from France to Spain would be deemed as shocking as a forced return of a German refugee to Hitler. We wait for speedy and full clarification."[26]

Clarification came in a cable from the French ambassador, who wired that General Menard had declared the order to be a fake. Reissig immediately made the cable public, but in the interim, a wholly contradictory series of communications had been sent out, without any authorization, on the letterhead of the Spanish Refugee Relief Campaign. These included copies of the Menard order itself, accompanied by pleas for a mass protest that featured the picketing of the French Consulate. In the belief that the demonstration was likely to do more harm than good, Reissig issued a statement dissociating the S.R.R.C. from the protest.

Not everyone agreed with Reissig. At the end of March, a letter was dispatched to the editor of the *New York Post* signed by Dr. Edward Barsky and others representing the Emergency Conference for Aid to the Spanish Refugees. The letter refuted the *Post*'s presumption that the order of General Menard was a fake and challenged its labeling of all those who acted in protest of the Menard order "as Communists whose only interest was to embarrass and discredit the French government."[27]

As the discord continued, it led to the banishment of other politically motivated members of the S.R.R.C., including John Sherman, once an organizer with the North American Committee. With the dismissals of Jacobs and Sherman, who were the leaders of the leftist element, the New York chapter took steps to repudiate the national organization. Even though Cannon played no official part in the national group, Reissig, out of habit and a sense of solidarity, kept him apprised of the external and internal difficulties at the S.R.R.C. Thus it was that Cannon came to hear about the factionalism at headquarters, which was bringing in its wake reductions in staff and defections among the local branches.

For Cannon, the troubles in New York suddenly explained the mysterious dissension that he had encountered in the Boston chapter. He and Dorothy Hickie continued to raise money locally for the Spanish refugees in France, but he had not heard from the Boston chapter in some time. When in the spring of 1940 he learned that the Boston group (or what was left of it) was joining with the dissidents, he decided to withdraw his name from the honorary chairmanship.[28]

Sending a copy of his letter of resignation to Dr. Barsky, Cannon explained where he stood in all this turmoil. To begin with, he simply could not support the actions of Sherman and Jacobs, which were, in his opinion, dishonorable. Second, he freely admitted that the refugees in France suffered grievously, but since the governments of England and the United States had paid no heed, it did not seem right to denounce unreservedly and exclusively the French government. Moreover, he agreed with Reissig and others that such denunciation was likely to do more harm than good. And third, he thought the rift in the organization in New York unfortunate, but hoped that the internal controversy would not continue and that attention be paid instead to alleviating the suffering of the refugees. Cannon regretted that he and Barsky must differ; nevertheless, he expressed warm appreciation for what the surgeon had done in the cause of the Spanish Republic.

In April, Jay Allen sent Cannon further information on the recent shake-up in the Spanish Refugee Relief Campaign and divulged his own feelings. "I well realize that without the left-wing boys to do the Jimmy Higgins job, we won't have an easy time," he wrote. There were at least two ways of betraying the refugees, he believed: "one is to abandon them to the French authorities, and the other is to hoist over them a hammer and sickle. We have refused to do either. I'm proud of the fact that we got through this crisis with no red-baiting, and the only baiting has been done by them."[29]

The commotion at the center of the S.R.R.C. had scarcely died down when further ructions came from, what was for Cannon, a most unlikely source. The Committee for Cultural Freedom (C.C.F.) issued in April 1940 a report on "Stalinist Outposts in the United States," listing among organizations controlled by Communists, influenced by Communists, or closely cooperating with Communists, the Medical Bureau to Aid Spanish Democracy, both before and after its affiliation with the North American Committee, as being under Communist influence. Cannon, as a member of the C.C.F.—an organization headed by John Dewey that was established in 1939, ostensibly in opposition to totalitarianism—received a copy of the report, and when he saw his friend Harlow Shapley's name as a supporter of an organization on the list (the National Emergency Conference) charged with being under outright Communist Party control, he wrote to Shapley about it. "As a rule," Cannon explained, "I take the attitude of the Chicago reformer who said he did not hitch his wagon to a star, but hitched it to anything going his way." He said he had approved of the goals of many of the organizations whose names appeared on the list and he was not about to be scared off by name-calling.

"On the other hand," Cannon continued, "my conviction has become fairly well settled by recent events that the communists have no real interest in maintaining a democracy, or at least American democracy. They have defended Russian terrorism through thick and thin, and are open apologists for the brutal attack of

Russia on Finland and the connivance of Stalin with Hitler and Hitler's perpetration of hideous cruelty."[30]

Shapley replied that growing doubts made him "half inclined to pull out of all the relief and policy committees." In a recent letter to Rev. Reissig, he told Cannon, he had "probably been nasty and cruel" even though he had written many an appeal in behalf of the Spanish refugees. Of course, "the cheapest way to blacken any project was to scream about the Communists being mixed in with the Professors, the New Dealers, the Philanthropists." However much he knew that, he was left feeling confused by the whole situation. Meanwhile, he salved his conscience by working "like a fiend" on refugee problems.[31]

When Cannon subsequently requested evidence from the Committee on Cultural Freedom for the charges against the Medical Bureau, he discovered that the guiding light behind the report on Stalinist outposts was Sidney Hook, a professor of philosophy at New York University and chairman of the C.C.F. executive committee. For Hook, a well-known interpreter of Marxist principle, the primary purpose of the C.C.F. was to sustain an all-out attack on the Soviet Union and its American sympathizers. He lumped the communists together with the fascists as totalitarians and saw a need to warn the public about seemingly worthy organizations the former had infiltrated in order to use non-communists for their own ends. Such was his animus against the Stalin regime that he became the patriarch of the neoconservatives who would carry the torch of anticommunism throughout the long years of the Cold War.

Cannon entered into a brief but intense correspondence with Sidney Hook that lasted from May through August 1940, with Hook's letters often running to several pages. Initially, Hook made a courteous reply to Cannon's inquiry, supplying lengthy details about the accusations. He even wrote that he admired Cannon's work on the emotions and was sorry that their correspondence should have got off on the wrong foot.[32]

According to Hook, the claim that the Medical Bureau was dominated by Communists came from two sources: testimony about American medical aid in Spain given by Dr. John Jacob Posner and his assistant, Miss Mitchell, and statements about the national office made by Francis Henson. Both Cannon and Reissig made an earnest effort to get at the root of the charges, the former writing to many of the doctors who served in Spain, the latter trying to recall the people and events in the New York office. Of those who had served in Spain, only Dr. Donald Pitts supported the charges of partisanship and incompetence that had been made by Posner and Mitchell. Drs. Leo Eloesser, Abraham Ettelson, Edward Barsky and Jose Puché (who had been in charge of the medical services of the armed troops for the Spanish Republic) repudiated the accusations and viewed the C.C.F. listing as yet another effort to discredit a worthy undertaking with which they were proud to be associated.[33]

When Cannon reported the conflicting testimony he had gathered, Hook objected to the tone and spirit of his letter and bristled at the counter-charges hurled at the C.C.F. "Certainly between two scholars who are searching for the truth," Cannon replied, "there should not be any feeling that one is engaged in insulting the other." But argumentation and Marxist dialectic were meat and drink to Hook and he used those tactics in an attempt to beat down his opponents.

Hook in essence wanted nothing more than to persuade Cannon to recant. It was a bad word to use with Cannon who believed he had nothing to recant. For him, the listing in "Stalinist Outposts" was an effort, a rather cowardly one, to smear a humanitarian organization that no longer existed. In a vigorous defense of the Medical Bureau, he indicated that he knew exactly what he was doing in his effort to aid the Spanish Republic, that he was more than a "nominal head" of the Bureau during its active years, and that he knew all about the organization's work for Spain. "I was well acquainted with the fact that there were communists in the organization," he averred. "That made no difference to me because we were all striving to bring help in a wholly honorable and, in my opinion, laudable manner to a people striving for freedom. There is no more reason, in my judgment, for supposing that the communists were using us who were non-communists than there was in supposing that we were using the communists."[34]

While Hook later offered an amended version of the statement about the Spanish Medical Bureau, he did not give an inch on his stance in the debate with Cannon. He would have gone on *ad infinitum* and *ad nauseum* but Cannon, tired of all the snarling and bickering, the fuss and the fury that seemed to have taken over the entire Spanish aid movement, brought the correspondence to an abrupt halt with a letter written on August 28. After Hook proposed that Cannon hear the charges of Francis Henson against the

Medical Bureau, Cannon, pleading commitments to his war work as well as to the refugees in France, said he did not know Henson and was unable to credit his testimony. In an effort to redirect Hook's bulldog persistence, Cannon suggested that he contact Roger Baldwin or Herman Reissig, who knew better than anyone else "whether Communists in fact gradually gained control of the organization."[35]

In response to the German occupation of France, the Emergency Rescue Committee (E.R.C.), with Frank Kingdon as chairman, was founded in New York for the purpose of helping all of the anti-fascist refugees in war-torn Europe. Several relief organizations worked through the Emergency Rescue Committee, including the Spanish Refugee Relief Campaign. After participating in a meeting at the American Academy in Boston in the fall of 1940, Cannon agreed to join with Freda Kirchwey of *The Nation,* Alvarez del Vayo, and Samuel Guy Inman to visit the offices of Nelson Rockefeller and other prominent government officials in Washington in behalf of the Spanish refugees in France. Although the delegation was received courteously, they learned that the problem of transport could not be solved by the use of U.S. ships because of the Neutrality Act, which would require an act of Congress to amend.[36]

In March 1941, Dorothy Salinger asked Cannon to serve on an advisory committee of a newly organized Boston branch of the E.R.C. She assured him that she and Dorothy Hickie would do the work but they needed advice and help with fundraising. He responded that he was so overwhelmed by numerous obligations that he must not take on any further burden. Not surprisingly, he left the door open. "If my name would be useful to you on the Advisory Committee and if I could be spared any time-consuming service, I am willing to cooperate with you. I hate not to join your effort with enthusiasm but you can readily understand that there are limitations which cannot be passed without danger." Later, when he became concerned about competing with the S.R.R.C. in fundraising, Reissig reassured him that the Emergency Rescue Committee was cooperating in a joint campaign conducted by the S.R.R.C., the New World Resettlement Fund, and the International Relief Association.

Even so, it wasn't long before both men acknowledged that the E.R.C. was turning out to be "pretty much a committee to rescue

Germans and other Central Europeans and has never had much stomach for trying to get Spanish Republicans out." Late in 1941, Reissig wrote Cannon that the S.R.R.C. was "nearing the end of its independent existence" and would continue to collaborate with the E.R.C. until it stopped operation, probably early in the new year.

Reissig, planning to return to the ministry, asked Cannon for a recommendation to help him find another position. He took a moment to look back at the enterprise which had brought them together and tried to express what it meant to both of them. "You will, I am sure, always be glad that you accepted a place of leadership in a cause that deserved the support of all good men. Every day brings new vindication of what you and others stood for. And in this work you had not a little to do with preparing our country for the tasks that now confront it."[37]

Some weeks later, after the Japanese attack on Pearl Harbor, Cannon, too, was in a reflective mood as he tried to fit the fate of the Spanish Republic into the series of events then rapidly coming to a head. Writing to his friend Carrasco-Formiguera (who had found sanctuary in Mexico), he looked to the future as well as the past:

> Although the United States is now involved in this vast martial conflict I do not find myself saddened, but rather, elated. It is a war of *production* and in the long run we can go far beyond the Axis Powers in making the given machinery for killing. I have no doubt that we shall be victorious and that the hideous experience of the past years will force mankind, for its own welfare, to organize against the renewal of these immense catastrophes.
>
> What a strange world! The very people who called me a "Red," a "Communist," a "candidate for an insane asylum," because I supported the Spanish Republic (aided by unspeakable Russia) are now ardently shouting for yielding everything possible to the Russian Government. I shout with them! But now I am not regarded as a queer one. How few of them realized that the war in Spain was the first act of the present world tragedy. When it is all over and peace comes again, may we hope that Spain will have the freedom from oppression which she sought.[38]

23 *The End of a Career*

AS THE 1930s drew to a close, Cannon met with Dean Sidney Burwell to discuss the future of the physiology department. He especially wished to remind the dean that the Rockefeller Gift would not continue beyond August 31, 1942, by which time he would no longer be head of the department, having reached the retirement age of seventy. Two problems thus confronted the Medical School. One was the appointment of a successor to head the physiology department, and the other, securing funds to support its future research.

Afterward, Cannon made a memorandum of the conversation. He ended by stating: "I expressed the intent of not participating in the selection of my successor. It seemed to me that the younger men who are responsible for the future of the School may look forward to a development of the Department different from the past and that they should not be affected by my views." Soon after the new year, he wrote Burwell that he had shown his memorandum to Drs. Forbes, Davis and Rosenblueth before sending it on to him.

With the beginning of the new decade, the reality of Cannon's approaching retirement began to seriously affect the physiology department. Hallowell Davis, based on his past experience with President Conant, cast about for a job offer to use as a buffer in negotiating his future position in the department. For his part, Cannon decided not to assume responsibility for any more Ph.D. students because he would not be able to see them through to the end of their training.[1]

The future of many members of the department was in doubt, and not just because of Cannon's approaching retirement. In the autumn of 1940 came the distressing news that Elizabeth Bright had been examined (at Cannon's suggestion) by the neurologist Houston Merritt and diagnosed as having multiple sclerosis. Although she remained on the payroll while she received care at Boston City Hospital, under personnel policies then in force Cannon had to inform her that her checks would cease as of October 1. Unfortunately, Miss Bright had little or no savings and retirement plans were few and far between at this time. To soften the blow, Cannon could only express his appreciation for the admirable contributions to physiological reporting that she had made for more than a score of years. Her preparation of records and graphs had, he wrote her, "set a standard which in the early days contrasted strikingly with the methods used elsewhere . . . I only wish that there was an immediate prospect of your continuing in this fine service." When Bright said she expected to get well enough to return to work, Cannon suggested to Dr. Merritt that he inform her of the seriousness of her disorder. Merritt, who shared Cannon's dilemma, replied, "I dislike to tell these patients the cold truth, but in some cases it is necessary and I believe this is true in the case of Miss Bright."[2]

The tragic case of Betty Bright was soon overshadowed by the turmoil of a world at war and the many changes wrought as the country geared up for supreme effort. Over the course of 1940, increasing numbers of individuals became engaged in activities in behalf of national defense, and not the least Cannon. In the spring, his chairmanship of the Shock Committee was added to his other obligations in behalf of the Rockefeller Foundation, the National Academy of Sciences and the National Research Council, the American Association for the Advancement of Science, the Spanish Refugee Relief Campaign, and a new appointment to the President's Polio Foundation—all requiring his attendance at a regular round of meetings away from Boston. "Sorry to miss you in Washington and in New York. We both seem to spend more time there than here!" wrote Hal Davis, who was working on projects investigating problems of sound and hearing for the National Defense Research Committee, as well as for a Committee on Aircraft Soundproofing of the N.R.C., in a postscript hastily added to one of his letters to Cannon.

By 1941 Alex Forbes had left the department for active duty as a Lt. Commander in the U.S. Navy—his second enlistment in a world war. In April, when Cannon wrote Forbes about continuing financial support for Dr. von Brücke's stay at Harvard, he added a wistful personal note: "With Dr. Davis away in Washington not infrequently and my own resort there not infrequently and your absence from the Laboratory, the place seems most unhappily inactive. I know that the services which are being performed are important but I cannot help feeling that science is suffering in consequence."[3]

Whatever gave rise to Cannon's forlorn lament, activity in the physiology department was not quite so bleak as he painted it. The biologist Hudson Hoagland, for one, spent a considerable part of 1940–41 in Cannon's laboratory, where he learned the techniques of mammalian physiology and completed two investigations. "Hoagland fitted into the group here admirably," Cannon later wrote Harlow Shapley. "There is no question in my mind about his excellent personal qualities." Oscar D. Ratnoff, a teaching fellow in physiology, for another, later recalled that he conducted experiments on the excretion of epinephrine after stimulation of the efferent nerves to the adrenal glands. He depicted Cannon at that time as "an inspiring, encouraging, thoughtful man, someone to whom I related instantly as a friend, if somewhat distant and much distracted by outside interests."[4]

That spring Cannon was faced with another loss when he learned that Harvard planned to retire Mrs. Agnes McLeod, a longtime laboratory assistant in his department, even though she preferred to go on working. Cannon's response to the news was heartfelt: "After all, some people are senile at 55 and others do not show limitations which spoil their efficiency even when they are 80! Why should we insist on chronological age instead of physiological age in the Department of Physiology?" If it had been up to him, he would have kept her on, but rules were rules and the Administrative Board was firm in holding to them.

Though Cannon had to go along with the decision, privately he wrote Mrs. McLeod that he had hoped they could retire together. "We are surely the only remnants of the old Department in the School on Boylston Street. When we go, that memory will be entirely lost." He could not forego thanking her "with the deepest feelings of gratitude for all the long years of faithful help which you have given to the very many persons who have worked here and who have needed your help."[5]

1940 was the last year that Cannon published a profusion of papers. They are an impressive lot, demonstrating that he maintained a range of interests although nearing the end of his career. He was co-author of a pair of articles on the early and late stages of neuromuscular transmission that followed up the work of Arturo Rosenblueth with R.S. Morison (1937) and J. V. Luco (1939). The first paper, by Rosenblueth and Cannon, reported their study of the action potentials of motor nerves in cats. The animals had received tetanic stimulation, by means of measuring the tension developed in striated muscles (recorded on electromyograms) and with injections of atropine, prostigmin, acetylcholine and potassium. The second paper, by Cannon and Rosenblueth, reported their study of muscular response in the late stages of fatigue and recovery, then considered the results in relation to the chemical and electrical theories of neuromuscular transmission.[6]

A paper on "Autonomic Control of Thyroid Secretion" reported the experiments that Harry Friedgood had been conducting in the physiology department during the 1930s—work for which Cannon had high hopes but which ultimately fell short of his expectations. Friedgood and Cannon, believing that the discovery of the anterior pituitary hypophyseal thyrotropic hormone offered a new experimental approach to research on thyroid function, had returned to the subject by repeating earlier experiments, as well as conducting new ones. Unfortunately, the relation of the sympathetic nervous system to thyroid function remained unclear, and they concluded that adequate explanation was still lacking to account for the difficulty encountered at Harvard and elsewhere in reproducing the results originally reported by Cannon, Binger and Fitz (1915).[7]

Cannon's address on "The Adrenal Medulla," delivered at the New York Academy of Medicine, summarized his work on the sympathico-adrenal system, but at its core was yet another refutation of the claims of Stewart and Rogoff. Cannon offered no apology for dishing up the old controversy because, he maintained, misleading statements still appeared which seemed to indicate that the matter had not been conclusively settled. Another speech, "The Role of Chance in Discovery," delivered to the

American Science Teachers Association, followed up on a previous lecture on serendipity. And a third one, a commencement address delivered at the Washington University of St. Louis, where Cannon added another honorary degree to his collection, outlined changes in medical education and advances in medical knowledge.[8]

These were years when honors continued to come Cannon's way. At his wife's urging, he accepted an invitation from the University of California to serve as its Hitchcock Visiting Professor in February 1941. That spring his early researches were acclaimed anew when he became the first recipient of the Friedenwald Medal of the American Gastroenterological Association. The citation read, "in recognition of his pioneer utilization of the X-rays in gastroenterology, and his important contributions to the mechanics of digestion, to the elucidation of the sensations of hunger and thirst, and to the development of the science and practice of gastroenterology."

No sooner had Cannon received the Friedenwald award than he was on his way to California again, this time as one of the honorary fellows at the 50th anniversary celebration of Stanford University. For a symposium on "The University and the Future of America" he delivered a speech on problems confronting medical investigators. Forecasting increasing numbers of elderly in our population, he urged that a medical attack be made on the diseases and disabilities of old age. He called for future studies to consider such subjects as the mechanism which keeps the body's life fluids in delicate balance, the increase of nervous disorders, and the use of drugs in healing.[9]

Upon his return home, the summer of 1941 was largely given over to reparative work on Cannon's hands. Ever since the early attempts in 1933 by John Homans to deal with the damage caused by old X-ray burns, Cannon had from time to time needed further operations. These had been performed either in Boston by Homans or Elliott Cutler or in St. Louis by Vilray Blair and Barrett Brown. In mid-July, Bradford operated on Cannon under local anesthesia for more than two hours, putting on five grafts, skinning bad places, and (to his mother's mind) generally making a new man of his father. Post-operatively, she reported to her family that he was comfortable and cheerful, no pain and no fever. When Brad removed the dressings, his hands looked fine, and overall both father and son were pleased that all had gone well. Dr. Aub, who attended the de-bandaging, told Cannon, "That's the best work that has ever been done on your hands."[10]

After recuperating in New Hampshire, Cannon returned to the Medical School for his final year as George Higginson Professor of Physiology. Despite the war's impact, the year began well. It was marked in the autumn by the arrival of his son-in-law, Douglas Bond, who had a National Research Fellowship for a year's training in physiology before going on with his career as a psychiatrist. There being very few graduate students this year, Douglas received a goodly share of attention. "He has developed a great admiration and fondness for Dr. Rosenblueth," Cornelia remarked, "with whom he is working, which pleases Father whose beloved pupil Dr. R. is, and also makes Douglas's work more interesting and run more smoothly."

The research that came out of this collaboration was published as "The Control of Clonic Responses in the Cerebral Cortex," by Rosenblueth, Bond and Cannon. It followed on the heels of an earlier report on "Cortical Responses to Electric Stimulation," by Rosenblueth and Cannon—a paper of such length and complexity that it required a table of contents. In brief, Rosenblueth and Cannon first studied the motor and the electric cortical responses to single and repetitive cortical stimulation, chiefly in rhesus monkeys but also in dogs and cats. Originally, the investigation had intended to elucidate some of the interesting problems offered by the continuing after-effects of electrical stimulation of the motor cortex (experimental epilepsy). But as it developed, the purpose was enlarged to include several different types of cortical responses that could be elicited by stimulation of any cortical area.

The second paper of this pair refers to the first, which reported some features of the tonic-clonic, self-sustained responses of the cerebral cortex to electric stimulation. Here Rosenblueth, Bond and Cannon attempted an analysis of the mechanism by which different cortical regions may become coupled during clonic activity, as well as an elaboration of the factors determining the rate of clonus, and the reason for the simultaneous abrupt end of the responses in several areas. To do so, they used young rhesus monkeys anesthetized with chloralose and applied stimulating electrodes by hand. The cortical electric responses were then recorded by means of a

Grass six-channel set of ink-writing, moving-coil galvanometers and associated resistance-capacity coupled amplifiers.[11]

The apparatus used in these studies was provided by Albert Grass. In 1935 Grass, just out of M.I.T. with an electrical engineering degree, had been taken into the Forbes-Davis section of Cannon's department to develop recording apparatus for detection and measurement of electric phenomena. He was so adept at inventing, designing and building such equipment, for which there was a great and growing demand, that he and his wife Ellen (whom he met in the physiology department) established their own company as makers of electroencephalographs and electrophysiological instruments.

In October 1941, Cannon's seventieth birthday was greeted with fanfare. First, there was a party in Washington arranged by his daughters Linda and Wilma, then a family dinner in Cambridge that included the Pierces, and a third celebration with members of his department, organized by his secretary, Minna Weiner. "There were Brazilians, Mexicans, women students and men, a nice international democratic group," Cornelia fondly reported, "all listening respectfully while the great man reminisced about the good old days of no ventilation, cat fleas and mice smell in the Medical School down on Boylston Street and told amusing tales of former students in the School."[12]

Cannon had scarcely rung in the new year of 1942 when he sent a reminder to the dean about his approaching retirement. He wondered whether Burwell would like a more definite statement of his resignation than the one he had sent to President Conant when he reached the age of sixty-six. "Since the autumn of 1940 the world has greatly changed and the demands on everyone in the United States have greatly increased," he wrote. "I wish it to be understood that my services are at the disposal of the Medical School so long as my health and vigor persist, even though I am on the retired list."

Acting on the advice of Burwell, Cannon sent a similar letter to Conant. When after a reasonable time he had not received a reply, he wrote Burwell again to tell him of the uneasiness in the department of physiology as to its future. No sooner had Cannon sent his letter to Burwell than he received the president's reply. Conant wrote that he had discussed the matter with the Corporation, and he expressed their appreciation not only for Cannon's offer but also for "your distinguished service and how much the medical profession will lose by your retirement." "Nevertheless," he stated, "considering all the factors which you mention in your letter, I believe that it would be well to continue on the plan which you have outlined, namely to retire at the age of seventy at the end of this academic year."[13]

Conant's letter had a salutary effect on Cannon, or so it seemed to Cornelia. She told her family he was like a boy released from school, so relieved after showing his willingness to help yet thankful that it was not asked of him. Moreover, he was in fine condition, rested and untroubled by the upheavals in the laboratory and the large class of medical students. Determined to make his last year a good one, he was spending more time than usual with the students.

Although Cannon kept up a good front, he could not help drawing comparisons between then and now. "The Laboratory is a very different place from what it was ten years ago," he wrote one of his former fellows, Abe Ravin. "I look back on that time as the acme of my experience—then everything was going at top efficiency and groups of investigators from various countries were working happily together on interesting problems. On August 31 I shall end my Professorship. It is not altogether pleasant to go out leaving the Laboratory in a relatively inactive state."[14]

Astonishingly, with Cannon's retirement fast approaching, no decision had yet been reached in regard to his successor. An anxious Hal Davis told the dean that he hoped the situation would be clarified in the near future, both for the sake of the other members of the department and for himself. Not only was Davis negotiating for another job, but so was Arturo Rosenblueth, who had been asked if he would consider taking a post in Mexico City. "As I have told you," Cannon wrote Burwell, "I should regard the departure of Dr. Rosenblueth . . . as a very serious loss to the School and especially to the Department of Physiology. If there is any sort of statement which can be made to Dr. Rosenblueth which will encourage him to stay here I hope that it will be made."[15]

Even though Rosenblueth held a privileged place as Cannon's close collaborator, his years at Harvard had been clouded by the insecurities of an outsider. The budget of the physiology department allowed for only two tenured full professorships—those of Cannon and Forbes. When it came to job security, even Davis was skating on thin ice, having constantly to come up with outside offers to be assured of promotion or reappointment. Rosenblueth's appointments as assistant professor (in line with Conant's "up or out" policy) had been limited to two three-year terms and his salary was patched together with funds from the tutorial budget, thus adding financial insecurity to job insecurity.

In 1939 Rosenblueth, seeking a more permanent foothold at Harvard, had applied for a position at the Howe Laboratory, an affiliate of the Medical School. At the time, Cecil Drinker, acting in behalf of the ad hoc committee, wrote Phil Bard a confidential letter in which he requested a frank opinion of Arturo. Bard wrote back that he was not sure what he thought of Rosenblueth; nevertheless, he proceeded to put his thoughts down on paper. While Rosenblueth had tremendous energy and an active mind, he was also "just a bit of a physiological Jesuit," Bard began—but hastened to add that was the worst he could say of him. He was certain that Rosenblueth had never engaged consciously in any active intellectual dishonesty.

And yet Bard continued to qualify whatever praise he had for Rosenblueth. If Arturo could "think circles around most people," the course of those circles was "not entirely even." He was, on the one hand, extraordinarily good at picking up and adapting for his purpose various techniques; on the other hand, he was poorly adapted for scientific investigation requiring time and patience. Though Bard's relationship with Rosenblueth had always been pleasant and he was really quite fond of him, he acknowledged that a good many people did not like him and some did not trust him. "I think you can be sure," Bard told Drinker, "that if his position is such that he has to deal with a variety of individuals there will sooner or later be trouble." Bard ended his letter by admitting that Rosenblueth was a complex fellow and that his impressions might not have been fair. "I think he is genuinely devoted to Dr. Cannon," he concluded, "and I have seen evidence that he is willing to put himself in the background on several occasions for the sake of Dr. Cannon's feelings. On the other hand, I feel quite sure that on many occasions he has done Dr. Cannon no good. And I, for one, rather deplore that scientific partnership."[16]

Though Bard's letter may be viewed by some as "back-stabbing," there is no denying that Rosenblueth's strong personality frequently clashed with the standard of behavior then prevailing in the scholarly community. Not only did his sometime arrogant and patronizing manner cause others in the department to resent him, he was known to belittle the medical students and called their teachers "goddamnfool clinicians." Besides being a volatile Latin, Rosenblueth was also of Jewish extraction—two strikes against him at an institution like Harvard with its inbred ancestry. There was a third strike against him as well. Although he had filed preliminary papers, Rosenblueth had not taken the final step to become an American citizen. In part, it was because he did not wish to sever the ties that bound him to Mexico and, in part, it was because he did not yet have a permanent position in the U.S. with an adequate salary.

This is not to say that Rosenblueth lacked supporters. It was no small matter that the two top men in the physiology department—Cannon and Forbes—had nothing but praise for his abilities. "Dr. Rosenblueth is one of the most brilliant and productive physiologists in this country, if not in the world," Forbes wrote in 1939 when nominating him for membership in the American Physiology Society. Indeed, Forbes, though a member of the supposedly stodgy Boston establishment, got along famously with Rosenblueth.

Another person who got along famously with Arturo was Norbert Wiener, the very exceptional and somewhat eccentric M.I.T. mathematician. In the introduction to his 1948 book *Cybernetics,* Wiener recalled attending the monthly seminars that were organized and run by Rosenblueth beginning in the late 1930s. A sort of after-dinner roundtable held in Vanderbilt Hall (the medical students' dormitory), each seminar was a lively presentation and discussion by young scientists and students of a single topic that related philosophy to scientific methodology. Wiener became a regular member of the seminars; he also entered into an exciting new collaboration with Rosenblueth on feedback systems, which would eventuate in the new science of cybernetics.[17]

The future presented a host of problems for Cannon as he neared the end of his active years at the Medical School. Not only did he worry about the prospects of Rosenblueth, Davis and others in his department for whom he felt responsible, he also wondered about a place for himself in the School after retirement. His own situation was complicated. When Cannon asked the dean if he might have assigned to him "a room which, so long as I am fit, I could use for my purposes in the Harvard Medical School," Burwell assured him that he had already discussed the matter with one or two people "because I knew it would be a great asset to the School to have you here." Cannon, expecting that he might soon be going abroad, preferred to postpone his move out of the department—though he was prepared to leave the rooms as soon as he received word that his successor was moving in. "The long delay in appointing someone to take my place," he told Burwell, "has been the main source of difficulty and I do not know that anyone is to be blamed for it."

For all the dean's assurances, Cannon was not assigned a place to work within the Medical School after his retirement. Instead, Burwell telephoned him with an alternate plan: Elliott Cutler had extended the hospitality of the Brigham Hospital, which offered to provide space in the Surgical Research Laboratory. If Cannon was disappointed, he didn't show it; rather, he wrote Cutler that he was deeply touched by the generous offer. One bright spot was that he would not have to sever his relations with the group at the Brigham, where he had happily served as consulting physiologist for so long.[18]

If there were problems, there were also satisfactions as Cannon looked forward to life in retirement. In March he sent Alan Gregg his elaborate study with Rosenblueth on self-perpetuating responses from stimulation of the cerebral cortex. It was, he said, "perhaps the most important contribution of this final year of the prolonged period during which the department of physiology has had the great advantage of support from the Rockefeller Foundation." Cannon told Gregg about his service as chairman of the Committee on Shock and Transfusion and on other boards and committees "where perhaps my experience may still count for something." And he promised that when he could find the time, he would begin writing a monograph on "The Increased Sensitiveness of Denervated Structures"—an outcome of work which the Foundation has supported.

Gregg, in reply, assured Cannon that the Foundation was well satisfied with the grants made in support of his work:

> You have added to knowledge steadily ideas as well as facts of importance, you have found, interested and trained a large number of valuable workers, you have been a peculiarly helpful friend to the best interests of your own school and medicine everywhere.
>
> If in your opinion the Foundation's aid has been helpful in enabling you to do these things, imagine the satisfaction and gratitude we bear you for having used such aid so effectively.[19]

Yet another satisfaction for Cannon lay in the publication of his article on "'Voodoo' Death" that spring in the *American Anthropologist.* Not only did "'Voodoo' Death" arouse widespread interest (and continues to do so today), it was also the last of his pronouncements on the bodily effects of strong emotion. Moreover, its publication coincided with the end of his career.

Cannon's interest in the practice of sorcery and witchcraft in primitive tribes had been sparked nearly a decade earlier, after one of his research fellows, Norman E. Freeman, found that intense excitement could cause a progressive reduction of blood volume until a shock-like state developed that resulted in death. In 1934, Cannon began to gather reports from anthropologists, practitioners and other observers of cases of voodoo death among natives in South America, Africa, Australia, New Zealand, Haiti and several of the Pacific Islands. He also sought out information from old friends like Walter Alvarez and Daniel de la Paz, among others. In 1941, while serving as visiting professor at the University of California, he had presented a preliminary form of his paper to a group with an interest in anthropology.[20]

The first part of "'Voodoo' Death" cites, from the case records Cannon gathered, instances of death resulting when natives were subjected to spells or sorcery or black magic. While ethnologists, on one side, admitted that the belief that one has been subjected to sorcery and inevitably condemned to death could actually end in death, physiologists and physicians, on the other side, were skeptical and considered the phenomenon impossible. Rather than deny that voodoo death was within the realm of possibility, Cannon looked closely at its features, the most prominent being an ominous and persistent state of fear and the profound physiological disturbances associated with it.

He referred to the evolution of the fight-or-flight response and summarized the action of the sympathico-adrenal division of the nervous system that controls "the 'internal environment' of our living parts." He described the experiments of Philip Bard and of Norman Freeman, as well as findings on shock in World War I, which were newly pertinent to his discussion because they showed that a disastrous fall of blood pressure—whether brought on by prolonged and intense emotion or by severe wounds—ends in death. Persons dying as a consequence of sorcery go without food or water as they wait in fear for their impending death, and this factor augments the damaging emotional effects to induce the fatal outcome. After recounting further instances of shock brought on by fear, fatigue, and lack of food and water, Cannon concluded that voodoo death may indeed be real. It may be explained as due to shocking emotional stress—in other words, to obvious or repressed terror.

Fifteen years later, "'Voodoo' Death" was reprinted in an issue of *Psychosomatic Medicine* that contained papers presented at a meeting of the American Psychosomatic Society held in 1957 to memorialize Cannon. Among the contributions was one by Curt P. Richter of Johns Hopkins, which followed up on Cannon's article and which he summarized: "A phenomenon of sudden death has been described that occurs in man, rats, and many other animals apparently as a result of hopelessness . . . In this instance as in many others, the ideas of Walter Cannon opened up a new area of interesting, exciting research."[21]

In March 1942, just a few months before Cannon's retirement, B.A. Houssay, who ran the most productive laboratory of physiology in Latin America, wrote him a letter that offered the alluring prospect of an invitation to visit Argentina. Cannon and Houssay were old friends. Not only had Houssay sent students like J. T. Lewis, Teodoro Schlossberg and Oscar Orias to Cannon for advanced training, he also had visited Harvard himself as a Dunham Lecturer in 1935 and as a delegate to the Tercentenary celebration in 1936. He held Cannon's research in such high regard that he joined with others in nominating him for a Nobel Prize in medicine or physiology. Cannon, in turn, believed Houssay to be the premier medical scientist in South America. He proposed him for honorary foreign membership in the National Academy of Sciences and also nominated him for a Nobel—a prize that Houssay would ultimately win in 1947.[22]

Houssay in his letter told Cannon that some of his former

students in Argentina had already written to the Division of Cultural Relations of the State Department about the prospect of his visit and if he would indicate that he was willing, they would gladly sign a collective invitation asking him to come. Cannon, though intrigued by Houssay's proposal, scarcely knew how to answer him because of the various obligations confronting him, especially his war work in conjunction with the Committee on Shock and Transfusion. But he left an opening: if the State Department should ask him to visit some of the countries of South America in the autumn, that is, after he had retired from his post at Harvard, he might be able to do so.

A few weeks later, Lewis Hackett, a representative of the Rockefeller Foundation in Buenos Aires, reported to his home office that the most impressive group of investigators in Argentina, Chile and Peru was that of the physiologists. "They have not only introduced the spirit of investigation and the scientific method in their own field," he continued, "but have had extraordinary influence in many other fields of scientific teaching in the Universities. The originator of all this was Dr. Walter B. Cannon of Harvard University who now has students in influential University positions throughout South America." Wherever Hackett went, he had been asked whether there was any likelihood of a visit from Dr. Cannon to South America after his retirement. In Hackett's opinion, the Moe-Keppel-Stevens committee (i.e., the Committee for Inter-American Artistic and Intellectual Relations of the Rockefeller Foundation) could not hope to do anything better with their money than to send Cannon to South America, provided he was willing to go.[23]

Still other forces contributed to the drive to send Cannon to Latin America. That spring he was nominated to fill the post of Foreign Secretary of the National Academy of Sciences, which had been left vacant by the death of his longtime friend and colleague, Lawrence J. Henderson. Though Cannon thought the position should go to a younger man and had even suggested Harlow Shapley as a better choice, a short time later his election to the office was formally announced. As a component of Cannon's new duties, he automatically became head of the Division of Foreign Relations of the National Research Council. As such, the prospect of bolstering relations between the United States and South America fit very well with a possible goodwill tour. Accordingly, he wrote Ross G. Harrison, Chairman of the N.R.C., to tell him about the tentative invitation, though he supposed that going on an extensive journey to South America would perhaps not be the best way of performing the services of Chairman of the Foreign Relations Division.[24]

Dr. Harrison raised no objections, however, and plans for the trip proceeded apace. Late in May, Henry Allen Moe (of the Guggenheim Foundation and the Rockefeller committee) asked Houssay to send him a specific program for Cannon—itinerary, lectures, consultations—and also, the transportation situation being what it was because of wartime exigencies, a statement about Mrs. Cannon as well. And there lay the rub. Although no one could doubt that Cornelia, were she to accompany Cannon on the Latin American tour, would make a fine goodwill ambassador in her own right, getting air priority for her was no easy matter. She told her family she had suggested to Cannon "that if we could not both go, he go alone, feeling pretty noble as I said it. 'No,' said Father cheerily, 'I don't want to die alone.'"

All joking aside, Cannon was perfectly serious about his wife going with him. "I have reached the time of my life when I feel unwilling to be long separated from her," he wrote Mr. Moe in early June. "I should not plan to go to South America unless she should go with me. If arrangements could not be made for her to accompany me I fear that the plan for a South American trip is definitely *out*."[25]

Nevertheless, plans did go forward for a visit in the fall of 1942, especially when Cannon found that he could give his lectures in English, that Dr. Weed approved his absence from the Shock Committee, and that Moe's committee's funds would not be available if he put the trip off until 1943. Though the sticking point still remained the question of obtaining air priority for Mrs. Cannon, Henry Moe worked out a schedule of the tour and wrote Houssay it was all but settled that the Cannons would make the trip.

Late in June, after discussing the situation with a colleague in the State Department where she was then working, Wilma wrote her parents that for the duration of the war there was no possibility of their going to South America together. She had been told that with the heavy traffic to Africa and points east via South America, it was all the Department could do to get air priority for a single individual and no wives were being sent back and forth from either direction. Wilma was indignant that Mr. Moe (or whoever it was) had not told them as much quite definitely so that they would not be expecting to go at any moment. Her mother took the news in stride. If Wilma's pessimism about air priority was justified, they wouldn't be desolated if they did not get to go; but Cornelia thought it would be nice for Walter to have a complete break after his retirement.[26]

Although others stepped in and tried to pull strings, they did not get very far. Cannon was furious, but Moe urged him not to express his anger officially. "Wilma writes that it is all off," Cornelia informed her family early in August, "but Mr. Moe asks Father to wait a minute . . . so we wait, muttering Spanish by the hour meanwhile."

At last, it became clear that the visit must be indefinitely postponed, and Cannon drafted a letter of regret to the various Latin American scientists. "If normal circumstances were prevailing," he wrote, "I am assured that responsible authorities in the Department of State would feel justified in urging that we go now; but present conditions and the prospect for the immediate future are not favorable. May I express the wish that at a later—and happier—time we may realize our hopes."[27]

On August 31, 1942, the official date of his retirement, Cannon marked the end of his career by sending a letter to President Conant:

> This is the last day of my formal connection with Harvard. I do not wish it to pass without a word of tribute from me to express my great debt to the University.
>
> Fifty years ago this September I arrived in Cambridge from St. Paul, Minnesota. A remarkable teacher of English literature in the High School had encouraged me to try to enter Harvard College. There was little possibility of family support, but Price Greenleaf aid and about $180 from my father during the first year gave me a start. With that beginning and with outside work and scholarships I was enabled to finish the eight years of the College and the Medical School. Life in the University during that time was extraordinarily rich with thrills from new interests and enthusiasms.
>
> Forty-three years ago I began to teach: first as Instructor in Zoology, then as Instructor in Physiology at the Medical School, and in 1906 as Dr. Bowditch's successor. In teaching and in conducting research I have enjoyed for more than four decades the marvelous experience of doing what I had rather do than anything else in the world, in the shelter and care of the University.

> Harvard has been to me truly an *alma mater.* Everything I have been able to accomplish during my career I owe to her. I look upon her, as on a bounteous and benevolent giver to all that is worthy in human striving, with reverence and affection; and I shall continue to do so to the end of my days.

Conant, in a belated reply, expressed the belief that it was "the University who is indebted to you, not the other way round." He called Cannon's tribute to his alma mater "a document that no one can read without emotion."[28]

By the time Cannon received Conant's rejoinder, however, he was in the grip of an unusual manifestation of his old ailment. "About two weeks ago I was afflicted with a Bell's palsy on the right side—a total paralysis of the facial muscles," he wrote his friend Robert Yerkes in September. "Fortunately the damage is peripheral and is commonly recovered from in four to eight months. Meanwhile the troubles of speech and eating and drinking are unpleasant. Added to my palsy has been a sudden and distressing deafness." Cannon told Yerkes that he was counting on the *vis medicatrix naturae* to pull him out of his difficulties; curiously enough, he ended, "You can understand, however, that I, who have enjoyed health and vigor for decades, take these limitations with ill grace."

Although Cannon was subject to occasional blue moods, it was typical of him, in times of trouble, to emphasize the positive and be optimistic about the future. Ever since he had gone into remission from his attack of mycosis fungoides in the early 1930s, he had considered himself in relatively good health. While others were wary and believed his physical condition to be precarious, at best, Cannon's cheerful outlook had allowed him to function in a manner that was the envy of men in far better health.

Even so, he had had to live with painful keratoses on his hands from old X-ray burns, red, itching skin, and bouts of sleeplessness, headaches or digestive upsets, especially in times of extreme fatigue or stress. More than once, his friend and physician, Joe Aub, had interceded so Cannon might be relieved of onerous duties that exacerbated his distress. The hectic days of heading the Spanish Medical Bureau was such a time and in the summer of 1938, Cannon decided to head for the restful hills of New Hampshire rather than attend the International Physiology Congress in Zurich. Among his reasons, he wrote Phil Bard, was the condition of the skin on his face and his hands, which were covered with surgeon's plaster. "I hate to be conspicuous, and I hate to be eternally explaining. That's a dreadful bore. So I decided to stay in Franklin."[29]

The following spring, Cannon detailed the course of his illness to Peyton Rous, with whom he served on the Board of Scientific Directors of the Rockefeller Institute, and offered himself as a guinea pig. In a letter of May 1939, he told Rous that within the past year, the redness had reappeared on his face, and there were also areas with red blotches over the lower abdomen and on his arms and thighs. He continued:

> You probably know what the later stages of this disorder are likely to be. The only cheering feature of the disease is that in my case it appears to be developing very slowly, and of course at my stage of life may not reach its ultimate manifestations.
>
> I am sure that you will regard this letter as confidential. I am telling you about my experience because it may be related to your interest in the lymphatic system and also to the other research activities in which you have been engaged. I am at your service if there is any remote chance of this rather rare condition being used for study.

Other than the condition of Cannon's skin, to his physician he looked very well and was quite cheerful—especially when he was pleased with the way his research was going. Moreover, Cannon was fully convinced that the combination of fish oil and vitamin D in the Navitol drops he took daily was especially efficacious. Despite recent setbacks, noted Dr. Aub in 1940, Cannon had been improving steadily for some years.[30]

Late in September 1942, Cannon alerted Philip Bard that because of his attack of Bell's palsy, he might need help at the upcoming conference on shock in New York that he had been planning. Even worse than the facial paralysis was the deafness that had stricken him, for it would render him unable to record the discussion for the conference report. At the same time, he wrote Magnus Gregersen and asked him to cooperate with Bard in preparing an account of the afternoon session, and then asked Alfred Blalock to take charge of the conference. "I cannot tell you how deeply I regret my inability to participate in the meeting next Thursday," Cannon wrote Bard. "I have been looking forward to it, eagerly expecting to promote progress in our researches by discussing moot points. Now I must count upon the men whom I have selected to lead the groups and also upon you and Dr. Blalock to help me in my predicament."[31]

There was worse to come, however. One morning in October, Cannon woke up with the muscles paralyzed on the left side of his face as well as on the right side—a case of bilateral Bell's palsy. Cornelia was horrified when she saw him, but it was not the palsy that he minded most. "He never had any personal vanity so his paralysis is not so hard to endure," she wrote her family, "but he says deafness makes him feel 'so alone' and it has been the basis of his withdrawing from his job with the shock workers and from any activity which requires committee meetings." Cannon planned to spend the next several months largely in their Cambridge apartment, writing on his book, walking in the dusk, listening to the radio and, if his eyes allowed, going to the movies.

The affliction interrupted Cannon's activities in behalf of foreign relations as well. Ross Harrison, writing for the National Research Council, told him that he had enough trouble with Bell's palsy and was not to worry about further affairs for the time being. Henry Dale, when he heard the news, asked A.N. Richards, then working with the Shock Committee, to be sure and give Cannon "a message of my sincere sympathy on his misfortune, which I hope and believe will only be temporary."[32]

By the time of Cannon's seventy-first birthday on October 19, he had begun to improve considerably. Cornelia, overjoyed, remarked that he was a different man, able to hear and almost normal in appearance. "Isn't it wonderful that one can recover from something which seems so menacing at the beginning?" she exclaimed.

Cannon was feeling so much better that he decided not to delay his move out of the Medical School any longer, though his successor still had not been named. With his wife's assistance, he cleared the accumulation of forty years out of his office—a task that exhausted him both physically and emotionally. Nonetheless, Cornelia was thankful that he was making a complete break for, as she saw the Medical School, "it will soon be over there a land that knew not Joseph and he will feel a back number and painfully set aside." As Cannon prepared to leave, Dr. Rosenblueth hovered affectionately around him, looking pretty sad at the changes taking place.[33]

24 *Medical Aid to China; American-Soviet Friendship*

IN THE SUMMER of 1937, the Sino-Japanese War, which everyone had for the most part expected, finally erupted. "Our neighbors are very active again," H.C. Chang wrote Cannon a day after the Japanese army entered Peiping. "It is a serious business this time. The city is quiet, although it is in impending danger at any moment." Then, as if to deny existing events, he continued, "the experiments on humoral transmission are going on under Cannon's light without interruption even during this critical moment." Some time later Cannon learned that Robert K.S. Lim, who had been visiting his father in Amoy before the war, did not return to Peiping and instead took a post in Chungking as director of the National Red Cross Society of China.

Cannon kept in touch with his Chinese friends and students and was well aware of the parallel between the situations in the Far East and in Europe. His letter of November 1937 to one friend, for example, addressed the efficacy of applying economic pressure to aggressor-nations: "After all, the aggressors at the present time are Germany, Italy and Japan. All of them are in financial straits. If England, France and the United States should refuse to have commerce with them when they break treaties and attack peaceful nations, a position of leadership in decency and respect for the given word might be restored. At present the world is in a turmoil because treaties are merely 'shreds of paper.'"[1]

Almost as a reflex to the announcement of the resumption of war in China, a number of committees for Chinese medical relief were organized in the United States that fall. One, like the Emergency Medical Committee for China, was a branch of the long-established Foreign Missions Conference of North America. Another, the Chinese Aid Council, was backed by the American League for Peace and Democracy. There were several others. Cannon, for his part, joined the New England Committee for Relief in China, which cooperated with the recently organized American Bureau for Medical Aid to China (ABMAC).

Despite the enormous good will of these various committees to help the embattled Chinese, few had defined the boundaries of their operations or developed methods for cooperating with one another or, for that matter, sorted out ways of sending the funds and supplies they had collected for distribution to the wounded and needy. Cannon was so appalled at the confusion that existed that he continually apologized in his letters to Robert Lim and tried to reassure him that prospects for the delivery of medical aid to the Chinese Red Cross were slowly but surely becoming more favorable.

"I am so glad to hear that you are helping out our Red Cross," Lim wrote Cannon in June of 1938. "We have received valuable supplies from the U.S.A. and hope that the flow will be maintained." He went on to say that he had been in contact with Dr. Co Tui of the American Bureau for Medical Aid to China about supplying further materials; apart from that, funds and more especially motor transportation were desperately needed.[2]

In the spring of 1939, after the fall of the Spanish Republic, Marian Exter, who had been active in the Boston branch of the Medical Bureau, wrote Cannon that the time seemed appropriate to establish a branch in Boston of the American Bureau for Medical Aid to China. Cannon, who characterized ABMAC as "a thoroughly responsible organization associated with the Chinese Red Cross," agreed to become a sponsor. His willingness to join the Bureau rested on his friendship with and confidence in Dr. Lim, to whom they were providing funds and supplies.[3]

At approximately the same time, G. Canby Robinson on behalf of the Rockefeller Foundation invited Cannon to join with Alfred Cohn and himself as medical members of the China Medical Board, which acted in cooperation with the Peiping Union Medical College (P.U.M.C.) to promote medical education in China. The last thing Cannon needed was another committee assignment, but he promptly accepted Robinson's invitation. For one thing, he felt indebted to the Foundation because of its generous support of his research; for another, he saw membership on the Board as an opportunity to promote the development of scientific medicine (as opposed to folk medicine) in China. Until his resignation in the spring of 1944, he would make every effort to attend the meetings of the Board held each spring and fall.

It must be said that when Cannon indicated he would join the China Medical Board, he really didn't know what the internal problems of the Board were. On his visit to the Peiping Union Medical College in 1935, everything seemed to be running smoothly. Since that time, much had happened of which Cannon knew little. There were, for example, endless debates between the Chinese trustees of the school in Peiping and the China Medical Board, as well as a great rivalry between the foreign physicians at the College and the native physicians. Nor did he know about the rampant inflation that had caused a decline in monetary value of the grants given by the China Medical Board to the P.U.M.C. In truth, Cannon's picture of the Peiping Union Medical College, a rosy one, dated from the time of his visit there and he remained convinced that the school was a model for scientific medicine.[4]

Henry Houghton, who had succeeded Roger Greene as director of the College, proved to be a very able administrator and managed to keep the school going for a number of years during the conflict with Japan. However, Houghton and John Layton Stewart, the head of Yenching University, had diametrically differing viewpoints about what was happening in China, and their letters to the home office that came across Cannon's desk often bewildered him.

In 1940, Cannon attended an informal conference held in New York, at which the dean of P.U.M.C. met with E.C. Lobenstine, chairman of the China Medical Board, and the members of the medical committee, and told them that the administrative and financial organization of the College was unsatisfactory and confusing. This situation was due, he said, "partly because of the division of authority and financial responsibility between the acting director, the trustees of the College, the China Medical Board and the Rockefeller Foundation, and partly because of the disturbed political and economic conditions which tend to make effectual cooperation between the various bodies especially difficult."[5]

As the situation in China grew more dire, particularly in 1940–

41, the question of what to do with the Peiping medical school became urgent. The position of members of the China Medical Board in New York was unanimous: they wanted to keep the College going, as far as they could, no matter what happened. Yet when this recommendation was presented to members of the Board of Trustees in China, they could not go along with it. They had no other plans, but as they saw it, the future of the Peiping Union Medical College rested in the hands of the Japanese. And, as it turned out, they were right.

Under deteriorating Japanese-American relations and with the prospect of open war between the two countries, the status of the P.U.M.C. became perilous. On December 8, the day after Pearl Harbor, the College was taken over by the Japanese army and closed. Everyone was imprisoned, and while most were freed a few months later and allowed to leave China, Dr. Houghton and two others were interned for the duration of the war.[6]

After the closing of P.U.M.C. by military authorities, funds were initially diverted to medical colleges in Free China at Chengtu, Chungking and Kweiyang. The China Medical Board, after much soul searching and heated debate, decided in 1942 to send Claude Forkner (who as a representative of the China Aid Council served on ABMAC's Medical Division) to China to ascertain how the Rockefeller Foundation could help these other medical colleges as well as how best to sustain the remnants of the P.U.M.C. staff.

Though the China Medical Board had a proprietary interest in the P.U.M.C., no longer could it hope that the College would remain a going concern. Rather, the Board members decided to focus on helping Western medicine in China survive the war. Thus it was that Dr. Forkner, who had very definite ideas about what relations between P.U.M.C. and the other medical schools should be, was invited to be the C.M.B. representative in China in conjunction with his visit there for the Cultural Relations Division of the State Department, due to take place early in 1943.[7]

In 1941, when relations between the United States and Japan worsened, the hearts of Americans went out to the Chinese. Even before the Lend-Lease Act officially granted aid to the Nationalist government of Chiang Kai-shek, a major effort to provide more substantial help from the private sector had led to the formation of United China Relief, Inc. (U.C.R.). Founded as a temporary agency for the purpose of launching a large-scale fund drive, the new agency's goal was to raise five million dollars during the spring and summer months. It aimed to do so by consolidating the efforts of several China aid organizations—ABMAC, the American Committee for Chinese War Orphans, the American Friends Service Committee. the Associated Boards for Christian Colleges, the China Aid Council, the China Emergency Relief Commission, the Church Committee for China Relief, INDUSCO, the American Committee in Aid of Chinese Industrial Cooperatives.

In April, Cannon agreed to accept the chairmanship of the Medical Division of U.C.R. The Division was to serve in an advisory or sponsoring capacity, and he took the post on condition that his responsibilities be made as light as possible.[8] His decision was in no small measure influenced by a letter written by Theodore H. White, correspondent for Time-Life and a former student of John Fairbank, which had recently come into his hands. White's detailed letter of February 20, 1941, concerned itself with the medical needs of the Chinese armies—a story, he believed, that had never been told.

In telling of the horrendous difficulties involved in getting materials to the people who needed them, Teddy White emphasized the persistent corruption, the lack of transport, and the problem of distribution. He reported that the greatest enemy of the Chinese soldier was not the Japanese but disease, and every kind of disease imaginable. As many as 80 percent of the soldiers had scabies. Malaria, skin diseases of various sorts, dysentery and epidemic diseases were ubiquitous. And what was available to treat them? Precious little; there was a lack of proper vitamins, not enough medicine, and no quinine.

"The American program of all-out aid to China, if it is to be a wise move, must be founded on fact," White ended his tale. "*The hundreds of planes we have promised the Chinese government cannot supply China with an offensive sting. China's strength lies in her infantry, and if the infantry is sick, China's strength is gone. In terms of value to America, every dollar given to medical aid to China is as important as a dollar spent on arms and munitions.*" There is no denying that White's letter made a tremendous impression on Cannon.[9]

Almost at once, an appeal letter signed by Cannon was sent out, which elicited a wide response throughout the country. In addition, he invited other doctors to serve as members of the

Medical Division. One of those called on was the Chicago physiologist, A.J. Carlson, who promptly agreed to Cannon's request. However much they may have disagreed in the past, Carlson held Cannon in high esteem. "I will be glad to work on any committee with you," he wrote, "because you are still a scientist and a humanitarian, first, last and all the time, and you will not consent to be a front for any political propaganda. That is more than can be said for some of the persons in the new organization."

As chairman of the Medical Division of U.C.R., Cannon automatically became a member of ABMAC's Medical Committee and was sent materials about its activities by Helen K. Stevens, its executive director. In June he was informed by Donald van Slyke, head of the Executive Committee, that he had been elected a director of ABMAC. When the fund drive fell far short of its goal, Dr. van Slyke joined with others to urge a complete reorganization of United China Relief on a permanent basis.[10]

In the fall of 1941, van Slyke asked Cannon to become one of the directors of United China Relief, and wrote him: "I sincerely hope that you will accept; considering the fact that the relief to China for some time to come must, on account of transportation limitations, be chiefly limited to medical supplies, it is important to have members on the Board who know something about medical problems. As made up now, you and I are the only ones that would know anything at all about such problems, and you are the only one that is really qualified from the medical standpoint."

Cannon replied that he liked the new organization of U.C.R. "It avoids the overlap and confusion of multiple drives and yet leaves the administration of funds in charge of organizations which know the needs and possibilities much better than less experienced general managers could possibly know them." Then he added, "As usual I hesitate to take on more responsibilities than I am carrying at present. Nevertheless, the possibility of aiding China in this time of pressing crisis appeals to me so deeply that I cannot refuse the invitation to become a member of the group of Directors."

During 1942, Cannon continued to lend his name to the China aid effort through ABMAC and U.C.R., though rarely was he able to attend meetings or take an active role. In consequence, he tried to resign as a director of ABMAC and member of its Medical Committee, but without success. Both Dr. van Slyke and Helen Stevens assured him it didn't matter whether he was able to attend meetings regularly or not, so long as they could call on him for advice and counsel. Besides, as van Slyke told him, "Having your name on the list of our Directors is a source of strength and inspiration."[11]

Much as Cannon disliked the pressure of duties that took him away from the laboratory and consumed his waning energy, his genuine concern for the people (including many old friends) who suffered the privations of war made it difficult for him to say "No" and mean it. First Spain, then China. Next came Russia. All made claims on him in the dwindling years of his life that he found hard to resist.

At the very time the China aid effort was gathering steam in 1941, there came the shocking news that Hitler had broken his pact with Stalin and ordered German troops to invade the Soviet Union. Within days of the invasion, John Kingsbury of the American Council on Soviet Relations sent Cannon a telegram, asking him to speak to a large public gathering in Madison Square Garden "in support of American friendly cooperation with the Soviet Union in resisting Fascist aggression." Minna Weiner, Cannon's secretary, replied that he was out of town and would not be able to participate in the rally.[12]

Some weeks later Henry Sigerist, on behalf of Kingsbury and others who had been discussing the organization of a committee for medical aid to the Soviet Union, invited Cannon to join the new group, telling him, "Your experience in the Spanish Committee would be of greatest value to all of us." Sigerist, a professor of the history of medicine at Johns Hopkins, had long been interested in the development of medicine and public health in the Soviet Union. The new committee, he assured Cannon, would devote itself to collecting and shipping medical supplies to Russia. It had no political affiliations of any kind and would cooperate closely with similar organizations sending medical aid to England—in short, it would serve purely humanitarian purposes.

Although Cannon was sympathetic to the idea of supplying medical aid to the Soviet Union, he refused Sigerist's request because of his prior commitment to China. "You can readily see," he explained, "that if I should join a Committee for aid to the Soviet Union I would at once be regarded as not wholeheartedly working for medical aid to China." Adding weight to his decision, Cannon continued, was his obligation to the Shock Committee as well as his need to wind up problems in the laboratory during the coming academic year—his final one at Harvard.[13]

The importance of Cannon's name to shore up support of the Soviet Union in the United States was not lost on Russian medical scientists. N. I. Propper-Grastchenkov at the A.M. Gorki All-Union Institute of Experimental Medicine, for instance, who had visited Cannon's laboratory during the fall of 1936, asked him to raise his voice to promote assistance to the peoples of Great Britain and the U.S.S.R. in their fight against fascism and to support closer relations between Russian and American scientists and scientific institutions. "It is my hope," Cannon replied, "that when this dreadful war is over, Russia, Great Britain and the United States will be drawn closer together and that there will be made possible an interchange of students and new science between your country and mine. Nothing would give me greater pleasure than to see an organization of this sort of interchange."

Not that Cannon needed to be reminded of the importance of rallying support and aid to the Soviet Union. On October 8, 1941, weeks before he received Propper's letter, he had sent Professor Orbeli, Pavlov's successor, a message that he hoped would be transmitted to the Congress of Russian Scientists then being held in Moscow:

> May I express through you my great admiration for the firm stand which your Government has taken against Nazi barbarism and aggression. We workers in science strive to learn and tell the truth; the Nazi leaders have established lying as a principle of public policy. We strive to be honest and direct; they rely on deception and deceit. By our researches we strive to advance the welfare of mankind everywhere; they arrogantly assume that Germans are superior to other peoples and with sadistic malevolence imprison and torture and attempt to crush any who do not bow down to them. We know that liberty is a prime requisite for acquiring new knowledge; they abolish liberty. We are sure that in peaceful pursuits science thrives and provides for enrichment of human living; they extol war—war which inflicts cruel mutilation and violent death on innocent men, women and even little children.
>
> In every one of these respects men of science must be inflexibly opposed to the despicable preaching and practice of the German tyrants. As a group devoted to liberty of learning and to the service of mankind we cannot resist too vigorously those who would thwart our endeavors.

Though Cannon's message was printed in the November issue of *Soviet Russia Today,* he did not know if it actually had been read at the Congress as he had intended.[14]

The Soviet Union, while accepting a variety of medical supplies from the United States, drew the line on certain forms of aid or sharing medical information. When an American Red Cross delegation to Russia in August and September of 1941 suggested that it might be useful to send doctors and scientific workers from friendly countries to help meet incipient epidemics there, the proffered aid was rejected. Several months later when Cannon in his capacity as chairman of the N.R.C. Committee on Shock wrote the U.S.S.R. Embassy in Washington for information of how Russian surgeons, under conditions of mobile warfare, provided transfusion in cases of hemorrhage and shock in the field, Victor Kallistratov, an attaché of the Embassy, blandly replied, "We do not have any material on the subject you write about at the present time."[15]

Despite these rejections, the number of agencies interested in providing medical relief to the Soviet Union proliferated. Early in July 1942 Hugh Cabot in behalf of the Massachusetts Committee for Russian War Relief asked Cannon to join its executive committee. Cannon, not liking to refuse an old friend, indicated that if he could be absent from some meetings during the fall because of prior commitments, he would join. Cabot was overjoyed. "Quite frankly," he confided to Cannon, "your presence on the Committee would make me feel a little less lonesome since I too have been declared on the high authority of Morris Fishbein and our own Roger I. Lee, to be a Communist. I should feel more comfortable therefore, to have another Communist of the same type to hold my hand."

In October, Cannon informed Cabot that he was out of commission because of his bout of Bell's palsy. "I am much distressed by your attempt to validate the conclusions of Sir Charles Bell," was Cabot's sympathetic rejoinder. "I am also struck by the fact that you take it in your stride with your usual philosophy. We shall, of course, not expect anything of you until the situation quiets down, but it confirms us in the view that when you are again fit, we shall need you very badly."[16]

By 1943, seasoned China hands had become outspoken in their complaints about the Kuomintang (the Nationalist government of

Chiang Kai-shek). The consensus of most American officials and correspondents working in China—Theodore White and Brooks Atkinson among them—was that it was "incompetent, corrupt, oppressive, unrepresentative, riddled by internal weakness and unlikely to last." Still others believed that civil war between the Nationalists and the Communists was inevitable, once the Japanese had been dealt with.

Cannon did not take sides. He was only concerned that the flow of medical aid would reach those who needed it most. Because of this concern, he agreed to serve on two subcommittees of the U.C.R. Advisory Committee on Medicine and Public Health; he also, in response to a request from Dr. Frank Co Tui, became a member of the ABMAC Committee on the Chinese Army Medical Service. Through these assignments, as well as his continued service on the China Medical Board, Cannon was able to keep informed of the activities of his friend, Robert K.S. Lim.

When in 1937 China began to resist further Japanese aggression, the country had 6,500 doctors to take care of a population of close to 450 million. Most of these doctors were in occupied territory, and there were perhaps only a thousand or so available to serve with the army. In order to address the shortage of qualified medical personnel, Dr. Lim had established a system of emergency training schools for doctors, nurses and paramedical help. These schools were financed by the Rockefeller Foundation, the British Boxer Indemnity Fund and private donations. Moreover, they had the support of Chiang Kai-Shek, who ordered staff of all army hospitals to send personnel to the branch schools in various war zones for training.[17]

In 1942, Cannon, in his capacity as Foreign Secretary, had proposed Lim for honorary foreign membership in the National Academy of Sciences—an honor, he was soon proud to announce, which was duly voted. Lim was then planning a visit to the United States, at which time Cannon intended to inform him of his election, but the visit had to be postponed because of political pressure that brought about a reorganization of the Chinese Red Cross.

Early in 1944, a flurry of letters and telegrams, initiated by Lin Yutang, urged that Robert Lim be appointed an ABMAC director as soon as possible and brought to the United States in order to protect him from Chinese politics. After considering the wisdom of such a ploy for Lim's future service and for ABMAC's, and after the War Minister stated that he understood Lim perfectly and approved the appointment, and after the cooperative nature of the arrangement was clarified, Lim accepted an invitation from Donald van Slyke to make a brief visit to New York for consultation with ABMAC, U.C.R., and the Rockefeller Foundation on future policies. The maneuver worked admirably; after his return to his homeland, Lim the following year became surgeon general of the Chinese Army.[18]

A large part of Cannon's interest during this time became channeled through the China Aid Council, which was trying to get supplies to the Communist guerrillas in Northern China through a blockade that had existed for some years. Mme. Sun Yat-sen, on learning of Cannon's outspoken support of medical aid to the blockaded rebel forces, sent an expression of appreciation to him, writing on May 3, 1943, "It is good to know that we have with us such a veteran and eminent fighter, on another front, for the health and strength of man." She pointed with pride to the role of her China Defense League in helping the people's army for five years—and not least, the medical service which had made their sustained effort possible "in the face of an embargo on medical supplies enforced not only by the enemy but by our own government." Cannon was especially interested in Mme. Sun's praise for Norman Bethune, the Canadian surgeon who had helped to build the medical service in North China before his death in 1939, and expressed a desire to see his papers. "We were associated in our support of the Spanish Loyalists," he wrote Mildred Price of the China Aid Council, "and I had great admiration for him."[19]

In September Miss Price, passing along a report in the *New York Times* that Chiang Kai-shek and the Kuomintang had made certain charges against the Communists in China, asked Cannon to write a letter to the editor about the "cruel blockade," whose existence the Nationalist government flatly denied. His reply was that writing letters to newspapers was futile and that the agencies rendering aid to China needed to take a strong stand. Cannon wrote a similar letter to James McConaughy, the president of United China Relief, urging that U.C.R. make such a stand. He concluded his letter, "By showing our interest in *all* Chinese who are battling the aggressive dictation of Nippon, I believe we can help to prevent a rift—a disastrous rift—in the Chinese armies."

Several weeks later, Mrs. E.C. Carter, the president of the China Aid Council, sent Cannon two letters from Mme. Sun about the situation in the North, and asked his advice. Cannon replied by alluding to information received from McConaughy, which made him see how really complicated the situation in China was and how easily one might misjudge it. He suggested that getting medical supplies into China and distributing them equitably might best be done through the agency of the Red Cross. "If that could be achieved, it would be best of all, and I would advise trying it."[20]

While a visiting professor at New York University in the winter of 1944, Cannon attended a meeting of the China Aid Council where plans were discussed for a protest of the blockade to be signed by eminent physicians. Troubled that he really didn't know if the blockade was in fact still in effect, Cannon first asked his son-in-law, John Fairbank, and various people in the State Department for up-to-date information. The responses he received, along with recent reports filed by Ted White of Time-Life, convinced him to participate in drafting a letter to the Chinese Ambassador. With the aid of Mrs. Carter and Miss Price and after many revisions, Cannon was satisfied that the letter he produced, instead of being a political protest, emphasized the humanitarian tradition of the medical profession.[21]

A roster of distinguished physicians signed the letter, with Cannon serving as agent for the group. The Ambassador's response in June was that neither guerrilla fighters nor regular troops had been deliberately denied medical supplies, but that insufficient amounts for total needs was the real problem. Rather than address the issue of the blockade *per se,* he called for an increase of medical shipments to China. The answer, although disappointing, was not unexpected. The next step, as far as the China Aid Council was concerned, was to release the physicians' letter to the press.

Cannon again turned to his son-in-law for advice. John Fairbank told him that the Chungking government was getting more and more reactionary and advised him to publish the letter as a public service. Still, Cannon had reservations about going public, in part because he did not wish to put the Chinese government in a bad light in the eyes of the world, and in part because he had not asked the doctors who signed the letter for permission to publish. Before long, however, word came from Mme. Sun and others that supplies were getting through on a regular basis, so there was no longer any need to publish the letter.[22]

In the two years after Cannon's retirement from his Harvard professorship, concurrent with his interest in providing medical aid to China was his effort to improve communication between American and Russian scientists. While the aid to China movement involved a number of organizations and projects that vied for Cannon's support, his effort in behalf of Soviet-American relations was much more focused.

During the early winter of 1943, Henry Sigerist after months of discussion with friends and colleagues laid down guidelines for the organization of an American-Soviet Medical Society. The prime function of the new society would be to publish in a journal, the *American Review of Soviet Medicine*, translations of articles on various phases of medical science and news of war medicine in the U.S.S.R., as well as reviews of Soviet medical books. Inasmuch as medical journals in the U.S. reported very little about Russian achievements, and since the language barrier made it difficult for American doctors to keep in touch with Russian medical literature, there was a decided need for an exchange of medical information with the Soviet Union. This was what the new journal hoped to do under Sigerist's editorship.

A great deal, he wrote Cannon in March, depended on having a good president "who enjoys a high reputation in this country as well as in the Soviet Union. We could think of no one more qualified for that position than you." Assuring Cannon that the Society had no political connections of any kind, Sigerist added, "It is a medical society and serves only one purpose, the exchange of medical information and the promotion of a closer cooperation between the medical corps of the two countries."[23]

When Cannon replied that he was deeply involved in a number of projects and had little time to take on anything else, Sigerist hastened to assure him that the presidency would not involve any office work; his own secretary would handle all the correspondence and business details, while he and his staff would take care of the editorial work. Against his better judgment, Cannon succumbed to Sigerist's artful cajoling—in no small part because he learned that Leo Eloesser, who was one of his allies during a trying controversy over the administration of medical aid to Spain, had been chosen by Sigerist to be vice president.[24]

A month after accepting the presidency, Cannon gave a talk at a gala dinner held in New York in honor of Professor Lebedenko, the representative of the Soviet Red Cross in the United States. Cannon's speech was so interesting and stimulating that nearly everyone attending the dinner joined the Society. In short order, he was invited to give other talks and complained to Sigerist of this new onus. Sigerist was all commiseration. "For goodness sake," he replied, "take care of yourself and do not let yourself be bullied into giving lectures . . . I always insisted with the Committee in New York that they should not impose upon you for lectures here and there. You are doing your share for the Society by being its president and by attending some really important meetings, but you must not allow yourself to be blackmailed into attending meetings that are not important at all."[25]

There were, however, two occasions that were very important indeed as far as Cannon was concerned. In the summer of 1943, he interrupted his vacation in New Hampshire to participate in a reception at the Soviet Embassy in Washington presided over by Andrei Gromeyko, Soviet chargé d'affaires. The reception was held to applaud three Americans—Cannon, G.N. Lewis, and E.O. Lawrence—who had been made honorary members of the Academy of Sciences of the U.S.S.R. The following autumn, Cannon became a sponsor of and participated in the Science Congress of the National Council of American-Soviet Friendship.[26]

Under Sigerist's leadership the American-Soviet Medical Society flourished. Within weeks of its birth 500 physicians had joined and within a year it had 2,000 dues-paying members as well as 1,000 separate subscribers to the *Review.* In large part, interest in the Society stemmed from the content of the *Review,* which featured articles by both American and Soviet physicians and medical scientists. Besides the bi-monthly publication, the Society undertook to collect and exchange books and other materials to be sent to the Soviet Union. Sigerist was especially pleased when he had a regular flow of books, periodicals, and reprints going to the Soviet Union, as well as an exchange of the *American Review of Soviet Medicine* with a number of other English-language journals.

At the end of March 1944, Cannon wrote Sigerist about an article that Dr. Lena Stern, with whom he was on friendly terms, had recently submitted to the *Review* on the treatment of shock. Cannon was perturbed by the submission on several counts, among them that it had been previously published in *Lancet* and worse, that it was poor. "I do not think," he concluded after a lengthy explanation, "that publication—or really, re-publication—in English of Lena Stern's ideas and methods would be regarded as a credit to Soviet medical science. My own judgment would be to disregard this particular contribution."[27]

Sigerist, in reply, confessed that he had never intended to be editor of the *American Review of Soviet Medicine* for more than a year or two and detailed the unusual difficulties that he faced. Not only were Russian periodicals not available in Baltimore, it was hard to find American physicians who were expert in their fields and also fluent in Russian. The *Review* had not yet constituted an expert editorial board whose members could read and survey Russian medical periodicals, then decide which articles particularly relevant to the medical problems of the war should be translated and published.

Having said that, Sigerist turned to some of the criticisms that had been leveled at the *Review.* One involved exaggerated press reports following publication of a series of papers by Dr. Alexander Bogomolets and his associates about a serum developed in their laboratory that they claimed had extraordinary qualities of curing disease. Another was presented by the unconventional way in which the Russians reported the results of their research, with a bold and careless style that often shocked American readers.

"All in all," he concluded, "I think the *Review* so far has been as good as it can be under the circumstances but it is obvious that it must be improved, and I feel confident that after a year's experience, we shall be able to improve it considerably." The criticisms that Cannon and others brought to Sigerist's attention were to have a telling effect and in consequence, the quality of papers published by the *American Review of Soviet Medicine* improved thereafter.[28]

In the autumn of 1944, as Cannon determined to shed most of the obligations he had undertaken on behalf of China and the Soviet Union, he found himself in conflict with Henry Luce, the powerful publisher of *Time* and *Life* magazines. The root of the conflict lay in the publication of an article in the September 4 issue of *Life* written by William Bullitt, the former U.S. Ambassador to France and the Soviet Union. Briefly, Bullitt maintained that the United States and England, in failing to stand up for the independence of Italy, had contributed to the widespread fear in Italy that the country might become communistic or dominated by Soviet Russia. Even more alarming, Bullitt predicted that within fifteen years there would be a third world war between the Soviet Union and Western Europe backed by England and the United States.

Cannon was upset by the article and joined with Albert Einstein, William Rose Benet, Serge Koussevitsky, Ralph Barton Perry and others in writing an "Open Letter to Henry Luce" in behalf of the National Council of American-Soviet Friendship. They asked two questions. First, whether it was *Life* magazine's wish to split the United Nations coalition just as victory over the Axis was approaching? And second, whether Luce, as one of the most powerful and influential editors in the country, honestly wanted to line up *Time* and *Life* on the side of anti-Soviet slanders and the possibility of war against Soviet Russia after the present conflict was over?[29]

The executive director of the National Council sent a copy of the open letter to Luce. To the first question Luce quickly answered "No. Nor in the opinion of the Editors of LIFE is there any serious danger of splitting up the anti-German coalition at this time by serious discussion of our common problems." To the second question he again replied "No." adding, "And if the foreign affairs of the United States are competently managed, there is no reason whatever why the United States should, ever again, be in two big wars in one generation." Then, in a postscript, he posed a question of his own, asking to be supplied "with objective information as to any of the past or present Communist affiliations and connections of the signers of the open letter."

When Cannon learned of Luce's rejoinder to the open letter, with its politically intimidating postscript, he wrote privately to Luce. Describing himself as an independent who more frequently voted the Republican than the Democratic ticket, he said he had supported the Spanish Republic against Hitler, Mussolini and Franco, and was not a pacifist, having worked on cases of wound shock in France in 1917–18. "I signed the protest," he explained, "because our people and the peoples of the Soviet Union have to live together in the same world. Until the cause is proved to be hopeless (and there is no sign that it is) I intend to do what I can to promote friendship and understanding between the two populations. To arouse suspicions and fears and hatred, which might lead to war, seems to me abominable conduct."[30]

The open letter to Henry Luce that Cannon joined in signing, like the letter that was sent to the Chinese Ambassador a few months earlier, was among his last public statements urging that humanitarian motives guide the future of international relations.

A chapter on "Being a Citizen" in his memoir, *The Way of an Investigator,* reflects on his experiences in the political arena; here he was able to look back with evident satisfaction: "I am grateful that fortunate conditions gave me opportunities to test democratic methods by opposing the efforts of antivivisectionists at destroying freedom of medical investigation; to attempt aiding republican forces in Spain and in China as they struggled against oppressors; and to promote deeper sympathy and understanding between the Americans and the Russians as they look forward to a better world."[31] Cannon modestly understated his own valiant role in the struggles. Though his critics believed he had made the mistake of letting his sympathies master his judgment, he clearly had no regrets about his efforts in the cause of liberty, democracy, and humanitarianism.

25 *On the Wings of Time*

Oh, that I had wings like a dove!
for then would I fly away,
and be at rest.

—(*Psalms 55:6*)

THE DAWN of 1943 saw Cannon on the slow road to recovery from his facial paralysis. During the winter months, what had become an annual hegira to Florida not only helped him to regain his old energy and strength, it also offered a respite from the harsh New England weather, for he had become extremely sensitive to cold and was easily chilled. The visit was just what he needed. When he returned to Cambridge, his face was so much better that he no longer felt embarrassed about his appearance and was soon his old self. After he was fitted with a hearing aid, he expected to find it easier to pick up his social life as well as his committee work.

By May, Cannon and Cornelia were in residence at their New Hampshire farm, their trips to and from Boston curtailed by wartime gas rationing. Although he felt well enough to travel frequently to Washington and New York to attend meetings, Cannon was trying out a new diet to control his itching, which had returned with a vengeance. While attending a meeting of the Canadian shock group in Ottawa, he was in torment. Cornelia thought he was wonderfully courageous about his affliction but privately called it a martyrdom. She vowed that she was ready to try Christian Science, osteopathy or anything else, but Walter was suspicious of any unorthodox promise of cure.

Still, he was able to stay active and fulfill his commitments—though not without difficulty. Scheduled to welcome the head of the Russian Red Cross in New York, Cannon became so distressed at the way things were handled that, in Cornelia's words, he got a headache and an upset stomach in the service of international relations. On top of everything, he woke up one morning with a stiff neck. He turned to his wife and said (only partly in jest), "If I had had all the troubles I have now fifteen years ago, I would never have written *The Wisdom of the Body*."[1] Somehow, he managed to keep cheerful while he busied himself with his prodigious correspondence and found solace in tending his victory garden. Best of all, he had the company of his grandson and namesake. Little Walter had a wonderful time with his grandfather, who delighted in finding a new enthusiast when he set up the old railroad system he had constructed many years before for Bradford.

Early that summer, an article in the Franklin newspaper reporting that the war had made inroads on the New Boston Road colony gave a rundown on the whereabouts of the Cannon family. Wilma was with the State Department in Washington, D.C., working in a division devoted to cultural relations with China, while her husband John Fairbank was with the Office of Strategic Services (O.S.S.), spending some of his time in China. Linda and Marian were also in Washington with their husbands. Harry Burgess was working under Donald Nelson in charge of production and priority of aluminum, and Arthur Schlesinger, Jr., was with the O.S.S. Marian expected to spend part of the summer in Franklin and planned to illustrate a children's book that her mother was writing about Abbé Huc's trip from China to Lhasa, Tibet, in 1844. Helen was in Texas with her husband, Douglas Bond, a lieutenant in the U.S. Army Medical Corps doing research on aviation fatigue at Randolph Field. And Bradford, who had been delayed from entering the service because he was needed in Boston for emergency work with survivors of the Cocoanut Grove disaster (one of the worst fires in the city's history), was a lieutenant stationed at the U.S. Army hospital in Valley Forge, Pennsylvania, where he specialized in plastic surgery. His wife, Ellen, with her family in Lincoln, Massachusetts, would join him there after the birth of their expected child.[2]

Throughout the summer months, Cannon suffered a bout of the worst itching he had endured in a decade. He was in such misery that he could not concentrate and had to give up any thought of sustained mental work. "I hoped that I might finish this summer a monograph on hypersensitivity of denervated structures," he confided to Robert Yerkes. "All the material was at hand, but I have been so wretched with itching and little sleep that answering letters and attending to Washington conferences have been my main activities." He had "a very strong hunch that a chemical stimulation of sensory nerves" was at the root of his condition, and wrote his physician Dr. Aub that he would try subcutaneous injection of acetylcholine when he got back to town.

In September, Mrs. Cannon persuaded him to see Dr. John Stokes, a dermatologist in Philadelphia—something that Douglas Bond had been urging for some time. "I took the liberty, and I realize it was a liberty," Dr. Stokes wrote Cannon after examination, "in calling your attention to the seriousness of the nervous tension factors in persons who have a serious itch." Of course, he continued, Cannon recognized these elements in his own case and it was up to him to correct them—starting first by giving up all work that caused him anxiety and worry.[3] Afterward, Cannon confessed to Joe Aub that he had consulted Stokes because the itching had become unbearable. He did not want Aub to think, however, that he had been "shopping around" or that he had lost any of his "supreme confidence in your wisdom."

Be that as it may, Stokes gave Cannon a fresh point of view about his troubles, and he immediately began to improve. Cannon was so sure Stokes was right about worries and anxieties, he not only told Dr. Harrison that he was overtaxed with his duties as head of foreign affairs at the N.R.C. and needed assistance, he also informed Dr. Weed that the lack of coordination in shock research was a constant source of distress to him and asked that someone else take his place as chairman of the committee.[4]

Despite Cannon's intention to give up the shock work, he continued to serve as head of the N.R.C. committee. In March 1944, he appointed Joseph Wearn to replace Robert Loeb (who had replaced Sturgis) as head of the Subcommittee on Blood Substitutes. Throughout 1943 and 1944, he continued to rely on Philip Bard and Magnus Gregersen to produce reports and help him with duties that involved travelling and conferences. A.N. Richards, in turn, continued to count on Cannon, as a consultant to his Committee on Medical Research, for advice on shock. Above all, Cannon was excited about the work that Edwin Cohn was doing at Harvard on plasma fractionation and saw promise in the researches of Joe Aub's group at the Massachusetts General Hospital, which brought back into the picture of shock the concept of traumatic toxemia.[5]

Meanwhile, the matter of finding a successor to Cannon as head of the physiology department at the Medical School had not proceeded smoothly. Although Dean Burwell recommended, and President Conant appointed, a search committee chaired by William B. Castle well in advance of Cannon's retirement, the committee had been disbanded without reaching a decision by the end of the academic year in 1942. Burwell, who maintained that the problem of Cannon's successor was not dead but sleeping, arranged to keep the department running by asking Hallowell Davis to serve as acting head for a year and enlisting the aid of Drs. Drinker and Aub, if needed. Aub was not happy with the compromise. For him, Burwell's solution represented "a lack of confidence in the nucleus of congenial young men who are among the most stimulating teachers and thinkers in the Medical School." Besides, he told the dean, it did not make use of Dr. Cannon in a time of emergency. Aub suggested that Cannon be invited to continue teaching for another year (and believed he was likely to accept since his South American trip had been cancelled), but Cannon's subsequent attack of Bell's palsy put Aub's suggestion on the shelf.[6]

A few months later Cecil Drinker was on a visit to the University of Virginia where Eugene Landis, who was on the short list of possible candidates, was Professor of Medicine. Drinker asked Landis outright whether he would accept the position at Harvard if offered. "When I put the question to Landis," Drinker wrote Burwell upon his return, "I was much surprised at his answer, which was, 'Yes!'" On receiving Drinker's letter, Burwell, who had reviewed Landis's career and researches with care and discussed his candidacy with a number of men at the School, laid the matter before Conant. More than that, the medical dean urged Landis's appointment and asked that it proceed as rapidly as possible.[7]

In short order, a new committee was appointed and voted unanimously to recommend the appointment of Dr. Eugene Markley Landis as the George Higginson Professor of Physiology beginning July 1, 1943. But Otto Krayer, Hermann Blumgart and Edwin Cohn at a meeting of the Full Professors raised objections to the selection of Landis that perturbed President Conant, who was not convinced that they had indeed got the right person. Conant worried that it might be a step backward if they appointed a man who was essentially a clinical investigator to a chair in a preclinical subject. He was therefore inclined to postpone the whole matter until after the war—when there would be a new crop of candidates on the horizon.

Now it was Burwell's turn to be perturbed. He believed delay in filling such a key position would be "an extremely dangerous and unwise thing to do" for it would cause a host of problems, both within the physiology department and affecting the entire Medical School. Near the end of 1942, he called together a joint meeting of the old and new ad hoc committees, which proceeded to give a vote of confidence for the appointment of Dr. Landis.[8]

A holder of both the M.D. and Ph.D. degrees from the University of Pennsylvania, Landis had spent the early part of his career there, except for fellowships with August Krogh and Thomas Lewis abroad, before taking the post at Virginia. His research began with the study of the physiology of capillaries, went on to exploration of the disturbances of vascular function, and most recently focused on pressor effects of kidney extracts and pathogenesis of hypertension. Moreover, he was reputed to be a person of good character, of scientific and personal integrity, and of effective and driving intelligence.

After the two committees made their report, Conant authorized Burwell to offer the Higginson Professorship to Landis. It gave the medical dean a great deal of pleasure to bring the difficult problem to a close. It pleased him even more when Landis promptly accepted the offer.[9]

An immediate casualty of the new regime in the physiology department was Arturo Rosenblueth. "Due to the changes in the Department resulting from Dr. Cannon's retirement," Rosenblueth announced in a March 1943 letter to Henry Allen Moe at the Guggenheim Foundation, "the Medical School finds that they don't need me any longer and I will leave next year. Indeed, I would like to leave sooner if I can." Rosenblueth, at the age of forty-three, was embarked on an anxious and arduous job hunt, supported in part by both Cannon and Moe. In an effort to help his former collaborator find a post, Cannon wrote a great many letters in his behalf, but without success. Though Arturo for a time pinned his hopes on the slim chance of a position at the University of Pennsylvania, it all came to naught.

Rosenblueth did have one job offer, but it was in Mexico City where Ignacio Chavez invited him to organize a department of physiology in his new Institute of Cardiology. For a time, Rosenblueth kept Dr. Chavez dangling in hope of something opening up in the U.S. His hope appeared to be realized in the summer of 1943 when Warren McCullough, one of Rosenblueth's colleagues in his work with cybernetics, invited him to join the department of electroencephalography at the University of Illinois. Arturo had difficulty choosing between his two offers but ultimately decided to go to the medical school in Chicago. In negotiations for the post, however, it soon became clear that he would be blocked from further advancement because he was not an American citizen.[10]

Some years previously, Rosenblueth had decided not to change his citizenship until he had secured a permanent post in the United States. He believed that if he relinquished his Mexican citizenship, he would not be able to find a position there, should things not work out in the U.S. "I suppose I am hypersensitive about my future since my recent nasty experience at Harvard," he explained to McCullough in October. "Their dismissing me, together with the fact that the only offer of a position which I have had in this country is yours, leads me to conclude that there are a very limited number of possibilities for me here." Unable to reach an agreement with the University of Illinois, Rosenblueth had no recourse but to accept the offer from Chavez. Cannon was reported to have buoyed his spirits by telling him, "Go put Mexico on the map of science."[11]

In the spring of 1943, after recovering from his bout of Bell's palsy, Cannon resumed the duties he had undertaken for the N.A.S. and N.R.C. in behalf of foreign relations. Because wartime priorities interfered with international scientific efforts of a non-military nature, there was not a whole lot going on save for a few meetings and reports. At the National Academy, his efforts were largely confined to the election of foreign associates and distribution of the N.A.S. Proceedings; at the N.R.C., he soon found there was considerable overlap between the scientific divisions and the Division of Foreign Relations.[12]

Perhaps it was this lull in activity that persuaded Cannon to add yet another committee assignment to his list at the National Research Council. In the autumn, William deB. MacNider, a pharmacologist at the University of North Carolina who had recently been appointed chairman of the N.R.C.'s Committee on the Biological Processes of Aging, wrote Cannon and asked for his help. In his letter, MacNider outlined plans for reorganizing the committee's efforts in providing support for research on aging. "In other words," he explained, "this Committee would operate in much the same fashion in which your most valuable Committee on the Endocrines carries forward its very effective work." Cannon agreed to become a member of the reconstituted committee and attended its first meeting, where it was announced that the organization formerly known as the Club for Research on Ageing would henceforth cease to exist, its activities taken over by the N.R.C. Committee.[13]

Both Cannon and MacNider had been involved in the Club for some time. In the summer of 1939, following a visit to the United States by Dr. V. Korenchensky of the Lister Institute in London, MacNider had agreed to serve as organizing chairman of the American branch of an international Club for Research on Ageing and Cannon agreed to become a member. "I have been much interested in the problems presented by the passage of years," Cannon had written MacNider in October of that year, "an interest possibly intensified by the fact that I am becoming more and more conscious of limitations in the passage of years, for today happens to be my birthday!" He assured MacNider that he would be glad to participate in the efforts "to know more about the old man," and might even be willing to present himself as a subject.

Cannon had not only attended the Club's initial meeting early in 1940, he subsequently sent MacNider a copy of his Welch Lecture on homeostasis in senescence. Once the Club was officially launched, MacNider, in appreciation for Cannon's advice and support, made him head of his executive committee. Soon after, the Club and the Committee on Biological Processes of Aging, then chaired by E.V. Cowdry, decided to coordinate their efforts, with the N.R.C. responsible for distributing funds for their projects.[14]

Despite his first rush of enthusiasm, later that year Cannon voiced doubts about his ability to take an active role in the Club. When A.J. Carlson, who had succeeded Cowdry as chairman of the N.R.C. Committee, invited members of the Club to meet with his committee in Washington, Cannon asked to be excused, complaining that he was spending too much time on the road. Not only that, but membership in the Club was supposed to be limited to those investigating or supervising problems of aging. Although Cannon planned to assign a project on the effect of old age on homeostasis to one of the workers in his laboratory, much to his chagrin the plan had to be dropped. He came to realize that research on aging was a long-term commitment, something he could not undertake because of his own approaching retirement.

Believing that his service as chairman of the executive committee had been "utterly null and void," Cannon in 1941 asked MacNider to find some younger man "actively engaged in research *de senectute*" to replace him. By then the U.S. Public Health Service had undertaken to make problems of aging part of its program; but the demands of stepped-up defense activities and the exigencies of world crisis made it increasingly difficult to obtain funding for extra-military research. Still, MacNider assured Cannon that research on aging was alive and well and dutifully kept him informed of the Club's activities.[15]

During the war years when other groups were forced to give up regular meetings, the Macy Foundation, through the good offices of Frank Fremont-Smith, provided funds for the Club to hold annual meetings. Its roster of distinguished members, Cannon among them, made every effort to attend these meetings, which they thoroughly enjoyed and found stimulating. Though he did little in the interests of its successor, the N.R.C. Committee on the Biological Processes of Aging, Cannon served as an inspiration to MacNider, who used as his model what had already been done "in such excellent form for research on sex and on the endocrine glands."[16]

While Cannon acted in an official capacity as head of foreign relations at the N.A.S./N.R.C., he responded as a private citizen when he found himself once more drawn to affairs in Latin America, especially as they affected his friend, B.A. Houssay. In November 1943, a young physician named Christiane Deane, working in Chile with Dr. Lipschutz, wrote her sponsor, Herbert M. Evans, at the University of California that the revolution under way in Argentina had turned ugly. At first, she informed him, everyone had hoped for the best under the government of Ramirez, as no one could easily imagine a change for the worse. But after a number of citizens signed a manifesto demanding effective democracy, mostly to let the new government know what the people really wanted, a resolution denouncing the signers was issued—which came as a complete surprise, soon to be followed by indignation.

"Amongst the ones you and I are interested in, Houssay,

Castex, Orias and Lewis had signed and they were thrown out into the street," Dr. Deane told Evans, adding that what happened at the Institute of Physiology was especially distressing:

> Prof. Houssay gave his last class in physiology in the presence of 2,000 people in the "aula" and "patio" which you well know and in spite of the cheering and applause, refused to say anything that wasn't physiology. Then there was a very sad farewell in his office. Now, what you may not realize to its full extent is that Prof. Houssay is profoundly affected by this situation, and especially by the very bad conduct of some of his disciples; he has taken all of his books, pictures, etc. to his house and has said that he will never return to his Institute—and remember that the latter is his life, he founded it, the only one of its kind in South America, and it is now going to waste.

To make matters worse, Ramirez had convinced the masses that Argentina must stay neutral at all cost, and not let any other country interfere in their business. Nevertheless, Dr. Deane implored Evans to see what could be done in the United States, noting, "The Rockefeller and other foundations which had been helping Houssay and Lewis and Orias cannot do anything now because of the political angle to the whole situation, so that help must come directly from outside sources."[17]

Almost immediately after receiving the letter, Evans got in touch with Cannon to apprise him of the situation in Argentina. Houssay's greatest problem appeared to be his need for the journals which he had gathered in the Institute of Physiology and to which he no longer had access. In order to tide him over this critical situation, Evans suggested circularizing the American Physiological Society and other groups to raise private funds for subscriptions to the most important journals in medicine and science published in the United States.

Before long, Houssay, with continued aid from the Rockefeller Foundation, was able to set up an independent laboratory where he was joined by Lewis, Orias and others. A committee in the United States, chaired by Evans and with Cannon, John Fulton and Carl Wiggers as members, sent out circular letters to a number of their colleagues and succeeded in collecting funds and journals for Houssay. He later regained his University position and, in 1947, shared a Nobel Prize with the Coris "for his discovery of the part played by the hormone of the anterior pituitary lobe in the metabolism of sugar."[18]

Less than two years after his retirement, Cannon returned to the classroom when he began a three-month term as Visiting Professor at the New York University College of Medicine early in 1944. The post, supported by the Commonwealth Fund, he wrote Robert Yerkes, was to be "an experiment both on the part of the College and on my part too." He would have an entirely free hand, his sole obligations being weekly lectures to the medical students and two public lectures. "Apart from this I wander about and converse with persons carrying on researches and professors holding staff meetings in clinics and etc. Whether my services can possibly be worth what is being paid for them, I don't know." As far as Yerkes was concerned, Cannon was being too modest. "When one has acquired as much and as varied experience and such precious wisdom as yours," he gently rebuked him, "there isn't much excuse for worrying about the value of one's presence, accessibility and activities to institutions and individuals. So, I shall not ask you whether you feel that you have given value received! Instead I am delighted that you can spend a few months in New York to the advantage of your friends and profession."[19]

Otto Loewi could hardly wait for Cannon to arrive, telling him, "We will have a splendid time." And they did, too, with plenty of opportunities for reminiscences and conversations about their mutual interests. Dean Donal Sheehan couldn't do enough for the Cannons, accommodating his schedule to theirs and arranging for Cannon to have every convenience, including not one but two offices. Cannon gave public lectures on wound shock and self-sustained cortical activity, and then gave an additional talk at Mt. Sinai Hospital on the emotions. His lectures to the medical students were well-attended and ranged in subject matter from bodily changes in emotional excitement to sympathin and neurotransmission, from hunger and thirst to homeostasis.

While Cannon was in New York, a dinner honoring him as president of the American-Soviet Medical Society was held on March 8. Dr. Sheehan participated in the occasion, speaking of Cannon with admiration and real affection. Several of Cannon's Harvard colleagues came down for the occasion—Roy Hoskins and

George Wislocki among them. Mrs. Cannon sat at the speakers' table beside Dr. Landis, whom she found very likeable.

All in all, Cannon fulfilled his commitment at N.Y.U. to everyone's great satisfaction. For his part, he felt privileged to have included the experience in his career, one that he would look back on with a very deep sense of gratitude.[20]

No sooner was Cannon done with his obligations in New York than he was in the hands of the surgeons—this time it was Barrett Brown assisted by Brad Cannon at the Army hospital in Valley Forge. On March 24, a growth on Cannon's upper lip (a basal cell carcinoma) was removed in an operation under local anesthetic. A few days later, Cornelia reported that the graft had taken 100 per cent and Bradford thought it looked very well. Dr. Brown kept close watch on him and he expected to be through in another week. Cannon was heartened by the enthusiasm of all concerned, his face more comfortable with each passing day.[21]

During the weeks of his recuperation, Cannon's thoughts were on the future. While at Valley Forge, he learned that Sen. Kilgore had indicated it would be helpful if a group of prominent scientists could identify fundamental problems concerning the relations of science and government, with particular reference to conditions arising out of the war. But even before that, a report that Cannon had fashioned with Harlow Shapley for the Division of Foreign Relations of the N.R.C. proposed that it was time to prepare for a future without war.

While still at N.Y.U., Cannon had asked Shapley, "Should not the Division of Foreign Relations, along with commerce, industry, education, et al., be making an attempt to determine what attitude we should have and what action we might take after the war?" He then sent the geologist Richard Field a copy of the letter he had written Shapley about the need for postwar planning. Expecting to be *hors de combat* while recuperating from his anticipated surgery, Cannon asked Field if he would design a questionnaire on postwar cooperation involving international scientific unions. Field was eager to carry out the commission and plunged right in, consulting with Cannon by mail and in person. Out of the questionnaires came "A Memorandum on the Activities and Future Plans of International Science Organizations, 1919–1944," which Field subsequently prepared with Cannon.[22]

The trend of Cannon's thinking in 1944 was evident in a telegram he sent that summer to Thomas B. Appleget of the Rockefeller Foundation, in which he wrote: "If there is any nucleus of international good will and understanding left in this world, it resides, I think, in scientific personnel. They will be the first to mend the broken wires of communication, and I hope this time all the world will realize, whether we like it or not, we have to live together on a globe which science has made too small for war." After it was learned that the Division of Cultural Relations of the State Department planned to set up a network of cultural attachés in many large countries, a joint invitation was extended by Cannon and Ross Harrison to a select group of scientists to attend a Conference on Problems of Restoring International Scientific and Scholarly Cooperation, to be held at the end of the year.[23]

Following a whirl of meetings in New York and Washington in the spring of 1944, Cannon returned to Cambridge—an unfamiliar place with thousands of soldiers and sailors milling around. Self-conscious of his changed appearance, now markedly altered by the ravages of failing health and rapid aging, he was scheduled to have X-ray treatments for his lip. In May, when he was up for renomination to the Board of Scientific Directors of the Rockefeller Institute, he offered to resign even though he enjoyed the associations it provided. Herbert Gasser, the director of the Institute, suggested that he stay on for another term, with the option to quit at any time should the duties become onerous. Gasser continued with a more personal observation,

> Ever since you joined our Board of Scientific Directors, your presence has afforded me both pleasure and reassurance. At times, I admit, I have had almost a feeling of guilt at the thought of imposing upon you; but as my admiration has been aroused at the heroic way in which you have been able to surmount your difficulties, the thought has always come to me that the same spirit that has made this possible would probably make you desire to continue. Certainly, I have not thought of you as old, rather as someone who has had time to accumulate wisdom; and that is what we need on the Board.

Cannon subsequently agreed to continue as a member of the Board, his new term running until 1947.[24]

The summer of 1944 was hectic but happy, with several grandchildren visiting the farm at one time or another. Cannon, free from itching or any manifestations of ill health, was at last able to get down to some serious writing. A large room had been built under the barn, where he could retire to work on the two books he had long had in mind. He began first with the autobiographical memoir that would become *The Way of an Investigator.* Mrs. Cannon, who took down his dictation on her typewriter, later recalled, "W.B.C. had long wanted to record his experience as an investigator as a possible help to young men starting on the job of adding to the world's knowledge. He had watched his students struggling with the problems common to all searchers for the truth, and he was eager to put his experience at their service."[25]

In a matter of weeks the memoir was finished, and they began on the second book, now renamed "The Supersensitivity of Denervated Structures." While working on this book in Franklin, Cannon received a visit from one of his former fellows, Joaquin Luco of Chile. Recalling Luco's work with Arturo Rosenblueth on the Vulpian phenomenon in 1936, Cannon decided to give him one of the treasures of his library. It was a book written by Vulpian, which William James had purchased the year of its publication and presented to Cannon when he was a medical student. Luco was thrilled with the gift, not least because it bore the signatures of both James and Cannon, as well as several more recent notes made by Cannon on the last page.

After Cannon had worked through all of his notes and preliminary materials, the scientific monograph still lacked some additional experimental data. This he hoped to supply by means of further work with Dr. Rosenblueth, which was planned during a visit to Mexico he and Cornelia were preparing to make early in 1945.[26]

Cannon spent the autumn in Cambridge. Harvard's Biological Laboratories, not far from his apartment, provided him with ample rooms and facilities for research, the Macy Foundation supplied a modest grant of money, and Dean Burwell offered secretarial help should he need it. Using himself as guinea pig and recording his data in a small notebook kept in his pocket, Cannon observed the effects of nervous fatigue, wakeful nights, indigestion, etc., on the number of steps he took per minute in going from his home to the biological building.[27]

By November, the Cannons were ready to head south for the winter months, having arranged for Marian and her twins to occupy their apartment while they were gone. Their plan was to travel by train, first to Florida, where they would spend the holidays with the Pierces, and then to Mexico for the long-awaited reunion with the Rosenblueths. The first leg of their journey was accomplished without incident, but once in Florida, they discovered that making arrangements to cross the country was not so easy. The wartime travel crunch caught them by surprise and they spent some anxious moments before Cannon at last managed to obtain tickets and reservations to Mexico City.

In the meantime, Wilma at the State Department had prepared a memorandum about the visit to Mexico of Dr. and Mrs. Cannon. It was in fact a formal request to government officials to extend to her parents the opportunity to meet persons in Guatemala and Mexico concerned with cultural cooperation. Dr. Cannon, she wrote, would be working in the laboratories of the Institute of Cardiology with his former student and longtime associate at Harvard, Dr. Arturo Rosenblueth. They were leaving on January 13 and planned to stay in Mexico during February, March, and possibly April.[28]

After arrival in Mexico City, the Cannons had to wait three days before they could get air passage to Guatemala, where they visited the *finca* of their friend, Carmen Petterson. They had an equally difficult time getting a flight back to Mexico City, as Cornelia testified in one of her family letters: "At last we arrived safely to be met by our friends, who had been waiting as plane after plane came in. When I complained of the difficulty of getting from one country to another, Arturo stole my thunder by saying that it is much worse getting into the United States than into any other country and I thought of the poor European refugees bombarding the U.S. consulates everywhere."

They were taken in hand by Arturo and Virginia Rosenblueth, who insisted that they stay with them in their comfortable house just steps away from the laboratory. At the end of February, when the Rosenblueths would not hear of them moving to a pension, they were only too happy to stay where they were. "Father especially is near his work, has Arturo to talk with about their old nervous system, and is being beautifully cared for," Cornelia reported. "I feel perfectly easy in my mind about going to Taxco for a week, remembering that he is in Virginia's thoughtful care and can want for nothing."[29]

During his visit, Cannon took time to write Camille Soula in Toulouse and Edward Barsky in New York to express his appreciation of all they had done for the Spanish refugees who were relocated in Mexico. He even accepted the honorary chairmanship of a fund appeal for the Boston chapter of the Joint Anti-Fascist Refugee Committee, which was channeling money through the Unitarian Service Committee for distribution to the Spanish refugees remaining in France.

In a cordial exchange of letters with Dr. Barsky, Cannon wrote that he had been favorably impressed when he visited the hospital in Mexico established in Barsky's name.[30] This hospital had been greatly helped in 1943, after Cannon made the acquaintance of Joseph Ford, a Boston manufacturer of ladies' underwear who supported the cause of the Spanish refugees. Ford had been easily won over by Cannon and came to depend on him for advice regarding his donations. When Cannon suggested that his money could be put to good use in providing equipment for a hospital that was taking care of the Spanish refugees in Mexico, Ford readily agreed. Cannon was filled with admiration for the hospital, calling it a monument to Barsky's devoted service to the outcasts fleeing from the tyranny of European dictators.

Everyone was pleased with Cannon's lectures and the young scientists who came to hear him seemed almost struck dumb with awe. A dinner given in his honor by the university officials, at which he made a speech complimentary to Mexico, was attended by a flock of dignitaries. The Spanish doctors in Mexico asked him to meet with them later in March. He planned to speak to them about his experiences studying wound shock in 1917–18 and his service as chairman of the N.R.C. Committee on Shock and Transfusion since May 1940.[31]

In mid-March, when Cornelia returned from a sightseeing trip to Taxco with her friend Elinor Lord, she found Walter much depressed by her absence and his researches going poorly. A few weeks later, he had recovered and his researches were again going well. She remarked as she left on another excursion with Elinor that she hoped, in spite of his occasional itching and stomach upsets, he would feel the Mexican experience was a triumphant success.

Soon, however, their Mexican experience, far from triumphant, took a turn for the worse. The two of them, sick and in bed, made a pathetic picture—she with nausea, then fever and sore throat, he with more serious ailments. As Cornelia recounted in a family letter, Cannon had been for weeks protesting against her insistence that he follow Dr. Stoke's advice to soak his hands and feet. The result was that his affliction crept back until there were so many cracks in his hands and feet that he refused the soaking on the score of the pain it caused. Then his old itching came back full force and he scratched his legs; every place he rubbed became infected. The worst discomfort came from the soles of his feet, which had great open sores so that standing was agony.

In spite of everything, Cannon still had to go on with his lectures and the elaborate dinner planned in his honor by the Spanish refugees. Cornelia unhappily felt too ill to attend the dinner. When it was over, Walter went back to bed and, as she put it, gave up trying to be anything but sick.[32]

Treatment with sulfa gave him temporary relief from his misery. But after a week of fever and development of pus pockets, Dr. Chavez took charge of Cannon's case, had him taken to the hospital just across the yard, "and there he has been for three days being stuffed with sulfa drugs and penicillin, which are already beginning to help him," Cornelia related. "Dr. Chavez is certainly one of the loveliest men who ever lived and a very expert doctor as well."

She had nothing but praise for the skillful attention Walter was getting, nothing but gratitude for the tender care of the Rosenblueths. But his general physical state began to fail, the fever and the infection kept returning and he became low in spirits, his sole cry being, "I want to go home." Chavez, Arturo and Cornelia talked matters over and decided that Cannon could not tolerate the long, arduous train trip. With Chavez in charge, they were soon able to get air priority to carry him through to Boston, his wife following by train.[33]

On arrival in Boston, Cannon entered the Massachusetts General Hospital for treatment—but not before he had mailed off the page proofs of the autobiographical memoir he had written. A few days later, his sister Ida wrote W.W. Norton, "He is worried now about the index which he seems to feel *must* be done by the first of May. Personally I do not see how he can possibly do it. He is having chills and temperature and has a great deal of discomfort most of the time as he is not sleeping well." But when Mrs. Cannon arrived, she helped him with the index and it was finished in a matter of days.

Cornelia was shocked to see Walter when she finally caught

up with him in Boston. His face, covered with red spots, was so swollen that his eyes were closed, and the skin was peeling off his feet and hands. The itching continued unabated and he was finding it hard to endure until spray irradiation was used to lessen it. In a few weeks, the care he received at the M.G.H. under Dr. Aub's direction had him well enough to be sent home.[34]

When the infection began to recur within a short time, however, Aub was beside himself and asked his young friend, the surgeon Oliver Cope, to help with the case. Cope put Cannon back in the hospital, where the source of infection could be better monitored, and there he remained for the better part of six weeks. After due consultation with Cannon's physicians, Cope began by taking cultures and found he was battling the worst staph and strep bugs that grow. An all-out attack on the germs, employing transfusions, antibiotics and supplementary measures, brought the infection under control and the itching, which had persisted, was treated with a course of brief X-ray exposure.

Cornelia kept the family informed of Cannon's progress at the M.G.H. and Bradford, as well as Linda, Helen and Douglas (back from England), were able to visit him there. Marian, who was holding down the fort in Franklin, wrote frequently to her mother to assure her of her children's love and support. "As a family, we have certainly lived charmed lives and this is the first real tragedy to strike at us," she observed. "The world is too full of sorrows these days to have this one added to the rest!"[35]

In the meantime, Norton had announced the publication of Cannon's memoir, *The Way of an Investigator,* which was chosen a principal selection of the Scientific Book Club. Robert Yerkes, one of the first to receive a copy, wrote Cannon:

> As I review my preprofessional years, I realize that had such a book fallen into my hands during my years in college, I should have read it with avidity and been immeasurably helped by it. This, perhaps, is the most significant attempt at evaluation I can offer. For I have no doubt that in planning and writing the book you had chiefly in mind future generations of scientists and investigators, instead of your students and colleagues who have had the benefit of profiting by your example and precept.

By the end of June, a relieved Cornelia wrote her children, "Father is so much better it would do your hearts good . . . He has dictated lovely inscriptions in his books for friends, which I write and he painstakingly signs."

A setback kept Cannon in the hospital for the first half of July. During that time, he received a letter from Wilma, who had been sent to Chungking by the State Department. She was looking forward to having a copy of his book, and said that many others were anxious to see it as well: "I am proud to see how many friends and admirers you have here—a truly international figure. I am privileged to live in your reflected glory." Cannon, for his part, was equally proud of Wilma, who had advanced during the war years to a position of great responsibility.[36]

On July 19 Cannon, clinically much improved, was released from the hospital to spend what would be the last weeks of his life in the place he called home. At Franklin Cornelia hoped he would benefit from days in the sun, far from the agitation and strain that so contributed to his suffering. Helen, Marian and Linda fixed up the room in the barn for their father so he could stay there in comfort and away from the noise of the grandchildren. But they all visited him and diverted his thoughts and kept his spirit young. The remainder of July and August passed pleasantly enough, with Cannon admitting somewhat sheepishly that he rather "liked to loaf." He was better in every way—more energy and interest in life and ideas—though he missed the presence of stimulating friends, especially the Pierces.

During these weeks when Cannon was free from suffering, he wrote Dr. Aub a poignant letter:

> I feel sure that you knew how deep my sense of gratitude was as we parted at the Hospital. I knew that if I attempted to express it in a single spoken word I should break down utterly. Indeed, I despair of ever expressing in any manner my profoundest thanks for the many acts of great kindness you have performed for me. And the sad feature of it is that I am so blind that I do not see any way of showing my gratitude. But do know, dear Joe, that it is in me—deep within me—and will be there till the end of my days.[37]

As September began, Cannon again had a spell of bad itching, then for a time he was free of it. The return of his affliction was accompanied by another blow—the death of George Pierce's wife, Florence. She had had an intestinal hemorrhage and was taken to the hospital for a blood transfusion but died there on September 3.

Just days after Florence Pierce's funeral, Cannon wrote a letter to Dr. Richard Hoffman, a refugee physician in Boston who had been recommended to him by Henry Christian. In it, he related the facts of his illness, then ended, "If you think there is anything that can be done to relieve my torment, may I have an appointment?" Cannon subsequently saw Hoffman, who supplied him with a new drug, Bellergal, which helped him through the night and gave him "the best sleep in months." The itching was still very bad during the day, and he had begun taking iron and injections of liver extract because he had a rather severe anemia.[38]

Cornelia kept her family posted of all this in her letter of September 21, as well as giving an account of their visit to town to see Cannon's doctors. She continued, "Father sits beside me trying to write letters between scratches. There are times when I almost lose my own nerve control when he scratches. It is curious how racking it is even for the non-itcher!"

Toward the end of September, Cannon's condition deteriorated and he became comatose. Dr. Aub, who rushed to Mrs. Cannon's side, recorded his patient's last few days of life in his case notes. Cannon was admitted to Franklin Hospital the afternoon of September 29 with a temperature of 102, which rose for the next few days until it reached 105 just before his death. Aub wrote, "When I saw him, he was having a chill and itching a good deal. Breathing fast and deep as though in acidosis. All movements free and normal, could open and close eyes—though they did not follow the finger very well, movements seemed o.k. Mouth and facial movements o.k. and obviously arms and legs moved freely. Reflexes not done—as he was so cold."

Cannon died on October 1, just weeks before his seventy-fourth birthday. On autopsy, Dr. Wolbach was "of the opinion that an acute febrile disease with pneumonia terminated his life. His disease of long standing at the time of death was in a period of remission."

Afterward, Bradford wrote to thank Aub for his devotion to his father and for the way he had handled the many problems that had arisen in the past fifteen years: "Your presence and your comforting words on Sunday night and Monday morning have given you undying gratitude from Mother and all of us." Ida wrote, too, and told Aub, "To me, your coming just when you were needed most is symbolic of the devoted service Walter has had all through the years." And Cornelia, from the bottom of her heart, acknowledged Aub's sympathy and strength in helping her through the hardest hours of her life, saying simply, "I cannot adequately thank you, I can only love you.[39]

At Cannon's death there was an outpouring of resolutions, memorials, and letters of condolence from near and far—all mourning the passing of a great man. Joe Ford's tribute was to establish a fund in Cannon's honor for scholarships in the physiology department at the Medical School. In Franklin, the immediate family (all except Wilma, who was with them in spirit) gathered in the old farmhouse to celebrate Cannon's life. Later they scattered his ashes over the land that meant so much to all of them. His daughter Marian put it best when she wrote in retrospect: "Though we mourned his death, there was a kind of exultation in recalling the richness of that life, and the quality of his person; his humor, his kindliness, his fairness, his fresh imagination that never seemed to dry up, his intellectual rigor, and the uncomplaining courage with which he bore the almost intolerable pain of his last years."

A funeral service was held at Mt. Auburn Cemetery in Cambridge on October 5, followed by two memorial meetings, one in New York City in late October and another at Harvard Medical School in early November. Otto Loewi was among those at the New York meeting, which Bradford and Ellen Cannon also attended. Chaired by Henry Sigerist, it featured speakers who focused on Cannon's international activities, especially with regard to American-Soviet relations and aid to Spain. Abraham Stone, who represented the American-Soviet Medical Society, moved that a W.B. Cannon fellowship be established for the exchange of students between the two countries.[40]

Joe Aub, who headed the committee for the meeting at the Medical School, worked closely with Cornelia Cannon on the plans. They both wrote to Arturo Rosenblueth about Cannon's last weeks and detailed the plans for the memorial service. But Rosenblueth replied that he would not be present. Much as they wanted Juan Negrin to speak at the service, they were unable to contact him. In the end, the speakers were Phil Bard, Ross Harrison, A. J. Carlson and "Tom" Perry, as well as Joe Aub, each remembering a particular aspect of Cannon's life.

Afterward, Cornelia recorded that the meeting at the Medical School was wonderful, something for her to cherish in memory. As they went into the Physiology Amphitheatre, Dean Burwell, who chaired the meeting, had said to her, "Remember, this is a happy day." His words both sustained and strengthened her, and she wrote that it was indeed a happy day for those who had not only admired the scientist but loved the man.[41]

ABBREVIATIONS AND LOCATIONS

END NOTES

INDEX

Abbreviations and Locations of Collections

A.A.A.S.	American Association for the Advancement of Science
A.A.S.W.	American Association of Scientific Workers
A.A.U.P.	American Association of University Professors
ABMAC	American Bureau for Medical Aid to China
A.E.F.	American Expeditionary Forces
A.F.S.C.	American Friends Service Committee
A.J.P.	*American Journal of Physiology*
A.M.A.	American Medical Association
A.P.S.	American Physiological Society
A.R.A.	American Relief Administration
B.E.F.	British Expeditionary Forces
C.C.F.	Committee for Cultural Freedom
C.C.S.	Casualty Clearing Station
C.M.B.	China Medical Board
C.M.R.	Committee on Medical Research
C.R.B.	Commission for Relief in Belgium
E.R.C.	Emergency Refugee Committee
FASEB	Federation of American Societies for Experimental Biology
F.D.R.	Franklin Delano Roosevelt
G.E.B.	General Education Board
H.M.S.	Harvard Medical School
Investigation	*Investigation of the Nature and Treatment of Wound Shock and Allied Conditions*
I.P.R.	Institute of Pacific Relations
"Letters Home"	bound volume of typed transcripts, "Letters Home from France and England, 1917–18"
Life and Contributions	*The Life and Contributions of Walter B. Cannon*, ed. C.M. Brooks et al., State University of New York, Downstate Medical Center, 1975
M.G.H.	Massachusetts General Hospital
M.I.T.	Massachusetts Institute of Technology
M.R.C.	Medical Research Committee (later Medical Research Council; National Institute of Medical Research)
N.A.S.	National Academy of Sciences
N.D.R.C.	National Defense Research Committee

N.I.M.R.	National Institute of Medical Research
N.R.C.	National Research Council
N.Y.U.	New York University
O.S.R.D.	Office of Strategic Research and Development
O.S.S.	Office of Strategic Services
P.B.B.H.	Peter Bent Brigham Hospital
P.U.M.C.	Peiping (or Peking) Union Medical College
SERE	Servicio de Emigración para Republicanos Españoles
"Servant of Science"	Unpublished biography of W.B. Cannon by Cornelia James Cannon (in Cannon Papers)
"Snatched from Oblivion"	bound volume of excerpts from Cornelia James Cannon's letters to her mother (in Cannon Family Papers)
Snatched from Oblivion	Marian Cannon Schlesinger's Cambridge memoir, Boston: Little, Brown, 1979
S.R.R.C.	Spanish Refugee Relief Campaign
U.C.R.	United China Relief, Inc.
Walter B. Cannon (1987)	Saul Benison, A.C. Barger and Elin L. Wolfe, *Walter B. Cannon: The Life and Times of a Young Scientist*, Harvard University Press, 1987
W.B.C.	Walter Bradford Cannon
The Way of an Investigator	Cannon's autobiographical memoir, N.Y.: W.W. Norton, 1945

J. C. Aub Papers
Countway Library of Medicine
Boston, Mass.

Birth Control League of Massachusetts Papers
Smith College Archives
Northampton, Mass.

C. S. Burwell Papers
Countway Library of Medicine
Boston, Mass.

W. B. Cannon Papers
Countway Library of Medicine
Boston, Mass.

Cannon Family Papers
Schlesinger Library
Radcliffe College
Cambridge, Mass.

H. A. Christian Papers
Countway Library of Medicine
Boston, Mass.

J. B. Conant Papers
Harvard University Archives
Pusey Library
Cambridge, Mass.

Harvey Cushing Papers
Sterling Memorial Library
Yale University
New Haven, Conn.

H. H. Dale Papers
Royal Society
London, England

Dean's Files
Harvard Medical Archives
Countway Library of Medicine
Boston, Mass.

Joseph Erlanger Papers
Washington University
School of Medicine Library
St. Louis, Mo.

Simon Flexner Papers
American Philosophical Society
Library
Philadelphia, Pa.

Alexander Forbes Papers
Countway Library of Medicine
Boston, Mass.

General Education Board Papers
Rockefeller Archives Center
North Tarrytown, N.Y.

Jerome D. Greene Papers
Harvard University Archives
Pusey Library
Cambridge, Mass.

Roger S. Greene Papers
Houghton Library
Harvard University
Cambridge, Mass.

Harvard Medical Archives
Countway Library of Medicine
Boston, Mass.

Harvard University Archives
Pusey Library
Cambridge, Mass.

W. W. Keen Papers
American Philosophical Society
Library
Philadelphia, Pa.

A. L. Lowell Papers
Harvard University Archives
Pusey Library
Cambridge, Mass.

Radcliffe Archives
Schlesinger Library
Cambridge, Mass.

A. N. Richards Papers
University of Pennsylvania Archives
Philadelphia, Pa.

Rockefeller Foundation Papers
Rockefeller Archives Center
North Tarrytown, N.Y.

G. N. Stewart and J. M. Rogoff
Papers
Countway Library of Medicine
Boston, Mass.

R. M. Yerkes Papers
Sterling Memorial Library
Yale University
New Haven, Conn.

End Notes

Unpublished materials cited below, unless otherwise noted, are in the Walter B. Cannon Papers, Countway Library of Medicine.

1. The Mystery of Shock, I: Acidosis

An account of Cannon's work on shock in World War I has previously been published; see Saul Benison, A. Clifford Barger and Elin L. Wolfe, "Walter B. Cannon and the mystery of shock: A study in Anglo-American cooperation in World War I, *Med.Hist.* 35(1991):217–249. See also Chapter 12, "A parenthesis of war," in Cannon's autobiographical memoir, *The Way of an Investigator.*

1. Thomas Johnson, tr., *The Collected Works of Ambroise Paré,* London, 1634; repr. Pound Ridge, N.Y., Milford House, 1968, p. 334.

2. For details concerning the formation of the National Research Council Physiological Committee, see Benison, Barger and Wolfe, *Walter B. Cannon* (1987), pp. 388–389.

3. Hermann Fischer, "Über den Shok," *Samml. Klinischer Vortr.* 1(1870):69–82; also Fischer's "Remarks on shock," *St. Louis Med. Surg. J.* 12, n.s. no. 2(1875):68–78.

4. Mitchell, Morehouse and Keen, *Reflex Paralysis,* Circular No. 6, Surgeon General's Office, 10 March 1864; also *Gunshot Wounds and Other Injuries of Nerves,* Philadelphia, Lippincott, 1864. For G.W. Crile, see *Blood Pressure in Surgery* (Philadelphia: Lippincott, 1903), pp. 401–402; also Peter C. English, *Shock, Physiological Surgery, and George Washington Crile: Medical Innovation in the Progressive Era,* Westport, Conn., Greenwood Press, 1980.

5. W.T. Porter, "Fat embolism as a cause of shock," *Bos. Med. & Surg. J.* 176 (1917):248; Yandell Henderson, "Acapnia and shock," *Am. J. Physiol.* 21(1908):126–156 and 27(1910):152–176; S.J. Meltzer, "The nature of shock," *Arch. Internal Med.* 1 (1908):571–588.

6. For Cannon's Shattuck Lecture, see "The physiological factors concerned in surgical shock," *Bos. Med. & Surg. J.* 176(1917):859–867. For Base Hospital No. 5, see *The Story of U.S. Army Base Hospital No. 5,* Cambridge, Mass., The University Press, 1919; Harvey Cushing, *From a Surgeon's Journal, 1915–1918,* Boston, Little, Brown, 1936. See also *Walter B. Cannon* (1987), pp. 388–392.

7. Cannon kept a running account of his war experiences in his diary for 1917–18 and in letters to his wife, Cornelia. For the latter source, reference here is primarily to the bound volume of typed transcripts, "Letters Home from France and England, 1917–18" [hereafter cited as "Letters Home" with the

appropriate page numbers]. Mrs. Cannon described the parting in a letter to her family of 9 May 1917 (in 1917–18 letter-book, Cannon Family Papers).

8. Cannon to Mrs. Cannon, letters of 9 and 10 May 1917.

9. For the voyage of the *Saxonia*, see diary entries for 11–22 May 1917; "Letters Home," pp. 1–16.

10. For Cannon's week in England, see diary entries for 23–30 May 1917; "Letters Home," pp. 17–30. *The Third Annual Report of the Medical Research Committee, 1916–17*, gives the background for the Committee's concern with shock as a subject of inquiry (pp. 77–79). See also "Memorandum upon surgical shock and some allied conditions, issued by the Medical Research Committee," *Br. Med. J.*, 24 Mar 1917, pp. 381–383.

11. "Letters Home," p. 38. For more on Camiers, see Cannon's diary notes for 1–25 Jun 1917; Cushing, *From a Surgeon's Journal* (n.6), pp. 142–159.

12. Ibid., p. 42. See also T.R. Elliott to Cannon, with copy of Elliott to Cuthbert Wallace, both 15 Jun 1917, and W.M. Bayliss to Cannon, 16 Jun 1917.

13. Diary entries for 24 and 25 Jun 1917; "Letters Home," pp. 46, 49, 52. Wolf went back to Cambridge University after the war, while Fraser became the Regius Professor of Clinical Surgery at Edinburgh. Fraser and Wallace wrote about their war experiences in *Surgery at a Casualty Clearing Station*, London, A. & C. Black, 1918.

14. See further E.M. Cowell, "The initiation of wound shock," in *Investigation of the Nature and Treatment of Wound Shock and Allied Conditions* [hereafter cited as *Investigation*], Report No. II, Special Investigation Committee on Surgical Shock and Allied Conditions of the M.R.C., 25 Dec 1917, pp. 58–66.

15. For Cushing's attempt in 1903 to persuade surgeons at Harvard of the value of taking blood pressure by the Riva Rocci method during surgery, see John Fulton, *Harvey Cushing: A Biography* (Springfield, Ill.: Thomas, 1946), pp. 213–215; also The Division of Surgery of the Medical School of Harvard University, *Report of Research Work, 1903–1904*, Bulletin No. II (Mar 1904):1–41.

16. Diary entries for 28, 29, and 30 Jun 1917.

17. For an expression of the view that accumulation of blood in the abdominal veins is not a feature of clinical shock, see Wallace and Fraser, *Surgery at a Casualty Clearing Station* (n.13), p. 8. For a more recent view of the problem, see C.R. Valeri and M.D. Altschule, *Hypovolemic Anemia of Trauma: The Missing Blood Syndrome*, Boca Raton, Fla., CRC Press, Inc., 1981.

18. This study eventuated in Cannon, Fraser and Hooper, "Some alterations in distribution and character of blood in shock and haemorrhage," in *Investigation*, pp. 27–39.

19. "Letters Home," pp. 63–64.

20. Ibid., pp. 66–67; Walter Fletcher to Cannon, 27 Jun 1917. Prominent among the English investigators working on shock were the physiologist William M. Bayliss and the pharmacologist Henry H. Dale.

21. Ibid., p. 67. For Henderson's publications on acid-base equilibrium, see the bibliography accompanying W.B. Cannon, "Lawrence Joseph Henderson, 1878–1942," *Biographical Memoirs of the National Academy of Sciences*, 23(1942):52–58.

22. Ibid., pp. 73, 76–77. The course of this research was noted in Cannon's diary entries for 6–12 Jul 1917.

23. Cannon, "Acidosis in cases of shock, haemorrhage and gas infection," in *Investigation,* pp. 47–60. For Wright's report on gas gangrene, see *Lancet,* 1(1917):1–9.

24. Mrs. Cannon's letter was written on 24 Aug 1917; in Cannon Family Papers. For Cannon's letter, see "Letters Home," pp. 89–90.

25. "Letters Home," pp. 92–93. Of his relations with Bayliss, Cannon concluded, "Bayliss has been so quiet and modest—no one would take him to be one of the world's foremost physiologists. It has been a great privilege to be associated with him during these days. We have done experiments together, gossiped about physiologists, talked about work ahead, discussed views of shock, and had a most delightful and friendly time getting better acquainted."

26. Bayliss to Cannon, 9 Aug 1917; see also Fletcher to Cannon, 10 Aug 1917. The Special Committee on Shock and Allied Conditions of the M.R.C. held its first meeting in London on 17 Aug 1917, with Ernest Starling in the chair and Henry Dale as secretary. The other members, besides Bayliss and Cannon, were F.A. Bainbridge, T.R. Elliott, John Fraser, A.N. Richards, C.S. Sherrington, and Cuthbert Wallace.

27. "Letters Home," p. 96. See also Cannon, "Acidosis in cases of shock" (n.23), pp. 54–57. Wallace and Fraser later paid tribute in their book (n.13) to Cannon for his work on the effect of anaesthesia during operation on patients in shock (pp. 20–21).

28. Bayliss to Cannon, 25 Aug 1917.

29. For Cannon's visit to Paris, see "Letters Home," pp. 119–125.

30. See Cannon, Fraser and Cowell, "The preventive treatment of wound shock," in *Investigation,* pp. 93–104; also Cannon's sections on the sugar content of the blood and on the effect of anaesthesia and operations on existent acidosis and low blood pressure in "Acidosis in cases of shock" (n. 23), pp. 53–57.

31. "Letters Home," pp. 106–107 and 137–139. Cannon received the letter from Boothby on 14 Aug 1917. The order to report for duty in the Gas Service arrived on September 15.

32. For Cannon's second visit to Paris, see "Letters Home," pp. 145–147; diary entry for 21 Sep 1917.

33. Dale to Cannon, 27 Sep 1917; Richards to Cannon, 22 Sep 1917. See also "Letters Home," p. 150.

34. "Letters Home," p. 175.

35. Ibid., pp. 180, 181. The investigations at Béthune were published as the second report of the Special Investigation Committee of the M.R.C. on 25 Dec 1917. Sir Walter Fletcher and the military censor consented to the subsequent reprinting of the papers in the *Journal of the American Medical Association,* 70(1918):520–535, 607–621, where they appeared as "The nature and treatment of wound shock and allied conditions," by W.B. Cannon, E.M. Cowell, John Fraser, and A.N. Hooper.

36. M.R.C. Special Committee on Shock and Allied Conditions, minutes of meetings for 2 and 9 Nov 1917 (on file at the Medical Research Council in London).

37. Gum acacia was proposed as an inexpensive and readily available expander of plasma volume. See further W.M. Bayliss, *Intravenous Injections to Replace Blood,* Report No. I of the M.R.C. Special Investigation Committee on Surgical Shock and Allied Conditions, 1917, revised 1919; also Bayliss, *Intravenous Injection in Wound Shock,* London, Longmans, Green, 1918.

38. H.H. Dale, P.P. Laidlaw, and A.N. Richards, "The action of histamine: its bearing on traumatic toxaemia as a factor in shock," in *Traumatic Toxaemia as a Factor in Shock,* Report No. VIII of the M.R.C. Special Investigation Committee on Surgical Shock, 1919; esp. p. 12.

The resumption of Dale and Laidlaw's work on histamine was given a prominent place in the "Memorandum upon surgical shock and some allied conditions issued by the Medical Research Committee" (published in *Br. Med. J.,* 24 Mar 1917, pp. 381–383; see also editorial, "The nature and prevention of shock," in the same issue, p. 397). When Laidlaw was drafted in the summer of 1917 to serve in the pathology department at Guy's Hospital, Dale, anxious to extend the research on histamine, recruited Richards, a young pharmacologist from the University of Pennsylvania, to serve in Laidlaw's place.

39. M.R.C. Special Committee on Shock and Allied Conditions, minutes of meeting for 27 September 1917; see also "Letters Home," p. 138.

40. Cannon's experiments with Bayliss may be followed in his 1917–18 diary entries; see esp. his notes for 22 Dec 1917, 14 and 18 Jan 1918.

41. "Letters Home," p. 259.

42. H.H. Dale, "Autobiographical note," pp. 159–161 (unpublished memoir in Dale Papers). See also M.R.C. Special Committee on Shock and Allied Conditions, minutes of meeting for 19 January 1918; "Letters Home," p. 254.

43. "Letters Home," pp. 262–263; M.R.C. Shock Committee, minutes of meeting for 8 February 1918.

44. Dale, "Autobiographical note" (n. 42), pp. 161–162. See further H.H. Dale, "The nature and cause of wound shock," in *The Harvey Lectures for 1919–1920* (Philadelphia and London: Lippincott, 1920), pp. 38–39.

2. The Mystery of Shock, II: Resuscitation

1. "Letters Home," p. 245. For Haldane, see also p. 181; for Barcroft, see further pp. 241–245, 248, 250.

2. Ibid., p. 273.

3. Reports of the monthly Red Cross Research Society meetings were published in their *Medical Bulletin* and its successor *War Medicine.* The session devoted to the subject of traumatic shock was held on 15 Mar 1918, and reported in the April issue; Cannon's remarks appear on pp. 424–428.

4. "Letters Home," 283–284, 286. Cannon began his enlistment in the U.S. Army Medical Service as a First Lieutenant, became a Captain in Aug 1917, then a Major in Feb 1918, and finally a Lieutenant Colonel in Oct 1918.

5. A.N. Richards to Cannon, 3 Apr 1918, in response to Cannon's letter of March 23; see also H.H. Dale to Cannon, 8 Apr 1918.

6. Cannon to Walter Fletcher, 12 Apr 1918.

7. Dale's letters to Cannon are dated 17 and 20 Apr 1917.

8. Cannon to Dale, 30 Apr 1918; see also his letter to Mrs. Cannon of the same date (not in "Letters Home").

9. Dale to Cannon, 7 May 1918.

10. Richards to Cannon, 10 May 1918, reviews the entire misunderstanding from its inception. See also Richards to Cannon, 14 May 1918, and Cannon to Richards, 21 May 1918.

11. "Letters Home," pp. 290, 287–288. Before long, a friendship had blossomed between Zinsser, Cannon and Yates that developed into a deep and lasting bond.

12. Cannon to W.M. Bayliss, 24 Apr 1918, in response to letters from Bayliss of 16 Mar and 3 Apr 1918. For a detailed discussion by Bayliss on acidosis, see his *Intravenous Injection in Wound Shock* (London: Longmans, Green, 1918), p. 60.

13. Mimeographed syllabus, "Course of instruction for preliminary training in resuscitation work for the wounded." For a more detailed description of the resuscitation training program, see Cannon's typescript report, "Historical Record of the Services of the Laboratory of Surgical Research, American Army, at Dijon, 1918" (esp. pp. 2–12), which was sent with letter of enclosure to General J.M.T. Finney on 2 Dec 1918. These and other W.W. I materials are in Box 136 of the Cannon Papers.

14. For the invitation to give the Croonian Lecture, see "Letters Home," pp. 260–261. Cannon cabled his wife to send his notes on thirst on 25 Feb 1918. For comments on preparing the lecture, see Ibid., pp. 327–328, 342–343, 347, 362–363.

15. Cannon's letter to his wife about his visit to London was dated 24 Jun 1918 (some parts not in "Letters Home"). On the way to the lecture, Fletcher told Cannon that it was the "blue ribbon of the biological turf."

16. "Letters Home," p. 337. Cannon's research with McKeen Cattell did not reach publication but is described in his "Historical Record" (n.13), pp. 13–14. For Cattell's previous work with Cannon, see *Walter B. Cannon* (1987), pp.326–329.

17. Dale to Cannon, 3 Jun 1918; see also Cannon to Bayliss, 21 May 1918, and Cannon to Dale, 25 May 1918.

18. For the differences between the "regulars" and the "volunteers" in the Army medical services, see "Letters Home," pp. 332 and 334.

19. The battle is described in Cannon's letter of 16–17 Jul 1918. See "Letters Home," pp. 396–405, esp. pp. 401–403 for the shock work at Écury; also *The Way of an Investigator,* pp. 138–140. Cannon's service in this engagement, referred to as the Battle of Champagne, as well as his service in the

Battle of St. Mihiel of September 1918 ("Letters Home," pp. 439–449), was later recorded on his military discharge.

20. See further Cannon's recommendations for improving work in resuscitation in his "Historical Record" (n.13), pp. 12–14.

21. T.R. Elliott to Cannon, 22 Aug 1918; in the archives of the Medical Research Council, London. For Cannon's interest in Robertson, see also their exchange of letters, 25 and 31 Jul 1918; Roger I. Lee to Cannon, 31 Jul 1918. Robertson's *Memdorandum on Blood Transfusion* was published as Report No. IV in the series issued by the Special Investigation Committee on Surgical Shock and Allied Conditions, of the M.R.C., 4 Apr 1918, revised 6 Sep 1918. The *Memorandum on Blood Volume after Haemorrhage* (with Arlie Bock) was published as Report No. VI in the same series, 8 Aug 1918.

22. Bayliss to Cannon, 27 Jul 1918, in reply to Cannon's letter of Jul 23. See further J.W. McNee, A.F. Sladden and J.E. McCartney, "Observations on wound shock, especially with regard to damage of muscle," in *Traumatic Toxaemia as a Factor in Shock,* Report No. VIII of the M.R. C. Special Investigation Committee, 1919.

23. Bayliss to Cannon, 24 Aug 1918; "Letters Home," p. 428. Cannon's report to the Commanding Officer of the Central Medical Department Laboratory was for the week ending 16 Aug 1918.

24. Cannon, "Historical Record" (n.13), pp. 15–17. Aub later took up this work under experimental conditions in the physiology laboratory at Harvard and published his results as "Studies in experimental traumatic shock. I. The basal metabolism," *Am. J. Physiol.* 54 (1920):388–407.

25. Dale to Cannon, 4 Oct 1918, in reply to Cannon's letter of Oct 2. See also Dale's letters to Cannon of 3, 13 and 20 Sep 1918. The memorandum was published as *Acidosis and Shock,* Report No. VII of the M.R.C. Special Investigation Committee Oct 1918.

26. Dale to Cannon, 11 Oct 1918, following up on his letter of Oct 4. For the invitation from Charles Richet, see "Letters Home," p. 430.

27. Cannon, "A consideration of possible toxic and nervous factors in the production of traumatic shock," *Ann. Surg.* 100 (1934):704–713; see esp. p. 708. An outline of Cannon's talk to the Société de Biologie was published as "Traumatic shock" in *War Med.* 2(1918–19):1367–1370; an abstract of the discussion appeared in *Med. Suppl.,* 1 Dec 1918, pp. 437–438.

28. Bayliss to Cannon, 1 Oct 1918. Some of the unfavorable opinions, as reported by C.G. Mixter, R.I. Lee and others, were later published in *War Med.* 2 (1918–19):1276–1278.

29. Paul De Kruif, then a captain in the Sanitary Corps, later published his results as "Experimental research on the effects of intravenous injection of gum-salt solutions," *Ann. Surg.* 69 (1919):297–311.

30. Robertson to W.B. Cannon, "Summary of observations and certain recommendations concerning the use of gum-salt solution and the treatment of shock," memorandum of 27 Oct 1918. See also Robertson's remarks in the discussion of gum-salt solution published in *War Med.* 2 (1918–19):1277.

31. "Letters Home," p. 482.

32. Ibid., p. 488. Cannon described his tour with Flexner and Lambert in a letter of 13 Nov 1918; see pp. 484–491.

33. "Conference at Boulogne of surgeons and pathologists of the B.E.F. on treatment of shock and haemorrhage, held on November 15th, 1918"; 6-page typescript in the A.N. Richards Papers. The comments recorded here are taken from this report.

34. Bayliss to Cannon, 12 Aug 1918. Bayliss' Oliver-Sharpey Lectures, delivered before the Royal College of Physicians in London, were published as *Intravenous Injection in Wound Shock* (n.12); see esp. p. 20.

35. For a description of recommended procedures, see *A Report upon Transfusion for the Recently Injured in the United States Army,* pamphlet published by the Medical Division of the American Red Cross Society in Paris, May 1918. Cannon's name appears as a member of the committee making these recommendations. The list of supplies is pitifully inadequate by later standards; for example, only four plain glass bottles of one-litre capacity "marked 100 c.c,, 400 c.c., and 700 c.c. at the proper levels" were recommended for each hospital.

36. Minutes of special meeting, 24 Mar 1919, held at the M.R.C. For later discussions of the case for and against gum-salt solutions, see Cannon, *Traumatic Shock* (New York and London: Appleton, 1923), pp. 179–183; Bayliss, "Acacia for transfusion," *JAMA* 78(1922):1885–1887.

37. "Letters Home," p. 496. The proceedings of the surgical conference held on Nov 22 and 23 were published in *War Med.* 2(1918–19):1251–1281.

There was some support for gum acacia, but the tenor of the criticism concerning its use was much harsher at this conference than at Boulogne. George Crile was one of the most critical and later wrote in his diary: "The physiologists were represented by . . . Cannon. They had their day first. They sang the hymn of praise! Then the practical surgeons of the Argonne fight sang the swan song! Gum was condemned! Exit gum! Exit physiologists! *Der Tag!*" (Quoted in Peter English, *Shock, Physiological Surgery, and George Washington Crile,* Westport, Conn.: Greenwood Press, 1980, p 205.)

38. "Letters Home," p. 498. Cannon's "Historical Record" was later incorporated into his chapter on "Wound Shock" in *The Medical Department of the United States Army in the World War,* vol. II, *Surgery* (Washington, D.C.: Government Printing Office, 1927), pp. 185–213.

39. Dale to Cannon, 12 Dec 1918. There was, however, an inkling or two that the British pharmacologist continued to smolder over *l'affaire Richards.* We leave it to some future biographer to reconcile the two faces of Dale. One side of him could smile and practice a public diplomacy that was both graceful and sensitive; then another side would take over and give vent in private to cruel and cutting comments.

3. A Lifeline of Letters

1. 30 Apr 1917, in "Snatched from Oblivion," a bound volume of excerpts from Cornelia Cannon's letters to her family. Mrs. Cannon's letters cited

below come from two other sources as well: manuscript letters to her husband and a bound volume of typescript excerpts from letters mostly to Cannon but also to her mother [hereafter cited as 1917–18 letter-book]. These and other letters and manuscripts of Cornelia James Cannon were on loan to the authors of this biography but are now located in the Cannon Family Papers.

2. See further 1917–18 letter-book for examples of the children's letters and drawings.

3. "Snatched from Oblivion," letter of 24 May 1917. The invitation from her sister is mentioned in the excerpt for May 14.

4. Ibid., 5 Jun 1917; see also Mrs. Cannon to Cannon, 4 Jun 1917. The family arrived in the Midwest on Jun 19 for a ten-week stay.

5. For the camping trip, see Mrs. Cannon's unpublished essay, "A Middle Aged Adventure," Aug 1917, 10-page typescript in the 1917–18 letter-book.

6. Ibid., letter of 8 Sep 1917. For Cannon's description of the bombing of American hospitals, see the bound volume of his "Letters Home," pp. 133 and 136–137; in Cannon Papers.

7. Excerpt from letter of 11 Sep 1917, in "Snatched from Oblivion"; see also Mrs. Cannon to Cannon, 18 Oct 1917. For the "radio boys," see J.L. Coolidge, "Naval Training at Harvard," *Harv. Grads. Mag.* 26(1917–18):202–208; President Lowell's Annual Report for 1916–17, p. 9.

8. 1917–18 letter-book, letter of 5 Oct 1917; also Mrs. Cannon's unpublished essay, "From a Hostess House, 1917–18," 10-page typescript.

9. Ibid., 22 Oct 1917 re "nice" boys.

10. Ibid., see letters of 6 and 28 Nov 1917.

11. Ibid., 11 Nov 1917. The reference to her sister Linda that follows is also in this letter.

12. Cannon's letter was written on 24 Jul 1918 (not in "Letters Home"), in reply to his wife's letter of June 7 ("I am sending you a copy of the pamphlet, 'Friend or Enemy,' published by the Y.M.C.A.").

13. Mrs. Cannon to Cannon, 13 Nov 1917. Robert Yerkes' letter to Cannon, 5 Aug 1917, is in the Yerkes Papers. Henry Lee Higginson, who gained his military rank in the Civil War, was a member of the banking firm of Lee, Higginson & Co., as well as a philanthropist and benefactor of Harvard. Among his many cultural and educational benefactions were the founding of the Boston Symphony Orchestra and the establishment of the George Higginson Professorship of Physiology at the Harvard Medical School.

14. Cannon to Mrs. Cannon, 23 Dec 1917 (not in "Letters Home"). For earlier letters, see Mrs. Cannon to Cannon, 11 Nov and 5 Dec 1917; also Cannon to Mrs. Cannon, 28 Oct 1917 (not in "Letters Home"). W.T. Porter's articles were published in a small volume, *Shock at the Front*, Boston, Atlantic Monthly Press, 1918. For Cannon's earlier relations with Porter, see *Walter B. Cannon* (1987), esp. pp. 112–114, 119–122, 348–349.

15. Cannon to Mrs. Cannon, 29 Dec 1917 (not in "Letters Home").

16. Cannon to Mrs. Cannon, 21 Apr 1918 (not in "Letters Home"). See also "Letters from Professor W.B. Cannon, '96," *Harv. Alum. Bull.*, May 30, 1918, pp. 664–666, which contains portions of six letters, 19 Dec 1917 to 30 Mar 1918. In 1935, the edited volume of Cannon's "Letters Home" was submitted (unsuccessfully) to the Atlantic Monthly Non-fiction Prize contest.

17. "Snatched from Oblivion,." excerpt of 14 Nov 1917. Still later, on Nov 18, Cornelia exclaimed, "I know now what it is to eat 'the bitter bread of poverty'—it is rye bread!"

18. Ibid., 24 Sep 1917.

19. Ibid., 29 Oct 1917.

20. Ibid., 4 Mar 1918.

21. See letters of 26 Jan and 2 Feb 1918, in the 1917–18 letter-book; also 2 and 8 Feb 1918, in "Snatched from Oblivion."

22. Cannon described his visit to Cambridge University in a letter of 5 Nov 1917; see "Letters Home," pp. 189–193. His follow-up letter of 13 Jan 1918 was addressed to "My dear Vice-Chancellor." For Shipley's article on "Hate," see *The Cornhill Magazine* (Sep 1915), pp. 289–299.

23. "Letters Home," p. 193.

24. After returning from a visit to the United States in May 1921, Walter Fletcher wrote a Memorandum for the University Grants Committee on the difference between English and American systems of medical education (11-page pamphlet in the Countway Library of Medicine). Years later, Cannon referred to his luncheon with Fletcher in a letter to T.R. Elliott, 22 Jan 1934.

25. "Letters Home," p. 321.

26. Cannon's letter to Mrs. Lathrop is dated 25 Feb 1918.

27. "Letters Home," p. 310.

28. 1917–18 letter-book, 22 Mar 1918; see also letter of May 9.

29. Ibid., 30 May 1918 re memorial service; see also 16 Mar 1918 re wives' meeting.

30. Ibid., see, e.g., Mrs. Cannon's letters of 5 Jun, 17 Jul, 22 and 27 Aug 1918.

31. Cannon to Mrs. Cannon, 20 Jul 1918 (not in "Letters Home"); see also excerpt of 21 Jul 1918, in the 1917–18 letter-book.

32. "Snatched from Oblivion," 2 Jul 1918. Mrs. Cannon had been deeply touched by an article, "The Gulf," in the Jun 1918 *Atlantic Monthly;* see also her letter to Cannon of Sep 10.

33. Cannon to Mrs. Cannon, 1 Sep 1918 (not in "Letters Home").

34. 1917–18 letter-book; see, e.g., 15 and 30 Sep 1918.

35. Mrs. Cannon to Cannon, 13 Nov 1918.

36. "Letters Home," p. 493; for the celebration in Paris, see also p. 492. Both descriptions are from Cannon's letter of 13 Nov 1918.

4. Postwar Adjustments, and a New Era at the Medical School

1. "Snatched from Oblivion," excerpt of letter from Mrs. Cannon to her mother, 22 Jan 1919; in Cannon Family Papers.

2. Cannon to Simon Flexner, 20 Feb 1919.

3. "Snatched from Oblivion," letter of 6 Feb 1919 headed "After Effects of War."

4. Ibid., 5 Mar 1919. See also Cannon to R.M. Yerkes, 13 and 25 Mar 1919; in Yerkes Papers.

5. Ibid., 30 May and 7 Nov 1919. For Cannon's review of the offer, see his letter to Franklin C. McLean, 7 Nov 1919; in Rockefeller Foundation Papers. Cannon explained that the situation had altered since McLean first approached him with the offer in the spring. With the departure of Dr. Redfield to Toronto, he did not wish to lay the burden of teaching the course on Dr. Drinker, who was undertaking the establishment of a new enterprise in industrial physiology.

6. Cannon wrote to Sen. Kellogg on 8 Dec 1919 and sent an identical letter to Sen. Knute Nelson, also of Minnesota.

7. "Snatched from Oblivion," excerpts of 17 Mar and 7 Aug 1919.

8. Ibid., 10 Jun 1919. When she consulted a Debrett's Peerage, Cornelia continued, "I ran my eye down through dukes, etc., and found 'Masters in Lunacy' and then 'Companions of the Bath'; so I didn't feel too dazzled." Her letter of Jun 30 reflects a change in attitude, and on Sep 17 she noted that Cannon had been receiving congratulations from English colleagues and that Pershing had received a C.B. as well.

9. See further W. W. Keen to Cannon, 7 and 15 Jan 1919; Cannon to Keen, 1 Feb and 17 Mar 1919. All in the American Philosophical Society Library. For Cannon's contribution on Traumatic Shock, see *Surgery, Its Principles and Practice,* ed. W.W. Keen, vol. VII (W.W. Saunders, 1921), pp. 296–315. In Dec 1919, Cannon signed a contract with his publishers, D. Appleton & Co., and spent the summer of 1920 preparing an account of the work done on shock in World War I. Publication of his monograph, *Traumatic Shock,* was delayed until 1923, however, because it was issued as part of a series on surgery.

10. Meltzer's article appeared in *Penn. Med. J.* 22 (Dec 1918):129–133; see esp. p. 130. Cannon wrote to Meltzer on 5 Feb 1919, and Meltzer replied on Feb 7; in American Philosophical Society Library.

11. Cannon's second letter to Meltzer was written on Feb 11. Meltzer's Letter to the Editor referring to Cannon's footnote (in "The course of events in secondary wound shock," *JAMA,* July 19, 1919, p. 178) was published in *JAMA,* Aug 9, 1919, p. 439.

12. Cannon to Mrs. Cannon, 2 Jul 1918 (not in "Letters Home"). He had received a letter from President Lowell informing him of Dean Bradford's resignation, the quest for a successor and the final selection of Dr. Edsall, and he confessed to his wife that he was relieved to be off the hook as a candidate for the deanship. See further *Walter B. Cannon* (1987), pp. 360–361; Joseph C. Aub and Ruth K. Hapgood, *Pioneer in Modern Medicine: David Linn Edsall of Harvard* (Harvard Medical Alumni Association, 1970), esp. Chapters 15 and 16 for Edsall's early years in the deanship; Henry K. Beecher and Mark D. Altschule, *Medicine at Harvard: The First Three Hundred Years* (Hanover, N.H.: The University Press of New England, 1977), pp. 192–195.

13. Confidential Statement of the Dean to the Faculty of Medicine, April 7, 1919; in Dean's Files, Harvard Medical Archives.

14. Minutes of Medical Faculty meeting, 7 Apr 1919; in Harvard Medical Archives.

15. For the fundraising campaign, see further President Lowell's Annual Report for 1918–19. Dean Edsall, in his Annual Report for 1927–28, looked

back over ten years to the precarious financial condition of the Medical School when he took over the deanship; see esp. pp. 208–210.

16. See further D.L. Edsall to A.L.Lowell, 19 Mar 1919, enclosing proposal to Henry S. Pritchett, president of the Carnegie Foundation; also Lowell's formal application to the General Education Board, 14 May 1920, requesting a share of the Teachers' Fund. Edsall wrote Lowell on 30 Sep 1919 that Jerome Greene had encouraged him to apply to the Rockefeller Foundation for aid to the Medical School, and correspondence continued between Greene and Lowell from Dec 1919 to May 1920. All in A.L. Lowell Papers, 1919–1922 series, Harvard University Archives.

17. See letters in DeLamar folder in the Lowell Papers. On 9 Dec 1918, Alfred Jaretzski informed Pres. Lowell of the bequest and sent him a copy of DeLamar's will. The portion affecting the Medical School was sent to Dr. Edsall on Dec 14.

18. See further obituary of Captain DeLamar in the *New York Times*, Dec 2, 1918, p. 13; biographical account in the *Johns Hopkins Hospital Bulletin* 34 (Apr 1923):135–137. See also C. Sidney Burwell to George P. Berry, 16 Nov 1962, with reply from Susan Scharff, 6 Dec 1963; both in C.S. Burwell Papers. The information on Alfred Jaretzski comes from his student file (Harvard College, Class of 1913, Harvard Law School, Class of 1916), courtesy of the Harvard University Archives. A discussion of the DeLamar bequest may be found in Aub and Hapgood (n.12), pp. 221–223.

19. H.A. Christian to Cannon, 3 Dec 1919; in Christian Papers.

20. Cannon's letter of 10 Mar 1920 was in reply to Abraham Flexner's letter of Mar 9. One of Cannon's first official acts after returning to the Medical School was to write Dean Edsall a lengthy letter about the financial needs of his department, probably in response to a request from Edsall for just such a statement. On 26 Mar 1919, Cannon laid out for the new dean what was necessary in the way of funds to run the physiology department in future.

21. For Bevan's comment, see *JAMA*, March 13, 1920, pp. 757–758.

22. Edsall to Lowell, 11 Mar 1920; in Lowell Papers, 1919–22 series.

23. For the offer to L.J. Henderson, see Edsall to Lowell, 29 Jan 1920; Lowell wrote to Henderson the very next day, Jan 30. Both in Lowell Papers. For the subsequent negotiations that resulted in Henderson's staying at Harvard, see further the Henderson file in the 1919–1922 series, Lowell Papers.

24. For Edsall's interest in industrial health and hygiene, see Aub and Hapgood (n.12), Chapter 12, esp. p. 184–189. See also Beecher and Altschule (n.12), pp. 195–197 (includes about Alice Hamilton as well).

25. Negotiations with Drinker may be followed through documents in fd. 644 of the 1919–22 series, Lowell Papers. See also Edsall to Hunnewell, 9 Jul 1920, with instructions for increasing Drinker's salary; in Budget folder, Lowell Papers. During the 1920s, Katherine Drinker edited the *Journal of Industrial Hygiene* at the Harvard School of Public Health.

26. L.G. Rowntree to W.B. Cannon, 14 Aug and 20 Aug 1920. The archivist at the Mayo Clinic was unable to find previous correspondence about the job offer, but he did find a reference to Cannon's visit to Rochester on Jul 14, at which time he gave a lecture.

27. For the previous concern about salary issues, see Report of Committee of the Division of Medical Sciences of the N.R.C., "An investigation of conditions in the departments of the preclinical sciences," in *JAMA* 74 (Apr 17, 1920):1117–1122, which followed up on the letter Christian wrote Cannon on 3 Dec 1919. See also exchange of letters between Christian and Cannon, 20 and 22 May 1920; in Christian Papers.

28. Christian to Cannon, 27 Aug 1920, enclosing Edsall's letter of Aug 25, to which he responded on Aug 26. Cannon had informed Christian of his offer from the Mayo Clinic about the same time that he told Edsall; see Christian to Cannon, 19 Aug 1920, in which he acknowledged the news. See further correspondence and related materials in fd. 10 of the Cannon Papers.

29. H.P. Walcott to Edsall, 29 Aug 1920, in reply to Edsall's letter of Aug 27; in Dean's Files, Harvard Medical Archives.

30. Cannon wrote his letters to Christian and Edsall on Sept 2. He told Christian about the formal offer to Otto Folin in a letter of Aug 31.

31. Edsall to Lowell, 28 Sep 1920. Lowell had directed Hunnewell to inform Edsall that he would have to withdraw money charged against the endowment fund; see F.W. Hunnewell to Edsall, 8 Sep 1920. Both in Lowell Papers.

32. Lowell to William Thayer, 1 Oct 1920, in reply to Thayer's letter of Sep 29; in fd. 719, 1919–22 series, Lowell Papers.

33. W.H. Howell to Cannon, 1 Oct 1920; see also Robert W. Lovett to Cannon, 30 Sep 1920.

34. Cannon to Charles W. Mayo, 11 Oct 1920.

35. Christian to Cannon, 6 Oct 1920. Edsall's note, written two days earlier, is dated Oct 4.

36. C.F. Adams to W.B. Cannon, 24 Dec 1920. Cannon responded to Adams' official note on Dec 29: "This present wholly unexpected increase of salary is most gratifying. It has given me the sense of being among friends, and the happiness of realizing that they think well of my services to the University." (In fd. 719, 1919–22 series, Lowell Papers.) Christian had written Forbes on Oct 6; see also W. Cameron Forbes to A. Lawrence Lowell, 19 Oct 1920; in fd. 749, 1919–22 series, Lowell Papers.

37. Excerpt of letter of 20 Oct 1920 in "Snatched from Oblivion."

5. The Crucible: Stewart and Rogoff Challenge the Emergency Theory

1. This chapter was previously published in part by A.C. Barger as "Cannon's emergency theory: The controversy with Stewart and Rogoff," *NIPS* 7(Apr 1992):80–88. See also two excellent undergraduate theses: Laurie Rae Green, "A Study of Scientific Controversy: Walter Bradford Cannon and the Emergency Theory of Adrenin Secretion," B.A. degree in History and Science, Harvard University, 1972; F.L. Holmes, "Walter Bradford Cannon's Research on Functions of the Sympathetic Nervous System," B.S. degree, M.I.T., 1954,

in which the Stewart and Rogoff controversy is discussed as part of a larger effort.

2. Cannon, "The movements of the stomach studied by means of the Röntgen rays," *Am. J. Physiol.* 1(1898):359–382; see also *Walter B. Cannon* (1987), pp. 62–63.

3. For a splendid review of the history of the secretion from the adrenal medulla, see further Horace W. Davenport, "Epinephrin(e)," *The Physiologist* 25 (Apr 1982):76–82. See also E.M. Tansey, "What's in a name? Henry Dale and adrenaline, 1906," *Med. Hist.* 39 (1995):459–476.

4. Cannon and de la Paz, "Emotional stimulation of adrenal secretion," *Am. J. Physiol.* 28(1911):64–70; see also *Walter B. Cannon* (1987), pp. 256–260.

5. See further Cannon and Hoskins, "The effects of asphyxia, hyperpnea, and sensory stimulation on adrenal secretion," *Am. J. Physiol.* 29(1911):274–279; Cannon, A.T. Shohl and W.S. Wright, "Emotional glycosuria," *Am. J. Physiol.* 29(1911):280–287; Cannon, J.C. Aub and C.A.L. Binger, "A note on the effect of nicotine injection on adrenal secretion," *J. Pharmacol. Exp. Ther.* 3(1912):379–385. See also *Walter B. Cannon* (1987), pp. 261–269.

6. For Cannon's article, see *Am. J. Physiol.* 33(1914):356–372, esp. p. 372. *Bodily Changes* was published in New York and London by D. Appleton & Company, 1915.

7. G. N. Stewart, "So-called biological tests for adrenalin in blood, with some observations on arterial hypertonus," *J. Exp. Med.* 14(Oct 1911):377–400; "The alleged existence of adrenalin (epinephrin) in pathological sera," *J. Exp. Med.* (Jun 1912):547–569. Biographical materials for Stewart, as well as his bibliography, may be found in the Stewart and Rogoff Papers.

8. For Macleod's methods, see Stewart's second article, esp. pp. 555, 559, 564; for Cannon and de la Paz, see pp. 562 and 569. For a discussion of Cannon's prewar relations with Stewart, see also *Walter B. Cannon* (1987), pp. 377–382.

9. Stewart, Rogoff and Gibson, "The liberation of epinephrin from the adrenal glands by stimulation of the splanchnic nerves and by massage, studied by means of the denervated eye reaction," *J. Pharmacol. Exp. Ther.* 8(May 1916):205–246.

10. Halsted to Stewart, 7 Feb 1914; Hough to Stewart, 30 Aug 1914; both in the Stewart and Rogoff Papers.

11. Stewart to Cannon, 6 Jun 1916, in reply to Cannon's letter of May 29; both in Stewart and Rogoff Papers. In his letter Cannon told Stewart that he had thought of making use of the iris deprived of its sympathetic innervation as an index of secreted adrenin; but the prompt dilatation of the pupil in fright led him to suppose that he was concerned with the nervous control rather than with adrenal secretion, and so he gave it up.

12. Stewart to Cannon, 24 May 1916, in reply to Cannon's letter of May 20; both in Stewart and Rogoff Papers.

13. Stewart and Rogoff, "The spontaneous liberation of epinephrin from the adrenals," *J. Pharmacol. Exp. Ther.* 8(1916):479–524; see esp. pp. 520–521.

14. See further Stewart and Rogoff, "Quantitative experiments on the liberation of epinephrin from the adrenals after section of their nerves, with

special reference to the question whether epinephrin is indispensable for the organism," *Proc. Soc. Exp. Biol. Med.* 14(1917):145–147; *J. Pharm. Exp. Ther.* 10(1917):1–48.

15. Cannon, "A note on the effect of asphyxia and afferent stimulation on adrenal secretion," *Science* n.s. 45(1917):463. Gasser and Meek's paper was "A study of the mechanisms by which muscular exercise produces acceleration of the heart," *Am. J. Physiol.* 34(1914):48–71.

16. See further the bibliography of Stewart's publications in the Stewart and Rogoff Papers.

17. J.J.R. Macleod, *Physiology and Biochemistry in Modern Medicine*, St. Louis, Moseby, 1918; see esp. pp. 745–746.

18. Macleod to Cannon, 26 Mar 1919, in reply to Cannon's letter of Mar 22.

19. "Is epinephrin indispensable to life?" editorial in *JAMA* 73(Jul 19, 1919):192.

20. "Studies on the conditions of activity in endocrine glands. V. The isolated heart as an indicator of adrenal secretion induced by pain, asphyxia and excitement," *Am. J. Physiol.* 50(Dec 1919):399–432; see esp. pp. 399–400 and 429. It is not easy to determine whether the experiments that Cannon describes and illustrates with supporting data in tables and figures were carried out in 1917 or in 1919. It seems likely, however, that they were done after Cannon returned from the war.

21. Cannon to Joseph Erlanger, 28 Jan 1920; in Erlanger Papers, Washington University School of Medicine.

22. Cannon to William Bayliss, 9 Dec 1919.

23. Stewart and Rogoff, "The relation of the epinephrin output of the adrenals to changes in the rate of the denervated heart," *Am. J. Physiol.* 52(Jun 1920):304–363; see esp. pp. 306, 310, 320, 359.

24. Stewart and Rogoff, "Essentials in measuring epinephrin output with further observations on the relation to the rate of the denervated heart," *Am. J. Physiol.* 52(Jul 1920):551–561.

25. Cannon and Rapport, "Studies on the conditions of activity in endocrine glands. VI. Further observations on the denervated heart in relation to adrenal secretion," *Am. J. Physiol.* 58(Dec 1921):308–337; see esp. p. 323.

26. Stewart and Rogoff's paper was "The relation of the epinephrin output of the adrenals to changes in the rate of the denervated heart" (n.23); see esp. p. 334.

27. Cannon and Uridil, "Studies on the conditions of activity in endocrine glands. VIII. Some effects on the denervated heart on stimulating the nerves of the liver," *Am. J. Physiol.* 58(Dec 1921):353–364. For the preliminary note by Cannon, Uridil and Griffith, see *Endocrinology* 5(1921):729–730.

28. Cannon and Griffith, "Studies on the conditions of activity in endocrine glands. X. The cardio-accelerator substance produced by hepatic stimulation," *Am. J. Physiol.* 60(May 1922):544–559.

29. Cannon and Carrasco-Formiguera, "Studies on the conditions of activity in endocrine glands. XI. Further evidence for reflex and asphyxial secretion of adrenin," *Am. J. Physiol.* 61(Jul 1922):215–227.

30. Stewart and Rogoff, "The supposed relation of the adrenals to reflex volume changes in the denervated limb," *Am. J. Physiol.* 63(1923):436–478; see esp. 469.

31. Roy Hoskins to Cannon, 28 Feb 1923, in reply to Cannon's letter of Feb 26. This is but one of a series of exchanges between Cannon and Hoskins (who served as editor of the journal, *Endocrinology*) during the early 1920s.

32. F.A. Hartman, H.A. McCordock and M.M. Loder, "Conditions determining adrenal secretion," *Am. J. Physiol.* 64(Mar 1923):1–34. As might be expected, the Clevelanders vigorously lambasted Hartman's paper; see their diatribe in the Oct 1923 issue of the *A.J.P.*, pp. 257–263.

33. Editorial in *JAMA* 80(1923):1007–1008.

34. S. Kodama, "Effect of stimulation of the sensory nerves upon the rate of liberation of epinephrine from the suprarenal glands," *Tohoku Journal of Experimental Medicine* 4(1923–24):166–242. Kodama published "A further report on the effect . . . " in the *Tohoku Journal,* also 1923–24; see pp. 465–493.

35. H.H. Dale to W.B. Cannon, 18 Oct 1923, in reply to Cannon's letter of Oct 9. Other examples are Cannon to Hartman, 17 Oct 1923; Cannon to Hoskins, 18 Sep 1923; Cannon to G.H. Simmons, editor of *JAMA,* 10 Oct 1923.

Cannon also wrote to Y. Sataki on Sep 17, telling him that Stewart and Rogoff regarded their caval pocket method "as thoroughly reliable because it was quantitative . . . Under the circumstances, it was highly interesting to find Dr. Kodama's article, and to learn that he had scrupulously followed Stewart and Rogoff and had secured results supporting the work from this Laboratory . . . Will you please convey to Dr. Kodama my hearty thanks and congratulations?"

36. Stewart's comment is in his article with Rogoff, "Effect of stimulation of afferent nerves upon the rate of liberation of epinephrin from the adrenals," *Am. J. Physiol.* 69 (1924):605–633; see esp. p. 628. See also J.M. Rogoff, "The output of epinephrin from the adrenal glands during cerebral anemia," *Am. J. Physiol.* 67 (1923–24):551–572.

37. T. Sugawara, M. Watanabe and S. Saito, "Effect of stimulation of the sensory nerves upon the rate of liberation of epinephrine from the suprarenal glands. The third report," *Tohoku J. Exp. Med.* 7(1926):1–79.

38. For Stewart and Rogoff's long-standing interest in the adrenal cortex, see Stewart's Mellon Lecture, "The adrenal glands," delivered at the University of Pittsburgh in 1926 and published four years later in *Arch. Int. Med.* 43(1929):733–766. See also J.M. Rogoff and G.N. Stewart, "Suprarenal cortical extracts in suprarenal insufficiency (Addison's disease)," *JAMA* 92(May 11, 1929):1569–1571. For Rogoff's continuing attacks, see Cannon's "Endocrinology by ukase: Some comments on recent statements made by J.M. Rogoff," in *Endokrinologie* 6(1930):24–26; Rogoff's "The adrenal medulla," *JAMA* 54(1935):2088–2092; Rogoff's "A critique on the theory of the emergency function of the adrenal glands: Implications for psychology," in *J. Gen. Psychol.* 32(1945): 249–268.

39. See further, e.g., note 18 on pp. 64–65, in Laurie Green's thesis; also F.L. Holmes's thesis, pp. 73–74 (both in n.1). In his autobiographical memoir,

The Way of an Investigator, Cannon referred to the controversy (but not to Stewart and Rogoff by name) in Chapter 9, "When Scientists Disagree."

40. For Pierce's memory of Carlson's talk, see *History of the American Physiological Society: The Third Century, 1937–1962,* ed. W.O. Fenn (American Physiology Society, 1963), pp. 142–143. For Cannon's comment, see his letter to Peyton Rous, 26 Mar 1926.

6. Antivivisection, Another Field of Battle

For Cannon's long-standing struggle with the antivivisectionists, see *Walter B. Cannon* (1987), esp. Chapters 10, 11, and 16. The documents on which this chapter is in large part based may be found in Boxes 32 and 33 of the Cannon Papers.

1. For House Bill No. 1130, introduced in the Pennsylvania Legislature, see, e.g., *Starry Cross* (formerly the *Journal of Zoophily),* Apr 1919, p. 53, and *Living Tissue* (published by the New England Anti-Vivisection Society), Sep 1919, p. 8. For the Washington, D.C. hearings, see F.S. Lee to Cannon, 12 Jun 1919.

2. The Myers bill, S.1258, was introduced on June 2, 1919, in the U.S. Senate, where it was referred to the Committee on the Judiciary. Cannon had been informed of the action on 3 Jun 1919, by Dr. G.W. McCoy, the director of the Hygienic Laboratory, who sent on a copy of the bill. Cannon promptly answered on June 5; requesting more information, he noted that a similar bill had repeatedly been defeated in the English House of Commons. See further Cannon to F.R. Green, secretary of the A.M.A. Council on Health and Public Instruction, 13 Jun 1919, and other correspondence in fd. 364 of the Cannon Papers; also Cannon's correspondence with McCoy, Simon Flexner and W.W. Keen in fd. 365. Still other letters about the Myers bill are located in fds. 366 and 367.

3. E.H. Baynes was president of the Meriden, N.H., Humane Society and had served as secretary of the Red Star Animal Relief. For a period of fifty years, English and American writers celebrated the extraordinary character of the dog and his warm and sensitive relations with man. See, e.g., Kipling's story about Garm, a bull terrier, Thomas Hardy's "Are you digging on my grave?," Steven Crane's "Little brown dog," and Jack London's *Call of the Wild.*

4. *Prohibiting Vivisection of Dogs,* Hearings before the Subcommittee of the Committee on the Judiciary, United States Senate, 66th Congress, first session, on S.1258, Washington: Government Printing Office, 1919. The above account is based on testimony in the printed transcript. The hearings were announced in *Living Tissue,* Oct 1919, p. 6. *Starry Cross* published the text of the Myers bill and gave an account of the hearings in the Nov 1919 issue, pp. 167ff.

5. Among the government scientists who testified at the hearings were Joseph Goldberger, Col. Charles W. Stile and Brayton Ransom. The Myers bill did not make it to the Senate floor for debate; nevertheless, there were repeated attempts to reintroduce dog bills in subsequent years.

6. Cannon to G.W. McCoy, 20 Feb 1920, in response to McCoy's letter of Feb 18.

7. Cannon to W.W. Keen, 10 May 1920; Cannon to Steven Paget, 10 May 1920, concerning the struggle in California and Dr. Whipple's efforts. Although Whipple from 1914 on had called on Cannon and the A.M.A. for help in coping with antivivisectionists in California [see *Walter B. Cannon* (1987), p. 373], by 1920 he actually needed little help. In the interim, he had not only become more adept in meeting their claims and charges in public debate, he also recruited such well-known practitioners and medical scientists to help him as William Ophuls, Karl Meyer, Ray Lyman Wilbur, and Walter Alvarez.

8. See, e.g., *California State Journal of Medicine* for Jul 1920, esp. pp. 246–247; also Walter C. Alvarez, "What are you doing to defeat anti-health and anti-medical legislation?" in the August issue, pp. 290–296.

9. Terhune's articles were published in *The Country Gentleman* of Oct 26, 1920, and in the *Ladies' Home Journal*, Oct 1920. The latter provided particularly gruesome details about what happened to stray dogs in St. Louis that were sold to medical establishments.

10. Cannon to Keen, 5 Nov 1920. For Cannon on Terhune, see also his letters to Simon Flexner, 2 Nov 1920, and to W.W. Keen, 25 Oct, 1 Nov, and 4 Nov 1920. All in American Philosophical Society Library.

11. For the defeat in California, see exchange of letters between Flexner and Cannon, 15 and 16 Nov 1920; also editorial, "The defeat of the antivivisection initiative in California," in *JAMA*, Nov 13, 1920, pp. 1344–1345.

12. Cannon to Keen, 17 Nov 1920: "It happens that while I was in Chicago last week, we seriously considered the establishment of a popular health journal which would be of interest not only to public-spirited medical men, but also to intelligent and interested laymen." For the lay society, see, e.g., Keen to Cannon, 31 Jan 1921; exchange of letters between Cannon and Flexner, 12 Jan, 14 Jan, and 19 Jan 1921. All in American Philosophical Society Library.

13. Cannon began to send out form letters to directors of laboratories on 24 Nov 1920. For a discussion of the laboratory code that was formulated in 1909, see *Walter B. Cannon* (1987), pp. 191–192; the rules are reproduced there on p. 450.

14. See Don R. Joseph (vice-dean, St. Louis School of Medicine) to Cannon, 10 Jul 1922, referring to a previous letter about the decision.

15. Flexner to Cannon, 14 Feb 1922; his earlier letter is dated 5 Oct 1921, in reply to Cannon's letter of Oct 3; all in American Philosophical Society Library. Cannon also wrote to E.H. Baynes about the open door policy at this time; see his letter of 7 Oct 1921. The open door policy was officially adopted in 1922; see *JAMA*, May 27, 1922, p. 1624; also Cannon's editorial in *JAMA*, June 10, 1922, p. 1820.

16. Baynes, "The truth of vivisection," in the *Woman's Home Companion*, Jul 1921, p. 9.

17. Ibid., pp. 9–10.

18. Cannon to Baynes, 20 Jul 1921. Baynes had first written Cannon on Jun 27, and enclosed the manuscript of his article for comment. The letters to

the editor appeared in the *Woman's Home Companion* of Nov 1921. See further Cannon's editorial about Baynes and his article, "A layman's defense of animal experimentation," in *JAMA,* Aug 6, 1921, pp. 429–430.

19. See, e.g., Cannon to Baynes, 20 Sep 1921; also their exchange of 19 and 20 Jul 1921. For further letters between Cannon and Baynes, see fds. 394–397 of the Cannon Papers. Cannon later commented on his friendship with Baynes in *The Way of an Investigator,* pp. 176–177.

20. Baynes to Cannon, 22 Mar 1922. Baynes's lecture was based on his article in the *Woman's Home Companion* (n.16) and was likewise called "The truth of vivisection." Cannon described Baynes's first lecture in Boston in a letter to W.W. Keen of 23 Oct 1921. Saying that Baynes had taken the point of view of the man on the street, he continued, "Altogether it was a telling lecture, and no doubt will be much improved . . . The audience at the end was thoroughly on Mr. Baynes's side and gave him a most hearty applause."

21. Cannon to Evarts Graham, 1 Dec 1921, in reply to Graham's letter of Nov 26; in American Philosophical Society Library.

22. For a full discussion of events leading up to the defeat, see further "Antivivisection campaign in St. Louis," *JAMA,* May 6, 1922, pp. 1412–1413.

23. See Cannon's letter to Joseph Erlanger of 22 Mar 1922, urging that an account of the St. Louis campaign be published: "We need to have such a precedent as you established made as widely known as possible." (In Erlanger Papers) The following year, after a similar success in Dallas, Cannon published an editorial, "Impounded animals for the service of medicine," in *JAMA,* Nov 10, 1923, pp. 1611–1612.

24. For an example of Hadwen's views, see "Grotesque medical fads: Fantastic theories of so-called modern science," in *Living Tissue,* Oct 1919, pp. 5–6. Cannon had tangled with the British antivivisectionist before and wrote about Hadwen in *Some Characteristics of Antivivisection Literature* (Defense of Research pamphlet 19, Chicago: A.M.A., 1911). See also *Walter B. Cannon* (1987), p. 275 and notes 5–7 on p. 473.

25. See further "Dr. Hadwen's tour," in *Starry Cross,* Oct 1922, p. 148; also letters in fd. 405 of the Cannon Papers.

26. Cannon's telegram and letter to Keen of 11 Oct 1922 mentions the letter received from Alvarez. See also May 1922 issue of *Calif. State J. of Med.* for editorial on chiropractic and osteopathic initiatives, pp. 155–156; also Oct 1922 issue, pp. 338–339, 363–367.

27. See further *JAMA,* Oct 21, 1922, pp. 1433 and 1448–1449.

28. Cannon to Woodward, 13 Nov 1922.

29. Cannon to Keen, 27 Feb 1922, in reply to Keen's letter of Feb 24 about Colorado.

30. Sewall had written to Cannon on 5 Jun 1922; George Simmons wrote Sewall on Jun 14 to say that Cannon had forwarded his letter to A.M.A. headquarters. This was followed by a letter from Woodward to Cannon of Jun 21, reporting that the Colorado situation was in a deplorable state. Woodward wrote Cannon again on Jul 28, enclosing a letter from Sewall of Jul 26 with his reply.

31. The Blücher editorial was in *Colorado Medicine* 19(Sep 1922):175–176. Editorial comments warning of the threat mounted by the antivivisectionists

led off the issues of *Colorado Medicine* for February, March and April 1922. On Aug 31, Woodward wrote Cannon that Dr. Simmons strongly endorsed their trip to Colorado

32. For the site visit, see Cannon to Flexner, 23 Sep 1922; Cannon to Keen, 29 Sep 1922; also "The antivivisection initiative in Colorado," *JAMA,* Jun 17, 1922, pp. 1895–1896. The following year, when Dr. C.S. Bluemel sent Cannon a full report of the campaign in Colorado, Cannon wrote Woodward (3 Apr 1923), "It is the most complete account of effective activities that I have ever had submitted to me . . . It will be of inestimable value in any future defense which we may be engaged in." See also Cannon to Bluemel, 3 Apr 1923.

33. "Antivivisection propaganda by camouflage," in *JAMA,* Oct 14, 1922, p. 1337. Cannon's earlier editorial about Hadwen, "Medical support for the antivivisectionists," was published in the Aug 13, 1921, issue of *JAMA,* pp. 559–560.

34. For the formation of the lay committee in Massachusetts, see, e.g., Cannon to Keen, 28 Nov 1921; Cannon to Simon Flexner, 16 Dec 1921; other letters in fd. 401 of the Cannon Papers. See also Samuel B. Woodward, "Legislative aspects of vaccination," *Bos. Med. & Surg.J.* 185 (Sep 15, 1921):307–310.

35. See, e.g., Cornelia James Cannon's editorial, "Is the woman vote sentimental?" in *The Woman Citizen* for Dec 30, 1922. Cannon, with some amusement, wrote W.W. Keen (6 Dec 1922) that the editor had hesitated to publish the editorial but his wife insisted, stating it was "high time for a first-class exhibition of back-bone."

36. Hadwen to Keen, 26 Dec 1922, published in *Starry Cross,* Feb 1923, p. 10. *Starry Cross* also published a letter from Dr. Keen to the Editor of 4 Jan 1923, pointing to the legislative defeat in California with barbed comment: "I trust you will import Dr. Hadwen again if this is to be the result of his campaign."

37. Cannon drafted a formal letter to Keen and others about the formation of a lay society, which had been agreed to at a meeting held in Philadelphia on Oct 17, 1922. "I was asked to draw up a plan of organization," he wrote on Oct 25, "that could be submitted and sent to the other members of the group. The simple form of constitution which is sent herewith is suggested for your consideration." He followed this up with a lengthy letter to Flexner, 2 Nov 1922, that detailed the status of plans for the lay society.

38. Exchange of letters between Victor Vaughan and Cannon, 11 and 12 Apr 1922; see also correspondence with Flexner and Keen in Feb 1922.

7. A Spirit of Fellowship: Ties with Pavlov and Yerkes

1. Cannon to Carl Tigerstedt, 8 Nov 1920. Cannon indicated that he was also writing to Richet, Johannson and Sherrington for information about Pavlov.

2. Simon Flexner to Cannon, 14 Jun 1921; Cannon to Flexner, 18 Jun 1921; Flexner to Cannon, 23 Jun 1921.

3. "Professor Ivan P. Pawlow still alive," Letters to the editor from Francis G. Benedict and Walter B. Cannon, *JAMA,* Sep 24, 1921, pp. 1039–1040.

4. See American Relief Administration *Bulletin* for Feb 1923; the entire issue is devoted to Russian relief.

5. Horsley Gantt, *A Medical Review of Soviet Russia* (London: British Medical Association, 1928), pp. 13–15. In 1928, Gantt also translated Pavlov's lectures on conditioned reflexes, for which Cannon wrote an introduction. He later joined the Phipps Psychiatric Clinic at Johns Hopkins.

6. See, e.g., "Starving Russian physicians," notice in *Bos. Med. & Surg. J.*, Jan 12, 1922, pp. 60–61.

7. Undated form letter headed "Medical Relief for Russia," signed by Richard C. Cabot, E.A. Codman, Francis W. Peabody, W.T. Councilman, Vincent Y. Bowditch, Alexander Forbes, and Walter B. Cannon.

8. I.P. Pavlov to Cannon, 15 Apr 1922.

9. Cannon to Frances Witherspoon, 14 Nov 1922, in reply to her appeal letter of Oct 27 for the National Campaign.

10. Cannon to Haven Emerson, 14 Mar 1923; Emerson to Cannon, 15 Mar 1923; H.O. Eversole (writing for the Russian Relief Commission of the National Information Bureau, Inc.) to Emerson, 19 Mar 1923.

11. Cannon to Mrs. Henry Villard, 16 Apr 1923, in reply to Mrs. Villard's letter of Apr 7. Cannon laid out a program for the National Campaign to appeal to the medical profession through their national journals and meetings, as well as through drug houses, for literature to send to the Soviet Union. For more on medical literature for Russian physicians, see, e.g., *Bos. Med. & Surg. J.*, Jul 13, 1922, p. 92; Gantt to Cannon, 25 Jan and 3 Apr 1923.

12. Gantt to Cannon, 16 Apr 1923; Cannon to Gantt, 9 May 1923; Cannon to Pavlov, 9 May 1923. In his letters, Cannon wrote that the fund of $2,000 sent to Helsingfors for Pavlov two years before was still available. He did not doubt that every contributor to the fund would think Pavlov well justified if he used it for the visit to America.

13. P.A. Levene to Cannon, 24 Jun 1923; Cannon to Levene, 25 Jun 1923. See also Cannon to Maurice C. Pincoffs, 25 Jun 1923; Jacques Loeb to Cannon, 26 Jun 1923.

14. Cannon to Gustav Eckstein, 14 Jan 1937. Dr. Eckstein planned to write a biography of Pavlov, for which Cannon sent him this lengthy summary of his relations with the Russian scientist, as well as some of Pavlov's letters. Almost at once, however, Eckstein reported that the Pavlov letters had been lost. See further correspondence in fd. 483 of the Cannon Papers. Eckstein never finished his biography, but did write an article, "Ivan Petrovich Pavlov," that was published in *The New Yorker,* Dec 2, 1950, pp. 158–170.

15. Cannon to Felix Cunha, 9 Apr 1936.

16. The story of Pavlov's 1923 visit to America has been published in several places; see, e.g., "Professor Pawlow's visit to America," *JAMA,* Jul 28, 1923, p. 308; also Cannon, *The Way of An Investigator,* pp. 185–186.

17. Mrs. Cannon quoted from the letter Cannon wrote to W.H. Welch; see her unpublished manuscript, "Servant of Science," p. 196, in Cannon Papers.

18. For Cannon's earlier relations with Yerkes, see *Walter B. Cannon* (1987), pp. 134–135, 197, 311. Bradford Cannon has referred to Cannon's friend-

ship with Yerkes in reminiscences of his father; see, e.g., his memoir in *Life and Contributions*, pp. 151–169.

19. James Reed, "Robert M. Yerkes and the mental testing movement," in *Psychological Testing and American Society, 1890–1930*, ed. Michael M. Sokal (Rutgers Univ. Press, 1987), pp. 75–94; see esp. pp. 83–85.

20. See John Burnham's article on Yerkes in the *Dictionary of Scientific Biography*, vol. 14, pp. 549–551; esp. p. 550.

21. Sophie D. Aberle and George W. Corner, *Twenty-five Years of Sex Research: History of the National Research Council Committee for Research in Problems of Sex, 1922–1947* (Philadelphia: W.B. Saunders Co., 1953), pp. 9–12.

22. Ibid., pp. 3–5. Aberle and Corner went on to note that biologists like Leo Loeb, Charles Stockard and Frank Lillie, building on work done in Europe, conducted experiments on the physiology of reproduction in mammals and looked on human sex behavior as the result of chemical, physical and mental factors.

23. Ibid., pp. 12–13, where the resolution is reproduced. For Cannon's relations with the N.R.C. Sex Committee during its formative period, see correspondence and related materials in Box 77 of the Cannon Papers.

24. Ibid., pp. 13–15.

25. Cannon to Yerkes, 11 Dec 1922.

26. *Twenty-five Years of Sex Research* (n.21), pp. 18–19. In the 1930s several of Cannon's students and colleagues, past and present, were recipients of grants from the Sex Committee—in addition to Bard, Sidney Britton at the Univ. of Virginia, Harry Friedgood and Jean Reboul in Cannon's laboratory; also Hallowell Davis and George Wislocki at H.M.S. and Gregory Pincus at Harvard's biology department.

27. Ibid., pp. 20–23 and 48. See also Bradford Cannon (n.18), p. 162. For Cannon's relations with the N.R.C. Sex Committee during the 1930s and '40s, see correspondence and related materials in Box 78 of the Cannon Papers.

28. Cannon to Victor Vaughan, 13 Dec 1921.

29. Cannon to Yerkes, 11 Oct 1922.

30. Cannon's enthusiasm for small foundations has been noted by Mrs. Cannon in "Servant of Science" (n.17), p. 201. For Cannon's views on the provision of fellows and funding for research, see *The Way of an Investigator*, pp. 51–55. The Elizabeth Thompson Science Fund was brought to H.M.S. by C.S. Minot and headed by him until 1911, after which time Cannon took over its direction. For the N.R.C. Medical Fellowship Board, see correspondence and related materials in Box 79 of the Cannon Papers.

31. Yerkes to Cannon, 10 Mar 1918.

32. For criticism of Yerkes' intelligence testing, see, e.g., Stephen J. Gould, *The Mismeasure of Man* (New York: Norton, 1981), pp. 192–233, and Allan Chase, *The Legacy of Malthus: The Costs of the New Scientific Racism* (New York: Alfred A. Knopf, 1977), esp. Chap. 11.

By the middle to late 1920s, geneticists such as T.H. Morgan, Raymond Pearl, H.S. Jennings, and H.J. Muller in the United States, as well as J.B.S. Haldane, Lancelot Hogben, and Otto Louis Mohr abroad, had advanced serious arguments against the scientific credibility of eugenics. Gradually, as such geneticists and biologists abandoned the movement, it became increasingly isolated from the scientific community. See further Garland E. Allen's essay on the American eugenics movement, "The role of experts in scientific controversy," in *Scientific Controversies: Case Studies in the Resolution and Closure of Disputes in Science and Technology*, ed. H.T. Engelhardt, Jr., and Arthur L. Caplan (Cambridge U. Press, 1987), pp. 169–202.

33. Cornelia J. Cannon to Robert M. Yerkes, 11 Jul 1921; in Yerkes Papers. It is not clear whether Mrs. Cannon referred to the volume Yerkes published with C.S. Yoakum in 1920, *Army Mental Tests*, or his official account, *Psychological Examining in the United States Army*, vol. 15, Memoirs of the National Academy of Sciences, 1921. The author of *Is America Safe for Democracy?* (1921) was William McDougall, who had left England to head Harvard's psychology department. For the influence on Cannon of McDougall's earlier work, *Introduction to Social Psychology*, see *Walter B. Cannon* (1987), p. 311.

34. See, e.g., R.M. Yerkes, "Testing the human mind," *Atlantic Monthly*, Mar 1923, pp. 358–370; also series of articles on the use and abuse of intelligence tests by Walter Lippmann in various issues of *The New Republic* in 1922. Among the books were *The Revolt Against Civilization* by Lothrop Stoddard (1922), *A Study of American Intelligence* by Carl Brigham (1923), *The Trend of the Race* (1921) and *Studies in Evolution and Eugenics* (1923), both by S.J. Holmes.

35. Mrs. Cannon's reply to Yerkes' letter of 4 Feb 1922 is undated; both in Yerkes Papers.

36. Reed (n.19), pp. 86–87.

37. Bradford Cannon, "Walter Bradford Cannon: Reflections on the man and his contributions," address before the centennial convention of the American Psychological Association, Washington, D.C., Aug 1992, *Intl. J. of Stress Management*, 1(1994):145–158; see esp. pp. 151–152. In his article, Bradford goes on to discuss his father's relations with Pavlov on pp. 152–154.

38. Ibid., p. 152; Reed (n.19), pp. 89–90. Reed concludes, "The patient friend of Julius the Orangutan and of Congo the Gorilla would be most often cited thirty years after his death as an exponent of scientific racism." (p. 90)

8. Homeostasis and the Sympathico-Adrenal System

1. A thorough discussion of Cannon's response to the criticisms of Stewart and Rogoff may be found in Chapter Five above. The Clevelanders continued their attacks after 1922; e.g., "The average epinephrin output in cats and dogs," *Am. J. Physiol.* 66(1923):235–236; "Effective stimulation of afferent nerves on the rate of liberation of epinephrin from the adrenals," *Am. J. Physiol* 69(1924):605–633.

2. Cannon to R. Carrasco-Formiguera, 27 Dec 1922. Cannon contributed a chapter, "The emergency function of the adrenal medulla," to the second volume of *Endocrinology and Metabolism*, ed. L. Barker et al. (D. Appleton, 1922), which not only contained a succinct exposition of his emergency theory but also lined up evidence that refuted the criticisms and arguments of Gley and Quinquad as well as Stewart and Rogoff.

3. Cannon, "Some general features of endocrine influence on metabo-

lism" (read at a symposium on internal secretion, Congress of American Physicians, New Haven, May 1925), in *Trans. Cong. Am. Phys. and Surg.* 13(1926):31–53; also in *Am. J. Med. Sci.* 171(1926):1–20. Cannon's editorial, "The endocrine glands—a caution," was in *JAMA,* May 28, 1921, pp. 1500–1501.

4. Cannon, "Organization for physiological homeostasis," *Physiol. Rev.* 9(1929):399–431; see esp. pp. 399–400 where Cannon cites Bernard's observation. Russell C. Maulitz, when an undergraduate at Harvard, wrote his senior thesis on homeostasis: "Walter B. Cannon: Scientific Explanation in Regulatory Physiology" (B.A. degree, 1966). Maulitz acknowledged assistance in drawing on F.L. Holmes's undergraduate thesis, "Walter Bradford Cannon's Research on Functions of the Sympathetic Nervous System" (B.S. degree, M.I.T., 1954), and his Ph.D. dissertation, "Claude Bernard's Concept of the *Milieu Intérieur*" (Harvard University, 1962).

5. Cannon, *The Wisdom of the Body* (New York: W.W. Norton, 1932); see esp. p. viii. See also Holmes's undergraduate thesis, p. 54.

6. See, e.g., Cannon's address to the American Association of Science Teachers, Dec 1939, "The role of chance in discovery," *Sci. Monthly,* Mar 1940; also published under title, "When luck aids the scientist," *Sci. Digest,* May 1940. See also Chapter 6, "Gains from Serendipity," in his autobiographical memoir, *The Way of an Investigator.*

7. Holmes, undergraduate thesis (n. 4), pp. 56–57. See further Aub, Forman and Bright, "The effect of adrenalectomy upon the total metabolism of the cat," *Am. J. Physiol* 61(1922):326; McIver and Bright, "Changes in metabolism following adrenal stimulation," *Am. J. Physiol* 67(1924):622; David Rapport, "The liberation of epinephrin in traumatic shock, *Am. J. Physiol* 60(1922):461.

8. W.B. Cannon and P.E. Smith, "Studies on the conditions of activity in endocrine glands. IX. Further evidence of nervous control of thyroid secretion," *Am. J. Physiol.* 60(1922):476–495. For their preliminary report to the Society for Experimental Biology and Medicine, see *Proc. Exp. Biol. Med.* 17(1920):88. For Cannon's earlier foray into the realm of the thyroid, see *Walter B. Cannon* (1987), pp. 323–329.

9. Dr. Carrasco has left an account of the occasion in his contribution to a symposium on insulin in retrospect; see further R. Carrasco-Formiguera, "From the preinsulin age to the Banting and Best era: Reminiscences of a witness and participant," *Israel J. of Med. Sci.,* 1972, esp. p. 484. Banting and Best's first published paper, "The internal secretion of the pancreas," appeared in *J. Lab. & Clin. Med.* for Feb 1922; see further Michael Bliss, *The Discovery of Insulin* (Univ. of Chicago Press, 1982), esp. pp. 94–95.

10. Cannon, McIver and Bliss, "Studies on the conditions of activity in endocrine glands. XIII. A sympathetic and adrenal mechanism for mobilizing sugar in hypoglycemia," *Am. J. Physiol.* 69(1924):46–66, provides references not only to Wilder, Boothby et al. and Boothby and Wilder, but also to Mann and Magath and Fletcher and Campbell. See further 1923 correspondence with Walter Boothby about his paper, "The calorigenic action of adrenalin chloride," in fd. 1725 of the Cannon Papers.

11. Ibid. See correspondence with George H.A. Clowes at Eli Lilly & Co. in fd. 1396 of the Cannon Papers. McIver, a graduate of H.M.S., was an N.R.C.

fellow attached to the department of surgery, while Bliss, a doctoral student in the Medical Sciences program, was a teaching fellow in biochemistry. This paper has a lengthy footnote in which Cannon refutes the continuing criticisms of Stewart and Rogoff.

12. Ibid., p. 66. See also Cannon, McIver and Bliss, "The effect of the blood-sugar level on adrenal secretion and sympathetic activity: a preliminary report," *Bos. Med. & Surg. J.* 189(1923):141

13. Cannon's earlier thinking was reported in his article with A.T. Shohl and W.S. Wright, "Emotional glycosuria," *Am. J. Physiol.* 29(1911):280–287, and in "The physiologic equivalent of war," *JAMA* 63(1914):1415. See further "The suprarenals in hypoglycemia," editorial in *JAMA*, Sep 6, 1924, p. 770.

14. J.B. Collip, "Hormones in relation to human behavior," in Harvard Tercentenary Publication, *Factors Determining Human Behavior* (Harvard University Press, 1937), see esp. pp. 21–22.

15. Cannon and Querido, "The role of adrenal secretion in the chemical control of body temperature," *Proc. Nat. Acad. Sci.* 10(1924):245–246; Cannon and Pereira, "Increase of adrenal secretion in fever," follows on pp. 247–248. In a letter of 26 May 1924, Cannon wrote W.W. Keen: "The general scheme which I have now in mind is that of working in a given field for some time and then putting my papers together in the form of a book. I did this in my 'Mechanical Factors of Digestion,' in 'Bodily Changes in Pain, Hunger, Fear and Rage,' and in 'Traumatic Shock.' Now I am working on the chemical control of body temperature, and if all goes well, I shall probably have another book on 'Temperature Control and Fever in Relation to Internal Secretion.' But that remains to be seen."

16. Cannon, Linton and Linton, "Studies on the conditions of activity in endocrine glands. XIV. The effects of muscle metabolites on adrenal secretion," *Am. J. Physiol.* 71(1924): 153–162. The brothers came to Boston from Scotland by way of Spokane, Washington. James later went into psychiatry and Robert became a highly-regarded surgeon at the M.G.H.

17. Cannon, "Some general features . . ." (n.3), p. 4.

18. Dr. Barker's comment in the discussion that followed Cannon's talk appears on p. 54 of the version published in the *Transactions* of the Congress (n.3).

19. Cannon, "Physiological regulation of normal states: Some tentative postulates concerning biological homeostatics," *À Charles Richet: Ses amis, ses collègues, ses élèves*, ed. Auguste Pettit par Les Editions Médicales, 22 Mai 1926, pp. 91–93. Cannon gave as his source for Richet, *Dictionnaire de Physiologie*, 1900, iv, 721, and for Bernard, *Leçons sur les Phénomènes de la Vie*, Paris, 1878, 121. Other contributors to the Jubilee Volume include Leon Fredericq, Eugène Gley, Graham Lusk, B.A. Houssay, E.A. Sharpey Schafer, C.S. Sherrington.

20. Donald Fleming, "Walter B. Cannon and homeostasis," *Social Research* 51(Autumn 1984):609–640; see esp. pp. 624–629. Holmes has pointed out that during the 1920s the ideas associated with Bernard's *milieu intérieur* attracted greater attention than at any time since Bernard himself had made it a central theme in his teaching. "It was an extraordinary occurrence that fifty years after Bernard's death, four of the most eminent physiologists [i.e., Haldane, L.J. Henderson, Cannon and Barcroft] of their generation should have found in

Bernard's generalization ideas so pertinent to their own times and work." (F.L. Holmes, "Joseph Barcroft and the fixity of the internal environment," *J. Hist. Biol.* 2(1969):89–122; see esp. p. 114.)

21. Cannon, Lewis and Britton (with the cooperation of J.R. Pereira, A. Groeneveld, J.R. Linton and R.R. Linton), "Studies on the conditions of activity in endocrine glands. XVII. A lasting preparation of the denervated heart for detecting internal secretion, with evidence for accessory accelerator fibers from the thoracic sympathetic chain," *Am. J. Physiol.* 77(1926):326–352. See esp. section on "The search for the unknown accelerator," beginning on p. 335 and footnotes on pp. 345–346. See also *The Way of an Investigator*, pp. 104–106.

22. Cannon, Querido, Britton, and Bright, "Studies on the conditions of activity in endocrine glands. XXI. The role of adrenal secretion in the chemical control of body temperature," *Am. J. Physiol.* 79(1927):466–507.

23. Cannon, Lewis and Britton, "The dispensability of the sympathetic division of the autonomic nervous system," *Bos. Med. & Surg. J.* 197(1927):514–515. About the same time Cannon published another article with Lewis deriving from the earlier work of 1925 and 1926; see further "The 'physiological maximum heart rate' as an artifact," *Am. J. Physiol.* 82(1927):67–74.

24. Cannon, Newton, Bright, Menkin and Moore, "Some aspects of the physiology of animals surviving complete exclusion of sympathetic nerve impulses," *Am. J. Physiol.* 89(1929):84–107; see esp. p. 87.

25. Ibid., pp. 106–107.

26. See, e.g., Robert Gaunt and W.J. Eversole, "Notes on the history of the adrenal cortical problem," *Annals N.Y. Acad. Sci.* 50(1949):511–521.

27. Cannon and Izquierdo, "Studies on the conditions of activity in endocrine glands. XXIII. Emotional polycythemia in relation to sympathetic and medulliadrenal action of the spleen," *Am. J. Physiol.* 84(1928):545–562. See also Cannon, *Bodily Changes in Pain, Hunger, Fear and Rage,* 2nd ed. (D. Appleton, 1929), pp. 178–192.

28. Cannon, "Die Notfallsfunktionen des sympathico-adrenal systems," *Ergebn. Physiol.* 27(1928):380–406. For Cannon's continued annoyance with the comments of European critics, see, e.g., his letter to Leon Asher, 31 Jan 1924; also Cannon to Graham Lusk, 5 Mar 1924, which fills in the details about Gley's trespasses.

29. Perhaps the continuing criticism of his European peers was on Cannon's mind when he wistfully wrote Cecil Drinker: "If I had only known that I was going into physiology and not neurology, I should have taken much more training in physics and chemistry than I actually had and less training in biology. Also, I should have made a point for which sacrifice should be made to go abroad early in my career in order to become acquainted with foreign investigators and to see various ways in which work is being carried on." (24 Sep 1925; in Harvard Medical Archives)

30. Cannon, "Reasons for optimisim in the care of the sick," *New Engl. J. Med.* 199(1928):593–597. For the diary note, see *Walter B, Cannon* (1987), p. 258. Unfortunately, after World War I Cannon gave up keeping diaries except for special occasions. Save for his early X-ray notebook and some scattered materials, he did not retain protocols of his experiments.

31. Cannon, "Organization for Physiological Homeostasis" (n.4). See also L.L. Langley, *Homeostasis: Origins of the Concept* (Benchmark Papers in Human Physiology, Dowden, Hutchinson and Ross, 1973). In addition to the Organization paper, Langley reprints Cannon's Croonian Lecture on thirst, his paper with McIver and Bliss (1924), the 1925 lecture to the College of Physicians, and the contribution to the Richet festschrift. In an earlier monograph, Langley wrote: "Cannon's word *homeostasis* embraces the fixed, or constant, internal environment, but then goes on to suggest dynamic, self-regulatory processes that serve to maintain the constancy or to return the internal environment to normal, should it get out of whack. This is the concept now referred to as *negative feedback,* that is, if there is a deviation in one direction, there is a reaction in the opposite direction." (Langley, *Homeostasis,* Reinhold, 1965, p. 9)

32. Ibid., pp. 401–402; for the definition of homeostasis, see also pp. 400–401. Cannon's knowledge of calcium storage was derived from the work of Joseph Aub with Walter Bauer and Fuller Albright.

33. Ibid., pp. 422–427.

34. Cannon, "The sympathetic division of the autonomic system in relation to homeostasis," *Arch. Neurol. & Psychiat.* 22(1929):282–294; also "Functions of the sympathetic system in maintaining the stability of the organism," *Proc. Inst. Med. (Chicago)* 7(1929):229–331. An editorial, "Physiologic homeostasis," published in *JAMA,* Nov 2, 1929, pp. 1385–1386, reviewed Cannon's pronouncements on the subject.

35. Fleming (n.20), pp. 630–631. Horace W. Davenport has noted that "Cannon was a 19th century rationalist for whom evolution had replaced the Divine Watchmaker, and it had produced a beautifully accurate watch. Cannon thought he had caught a glimpse of the wisdom of nature, and very likely he had." (Introduction to W.B. Cannon, *The Mechanical Factors of Digestion,* Science History Publications, reprinted from the original edition, 1986, p. xi.)

Despite such kudos, Cannon's protégé Arturo Rosenblueth felt compelled to respond to two latter-day critics of Cannon's theory, David L. Drabkin and Dickinson W. Richards, in his paper, "A critique of homeostasis" (*Perspectives in Biology: A collection of papers dedicated to B.A. Houssay on the occasion of his 75th birthday,* Amsterdam: Elsevier, 1963, pp. 323–331.)

9. An Alternative Theory of the Emotions

1. For Cannon's references to the work of Darwin and McDougall, see, e.g., his chapter on the utility of bodily changes in pain and great emotion in both the 1915 and 1929 editions of *Bodily Changes in Pain, Hunger Fear and Rage.* See also *Walter B. Cannon* (1987), p. 311. In a letter to Tom [Storey], 30 Jul 1923, Cannon advised him to read three books on the emotions: Darwin's *Expressions of the Emotions,* Crile's *Origin and Nature of the Emotions,* and his own book on *Bodily Changes.* In addition, he cited McDougall's *Social Psychology*—"an interesting discussion of instincts of emotions as a basis for judging social motives."

2. See further F. Goltz, "Der Hund ohne Grosshirn," *Pflüger's Arch.*

51(1892):570–614. For Sherrington, see *The Integrative Action of the Nervous System*, Yale Univ. Press, 1906; also R.S. Woodworth and C.S. Sherrington, "A pseudaffective reflex and its spinal path," *J. Physiol.* 31(1904):234–243. Decerebration and decortication are quite different procedures and should not be confused.

3. Both editions of *Bodily Changes*, 1915 and 1929, have frequent references to T.R. Elliott's papers published in the *Journal of Physiology* in 1905, 1912, and 1913. Though Cannon, in his 1914 paper on "The emergency function of the adrenal medulla," had previously cited Elliott's 1904 pathbreaking note "On the action of adrenalin," he did not allude to it again until 1930.

4. See, e.g., Head's *Studies in Neurology* (1920) and his paper, "Release of function in the nervous system," *Proc. Roy. Soc.* 92B(1921):184; also H. Head and G. Holmes, "Sensory disturbances from cerebral lesions," *Brain* 34(1911):109.

5. Cannon's address to the neurologists was published as "Evidence of nervous control of some internal secretions," *J. Nerv. Ment. Dis.* 54(1921):421–427; see esp. pp. 426–427. A discussion follows on pp. 428–433.

6. Cannon's address to the A.M.A. was published as "Some conditions controlling internal secretion," *JAMA* 79(1922):92–95.

7. Cannon's address to the psychiatrists was published as "New evidence for sympathetic control of some internal secretions," *Am. J. Psychiat.* 2(1922):15–30; see esp. 15–17.

8. Ibid., pp. 29–30. In an experiment that attracted Cannon's attention, Georges Marañon, a Spanish physiologist, had injected adrenalin into human subjects on grounds that it would evoke almost all the visceral changes that were characteristic of profound emotional disturbance and later noted the physiological effects: The subjects had turned pale, they had a tremor, their pupils were dilated, and occasionally their heads were moistened by a cold sweat, and so forth (pp. 28–29).

9. Sidney Britton, as a student at McGill, had written a paper on the effects of lowering temperature in fish, which his adviser sent to Cannon because of his interest in the bodily effects of temperature changes. After two years at Harvard as a fellow of the National Research Council, he went on to Johns Hopkins and, in 1928, to the Univ. of Virginia where he rose to be head of the physiology department. For Britton's work with Cannon on homeostasis, see Chapter Eight above, pp. 156–158.

10. Cannon and Britton, "Studies on the conditions of activity in endocrine glands. XV. Pseudaffective medulliadrenal secretion," *Am. J. Physiol.* 72(1925):283–294; see esp. pp. 284–286. For Woodworth and Sherrington, see note 2 above.

11. Ibid., pp. 292–294. A companion piece immediately followed in the same issue of the *A.J.P.*: E. Bulatao and W.B. Cannon, "The role of the adrenal medulla in pseudaffective hyperglycemia," pp. 295–313. This paper describes the high glycemic percentages associated with sham rage and shows that medulliadrenal secretion plays a major role in their production.

12. Cannon and Britton, "Studies on the conditions of activity in endo-

crine glands. XX. The influence of motion and emotion on medulliadrenal secretion," *Am. J. Physiol.* 79(1927):433–465; see esp. p. 463.

13. Philip Bard, "Ontogenesis of one physiologist," *Ann. Rev. Physiol.* 35(1973):1–16; see esp. pp. 6–10. For Bard's correspondence, see Box 110 in Cannon Papers; also reminiscences and photocopies in Box 199.

14. See further Bard's Ph.D. dissertation, "Diencephalic control of the sympathetic nervous system," Harvard University, 1927. The only time Bard ever collaborated with Cannon was on a study of the sympathetic innervation of the thyroid gland. But when they tried to repeat the experiments of Cannon, Binger and Fitz (1915), the results were negative and their study was not published. In a postdoctoral year at Harvard, Bard began to do sterile intracranial operations with a view toward preparing some chronically decorticate cats, in which he was instructed by Harlan Newton, one of Harvey Cushing's young surgeons-in-training. ("Ontogenesis of one physiologist," p. 11).

15. Cannon to Alvarez, 23 Aug 1927. For further correspondence between Cannon and Alvarez, see fds. 1514–1521 in the Cannon Papers.

16. Cannon, "The James-Lange theory of emotions: A critical examination and an alternative theory," *Am. J. Psychol.* 39(1927):106–124. A somewhat altered version of the paper appears as Chapter 18, "A Critical Examination of the James-Lange Theory" and Chapter 19, "Emotion as a Function of the Optic Thalamus," in *Bodily Changes in Pain, Hunger, Fear and Rage,* 2nd ed., N.Y. and London: D. Appleton & Co., 1929.

17. Cannon's talk to the psychologists, "Recent studies of bodily effects of fear, rage and pain," was published in *J. Phil., Psychol., and Sci. Methods,* Mar 14, 1914, pp. 162–165. The following month, an expanded version of Cannon's lecture titled "The interrelations of emotions as suggested by recent physiological researches" was published in *Am. J. Psychol.,* Apr 25, 1914, pp. 256–282. Here Cannon realized that he could not avoid discussing James's work. See further *Walter B. Cannon* (1987), pp. 314–315.

18. Cannon, "New evidence for sympathetic control of some internal secretions" (n.7), p. 16.

19. William James, "What is an emotion?" in *Mind,* 9(1884):188–205; see also his chapter on the emotions in *Principles of Psychology,* vol. 2, pp. 442–485. Lange's theory was first published in Danish and later translated into German and French. These three essays were gathered into a volume titled *The Emotions,* ed. Knight Dunlap, Baltimore: Williams & Wilkins, 1922.

20. Additional evidence that Cannon was mulling over the James-Lange theory during these years may be found in his letters, e.g., to Elliott Joslin of 7 Sep 1923, and to Edward J. Kempf, 7 Jan 1923.

21. R.B. Perry. *General Theory of Value* (N.Y.: Longmans, Green, 1926), see esp. pp. 295 and 299–300. Perry has written a lengthy biographical note on William James for the *Dictionary of American Biography* (vol. 9, 1932), pp. 590–600; see also his two volumes of James's unpublished correspondence and notes, *The Thought and Character of William James,* Boston: Little, Brown, 1935.

22. Cannon, "The James-Lange theory" (n.16); see esp. "A consideration of the visceral factors," pp. 107–114.

At the beginning of his paper, Cannon confessed to some trepidation in venturing to criticize a view of the nature of the emotions which had commended itself so generally to psychologists. Nevertheless, pertinent physiological facts had come to hand, "which were not available when James and Lange developed their ideas" and led him to consider alternative explanations of affective experience "before the James-Lange theory is granted basal claims in this realm of psychology."

23. Ibid., see section headed "A theory of emotion based on thalamic processes," esp. pp. 119–121.

24. Cannon, "Neural organization for emotional expression," in *Feelings and Emotions,* ed. Carl Murchison (Worcester, Mass.: Clark Univ. Press, 1928): pp. 257–269.

25. For Cannon's discourse to the Massachusetts Medical Society, see "The mechanism of emotional disturbance of bodily functions," *New Engl. J. Med.* 198(1928):877–884.

26. Philip Bard, "A diencephalic mechanism for the expression of rage with special reference to the sympathetic nervous system," *Am. J. Physiol.* 84(1928):490–513; "A central representation of the sympathetic nervous system as indicated by several physiologic observations," *Arch. Neurol. & Psychiat.* 22(1929):230–246.

27. Murchison to Cannon, 9 Feb 1928; Cannon to Murchison, 24 Mar 1928; Cannon to Murchison, 28 Mar 1928. Cannon congratulated Bard on a fine piece of work after reading his manuscript; soon after, he wrote Bard again that he had failed to mention "the admirable handling of the James-Lange theory. I think you did the job better than I did." (30 Apr and 2 May 1929). Bard's chapter in the *Handbook* is on pp. 264–311; his two-part paper, "On emotional expression after decortication with some remarks on certain theoretical views," is in *Psychol. Rev.* 41(1934):309–329 and 424–449.

28. Newman, Perkins and Wheeler, "Cannon's theory of emotion: A critique," *Psychol. Rev.* 37(1930):305–326.

29. Cannon, "Again the James Lange and the thalamic theories of emotion," *Psychol. Rev.* 38(1931):281–295; see esp pp. 292–294. Dana's paper was "The anatomic seat of the emotions: A discussion of the James-Lange theory," *Arch. Neurol. & Psychiat.* 6(1921):634–639.

30. Ibid., p. 295. In a letter to Bard of 28 Sep 1928, Cannon told of happening upon Dana's 1921 paper after he had expressed the very same idea regarding the seat of the emotions. He had received a friendly letter from Henry Head, who had written, "As you may readily understand, this theory [i.e., James-Lange] has never appealed to me and all my work has tended to run counter to it. You have summarized my conclusions perfectly accurately and I think your use of them is justifiable."

31. J.W. Papez, "A proposed mechanism of emotion," *Arch. Neurol. & Psych.* 38(1937):725–743; besides Cannon and Bard, Papez referred to the 1933–34 reports of Wilder Penfield and S.W. Ranson in advancing knowledge of the

functions of the hypothalamus. See also P.D. McLean, "Psychosomatic disease and the 'visceral brain': Recent developments bearing on the Papez theory of emotion," *Psychosomatic Med.* 11(1949):338–353. Melvin Konner comments on Papez in *The Tangled Wing: Biological Constraints on the Human Spirit* (Harper & Row, 1982), pp. 141–142; he discusses the James-Lange theory and Cannon's critique and alternative theory on pp. 136–141.

32. P.D. McLean, "Challenges of the Papez heritage," in *Limbic Mechanisms: The Continuing Evolution of the Limbic System Concept,* ed. K.E. Livingston and O. Hornykiewicz (N.Y. and London: Plenum Press, 1976), pp. 4–5.

33. "A proposed mechanism of emotion" (n.31); see esp. p. 729.

34. S.B. Nuland, *The Wisdom of the Body* (N.Y.: Alfred Knopf, 1997), pp. 333–335.

10. Friends and Associates of Medical Progress

1. The first number of *Hygeia,* announced in the issue of *JAMA* for Dec 2, 1922, appeared in April 1923. See also *Colorado Medicine,* Dec 1922, pp. 250–251; *Starry Cross,* Nov 1923.

2. Cannon to Olin West, 24 Sep and 17 Oct 1923; editorial in *JAMA,* Oct 27, 1923, pp. 1443–1444. See also Cannon to John M. Dodson, A.M.A. Bureau of Health and Public Instruction, 28 Sep 1923, with history of the lay society; pamphlet, *The Society of Friends of Medical Progress: Why It Was Founded,* Boston, 1923. Correspondence with and about the Society may be found in fds. 411–413 and 426 in the Cannon Papers.

3. Cannon quoted Dublin in a letter to W.W. Keen, 15 Oct 1923. After a founding meeting held in Boston, Cannon wrote Simon Flexner, "I am glad that you liked Barbour. He has many of the qualities we need for an Organizing and Active President. He is ardently interested in the protection and promotion of medical research . . . I think it highly likely that he will be selected for the post—perhaps within a few months." (15 Jan 1923; in the American Philosophical Society Library). For an earlier assessment of Barbour, see Cannon to Simon Flexner, 16 Mar 1923.

4. See Chapter Six above, pp. 113–116. See also Raymond Gorges, *Ernest Harold Baynes: Naturalist and Crusader* (Boston, Houghton Mifflin, 1928), esp. Chapter 11, "The Fight for Truth," pp. 194–223.

5. Keen wrote Cannon (13 Nov 1922) that he felt Baynes had sacrified a great deal for his antivivisection activities and asked if Cannon could stir up some paid work for him. Flexner also wrote Cannon about Baynes on the same date, with Cannon replying on Nov 14. Barbour wrote Frederick Bigelow (10 Jan 1923), "We have an excellent Field Secretary in Mr. Baynes who is ready to go to work at once."

6. Cannon to Flexner, 8 Feb 1924, in response to Flexner's letter of Feb 7. See also Mary Lee Thurman to Cannon, 7 Feb 1924, enclosing a letter from Regina Jenkins of Feb 6, which had been forwarded to the office.

7. Cannon to Flexner, 14 Feb 1924, in response to Flexner's letter of Feb 13. Flexner continued to be "nervous" about the Friends. After Miss Thurman sent him yet another complaint that had been addressed to Bishop Mann, Flexner wrote on Mar 11, "Who is Miss Thurman? And what particular qualification has she for conducting the affairs of the Friends of Medical Progress?" Cannon again addressed Flexner's apprehensions in a letter of 13 Mar 1924, to which Flexner responded somewhat sheepishly on Mar 14.

8. See, e.g., G.M. McCoy's letter to Cannon of 2 Feb 1923, and Cannon's lengthy reply of Feb 5.

9. McCoy to Cannon, 5 May 1923, in reply to Cannon's letter of Apr 26. The earlier exchange of letters between Cannon and McCoy was 9 and 19 Apr 1923.

10. Flexner to Cannon, 16 Apr 1923, in reply to Cannon's letter of Apr 11.

11. A reorganization at A.M.A. headquarters did away with the Council of Health and Public Instruction in the summer of 1923. Although it was subsequently decided to retain the Committee for the Protection of Medical Research, Cannon wrote William C. Woodward on Sep 28, "I am a little uncertain about my present relations with the American Medical Association." When John M. Dodson took over as executive secretary of the reorganized Bureau of Health and Public Instruction that fall, relations between the A.M.A., Cannon's committee, and the lay society were re-examined. On Oct 8, Cannon wrote Dodson that he would be glad to continue to serve as chairman of the new committee that was being appointed.

12. Cannon to Mary Lee Thurman, 11 Mar 1924.

13. George H. Simmons to A. Murat Willis, 21 Dec 1923; see also Cannon to Simmons, 14 Dec 1923. In the summer of 1924, Cannon became concerned about reports on animal experiments in psychological laboratories that might arouse the antagonism of the antivivisectionists and called the attention of the American Psychological Association to the situation.

14. For seeking a replacement for Barbour, see, e.g., Cannon to Flexner, 20 Nov 1924. While Barbour no longer directed the Friends, he continued to serve the organization as president.

15. See further Gruenberg's 7-page typescript "Memorandum to the Board of Directors," 25 Sep 1925, esp. p. 2.

16. Cannon to Benjamin Gruenberg, 30 Sep 1925. Gruenberg also prepared a printed Annual Report after his first year at the Association, which noted the death of Ernest Harold Baynes on Jan 21.

17. Gruenberg was especially alarmed by two editorials, "Putrid pus for vaccination," in the *Evening Graphic* of Dec 30, 1925, and "Experimenting on your child," in the issue of Jan 19, 1926. His mimeo letters to medical school deans and hospital superintendants were dated 21 Jan and 9 Feb 1926.

18. Gruenberg to the Board of Directors, American Association for Medical Progress, 9 Jan 1926, enclosing a letter from Lee K. Frankel, vice-president, to the Managers of the Metropolitan Life Insurance Company, 15 Dec 1925, and a press release from the Metropolitan Information Service.

19. Cannon reported the incident to Keen in a letter of 30 Mar 1926. See

also Gruenberg's letter to the Board of Directors of Mar 13, enclosing the letter from Mrs. Sanger together with printed statements from the A.B.C.L. Keen, replying to Cannon on Apr 2, was appalled at Gruenberg's action and began his letter, "Frankly I have lost all confidence in the good judgment of Mr. Gruenberg." And he ended, "If only Gruenberg had had the sense not to touch Birth Control in any way, we should have had no trouble. That is what irks me." (In American Philosophical Society Library.)

20. Resolutions of Board of Directors of the American Association for Medical Progress, Boston, April 1926, signed by Edward Wigglesworth, secretary.

21. Cannon to Keen, 8 Feb 1926. He had also written to Simon Flexner about his resignation on 15 Dec 1925. Cannon's final report for the Committee was published in *JAMA,* Mar 20, 1926, p. 862.

22. Sarah Cleghorn's letter to the editor of *The Nation* was dated Sep 21, but was not published until November 9, 1927. For Stewart and Rogoff's article, see *Am. J. Physiol.* 79(Jan 1927):508–535. It was the third in a series of their studies on adrenal insufficiency. The materials on which this account is based may be found in the Stewart and Rogoff Papers.

23. Cannon to Olin West, 28 Nov 1927.

24. Dr. West sent identical letters to G.N. Stewart and J.M. Rogoff, 30 Nov 1927; his letter to Cannon is of the same date.

25. Rogoff to West, 13 Dec 1927. Rogoff enclosed a copy of the reply he had written when the leaflet first appeared, which he had sent to the American Association for Medical Progress.

26. Cannon to West, 19 Dec 1927. West's reply to Rogoff was written on Dec 16.

27. See further Gruenberg's Memorandum of the Conference in Cleveland, February 14, 1928, that he enclosed in a letter to Rogoff, 23 Feb 1928.

28. Volume l, no. 1 of the new journal (the first and only issue) came out in March 1928. The monograph contained contributions from Haven Emerson of Columbia, Charles Stockard of Cornell and C.E.A. Winslow of Yale, among others. Both were published by the Association.

29. Cannon to McCoy, 14 May 1928, in reply to McCoy's letter of May 12.

30. *The Way of an Investigator,* p. 159. Keen's letter to Cannon had been written on 24 Mar 1926; in American Philosophical Society Library.

In recent years the debate on the subject of animal experimentation has heated up more than ever. See, e.g., Patricia Curtis, "New debate over experimenting with animals," *New York Times Magazine,* Dec 31, 1978, pp. 18–23; Andrew Rowan and Bernard Rollin, "Animal research—for and against: A philosophical, social and historical perspective," *Perspectives in Biology and Medicine* 27(Autumn 1983):1–17. The arguments for and against research using animals have become as polarized as those on the subject of abortion, the lunatic fringe as usual taking extreme positions. The comparison is all too grim, as attested in the *New York Times* for Jan 10, 1999. An article (page 10), "British researchers on animal rights death list," is accompanied by a photograph of Colin Blakemore, an Oxford physiologist, whom animal rights activists have threatened to kill because of his experiments.

11. One Man's Family

Cornelia James Cannon's letters and manuscripts, previously on loan to the authors, are now in the Cannon Family Papers, located in the Schlesinger Library. The letters to her family in St. Paul, Minnesota, were initially saved by Cornelia's mother, Frances Haynes James, and excerpted in a volume that she called "Snatched from Oblivion." Marian Cannon Schlesinger's delightful book of the same title, *Snatched from Oblivion,* also served as a source here, as did the bound volumes of letters (letter-books) for the years after Mrs. James's death.

1. See further H.H. Dale, "Prof. W.B. Cannon, For. Mem. R.S.," *Nature* 158(Jul 1946):87–88; "Walter Bradford Cannon, 1871–1945," *Obituary Notices of Fellows of the Royal Society* 5(Mar 1947):407–423.

2. *Snatched from Oblivion,* see esp. pp. 49, 52–53, 93.

3. "Snatched from Oblivion," 6 Oct 1922. For the deaths of Southard, Mrs. Jackson, and Miss Baldwin, see 11 Feb 1920, 9 Feb 1921, 14 Jan 1922. For more on the Cannons' circle of friends, see *Walter B. Cannon* (1987), esp. Chapter 8; also *Snatched from Oblivion,* pp. 98–99.

4. Ibid., 26 Oct 1922; see also 2 Nov 1920 and 11 Oct 1922.

5. Ibid., 29 Nov and 5 Dec 1922. For Mrs. Cannon's remembrance, see her unpublished memoir, "Servant of Science," p. 79; in Cannon Papers. Official notices of the citation were sent to Cannon by Archibald Campbell, 28 Nov 1922, and A.D. Tuttle, 9 Jan 1923.

6. 1919–25 letter-book, 14 Dec 1923; for Mrs. James's visit, see also Nov 1923 letters.

7. *Snatched from Oblivion,* esp. pp. 23–25.

8. Mrs. Cannon described the incident in a family letter of 14 Jun 1927; in 1926–28 letter-book, accompanied by newspaper accounts. For her continuing battles with Cambridge city officials, see, e.g., letters of Nov-Dec 1920 and Nov 1921 (in "Snatched from Oblivion") and 15 Apr 1924 (in 1919–25 letter-book); also *Snatched from Oblivion,* pp. 86–87. Three notebooks of press clippings about the Cambridge Public Schools Association, including her *Cambridge Chronicle* columns, may be found in the Cannon Family Papers.

9. "Snatched from Oblivion," 12 Nov 1921; see also letters of 14, 17, 19, and 21 Nov 1921. The editorial for *Woman Citizen* was published under title of "Freedom of Speech." Prof. Chafee was known as "probably the most respected civil libertarian in America."

10. 15 Dec 1921, in "Snatched from Oblivion"; 24 Mar 1924, in 1919–25 letter-book. See further Chapter Ten above, pp. 198–199; *Snatched from Oblivion,* pp. 103–106. The Birth Control League of Massachusetts, founded in 1916 under the leadership of Blanche (Mrs. Oakes) Ames, had been replaced by a Family Welfare Association during and after the war years. The decade of the 1920s was not a propitious time for either feminism or birth control, and the organization languished for lack of interest. In 1928, after the arrest of Dr. Antoinette Konikow, whose clinic in Boston had been raided by the police, Mrs. Cannon was one of the group who made the bold decision to reactivate the Birth Control League. A Planned Parenthood profile of her, written in the

1960s, delineates her role as a pioneer force in Massachusetts birth control (in B.C.L.M. Papers, Smith College).

11. "Snatched from Oblivion," 1 Nov 1920 and 1 Sep 1921. "Can our civilization maintain itself?" was published in the *Atlantic Monthly,* Nov 1920.

12. Sedgwick to Cannon, 11 Jan 1921; note of Mrs. Cannon, 3 May 1921; both in 1919–25 letter-book. See also 10 May 1923, in "Snatched from Oblivion." "Philanthropic doubts" was the lead article in the issue of Sep 1921; several folders of letters and printed rejoinders to the article may be found in the Cannon Family Papers. Cornelia never did publish the follow-up article, but another article appeared in the *North American Review* just one month later: "How much optimism do we need?" was in the Oct 1921 issue.

13. 30 Dec 1922, in "Snatched from Oblivion"; see also Chapter Seven above, pp. 141–142. The full title of the article is "Democracy in question: I. American misgivings."

14. 1919–25 letter-book, 27 Apr 1923 and 21 Jun 1923. "Can our public officials be our leaders?" appeared in the May issue of the *North American Review.* "The crabbing of youth by age," in the June *Atlantic Monthly* was followed by "Disassociated schools" in the November issue. A folder of letters from Ellery Sedgwick about Mrs. Cannon's articles may be found in the Cannon Family Papers.

15. Ibid., 13 Jun 1921; for the Radcliffe Funds, see also, e.g., 14 Jan and 18 Apr 1922. Cannon to Martin, 8 Nov 1923, in reply to Martin's letter of Oct 30; both in Cannon Papers.

16. Ibid., 16 Jan and 8 Aug 1923. Mrs. Jellyby is a character in Charles Dickens' *Bleak House.*

17. Cannon to Mrs. James, 23 Jan 1921, and Cornelia's letter of the same date accompanied by poems; both in "Snatched from Oblivion." For the trip west, see also 11 Jul 1922. For her public readings, see further *Snatched from Oblivion,* pp. 99–102.

18. Mrs. Cannon's articles include, e.g., "You and public health," *Hygeia,* Feb 1924; "Our public schools," *Child Welfare Magazine,* Jun 1924; "What college, daughter?" *Radcliffe Quarterly,* Jan 1925; "Prohibition and the younger generation," *North American Review,* Fall 1925; "Quantity production of ideas, *Atlantic Monthly,* Sep 1925.

19. Cornelia James Cannon, "The writing of *The Pueblo Boy,*" *Radcliffe Quarterly,* Feb 1958. See also 1919–25 letter-book, e.g., 2 Apr and 2 Jul 1925.

20. Cannon to Mrs. Cannon, 27 Jul 1925, in 1919–25 letter-book. Cornelia wrote her mother on Jul 21 that she had finished the book. *The Pueblo Girl* was published in 1929; *Lazaro in the Pueblos* in 1931; and *The Fight for the Pueblo* in 1934—all by Houghton Mifflin.

21. 1926–28 letter-book, 24 Oct 1926; for more on "Matts," see also letters of 8 Jun, 24 Jul, 18, 23, and 26 Aug 1926.

22. Ibid., 15 and 31 Jan, 9, 16, 17, 23, and 27 Feb 1927.

23. *Snatched from Oblivion,* p. 128; for more on *Red Rust,* see esp. pp. 125–128. See further *Book Review Digest,* 1928, pp. 131–132; also scrapbook of reviews and letters in Cannon Family Papers.

24. 1926–28 letter-book, see esp. letters of 30 Jun and 10 Jul 1927. *Heirs,* like *Red Rust,* was published by Little, Brown; see further *Book Review Digest,* 1930, p. 165. Mrs. Cannon wrote another novel which, like "The Clan Betrays," was never published. Entitled "Denial," it was a book of twenty chapters with birth control as its theme. Both of these, as well as other unpublished manuscripts, may be found in the Cannon Family Papers. (With gratitude to Maria Dietrich for sharing portions of her projected biography of Cornelia James Cannon.)

25. *Snatched from Oblivion,* p. 112; see also pp. 63–64. See further 1919–25 letter-book, 7 Mar 1923 ("She who raises a boy deserves the thanks of the nation, even if she does not raise him to be a soldier. It is a job!"), and 1926–28 letter-book, 10 May 1927. Another glimpse of Cannon family life is provided in "Introduction to a family memory, as told by Bernice, Ida, and Jane Cannon, to their niece, Linda Cannon Burgess, 1959"; in Cannon Family Papers.

26. For more on Bradford, see 1919–25 letter-book, e.g., letters of 6 Jan and 25 Dec 1925. See also *Boston Sunday Globe,* 2 Mar 192[?], in Cornelia James Cannon's biographical file, Radcliffe Archives.

27. 1926–28 letter-book, 6 Feb 1927; see also 25 Jan, 6 and 16 Feb, 25 Apr 1927. The story of the German exam is recounted in "Servant of Science" (n.5), p. 34.

28. 1919–25 letter-book, Mrs. Cannon to Bradford, 13 Jul 1925, to her mother, 10 Nov 1925, to Wilma, 14 Dec 1925. For Cannon's illness, see also her letter to her mother, 24 Oct 1926, in 1926–28 letter-book; Cannon to Simon Flexner, 22 Oct 1926 (in Cannon Papers).

29. "Snatched from Oblivion," 3 Aug 1922; see also 27 Mar, 7 Jun, and June-July 1922. Cannon wrote Lewis Hackett, 7 Nov 1922: "We saw Cannon Mountain in Glacier Park, but did not climb it again. It was between 1,000 and 2,000 feet higher than it was twenty-one years ago when we first ascended it. It is remarkable how rapidly the surface of the earth changes!" Years later the note in the bottle was found by two climbers, prompting Marian Cannon Schlesinger to write a short account of her parents' adventures, which was published in the Apr 1989 *American Heritage.*

30. Cornelia kept a detailed account of the trip, which she had bound into a separate volume; in Cannon Family Papers. Cannon described the trip for Philip Bard in a letter of 17 Sep 1928; in Cannon Papers.

31. Bradford Cannon, personal communication, February 1998.

32. 1929 letter-book, 29 Jun, 10 Jul, 21 Jul 1929. Bradford's first two research articles were published in *Am. J. Physiol.:* "The effect of progressive sympathectomy on blood pressure" in Jul 1931, pp. 592–596, and "A method of stimulating autonomic nerves in the unanesthetized cat, with observations on the motor and sensory effects" in Aug 1933, pp. 373–382.

12. The Physiology Department: An Extended Family

1. Cannon's talk, "The history of the physiology department of the Harvard Medical School," was published in *Harv. Med. Alumni Bull.* 1

(1927):12–19; see esp. pp. 18–19. For more on the physiology department under Dr. Bowditch, see further *Walter B. Cannon* (1987), pp. 40–44.

2. Alfred C. Redfield, "The laboratories of physiology of the Harvard Medical School," reprinted from *Methods and Problems of Medical Education* (N.Y.: The Rockefeller Foundation, 1925), see esp. pp. 1 and 4. The article features floor plans of Building C.

3. For Cecil Drinker, see letters in fd. 8, Cannon Papers, esp. E.H. Bradford to Cannon, 18 Mar 1918; also Chapter Four above, pp. 74–75. For the physiology department in the prewar era, see *Walter B. Cannon* (1987), e.g., pp. 121–122, 154–158, 245, 364–366.

4. Redfield (n.2) has a section on the faculty and staff of the physiology department on p. 15. See also descriptions of the physiology laboratories in *Harvard Medical School Announcements and Course Catalogs*, 1919–20, 1920–21,1921–22, 1922–23, as well as departmental correspondence in the Cannon Papers (esp. fds. 13, 19, 20). Cannon wrote Hugh Cabot, 4 Jan 1923: "I think I have the three best men in physiology with me. They are young, very well-trained, and already have a considerable body of investigation to their credit." He went on to describe the qualifications of Drinker, Redfield and Aub.

5. Cannon to W.O. Fenn, 27 Nov 1923. Redfield described the tutorial system in a section on teaching in his article, pp. 12–15. In 1923, Cannon put out a fourth edition of his laboratory manual (*A Laboratory Course in Physiology*, Harvard University Press, originally published in 1910)—the first revision in ten years. Subsequent editions appeared in 1926, 1927, and 1931.

6. Cannon to Joseph Barcroft, 13 Mar 1923. For Cannon's wartime thoughts on educational exchange, see Chapter Three above, pp. 54–55.

7. Cannon to Hal Davis, 20 Apr 1923; Cannon to D.L. Edsall, 10 Mar 1925. See further *The Professional Memoirs of Hallowell Davis* (St. Louis: Central Institute for the Deaf, 1991), p. 16; for more on Davis' years at Harvard, see also pp. 15–21.

8. *The Way of an Investigator*, p. 91; Cannon to Edsall, 21 Apr 1924. Alex Forbes's article appeared in *Physiol. Rev.* 2(1922):361–414. See further W.O. Fenn, "Alexander Forbes, 1882–1965," *Biog. Memoirs* 40(1969):113–141; Robert K. Frank and Judith H. Goetzl, "The Alexander Forbes Papers," *J. Hist. Biol.* 11(Fall 1978):387–393.

9. For Cecil Drinker's move, see Edsall to Cannon, 2 Aug 1927. Funding from the Rockefeller Foundation in 1921–22 had helped to provide the means for opening of the School of Public Health; the following year, Edsall wrote his first report as dean. As head of the two schools, he became the first full-time dean at H.M.S. A further gift from the Rockefeller Foundation in 1926–27 pledged "to pay to Harvard University for further development of the School of Public Health the sum of $1,090,000" (Jean Alonzo Curran, *Founders of the Harvard School of Public Health*, N.Y.: Macy Foundation, 1970, pp. 44–45).

10. For Roy Hoskins' appointment, see Minutes of the Harvard Medical Faculty, meeting of 1 Apr 1927, in Harvard Medical Archives. In 1926, a fund was established from the estate of Stanley McCormick of Chicago for the support of a Memorial Foundation for Neuro-Endocrine Research, with Mrs. McCormick serving as president. The Foundation's offices and laboratories were located in the department of physiology at Harvard, while the clinical

work was carried out at the Worcester State Hospital. Magnus Gregersen received his Ph.D. under Cannon and stayed in the department (working on thirst, plasma volume, and the salivary glands) for another ten years.

11. Harry Veach requested a $1200 scholarship in a letter to Cannon, 23 Aug 1922, with Cannon replying on Aug 24. Cannon later praised Veach in a letter to S.R. Meaker, 5 Apr 1924. For lists of assistants in the physiology department, see Cannon to Edsall, 6 and 10 Oct 1928; in Dean's Files, Harvard Medical Archives.

12. For Cannon's work day, see F.L. Holmes, "Walter Bradford Cannon's Research on Functions of the Sympathetic Nervous System" (B.S. degree, M.I.T., 1954), esp. p. 104. Holmes interviewed Cannon's family and associates for his description of Cannon's work and methods.

13. Fenn to Cannon, 19 Sep 1922. Fenn expressed remorse for not having collaborated with Cannon during his years at Harvard.

14. *The Way of an Investigator,* pp. 40–41; see also pp. 92–96.

15. Cannon wrote Mrs. George P. Baker, Dean of Radcliffe College, about Dorothea Smith on 26 Mar 1923. He wrote Dr. Martha Tracy of the Women's Medical College about Annie Minot on 25 May 1923. Minot received her Ph.D. from the Division of Medical Sciences in physiology, while Smith got hers in bacteriology. The only other women to get the Ph.D. through Radcliffe during this period were Abby H. Turner in physiology in 1926 and Olive Watkins (later Mrs. George Smith) in biochemistry in 1928. For later developments about women in the Medical Sciences program, see Chapter Twenty-one below, pp. 423–426.

16. Peyton Rous to Cannon, 21 Feb 1923, in reply to Cannon's inquiry of Feb 13; in American Philosophical Society Library. For Rapport, see, e.g., Cannon to Stuart Graves, 12 May 1923. Rapport later became professor of physiology at Tufts Medical School, while Menkin worked with S.B. Wolbach in the pathology department at H.M.S. until the 1940s.

17. D.L. Edsall to R.M. Pearce, 13 Jun 1924, in Dean's Files. Cannon had written Edsall, 19 Mar 1924, about his "apprehension lest the receiving of Rockefeller and other Fellows in the Department would run my expenses this year considerably beyond my budget." See also Edsall's annual report for the Faculty of Medicine, in *Annual Reports of the President and Treasurer of Harvard University, 1923–24,* pp. 198–199.

18. Cannon to Edsall, 10 Mar 1925, in Dean's Files; his letter to John Warren, 4 Dec 1922, is in the Cannon Papers. Eliot's comment has been quoted often; see, e.g., *Walter B. Cannon* (1987), p. 116.

19. L.J. Henderson wrote to John Warren as Chairman of the Space Committee on 27 Mar 1928; Cannon noted his approval on Apr 10; in Dean's Files. For the 1920 salary episode involving W.T. Porter, see fd. 738 in the 1919–22 series, A.L. Lowell Papers.

20. For Edsall's relations with the Rockefeller enterprise at this time, see further Joseph C. Aub and Ruth K. Hapgood, *Pioneer in Modern Medicine: David Linn Edsall of Harvard,* (Harvard Medical Alumni Association, 1970), pp. 281–284. Under Edsall's leadership, H.M.S. became a major beneficiary of Rockefeller grants, with awards going to the School of Public Health, the Lying-In

Hospital, new laboratories at the M.G.H., the departments of psychiatry and neurology, the tutorial system, and so forth.

21. Edsall to Pearce, 26 Apr 1927; his preliminary statement of needs was written on Apr 5. These letters, with Edsall's 20-page memorandum, may be found in the Dean's Files. While Edsall was in China, Cannon served as acting dean of the Medical School for the fall semester of 1926.

22. Edsall to Pearce, 4 May 1927; Flexner acknowledged Edsall's memo on Apr 30. Both in Dean's Files.

23. Edsall to Flexner, 22 Oct 1927; excerpt from Minutes of G.E.B. meeting, Nov 17, 1927. Cannon's letter to Edsall was 10 Oct 1927, enclosing Pearce's letter of October 6. All in Dean's Files.

24. Cannon to Edsall, 1 Oct 1928, re research activities. George E. Vincent notified Edsall of his appointment to the special committee of the Rockefeller Boards on Apr 17, 1928. Both in Dean's Files.

25. Cannon to Edsall, 6 and 10 Oct 1928, re students trained in physiology; in Dean's Files.

26. Pearce's dinner with Edsall was recorded in his diary note of 11 Mar 1929. For the appropriation awarded to Cannon in 1927, see also Pearce's Memo to the officers of the General Education Board, 13 Mar 1929. Both in Rockefeller Archives Center, as are some of the documents at Harvard.

27. Cannon to Edsall, 16 Jul 1929; Edsall to Lowell, 26 Aug 1929; both in Dean's Files.

28. Pearce to Cannon, 10 Dec 1929, in reply to Cannon's statement, 9 Dec 1929. See also Pearce to Cannon, 26 Nov 1929; Pearce's diary note of 7 Dec 1929; Edsall to Pearce, 9 Dec 1929. All in Rockefeller Archives Center.

29. Pearce's "Aid for a Leader" memo is undated. See further excerpt of letter from Pearce to Cannon, 10 Feb 1930, in Rockefeller Archives Center; also Edsall to Cannon, 6 Mar 1930, in Dean's Files.

30. Edsall to Cannon, 17 Apr 1930. Norma S. Thompson, secretary, officially notified Harvard of the Rockefeller resolution in a letter to A.L. Lowell, 16 Apr 1930. Both in Dean's Files.

13. Scientific Outreach at Home and Abroad

1. Further information may be found in the Dean's Files and the Dunham Lecture files, Harvard Medical Archives; photocopies of selected correspondence relating to Cannon's chairmanship are in Box 201, Cannon Papers.

2. Cannon to Mary D. Dunham, 8 Nov 1926. See also exchange of letters between Cannon and C.S. Sherrington, 2 Dec 1926 and 19 Feb 1927; W.B. Hardy to Cannon, 18 Mar 1928. For a report of the early Dunham Lectures, see *Harv. Med. Alum. Bull.*, Jun 1927, pp. 12–13.

3. Cannon sent letters to members of his committee on 30 Nov 1927, suggesting the selection of lecturers from those attending the International Congress. See further exchange of letters between Joseph Barcroft and Cannon, 29 Apr and 10 May 1928; Cannon to Louis Lapicque, 26 Mar 1929; Cannon to Mrs. Dunham, 26 Mar and 16 Oct 1929.

4. See W.H. Howell to D.L. Edsall, 18 Apr 1927, in Dean's Files; also file of correspondence and related materials concerning the 13th International Congress of Physiology. Both in Harvard Medical Archives.

5. For published reports of the Congress, see J.F. Fulton, "The 13th International Physiological Congress," *Harv. Med. Alumni Bull.*, Nov 1929, pp. 4–8; "The Thirteenth International Physiological Congress," *New Engl. J. Med.*, 201(1929):433 ff. See also printed programs, lists of members, and abstracts of the papers presented at the Congress.

6. *History of the American Physiological Society Semicentennial, 1887–1937*, Baltimore, 1938; an account of the Boston Congress is on pp. 139–140, including Howell's opening address and an excerpt from Krogh's oration.

7. 1929 letter-book, Mrs. Cannon's family letters of Aug 16 and 18 (in Cannon Family Papers); see also description of the Congress and Pavlov's visit in her unpublished manuscript, "Servant of Science," pp. 105–111 (in Cannon Papers).

8. Exchange of letters between I.P. Pavlov and Cannon, 14 and 24 Jan 1929. For Pavlov's earlier visit to America, see Chapter Seven above, pp. 131–133.

9. John Fulton reported the Pavlov incidents in his *Harv. Med. Alumni Bull.* article (n.5), as well as in *Harvey Cushing: A Biography* (Springfield, Ill.: C.C. Thomas, 1946), pp. 577–579. Cushing's quote was also published in an obituary article, "Ivan Petrovich Pavlov," *New Engl. J. Med*, Mar 5, 1936, p. 489. Pavlov's talk was titled "Inhibition in the normal activity of the cerebral hemispheres."

10. Ibid.; *New Engl. J. Med.* (n.5), p. 441, 447.

11. See Mrs. Cannon's unpublished manuscript, "The Friendship of Dr. Pavlov and Dr. Cannon," pp. 3–4; copy in fd. 508 of the Cannon Papers.

12. For more on the exchange professorship, see Chapter 11 in "Servant of Science" (n.7), pp.114–123. The story is confirmed in a letter from Dean Edsall to Pres. Lowell, 26 Nov 1928; in fd. 296, 1928–30 series, A.L. Lowell Papers. Cannon's official appointment was dated 12 Nov 1928; he outlined his plans in a letter to Alan Gregg, 7 Dec 1928.

13. Ibid., for the trip to Italy, see pp. 116–118. Another of Mrs. Cannon's unpublished manuscripts, "Art Awheel in Italy," may be found in the Cannon Family Papers. Marian Cannon Schlesinger has described the tour of Italy in Chapter 8, and the remainder of the trip in Chapter 9, of her memoir, *Snatched from Oblivion.*

14. Exchange of letters between Pierre Galpin and Cannon, 15 and 19 Oct 1929.

15. 1930–31 letter-book; see Mrs. Cannon to Bradford Cannon, 2 Jan 1930, to her father, 8 Jan 1930, and to Bradford, 14 Jan and 6 Feb 1930 (in Cannon Family Papers). Cannon described the dinner parties in his letters to Bradford of 9 and 12 Jan 1930; in Cannon Papers.

16. Ibid., Wilma Cannon to Ida Cannon, 7 Jan 1930.

17. The description of Cannon's lectures comes from his letter to Bradford, 9 Feb 1930, and from Mrs. Cannon's letter of the same date.

18. Cannon to Bradford Cannon, 24 Feb 1930; see also 1930–31 letter-book

for Mrs. Cannon's letter to Bradford of the same date. Cannon's speech before the Société de Psychologie was published as "Quelques observations sur le comportement maternel d'animaux privés du systés du système sympathique," *J. de Psychol. Norm. et Path.*, 1930, pp. 486–488.

19. Cannon to Bradford Cannon, 16 Mar and 4 Apr 1930. See also Mrs. Cannon to her daughters, 10 Mar 1930, and Cannon to the girls, 17 Mar 1930; both in 1930–31 letter-book. Cannon's speech at the Univ. of Louvain was published as "Le rôle du système sympathico surrénal au point de vue du maintien de la stabilité organique (homeostasie)", *Rev. Belge Sci. Med.* 2(1930):451–464.

20. Cannon to Bradford Cannon, 14 Apr 1930. For Carrasco at Harvard, see Chapter Five above, p. 99. Pi-Suñer's fellowship was in 1927; both he and his father, Auguste, attended the Boston Congress in 1929.

21. 1930–31 letter-book, Mrs. Cannon's letter to Bradford, 8 Apr 1930.

22. Mrs. Cannon to Bradford Cannon, 26 Apr 1930; Cannon to Bradford, 14 Apr 1930. For more on the visit to Spain (including menus signed by those who attended the gala dinners in honor of Cannon), see fd. 580 in the Cannon Papers

23. Cannon to Bradford Cannon, 10 May 1930; Mrs. Cannon to Bradford, 6/7 May 1930. Cannon had written Pres. Lowell, 23 Aug 1929, that Barcroft brought him the invitation to give the Linacre Lecture when he came to the Boston Congress. Cannon's lecture was published as "The autonomic nervous system: An interpretation," *Lancet* 1(1930):1109–1115.

24. See 1930–31 letter-book for Mrs. Cannon's letters from mid-May to mid-Jun 1930. For Cannon's battle experience during the war, see Chapter Two above, pp. 34–35.

25. Ibid., Mrs. Cannon wrote from Ragaz about Cannon's illness on Jun 13–20.

26. Ibid., see esp. Mrs. Cannon's letters from Jun 22 to Jul 16.

27. Ibid., undated note from London preceding Marian's letter of July 23. Wilma wrote her Aunt Ida about Bradford's jaundice on 13 Aug 1930.

28. See P.B.B.H. patient record for W.B. Cannon; the entry for 30 Jul 1930 includes a history in Cannon's own words, Dr. Quinby's note of the operation, and Dr. Wolbach's pathology report. See also Cannon's "Notes dictated after an operation on myself, July 31, 1930." Both in Box 8, J.C. Aub Papers.

29. See exchange of letters between W.W. Norton and Cannon, 24 and 28 May 1929; Cannon to Norton, 11 Dec 1930; Cannon to Norton, 7 Aug 1931.

30. Cannon to Norton and Norton to Cannon, 16 Dec 1931; Cannon to Norton, 9 Jan 1932; Norton to Cannon, 9 Feb 1932. A.J. Carlson in his review called Cannon's book "a model of accuracy and lucidity." Even so, he did not agree with Norton about the Epilogue, which he noted was "an anticlimax, an excursion into sociological speculation." "Argument by analogy may have some merit," he continued, "if the conclusions lead to experimental tests; otherwise it appears to the reviewer an infantile pastime, if not a nod." (*Am. J. Sociol.*, Jan 1933, p. 651)

31. Norton to Cannon, 13 Feb 1932, and to Wilma Cannon, 2 and 6 May 1932. Cannon first wrote Wallace Donham on 31 Mar 1931; see also Donham to Cannon, 11 Feb 1932, and Cannon to Donham, 13 Feb and 13 May 1932.

32. "The Ideal Scrapbook," in Box 165, Cannon Papers; see esp. pp. 17–21. Horace W. Davenport has written that Cannon's "prose flows in a limpid stream, unlike that of many scientists who write as though they were squeezing words like toothpaste out of a tube. His prose is clear, straight-forward and simple, but never monotonous. It carries conviction." (Introduction to reprint edition of Cannon's *Mechanical Factors of Digestion,* Science History Publications/USA, 1986)

33. Cannon's letter to Norton was written on Feb 8, 1933. His lecture on biocracy was published in *Technology Review,* 35, no. 6, Mar 1933.

34. For the article from the *Nottingham Evening Post,* Oct 12, 1933, see p. 17 of Cannon's Scrapbook. For a discussion of Cannon and social homeostasis, see Stephen J. Cross and William R. Albury, "Walter B. Cannon, L.J. Henderson, and the organic analogy," *Osiris* 2nd ser. 3(1987):165–192. See further Stuart Chase, *Technocracy,* N.Y., John Day, 1933; Jean Meynaud, *Technocracy,* N.Y., Free Press, 1969.

14. Chemical Mediation of Nerve Impulses: I. Searching for the Mysterious X-factor

1. For modern views, see, e.g., J.C. Eccles, "The development of ideas on the synapse," in *The Historical Development of Physiological Thought,* ed. C.M. Brooks and P.F. Cranefield (N.Y.: Hafner, 1959), pp. 39–66; O. Schmiedeberg and R. Koppe, "Pharmacology of the autonomic nervous system," in *Readings in Pharmacology,* ed. B. Holmstedt and G. Liljestrand (N.Y.: Macmillan, 1963), pp. 169–201; S.E. Mayer, "Neurohumoral transmission and the autonomic nervous system," in Goodman and Gilman's *The Pharmacological Basis of Therapeutics,* 6th ed. (N.Y.: Macmillan, 1980), pp. 56–90.

2. C.S. Sherrington, "The central nervous system," in Foster's *Textbook of Physiology,* 7th ed., London: Macmillan, 1897. Another pioneer was DuBois-Reymond, who suggested in 1877 that transmission might be produced either electrically or chemically.

3. For T.R. Elliott, see "On the action of adrenalin," *J. Physiol.* 31(1904):xx; also "The action of adrenalin," *J. Physiol.* 32(1905):401–467. For J.N. Langley, see "On the reaction of cells and nerve endings to certain poisons (nerve endings and receptive substance)," *J. Physiol.* 33(1905):374–413. Actually, Elliott enlarged the scope of earlier observations made independently by M. Lewandowsky (1898) and Langley (1901).

4. For W.E. Dixon, see "On the mode of action of drugs," *Med. Mag.* 16(1907):454–457; also Dixon and Hamill, "The mode of action of specific substances with special reference to secretin," *J. Physiol.* 38(1909):314–336.

5. For Reid Hunt, see "On the effects of intravenous injections of minimal doses of epinephrin sulphate upon the arterial blood pressure," *Am. J.*

Physiol. 5(1901):vi-viii; also R. Hunt and R. de M. Taveau, "On the physiological action of certain choline derivatives and new methods for detecting choline," *Brit. Med. J.* 2(1906):1788–1791. For W.H. Howell, see "Vagus inhibition of the heart in its relation to the inorganic salts of the blood," *Am.J. Physiol.* 15(1906):280–301.

6. Dale and Ewins, "Choline-esters and muscarine," *J. Physiol.* 48(1914):xxiv. See further H.H. Dale, *Adventures in Physiology, with Excursions into Autopharmacology* (Pergamon Press, 1953), esp. pp. x-xv; Wilhelm Feldberg, "Henry Hallett Dale, 1857–1968," *Biog. Mem. Fellows Roy. Soc.* 16(1970):126; W.M. Bynum's entry for Dale, *Dict. Sci. Biog.* XV(Supplement), pp. 104–107.

7. Dale, "The action of certain esters and ethers of choline, and their relation to muscarine," *J. Pharm. and Exp. Therap.* 6(1914):147–190; see esp. pp. 183 and 188.

8. Cannon's Kober Lecture, "The story of the development of our ideas of chemical mediation of nerve impulses," was published in *Am. J. Med. Sci.* 188(1934):145–159; see esp. p.148. About the same time, Dale also published a review; see his Linacre Lecture, "Chemical transmission of the effects of nerve impulses," *Brit. Med. J.*, May 12, 1934, pp. 835–831; reprinted in *Adventures in Physiology* (n.6), pp. 530–545.

9. O. Loewi, "Über humorale Übertragbarkeit der Herznervenwirkung," *Pflügers Arch. ges Physiol.* 189(1921): 239–242, and 193(1922):201–213.

10. Cannon's Kober Lecture (n. 8), pp.149–150. For further biographical and bibliographical information, see Dale's obituary notice of Loewi in *Biog. Mem. Fellows Roy. Soc.* 8(1962):67–89, esp. pp. 76–80; also G.L. Geison's entry for Loewi in *Dict. Sci. Biog.* VIII, pp. 451–457.

11. For the earlier account, see Chapter Five above, pp. 96–99. For Stewart and Rogoff's claims, see "The relation of the epinephrin output of the adrenals to changes in the rate of the denervated heart," *Am. J. Physiol.* 52(1920):304–363; also "Essentials in measuring epinephrin output with further observations on its relation to the rate of the denervated heart,"on pp. 521–561.

12. See Cannon and Rapport, "Studies on the conditions of activity in endocrine glands. VI. Further observations on the denervated heart in relation to adrenal secretion," *Am. J. Physiol.* 58(1921):308–337, esp. p. 323.

13. Cannon and Uridil, "Studies on the conditions of activity in endocrine glands. VIII. Some effects on the denervated heart on stimulating the nerves of the liver," *Am. J. Physiol.* 58(1921): 353–364.

14. Cannon's address to the neurologists was published in *J. Nerv. Ment. Dis.* 54(1921):421–427; see esp. p. 424. See also Cannon and Griffith, "Studies on the conditions of activity in endocrine glands. X. The cardio-accelerator substance produced by hepatic stimulation," *Am. J. Physiol.* 60(1922):544–559, esp. pp. 557–559.

15. Cannon's address to the psychiatrists was published as "New evidence for sympathetic control of some internal secretions," *Am. J. Psychiat.* 2(1922):15–30; see esp. 15–17. For an account of Cannon's first use of insulin in his experiments, see Chapter Eight above, pp. 148–149.

16. Cannon, Lewis and Britton, "Studies on the conditions of activity in endocrine glands. XVII. A lasting preparation of the denervated heart for

detecting internal secretion, with evidence for accessory accelerator fibers from the thoracic sympathetic chain," *Am. J. Physiol.* 77(1926):326–352; Cannon's Kober Lecture (n.8), pp. 151–152. For more on Cannon's work with Lewis and Britton, see Chapter Eight above, pp. 156–158.

17. Newton, Zwemer and Cannon, "Studies on the conditions of activity in endocrine glands. XXV. The mystery of emotional acceleration of the denervated heart after exclusion of known humoral accelerators," *Am. J. Physiol.* 96(1931):377–391, esp. pp. 388–390.

18. Cannon's Kober Lecture (n.8), pp. 152–153. Based on their research at Harvard, Zwemer and Newton published "Studies on the conditions of activity in endocrine glands. XXIV. Asphyxial stimulation of the denervated adrenal gland," *Am. J. Physiol.* 85(1928):507–511.

19. See further Z.M. Bacq, "Walter B. Cannon's contribution to the theory of chemical mediation of the nerve impulse," in *The Life and Contributions of Walter B. Cannon*, pp. 68–83; for his arrival at Harvard, see esp. pp. 68–69. See also Bacq's monograph, *Chemical Transmission of Nerve Impulses* (Pergamon Press, lst English ed., 1975), esp. pp. 36–37. Bacq wrote other reviews of chemical transmission that included his reminiscences; see, e.g., "Early work on the adrenergic mediator: How to go wrong," *TIPS*, Feb 1980, pp. 141–143; "Chemical transmission of nerve impulses," in *Discoveries in Pharmacology, Vol. I, Psycho- and Neuro-pharmacology* (Elsevier, 1983), pp. 49–103.

20. R.M. Moore, after graduating from Washington U. Medical School, spent two years at Harvard between his internship and residency at Barnes Hospital; based on his research at Harvard, he published "The effects of certain foods upon the rate of the denervated heart of the surviving unanesthetized cat," *Am. J. Physiol.* 89(1929):515–541.

21. Cannon to Bard, 19 Oct 1928. Cannon described his experiments with Moore in Cannon and Bacq, "Studies on the conditions of activity in endocrine glands. XXVI. A hormone produced by sympathetic action on smooth muscle," *Am. J. Physiol.* 96(1931):392–412; see esp. pp. 393–394.

22. For the 1929 experiments, see Cannon and Bacq (n.21), pp. 394–396.

23. Bacq, "Walter B. Cannon's contribution" (n.19), p. 69; *Chemical Transmission of Nerve Impulses* (n.19), p. 37.

24. Cannon and Bacq (n.21), pp. 396–397.

25. Ibid., pp. 399–405, 406, 407.

26. Cannon to R.L. Zwemer, 8 Nov 1930.

27. Bacq, "Walter B. Cannon's contribution" (n.19), p. 69.

28. Newton, Zwemer and Cannon (n.17), pp. 378–379.

29. Ibid., pp. 380, 381–386, 391.

30. Cannon and Bacq (n.21), pp. 407–408 and 410.

31. Ibid.; for the references to Loewi and Elliott, see pp. 407–409 and 411.

32. For Cannon's opinion of Bacq, see his letter to S. Burt Wolbach, 9 Feb 1931; also Cannon to F. Bremer, 21 Oct 1933. For Bacq on Cannon, see "Walter B. Cannon's contribution" (n.19), pp. 69–71.

33. Cannon, "Recent studies on chemical mediation of nerve impulses," *Endocrinology* 15(1931):473–480; see esp. p. 479.

34. Parker discusses the work of Cannon and Bacq on pp. 54–56.

35. Bacq, *Chemical Transmission of Nerve Impulses* (n.19), p. 12. Dale paid tribute to Elliott by dedicating his *Adventures in Physiology* (n.6) to his old friend; see also "Thomas Renton Elliott, 1877–1961" in *Biog. Mem. Fellows Roy. Soc.* 7(1961):53–74, esp. pp. 60–66.

15. No End of Emergencies

1. For the salary raise, see D.L. Edsall to Cannon, 25 Oct 1930, and Worth Hale to Cannon, 3 Nov 1930; for the Rockefeller grant, see Chapter Twelve above, pp. 238–242.

2. See further *Snatched from Oblivion,* p. 171; for the removal of the house, see also Cannon to A.L. Lowell, 10 Dec 1929; in fd. 574, 1928–30 series, A.L. Lowell Papers. W.W. Norton, writing to Cannon about publication of *The Wisdom of the Body* on 12 Feb 1932, referred to the "general desperate state of the book trade."

3. Cannon to Edsall, 18 Feb 1933, in reply to Edsall's letter of Feb 15. For the impact of the Depression, see also Edsall's form letters to department heads, 20 May, 20 and 23 Jun, 6 Sep 1932; J.C. Aub and Ruth K. Hapgood, *Pioneer in Modern Medicine: David Linn Edsall of Harvard* (Harvard Medical Alumni Association, 1970), pp. 320–322.

4. See correspondence in fds. 2328 and 2329 of the Cannon Papers; mimeo report of Mr. Kreuger's Speech in Chicago Office, Dec 1931. Among the biographies of Kreuger are William H. Stoneman, *The Life and Death of Ivar Kreuger,* Bobbs-Merrill, 1932, and Allen Churchill, *The Incredible Ivar Kreuger,* New York: Rinehart, 1967.

5. F.W. Sever to Cannon, 7 Jul 1932, in reply to Cannon's letter of May 9. There is in Jerome Greene's Papers a file of correspondence concerning the aftermath of the Kreuger fiasco. Greene, who had been a partner at Lee, Higginson since 1918, wrote F.P. Keppel on 28 May 1932 about the investment firm's relations with Kreuger and ended, "Some inscrutable fate had us picked out as his victims, a firm which possesses not only one of the oldest and most honorable records in American business but also one, as it happens, which pioneered in methods of thorough presentation of essential facts to investors through the present type of bond prospectus."

6. *Snatched from Oblivion,* p. 173; see also 1932–33 letter-book, Mrs. Cannon's family letters of 23 Jun, 21 and 25 Oct, and 6 Nov 1932 (in Cannon Family Papers).

7. See further *Walter Bradford Cannon: Exercises Celebrating Twenty-Five Years as George Higginson Professor of Physiology, October 15, 1931* (Cambridge, Mass.: Harvard Univ. Press, 1932), with Alvarez's remarks on pp. 16–22. For plans for the celebration, see correspondence in fds. 60 and 61 in the Cannon Papers; for letters collected by Alvarez, see fd. 1518.

8. Ibid., p. 65; Cannon's remarks are on pp. 59–65. At the end of the volume is a record of Cannon's war service, affiliations and honors, as well as a bibliography of his publications up to that time. A report of the celebration

was published in *Harv. Alum. Bull.*, Oct 23, 1931, and in the *New Engl. J. Med.*, Oct 15, 1931, pp. 779–781.

9. This account is based in large part on correspondence and related materials in Box 8 of the Aub Papers, "W.B. Cannon Case," as well as on Mrs. Cannon's letter-books in the Cannon Family Papers.

10. E.L. Oliver to J.C. Aub, 7 Oct 1931; in Aub Papers. Cannon, in a letter to R.M. Yerkes of 4 Mar 1932, told of Aub going to see Edsall; see also Cannon to W.C. Alvarez, 24 Mar 1932. He wrote Z.M. Bacq (17 May 1932) and others that although he had been looking forward to seeing his European colleagues at the Physiological Congress in Rome, he had decided to rest instead because his illness had been attributed to extreme fatigue.

11. Wolbach sent the biopsy report to Aub with a letter of 13 Jun 1932; see also Aub to Lane, 29 Jun 1932; Lane to Aub, 4 Aug 1932; Aub to Lane, 2 Sep 1932 (all in Aub Papers). For a retrospective report, see J.C. Aub, S.B. Wolbach, B.J. Kennedy and O.T. Bailey, "Mycosis fungoides followed for fourteen years: The case of Dr. W.B. Cannon," *Arch. Pathol.* 60 (Nov 1955):535–547. During the summer in Franklin, Cannon, as a distraction, took up modeling in clay and sculpted a bust of his daughter Helen, which later won a blue ribbon in a Boston physicians' art contest. (*The Way of an Investigator*, pp. 202–203)

12. For Mrs. Cannon's observations, see her 1932–33 letter-book, esp. 3 Oct, 13 Nov, and 17 Dec 1932; in Cannon Family Papers.

13. Cannon's letter to Aub was written on 15 Dec 1932, and he wrote a similar letter to Dr. Lane the same day. Rowe wrote Cannon on 30 Dec 1932 that he had transmitted his report to Dr. Aub. See also report of Rowe's meeting and letter from Aub to Rowe, 5 Jan 1933. All in Aub Papers.

14. See 1932–33 letter-book, 16 Mar, 19 Apr, and 22 May 1933. For a detailed examination of Cannon's early use of X-rays, see A.C. Barger's Sosman Lecture, "New technology for a new century: Walter B. Cannon and the invisible rays," *Am. J. Radiol.* 136(Jan 1981):187–195. Many pioneers in clinical radiology suffered injury and even death as a result of radiation damage; see further Percy Brown, *American Martyrs to Science Through the Roentgen Rays*, Springfield, Ill.: C.C. Thomas, 1936.

15. Ibid., 28 Sep 1933; see also letters of May 28, Jun 8, Jul 2, and Sep 24. Cannon's note of thanks was written to Charles D. Branch, 27 May 1933.

16. Ibid., Mrs. Cannon's letter to Bradford, marked "Private," is undated; it has been placed between Cannon's letters to Bradford of Sep 30 and Oct 10. Cannon also wrote to Bradford on Oct 18 and 30; the quotations are from his letters of Oct 10 and 30.

17. "Mycosis fungoides followed for fourteen years" (n.11); for Cannon's treatment in the fall of 1933, see esp. p. 536.

18. Cannon to Alvarez, 23 Nov 1933, in reply to Alvarez's letter of Nov 17. Writing to Robert Yerkes on 16 Oct 1933, Cannon broke the news that "Tom" Perry's wife, Rachel, was seriously ill with acute leukemia. In a postscript, Cannon confided that he and Cornelia had talked over their own situation and decided that "quality of life is more important than quantity." A few weeks later (27 Nov 1933), Cannon wrote Yerkes about his previous

attempt to resign from the Sex Committee: "At that time, I was in the midst of some perturbation because of what I had been told and also because I had not had time for any readjustment . . . I have had a considerable degree of satisfaction from work in our Committee. If you have not taken steps, therefore, to make new arrangements, I might be willing to reconsider."

19. For the epilogue to *The Wisdom of the Body* and Cannon's lecture on biocracy, see Chapter Thirteen above, pp. 261–264.

20. Carlson's letter to the editor appeared in *Science,* Mar 25, 1932, p. 338. Cannon wrote Hill on 10 May 1932, with Hill's reply on May 19. This account draws on Saul Benison's previously published Garrison Lecture, "Walter B. Cannon and the politics of medical science, 1920–1940," *Bull. Hist. Med.* 65(1991):234–251; see esp. p. 237.

21. Boas' petition to the National Academy was presented at a Council meeting of 23 Apr 1933; in fd. 1012 of the Cannon Papers. See also exchange of letters between Boas and Cannon, 24 and 26 May 1933. Around this time, the directors and faculty of the New School for Social Research in New York City, guided by its chairman, Alvin Johnson, began to set up a University in Exile for the displaced German scholars, as a protest against violations of academic liberty.

22. Exchange of letters between Harrow and Cannon, 31 May and 2 Jun 1933. Cannon's letter to Loewi was written on 21 Jun 1933.

23. Benison's Garrison Lecture (n.20), pp. 239–240. See further Marcia Graham Synott, *The Half-Opened Door: Discrimination at Harvard, Yale and Princeton,* Greenwood Press, 1979; also Stephen Steinberg, *The Academic Melting Pot: Catholics and Jews in American Higher Education* (McGraw-Hill, 1974), esp. Chapter 1, "The 'Jewish problem' in higher education."

Hans Zinsser, who was interested in getting contributions from non-Jewish academics for the relief of German scientists, asked Edsall (who had expressed sympathy for the ousted scholars) whether it might be possible to enlist the support of Pres. Lowell (15 May 1933). Edsall wrote to Lowell on May 17, but some weeks later reported that Lowell had not replied to his letter (16 Jun 1933). All in Dean's Files, Harvard Medical Archives. It should be noted that Lowell, on his way out of the Harvard presidency at this time, was in process of turning over the reins to J.B. Conant.

24. For the Committee in Aid of Displaced Foreign Physicians, see exchange of letters between Sachs and Cannon, 29 and 30 Oct 1933. This committee had taken over the purely medical applications from the Emergency Committee in Aid of Displaced German Scholars, which was headed by Livingston Farrand and Stephen Duggan.

25. For the Academic Assistance Council, see Dale to Cannon, 14 Oct 1933, and exchange of letters between Cannon and Hill, 3 and 15 Nov 1933.

26. George Baehr wrote to Cannon on 29 Nov, 22 Dec and 27 Dec 1933, with Cannon replying on Dec 1 and 28.

27. For Ernst Fischer, see Benison's Garrison Lecture (n.20), pp. 240–241; also Fischer to the Emergency Committee in Aid of Displaced German Scholars, 9 Jan 1934, Cannon to Baehr, 31 Jan 1934, and Hill to Cannon, 19 Feb 1934.

28. See further Cutler to Cannon, 9 and 12 Apr 1934; Cannon to Cutler and Cannon to Edsall, 12 Apr 1934; Edsall to Cannon, 13 Apr 1934.

29. Hill's Huxley Lecture was delivered at Birmingham on 16 Nov 1933 and published under title of "International status and obligations of science" in *Nature,* Dec 23, 1933; reprinted in A.V. Hill, *The Ethical Dilemma of Science and Other Writings* (Rockefeller Institute Press/Oxford University Press, 1960), pp. 205–221.

30. For Stark's letter to the editor of *Nature,* Feb 24, 1934, together with Hill's reply, along with the two further letters in *Nature,* Apr 21, 1934, see Hill's *Ethical Dilemma of Science* (n.29), pp. 222–226. For more on Johannes Stark and the Deutsche-Physik movement, see Mark Walker, *Nazi Science: Myth, Truth and the German Atomic Bomb* (N.Y. and London: Plenum Press, 1995), esp. Chapters 2 and 3.

16. Chemical Mediation of Nerve Impulses: II. Sympathin E and Sympathin I

1. *Libro Homenaje Arturo Rosenblueth,* ed. Juan Garcia Ramos (Mexico,D.F.: Pressa Medica Mexicana, 1971), contains a biography of Rosenblueth in English on pp. 22–37, as well as a bibliography of his publications on pp. 39–48. See also Louisa Barclay Benton, "Arturo Rosenblueth: Success or Failure?" undergraduate thesis presented to the Committee on History and Science, B.A. degree, Harvard University, 1986, esp. pp. 9–21.

2. Benton discusses Rosenblueth's Guggenheim fellowship and his arrival at Harvard on pp. 21–27; see esp. p. 27.

3. Ibid. For Cannon's experiments with Bacq, see Chapter Fourteen above, pp. 278–282.

4. A. Rosenblueth and T. Schlossberg, "The sensitization of vascular response to 'sympathin' by cocaine and the quantitation of 'sympathin' in terms of adrenalin," *Am. J. Physiol.* 97(1931):365–374. Schlossberg later published, with Margaret McKay Sawyer and others from the Harvard Laboratory of Physiology, a three-part article in the Apr 1933 *A.J.P.,* 104:172–194, under the heading "Studies of homeostasis in normal, sympathectomized and ergotaminized animals": I. The effect of high and low temperatures; II. The effect of anoxemia; III. The effect of insulin."

5. Cannon, "Recent studies on chemical mediation of nerve impulses," *Endocrinology* 15(1931):473–480; see esp. pp. 475 and 476.

6. Rosenblueth and Cannon, "Studies on the conditions of activity in endocrine glands. XXVII. Some effects of sympathin on the nictitating membrane," *Am. J. Physiol.* 99(1932)398–407.

7. Freeman, Phillips and Cannon, "An unsuccessful attempt to demonstrate humoral action of 'vagus substance' in the circulating blood, *Am. J. Physiol.* 98(1931):435–440. Norman Freeman, the grandson of W.W. Keen and brother of Walter, was interested in the physiological aspects of surgery. After serving as a surgical resident at the M.G.H., he went on to the University of

Pennsylvania, where his research interests included surgical shock and hypertension.

8. For bibliographic details, see the list of publications on p. 40 of *Libro Homenaje* (n.1). All of those papers were published in the *A.J.P.* For the Guggenheim renewal, see Benton (n.1), pp. 29–32. The selection committee wanted Rosenblueth to assure them that he would return to Mexico after his fellowship, but he refused to make any promises. The issue was resolved when two influential people rallied to Rosenblueth's support: Henry Allen Moe, the secretary of the Foundation, and Hans Zinsser, of the department of bacteriology and immunology at H.M.S.

9. See further Horace W. Davenport, "Signs of Anxiety, Rage, or Distress," *The Physiologist* 24(Oct/Nov 1981):1–5. Davenport has left a memorable portrait of Rosenblueth ("Cannon's disciple, collaborator and dark angel") in this article about Cannon's study of the sympathetic nervous system; see esp. pp. 3–4.

10. Interview of Bradford Cannon by Saul Benison and A.C. Barger, summer of 1987. Arturo Rosenblueth and Bradford Cannon published an account of their experiments, "Some circulatory phenomena disclosed by ergotoxine," in *Am. J. Physiol.* 105(1933):373–382.

11. For "The mode of action . . .," see *Am. J. Physiol.* 101(1932):149–165; for "The chemical mediation . . . , see *Am. J. Physiol.* 102(1932):12–38. In the first of these papers, Rosenblueth expressed his thanks to Dr. M. Vallarta of M.I.T. for help in the mathematical aspect of the problem and to Dr. H. Davis for his suggestions and criticisms.

12. For Rosenblueth, Leese and Lambert, see "Electrical potentials in smooth muscle," *Am. J. Physiol.* 103(1933):659–680, in which the authors give thanks to Dr. Alexander Forbes for his valuable advice and criticism.

13. For Rosenblueth and Rioch, see "The nature of the responses of smooth muscle to adrenin and the augmentor action of cocaine for sympathetic stimuli," *Am. J. Physiol.* 103(1933):681–685. Rosenblueth and Rioch published two other articles together in 1933, one on electrical excitation of multifibered nerves, the other on temporal and spatial summation in autonomic systems.

14. See Dale's 1934 Linacre Lecture, "Chemical transmission of the effects of nerve impulses," reprinted in *Adventures in Physiology,* (London, Pergamon Press, 1953), esp. p. 534. For a review of Loewi's work in his own words, see, e.g., "The Ferrier Lecture on problems connected with the principle of humoral transmission of nervous impulses," *Proc. Roy. Soc. London B.* 118(1935):299–316. For a bibliography of Loewi's publications, see "Otto Loewi, 1873–1961," *Biog. Mem. Fellows Roy. Soc.,* pp. 84–89.

15. See Dale's Croonian Lecture, "Some factors in the control of the circulation," *Lancet* 1(1929):1179, 1233, and 1285; Dale and Dudley, "The presence of histamine and acetylcholine in the spleen of the ox and the horse," *J. Physiol.* 68(1929)97–123. For Dale at the N.I.M.R., see also Gary R. Fick, "Henry Dale's involvement in the verification and acceptance of the theory of neurochemical transmission: A lady in hiding," *J. Hist. Med.* 42(1987):467–485, esp. p. 471. Fick discusses the two 1929 papers on pp. 472–474.

16. See Dale and Gaddum, "Reactions of denervated voluntary muscle, and their bearing on the mode of action of parasympathetic and related nerves," *J. Physiol.* 70(1930):109–144. See also Dale's Linacre Lecture (n.14), pp. 536–537; Fick (n.15), pp. 474–475.

17. Cannon and Rosenblueth, "Studies on the conditions of activity in endocrine glands. XXIX. Sympathin E and Sympathin I," *Am. J. Physiol.* 104(1933):557–574; see esp. pp. 557–561.

18. Cannon to Rosenblueth, 18 Nov 1932. Correspondence between Cannon and Rosenblueth may be found in fds. 1625–1626 of the Cannon Papers.

19. "Sympathin E and Sympathin I" (n.17), pp. 561–565.

20. Ibid., pp. 565–568; for Langley, see esp. p. 568. J.N. Langley summarized his previous work in *The Autonomic Nervous System*, Cambridge, England: Heffers, 1921. Years later, Zenon Bacq wrote about the two sympathins, "Write α instead of E and β instead of I and you have the modern language" ("Chemical transmission of nerve impulses," in *Discoveries in Pharmacology, Vol. I, Psycho- and Neuro-pharmacology*, Elsevier, 1983, p. 78).

21. Cannon, "Sympathin E and Sympathin I," *Trans. Assoc. Amer. Phys.* 48(1933):236–238; "Chemical mediators of nerve impulses," *Science*, Jul 21, 1933, pp. 43–48. Cannon's address in Cincinnati was delivered on April 12, while the meeting of the Association of American Physicians was held in Washington, D.C., May 9–11. According to a press release from Science Service, Cannon also announced his discovery to a group at the National Academy of Sciences on Apr 20, 1933.

22. Regarding the reception accorded to the sympathin E and I paper, Bacq wrote (*Chemical Transmission of Nerve Impulses*, Pergamon Press, lst English ed., 1975, p. 42): "No physiologist competent in his field could accept this theory." And again, in his review article in *Discoveries in Pharmacology, Vol. I* (n.20), p. 79, he noted that "many U.S. pharmacologists and physiologists who disliked the concept of chemical transmission were happy to voice severe criticism."

23. Bacq to Cannon, 7 Dec 1932, Cannon to Bacq, 12 Dec 1932, Bacq to Cannon, 28 Dec 1932; see further correspondence between Bacq and Cannon in fds. 748–752 of the Cannon Papers. See also Bacq's article, "Inhibition des mouvements des l'intestin par la sympathine," in *Compt. Rend. Soc. Biol.* 112(1933):211–212, in which he acknowledged aid from the Ella Sachs Plotz Foundation.

24. Bacq to Cannon, 3 Feb 1933. Bacq had applied for a C.R.B. advanced fellowship to work at Woods Hole during the months of July and August, and proposed spending September and October in Boston in order to discuss the sympathin matter. (Bacq to Cannon, 5 Jan 1933)

25. For Bacq's recollection, see *Chemical Transmission of Nerve Impulses* (n.22), p. 43. See also G. Barger and H.H. Dale, "Chemical structure and sympathomimetic action of amines," *J. Physiol.* 41(1910):19–59.

26. Davenport, "Signs of Anxiety, Rage, or Distress" (n.9), p. 4. We are grateful to Dr. Davenport for calling this important point to our attention.

27. Cannon to Bacq, 17 Apr 1934; Bacq to Cannon, 28 May 1934. See also Bacq and Rosenblueth, "The action of calcium and potassium ions on the

nictitating membrane, the adrenal medulla and the non-pregnant uterus of the cat," *Am. J. Physiol.* 108(1934):46–49.

28. Cannon acknowledged receipt of Bacq's review on 8 Jun 1934; his lengthy letter was written on June 20.

29. Bacq to Cannon, 18 Dec 1934; Cannon to Bacq, 18 Dec 1935. Earlier letters are Bacq to Cannon, 7 Aug 1934; Cannon to Bacq, 3 Oct 1934; Bacq to Cannon, 16 Oct 1934. Cannon ended his criticism of Bacq, Brouha and Heymans (5 Nov 1934): "I am sure you will take the remarks in this letter in good spirit. The worst effect of the influence of an investigator who associates with younger men is to have the younger men replicas of himself. The most important thing that he can do is to develop in his collaborators an independent spirit."

30. Dale to Bacq, 9 Feb 1934; in Dale Papers, Royal Society. For Cannon and Rosenblueth's comparative study, see *Am. J. Physiol.* 112(1935):268–276. Two other studies by Rosenblueth and Cannon were published in the *A.J.P.* in 1934: "Direct electrical stimulation of denervated autonomic effectors" and "A further study of vasodilators in sympathectomized animals."

31. Bacq, *Chemical Transmission of Nerve Impulses* (n.22), p. 23. See also Fick (n.15), pp. 475–476; Wilhelm Feldberg, "The early history of synaptic and neuromuscular transmission by acetylcholine: Reminiscences of an eyewitness," in *The Pursuit of Nature*, ed. A.L. Hodgkin et al. (N.Y.: Cambridge Univ. Press, 1977), pp. 65–83.

32. For Dale and Feldberg, see "The chemical transmitter of vagus effects to the stomach," *J. Physiol.* 80(1934):16P-17P, and "The chemical transmission of secretory impulses to the sweat glands of the cat," *J. Physiol.*, 82(1934):121–128. See also Feldberg, Minz and Tsudzimura, "The mechanism of the nervous discharge of adrenaline," *J. Physiol.*, 80(1934):15P-16P; Feldberg and Gaddum, "The chemical transmitter at synapses in a sympathetic ganglion," *J. Physiol.*, 80(1934):12P-13P.

33. H.H. Dale, "Nomenclature of fibres in the autonomic system and their effects," *J. Physiol.* 80(1933):10–11.

34. For Dale and Feldberg's communication to the Physiological Society, see "Chemical transmission at motor nerve endings in voluntary muscle?" *J. Physiol.* 81(1934):39P-40P. Other members of the Hampstead group included Martha Vogt, who worked with Feldberg and Dale to extend their research on neuromuscular transmission, and G. Lindor Brown, who conducted electrophysiological experiments which supported the neurochemical position. Dale's William Welch Lectures in New York, published in *J. Mt. Sinai Hosp.* 4(1938):401–429, are among several instances where he presented the experimental discoveries of his research group and explained their significance.

35. Fick (n.15), pp. 477–482. For the 1936 award, see G. Liljestrand's discussion in *Nobel, the Man and His Prizes* (2nd. and enl. ed., Elsevier, 1962), pp. 314–317.

36. Dale to Loewi, 12 Mar 1936 (in Dale Papers); see also Dale's Linacre Lecture (n.14), p. 536.

37. Cannon to Bradford Cannon, 28 Oct 1935; see also Bacq, *Chemical Transmission of Nerve Impulses* (n.22), p. 45.

38. For Greer, Pinkston, et al., see "Nor-epinephrine . . . as a possible mediator in the sympathetic division of the autonomic nervous system," *J. Pharm. & Exp. Therap.* 62(1938):189–227. See also R.L. Stehle and H.C.Ellsworth, "Dihydroxyphenyl ethanolamine (arterenol) as a possible sympathetic hormone," *Ibid.*, 59(1937):114–121; K.L. Melville, "The antisympathomimetic action of dioxane compounds (F883 and F933), with special reference to the vascular responses to dihydroxyphenyl ethanolamine (arterenol) and nerve stimulation," *Ibid.*, 59:(1937):317–327.

39. Ibid.; for their discussion of the two sympathins, see esp. pp. 191–194 and 223–225.

40. Cannon to Bacq, 27 Apr 1937. Pinkston was at Harvard from 1930 to 1935, at Vanderbilt from 1935 to 1937, then at the American University at Beirut, where he replaced Otto Krayer as professor of pharmacology until the war forced his return to America.

41. Preface to Cannon and Rosenblueth, *Autonomic Neuro-Effector Systems* (Macmillan, 1937), p. viii. Years later, Bacq criticized Rosenblueth for publishing after Cannon's death (in *The Transmission of Nerve Impulses at Neuroeffector Junctions and Peripheral Synapses*, New York, John Wiley & Sons, 1950) a defense of the two sympathins hypothesis despite the accumulation of evidence by von Euler in favor of noradrenaline; see further Bacq, "Early work on the adrenergic mediator: How to go wrong," *TIPS*, Feb 1980, pp. 141–143.

42. For Blashko and Smith's comment, see Volume 6 (The Adrenal Gland), introduction to Section 7 (Endocrinology), p. viii. For von Euler's comment, see p. 4 of his volume on noradrenaline; also Henry Dale's statement about sympathin on p. xii of the Introduction. In 1970, U.S. von Euler shared a Nobel Prize with Julius Axelrod and Bernard Katz "for their discoveries of the role played by certain chemicals in the transmission of nerve impulses."

43. Dale to Cannon, 15 Nov 1936. See further Cannon's notice, "The Nobel Prize in physiology and medicine," *Sci. Monthly*, Feb 1937, pp. 195–198.

17. Around the World in 1935

1. 1934 letter-book, Mrs. Cannon to her family, 7 Oct; see also letters of Oct 16 and 24 (in Cannon Family Papers). For the invitation from P.U.M.C., see Roger S. Greene to Cannon, 11 Sep 1934, and Cannon's reply of Sep 13; Cannon to D.L. Edsall, 3 Oct 1934. Cannon referred to Pavlov's invitation of Sep 7 when he accepted on 3 Oct 1934.

2. Ibid., 2 Jun 1934.

3. Ibid., 4 Nov 1934; for reading books about Russia, see letter of Aug 5. Despite Mrs. Cannon's reference to *Escape from Soviet Russia*, efforts to identify the book have not met with success. The visit of Mme. Samoiloff, the widow of a Russian physiologist, is described in Mrs. Cannon's letter of Oct 7.

4. Karl Compton's address in New York of Dec 1934 is quoted in Peter J. Kuznick, *Beyond the Laboratory: Scientists as Political Activists in 1930s America* (Univ. of Chicago Press, 1987), p. 126; for other accolades for the Soviet experiment in the early 1930s, see also pp. 106–128.

5. Cannon wrote to B.P. Babkin on 14 Nov 1934 and to V. Boldyreff on Nov 22; Babkin replied on Nov 23, Boldyreff on Nov 27. Both Babkin, who was in the physiology department at McGill University, and Boldyreff, at the Pavlov Institute at Battle Creek, Michigan, had been exiled from Russia after the Revolution. Babkin went on to write a definitive biography of Pavlov, which was published in 1949.

6. Cannon to Horsley Gantt, 14 Nov 1934; Gantt replied on Dec 5.

7. Cannon to A.A. Horvath, 19 Nov 1934, with Horvath's reply on Nov 20. Cannon wrote S. Lefschetz on 24 Nov 1934, after reading his report of the congress in *Science* (Nov 23, 1934); Lefschetz replied on Nov 28, adding: "One of the more serious difficulties that Soviet scientists have to contend with is the impossibility for them to obtain permission to go abroad . . . There is grave danger that a serious gap is being created between Soviet and non-Soviet scientists which is increased by the difference in the languages and difficulties in obtaining foreign publications."

8. Exchange of letters between Cannon and Babkin, 11 and 14 Feb 1935. According to Mrs. Cannon's family letter of 13 Feb 1935, Mme. Samoiloff (n.3) also went over Cannon's speech and told him, "Every word will be scanned by the politicians, for with all their bravado they care a lot what the Americans think of them." Before leaving on his journey, Cannon had gone to St. Louis to have a graft done on his hand by the plastic surgery team of Vilray Blair and Barrett Brown at Barnes Hospital.

9. See further 1935 letter-book for Mrs. Cannon's family letters about the trip; in Cannon Family Papers. For the visit to Shanghai, see her letter of Apr 13; the voyage with stops in Hawaii and Japan is described in letters from Mar 16 to Apr 2.

10. The 1935 letter-book, in addition to family letters from Mrs. Cannon, and a few from Dr. Cannon, contains photographs and memorabilia, as well as Mrs. Cannon's impressions of "Birth control around the world" at the end of the volume.

11. Robert K.S. Lim to Cannon, 17 May 1934, about receiving A.C. Liu at Harvard, with Cannon's reply on Jun 7. See also other correspondence between Cannon and Lim, 1932–1935, in fd. 511 of the Cannon Papers. The arrival in Peiping is described in Mrs. Cannon's family letter of 18 Apr 1935.

12. Cannon's assessment of Liu was written on 7 Feb 1935. See further Roger Greene to Alan Gregg and to H.S. Houghton, both 7 Aug 1934; Greene to Gregg, 13 Aug 1934; all in Rockefeller Archives Center. Greene, on learning of Cannon's interest in going to China, talked with him about accepting an appointment at P.U.M.C. See also exchange of letters between Cannon and Lim about plans for his visit, 11 Oct and 16 Nov 1934.

13. 1935 letter-book, Mrs. Cannon's letters of Apr 18 and May 22; also Cannon's letter to Bradford of May 14. For more on the educational and scientific program of the P.U.M.C., see John Z. Bowers, *Western Medicine in a*

Chinese Palace, New York:Macy Foundation, 1972. Chapter 13 in Mrs. Cannon's unpublished manuscript, "Servant of Science," describes the trip to China and Japan; in Cannon Papers. See also *Snatched from Oblivion*, in which Marian Cannon Schlesinger devotes a chapter to China; for her parents' visit, see esp. pp. 205–207.

14. Ibid., 22 May 1935; "Servant of Science," p. 136. Cannon's undated "Notes from China" concern the dirt and smells and noise and backwardness that he observed in China.

15. The story of Cannon testing his audience's English is recounted by Mrs. Cannon in "Servant of Science," p. 135. Cannon's note to Bard is dated 16 May 1935.

16. For Dr. Lim's research group, see his letter to Cannon of 16 Nov 1934; also Cannon to H.C. Chang, 1 Nov 1935, and other letters in fds. 515 and 516 of the Cannon Papers. Cannon enclosed a list of his lectures and special demonstrations with a report of his visit to P.U.M.C. to Roger Greene, 1 Jun 1935. See also S.C. Shen and W.B. Cannon, "Sensitization of denervated pupillary sphincter to acetylcholine," *Chinese J. Physiol.* 10(1936):359–371.

17. Lim to Cannon, 8 May 1936. For Cannon's Chinese name, see H.C. Chang to Cannon, 2 May 1935. Mrs. Cannon described his Chinese costume in her family letter of 22 May 1935; in 1935 letter-book. See also Roger Greene to Alan Gregg, 3 Jun 1935, enclosing copy of statement from Cannon re his visit to Peiping; in Rockefeller Foundation file, Rockefeller Archives Center. In a growing dispute among the Rockefeller Foundation, China Medical Board, and the P.U.M.C., Roger Greene had become the central figure. His ouster from his post as the regional director was only a matter of time when Cannon visited Peiping. See further Mary Brown Bullock, *An American Transplant: The Rockefeller Foundation and the Peking Union Medical College* (Berkeley: Univ. of California Press, 1980), pp. 63–72.

18. Copy of letter from Yas Kuno to Dr. Lim, 14 Jan 1935, enclosed with letter to Cannon of Jan 19, in which Lim wrote: "He tells me that Prof. Sataké of Sendai has extended an invitation to him on behalf of the Imperial Universities of Sendai, Tokyo and Kyoto." See also Sataké's invitation to Cannon, 13 Feb 1935; Kuno to Cannon, 25 May 1935, with itinerary for trip.

19. For the travels through Mukden, Manchuria and Korea, see Mrs. Cannon's family letters of June 11 and 17; also "Servant of Science," p. 139. For Kuno's election to the International Committee, see Cannon to Kuno, 4 Nov 1936.

20. For the trip to Korea, see "Servant of Science," pp. 140–141. For Cannon's later remembrance of Satake, see *The Way of an Investigator*, pp. 183–184; for his visit to China and friendship with Lim, see also pp. 182–183.

21. For the arrival in Japan and the visit to Tokyo, see Mrs. Cannon's family letters of 21 and 24 Jun 1935. Cannon had written Dr. Goto on 29 Oct 1934, recalling their work together in Agassiz's laboratory in Newport nearly forty years before and hoping that he would be able to see his old friend in Japan. Press clippings and memorabilia of the visit to China and Japan may be found in Cannon's Ideal Scrapbook; in Cannon Papers.

22. The journey on the Trans-Siberian Express lasted from July 10 to the

18th, when they arrived at Sverdlovsk (earlier known as Ekaterinburg). For details, see Chapter 14 in "Servant of Science," esp. p. 148. There are several pages of press clippings and memorabilia in Cannon's Scrapbook about his visit to the Soviet Union.

23. Ibid., pp. 149–150.

24. Mrs. Cannon's arrest is described in her family letter of 20 Jul 1935; see also "Servant of Science," pp. 151–156, and *Snatched from Oblivion*, pp. 214–217.

25. Mrs. Cannon to her family, 25 Jul 1935.

26. For the visit to Moscow, see "Servant of Science," pp. 157–160; *Snatched from Oblivion*, pp. 217–218.

27. For the visit to Leningrad and the 15th International Congress, see Mrs. Cannon's family letters of 8 and 10 Aug 1935 and Chapter 15 in "Servant of Science"; also *Snatched from Oblivion*, pp. 218–221.

28. Cannon's address was published as a pamphlet by the State Publishing House for Biological and Medical Literature, Moscow-Leningrad, 1935; see esp. p. 4. It was later published in the Proceedings of the 15th International Congress, put out by the *Sechenov Journal of Physiology of the U.S.S.R.* in 1938. See also Saul Benison's Garrison Lecture, "Walter B. Cannon and the politics of science, 1920–1940," *Bull. Hist. Med.* 65(1991):234–251; esp. pp. 241–242.

29. Ibid., pp. 9–11. For a description of the Congress, see K.J. Franklin, "A Short History of the International Congresses of Physiologists," *Annals of Science*, Jul 15, 1938, pp. 314–320; for the opening plenary session, see esp. pp. 315–316. Franklin noted that from the idea of international cooperation between scientists, Cannon made "an easy transition to the mode of action of autonomic nerves on their effector organs, for at least a dozen nations had contributed to the more recent advances made in that particular field." Cannon later described the reception of his speech and the reaction of the German delegates in a letter of 21 Dec 1937 to Edward I. Hartshorne at Harvard's sociology department.

30. Cannon's note about Pavlov is quoted in "Life with a Scientist," p. 161.

31. For the visit with the Pavlovs, see Mrs. Cannon's unpublished manuscript, "The Friendship of Dr. Pavlov and Dr. Cannon," pp. 6–8; copy in Cannon Papers. She expressed the opinion that part of the respect and consideration shown to Pavlov might well have been due to Lenin's putative remark on his deathbed, "Take care of Pavlov."

32. Mrs. Cannon to her family, 16 Aug 1935. See further D.Y. Solandt, "International Physiological Congress: Meeting in the U.S.S.R.," *Nature*, Oct 12, 1935, pp. 571–572.

33. After Pavlov's death, Cannon sent a short statement of appreciation to *Soviet Russia Today* (12 Mar 1936), and a longer obituary notice for inclusion in the Apr 15, 1936 issue of the *Research Bulletin of the Soviet Union*.

34. For the meeting in Moscow, see Mrs. Cannon's family letter of 17 Aug 1935.

35. Ibid.; also "Servant of Science," p. 174. In *Snatched from Oblivion*, Marian comments (p. 223): "I do not think my mother was ever able to rid

herself of her Alice-through-the-Looking Glass perplexity in face of the contradictions she saw at every turn." Among Mrs. Cannon's unpublished manuscripts in the Cannon Family Papers is a bound typescript of 138 pages titled "The U.S.S.R. and an International Congress," which is dated [1965] and dedicated to the memory of Dr. Cannon. Also there are two versions of her experience in the Soviet Union and an "Open Letter to Stalin," which were written after she returned from the trip, either in late 1935 or early 1936.

36. Mrs. Cannon's family letters from 18 Aug to 14 Sep 1935 describe the last leg of the trip.

37. Exchange of letters between Cannon and Babkin, 2 and 8 Oct 1935.

38. L. Cherniansky to Cannon, 5 Dec 1935, with Cannon's reply on Dec 20. See also Kuznick (n.4), pp. 160–163; an account of the International Congress, as well as of Cannon's relations with Pavlov and other Russians, may be found on pp. 150–158.

39. Cannon's report card of two pages, headed "My Strongest Impressions of Soviet Russia," was written at the request of M.J. Olgin for *Pravda* (15 Oct 1935), and enclosed with Cannon's reply of Oct 18. A more detailed outline of Cannon's Russian experience (5-page typescript) is in fd. 473 of the Cannon Papers.

40. Herbert Goldfrank to Cannon, 17 and 25 Oct 1935, with Cannon's replies of Oct 19 and 26.

41. Postscript in Cannon's letter to Mary Dalton, executive secretary of the Friends of the Soviet Union, 6 Dec 1935. On Nov 27, she had written Cannon about a series of articles in the *Saturday Evening Post* "slandering the American teachers, students, and others who were in the Soviet Union this summer, as well as the Soviet Union itself."

18. The Spanish Medical Bureau

1. W.A. Neilson to Cannon, 28 Apr 1937, in reply to Cannon's letter of Apr 27. William O. Field, Jr., invited Cannon to the presidency of the American Russian Institute in a letter of 21 Apr 1937.

2. The Earl of Listowel had written as chairman of the Congress of Peace and Friendship of the U.S.S.R. on 29 May 1937; Cannon replied on Jun 16. His refusal of the American Russian Institute presidency was written on 29 Apr 1937. For Cannon's growing doubts about the Soviet Union in the late 1930s, see Peter J. Kuznick, *Beyond the Laboratory: Scientists as Political Activists in 1930s America* (Univ. of Chicago Press, 1987), pp. 163–165 and 167–168.

3. For the 1930 visit to Spain, see Chapter Thirteen above, pp. 256–257. See also Cannon's 1945 memoir, *The Way of an Investigator,* pp. 159–160.

4. Cannon to G. Marañon, 17 Apr 1931; Cannon to Juan Negrin, 19 May 1933.

5. See, e.g., F. Jay Taylor, *The United States and the Spanish Civil War,* New York, Bookman Associates, 1956; Dante Puzzo, *Spain and the Great Powers, 1936–1941,* Columbia Univ. Press, 1962; Gabriel Jackson, *The Spanish Republic and the Civil War, 1931–1939,* Princeton Univ. Press, 1965; Raymond Carr,

Spanish Tragedy: The Civil War Perspective, London: Weidenfeld and Nicolson, 1977; Hugh Thomas, *The Spanish Civil War,* rev. and enl. ed., New York: Harper and Row, 1977. See also David T. Cattell, *Soviet Diplomacy and the Spanish Civil War,* Berkeley: Univ. of California Press, 1957; Verle B. Johnston, *Legions of Babel: The International Brigades in the Spanish Civil War,* Pennsylvania State Univ. Press, 1968; R. Dan Richardson, *Comintern Army: The International Brigades and the Spanish Civil War,* Univ. Press of Kentucky, 1982.

6. Cannon briefly discussed the outbreak of civil war in Spain and his subsequent effort in behalf of the Spanish Republic in *The Way of an Investigator,* pp.160–162. See also Saul Benison's Garrison Lecture, "Walter B. Cannon and the politics of science, 1920–1940," *Bull. Hist. Med.* 65(1991):234–251, esp. pp. 243–250.

7. R. Carrasco-Formiguera to Cannon, 23 Feb 1937; Roger Chase to Cannon, telegram of 12 Jan 1937, and Cannon's refusal of Jan 13; Chase to Cannon, telegram of 30 Mar 1937, with note of Cannon's acceptance on the same day.

8. Cannon's address at the Hippodrome on Apr 18, 1937, was published as a 15-page pamphlet by the North American Committee to Aid Spanish Democracy; see further fd. 620 in Cannon Papers. For Cannon's views on the situation in Spain at this time, see, e.g., his letters to Perry C. Baird, 13 Apr 1937, and to Fred T. Murphy, 28 Apr 1937. After reading that Dr. Negrin had become premier, Cannon wrote him about his efforts in behalf of the Medical Bureau (18 May 1937).

9. For Picasso's "Guernica," see Robert Hughes, *Culture of Complaint: The Fraying of America* (Oxford Univ. Press, 1993), p. 185. See also Cannon to Carrasco, 4 May 1937.

10. Marian Craig Wentworth to Cannon, 13 May 1937. Other letters of support may be found in fds. 633–634, Cannon Papers.

11. See further Sarah Leah Elson, "Transcending Politics: Physicians' Humanitarian Involvement in Politics in the Medical Bureau to Aid Spanish Democracy and the International Physicians for the Prevention of Nuclear War," senior thesis presented to the Committee on History and Science, Harvard University, 1993. Elson's thesis undertakes a comparison and contrast of the two humanitarian aid organizations that operated in very different times before and after World War II.

12. The letter from Hart and Curtiss was published in the *Harvard Crimson* for Sep 30,1937, and Cannon's rebuttal appeared on the following day, Oct 1. The newspaper in Portland was the *Evening Express,* May 22, 1937.

13. Exchange between Caspar Weinberger and Cannon, 22 and 26 Nov 1937. See also Felix Frankfurter to Cannon, 6 Oct 1937; Wiley Mayne to Cannon, transmitting report of the student committee, 8 Nov 1937, and Cannon's acknowledgment of Nov 12. For a previous complaint about fundraising for the Harvard ambulance, see Ruth Ludlow Hagler to Cannon, 24 May 1937, and Cannon's reply of May 26.

14. See 1937 correspondence with the Speakers' Bureau in fd. 617 of the Cannon Papers. At one point, Cannon suggested Dr. Edward Barsky for the post of executive director of the Medical Bureau but Barsky preferred service in Spain and returned there at the end of 1937 with Leo Eloesser.

15. For the speaking tour in the midwest, see 1938 correspondence in fd. 628, Cannon Papers. For Jay Allen in Spain, see Gabriel Jackson (n.5), pp. 268–269. For Dr. Bethune, see *The Politics of Passion: Norman Bethune's Writing and Art*, ed. Larry Hannart, Univ. of Toronto Press, 1998; Ted Allen, *The Scalpel, The Sword: The Story of Dr. Norman Bethune*, rev. ed., Toronto, 1989.

16. Cannon to Jay Allen, 1 Nov 1938; Benison's Garrison Lecture (n.6), pp. 246–247.

17. These names were among those on the roster of national sponsors of the medical committee as listed on the letterhead of the North American Committee and Medical Bureau.

18. Cannon to Herman F. Reissig, 11 Apr 1938; see also Reissig to Cannon, 30 Mar 1938, and Cannon to Reissig, 1 Jun 1938. After William J. Crookston, the executive secretary of the Medical Bureau, died in Aug 1937, Frank T. Woodbury was appointed to succeed him. With the merger early in 1938, however, Rev. Reissig, who had filled a similar post in the North American Committee, became the executive secretary of the new Spanish aid organization.

19. Cannon to Carrasco, 3 Aug 1938. His earlier letter to Negrin was written on June 10. Several pages of press clippings and memorabilia about the Medical Bureau, especially relating to the relief ship as well as the great "Lift the Embargo" rally held in Madison Square Garden, may be found in Cannon's Ideal Scrapbook.

20. For materials about "A Nutrition Unit for Spain" and correspondence with T. Addis, H. Borsook, and F.M. Pottinger, see fd. 645 in the Cannon Papers.

21. Cannon's message is undated; see also message to Juan Negrin from the executive secretary and chairman of a special committee of the Medical Bureau, 8 Nov 1938; Cannon to Reissig, 25 Nov 1938. Cannon, writing to A.A. Heller, reviewed his stance on the reports of pellagra and need for nicotinic acid: "As I pointed out to the headquarters of the Medical Bureau, it is rather futile to cure the disease and not to provide the proper food for keeping people well." (9 Dec 1938)

22. Letter of acknowledgment to Cannon with multiple signatures, Dec 1938; Carrasco-Formiguera to Cannon, 11 Jan 1939.

23. For Cannon's complaint, see his letter to Reissig, 7 Sep 1938; for the Boston Medical Bureau, see, e.g., Cannon's letters to Dorothy Hickie, 20 Oct 1938, to A.S. Coolidge, 8 Nov 1938, and to Hickie of 22 Dec 1938.

24. Cannon to Crane Brinton, 20 Apr 1938. For his views on the Catholic Church's opposition, see, e.g., Cannon to Herman F. Reissig, 9 Feb 1937; Cannon to Mrs. John Sherman Hoyt, 28 May 1937.

25. Cannon to Negrin, 25 Sep 1937. For Sen. Burke's investigation, see Cannon to F.W. Cook, 2 Jul 1937; article in *Cambridge Tribune*, Jul 9, 1937, with headline, "Senator Burke Again Flays Harvard University Heads"; Ruth Robinson to Cannon, 25 Aug 1937.

26. Cannon to Rep. Adolph Sabath, 30 Jan 1939; for his later views, see his letters to Franz Boas, 5 Jan 1940, and to Kirkley Mather, 3 Jun 1940. For the House committee's charges against the Spanish aid organizations, see Reissig to Cannon, 1 Sep 1938; Reissig to Rep. Martin Dies, 29 Nov 1938.

27. Richard Miller's letter to the editor of the *Harvard Alumni Bulletin* was published in the Oct 14, 1938 issue; Cannon's letter followed in the issue of Oct 21. The appeal letter that upset Dr. Miller in the first place was sent out from the Boston Medical Bureau, signed by Dorothy Hickie, executive secretary, 28 Sep 1938. The subsequent letters were Cannon to Miller, 15 Oct 1938; Miller to Cannon, 17 Oct 1938; Cannon to Miller, 19 Oct 1938. See also Benison's Garrison Lecture (n.6), p. 245.

28. Miller to Alex Forbes, 20 Oct 1938, in reply to Forbes's letter of Oct 14 (corrected Oct 17). Forbes and Miller had collaborated in a research together, "Detection with the string galvanometer of afferent impulses in the brain-stem and their abolition with ether anaesthesia," that was reported in the Proceedings of the American Physiological Society; see *Am. J. Physiol.* 40(1916):148–149.

29. Alexander Forbes to W. Cameron Forbes, 2 May 1938, enclosing a letter from Carrasco-Formiguera of 8 Jun 1937. For Ellery Sedgwick's trip to Spain, see Cannon to W.W. Norton, 10 Jan 1938, and exchange of letters between Norton and Cannon of 18 Feb 1938; "Franco's conquest of the Atlantic," *The Nation,* Feb 26, 1938, pp. 233–234; letter of protest to the editor of *The New Republic,* Mar 9, 1938. See also Benison's Garrison Lecture (n.6), p. 246.

30. Cameron Forbes replied on 3 May 1938, and there was a further exchange of letters between the two on May 7 and 9. Alex was the youngest of eight children of William Hathaway Forbes (president of the Bell Telephone Company) and Edith Emerson (daughter of Ralph Waldo Emerson); his brother Cam was formerly the governor-general of the Philippines, as well as ambassador to Japan.

31. Whitehill's letter was published in the *Harvard Alumni Bulletin* of Nov 4, 1938; Warren's was in the Nov 11 issue, and Cannon's rebuttal to both letters was in the Nov 25 issue.

32. A.N. Holcombe to Cannon, 21 Oct 1938; Cannon replied on Oct 25: "When there is a good deal of criticism coming my way, the approval which you, a friend whose judgment I prize highly, send me is most cheering."

33. Answering a "call to action" in March 1938, Cannon wrote letters to academic and medical colleagues, as well as to congressmen and the State Department, urging modification of the Neutrality Act and lifting of the embargo. As examples, see, e.g., his letters to Felix Frankfurter, 28 Mar 1938, and to Harlow Shapley, 9 May 1938; see also fd. 648, Cannon Papers.

34. Cannon's statement was enclosed with his letter to Herman F. Reissig of 4 Nov 1938. Cannon wrote Reissig again on Nov 22 about a meeting he had attended the previous evening at the Boston Bureau: "There was enthusiasm for lifting the embargo but also there was a strong feeling that in that effort we should not lose sight of the primary function of the Medical Bureau, that of caring for the wounded, the mutilated, the maimed and the sick among the fighting forces and civilian population of the Spanish Republic." On Nov 2, Reissig had sent Cannon a suggested resolution to be circulated to medical faculties and hospital staffs. When Reissig later asked Cannon to join with Thomas Addis, A.J. Carlson and Haven Emerson to sign a letter to President Roosevelt, Cannon readily agreed (exchange of letters, 6 and 7 Dec 1938). For more about the embargo at this time, see Edward Barsky to Cannon, 19 Dec 1938, with Cannon's reply of Dec 21; Cannon to Jay Allen, 25 Jan 1939.

35. Cannon to Cushing, 8 Mar 1939, in reply to Cushing's letter of Mar 7. Cannon later expressed a sense of satisfaction when he read Sumner Welles's *Time for Decision*. In 1944, years after the start of the civil war in Spain, Welles had written: "In the long history of the foreign policy of the Roosevelt administration, there has been, I think, no more cardinal error than the policy adopted during the civil war in Spain." (quoted in *The Way of an Investigator*, p. 162)

36. Norman Thomas to Cannon, 3 Apr 1939, with Cannon's reply on Apr 5. See further Benison's Garrison Lecture (n.6), pp. 248–250.

37. Roger Baldwin to Cannon, 6 Apr 1939, in response to Cannon's telegram of Apr 5; Thomas to Cannon, 6 Apr 1939. One wonders what Cannon might have said when he read the copy of Thomas's letter to Baldwin (7 Apr 1939) that was sent to him. "In the reconstruction of the Committee," wrote Thomas, "it must be borne in mind that from our point of view it is better to get a known and responsible member of the Communist Party . . . than to get the type of liberal, who either as a convinced fellow traveller or as a plain innocent, can be manipulated into taking the Communist Party line, while at the same time appearing before the public as a liberal. I'd rather work directly with Earl Browder than with a half dozen so-called liberals whom you and I know and whom I could name." See further correspondence in fd. 668 of the Cannon Papers.

38. Cannon to John Sherman, 15 Feb 1938; see also Cannon to Dorothy Hickie, 17 Apr 1939.

39. Cannon's resignation was addressed to Harold H. Lund, 15 Apr 1939. He wrote that his election to the presidency of the A.A.A.S. "has added to the demands on my time and energies until the total burden on me has been unbearable. I must have relief or break under it. In the circumstances my resignation must be regarded as irrevocable." See also Cannon to Reissig, 28 Apr 1939; Cannon to Hickie, 29 Apr 1939. Dr. Aub wrote Hickie on Apr 29, too, saying that Cannon hated to give up his Spanish activities, but he thought it was essential for him to do so. He added, "Could you and your committee please make his withdrawal from this work as easy as possible for him?"

40. Thomas to Cannon, 20 May 1939; Baldwin to Cannon, 26 May 1939, with Cannon's reply of May 27.

41. Juan Negrin to Cannon, 10 Jul 1939. Miss Hickie's letter was written on May 3 and Cannon replied on May 4. See also 1939 letter-book for Mrs. Cannon's family letter of May 10; in Cannon Family Papers.

19. Refugees from War-torn Europe

1. Lead articles in the *New York Times* and *New York Herald-Tribune*, Dec 30, 1935, with text of James G. McDonald's letter of resignation of Dec 26. McDonald charged that Nazi terrorism had spread beyond the Jewish phase, and he demanded concerted action to stop the persecution in Nazi Germany. See also appeal letter from the newly-organized Emergency Committee in Aid of Political Refugees from Nazism, 16 Jan 1936, which listed Cannon's name among its sponsors.

2. Johannes Stark, "The pragmatic and the dogmatic spirit in physics," *Nature*, Apr 30, 1938, pp. 770–772. For an earlier discussion of Stark, see Chapter Fifteen above, pp. 306–307.

3. Cannon enclosed the preliminary draft statement in a letter to Harvey Cushing, 10 Jun 1938. His "Ideals of the man of science" had originally been published in *The Christian Register* in 1905 (vol. 84, pp. 1192–1195); see also his statement on the relation of Jewish physicians and investigators to the development of medical knowledge and practice, enclosed in a letter to Franklin Folsom at the League of American Writers, 29 Nov 1938.

Another version of a response to Stark's article was prepared by a group of signatories headed by Wesley C. Mitchell, president of the A.A.A.S. (in fd. 1185 of the Cannon Papers). The replies to Stark led by the end of 1938 to formulation of a "Manifesto on Freedom of Science," which declared that scientists had "a moral obligation to educate the American people against all false and unscientific doctrines, such as the racial nonsense of the Nazis." See further Peter J. Kuznick, *Beyond the Laboratory: Scientists as Political Activists in 1930s America* (Univ. of Chicago Press, 1987), pp. 184–186.

4. See exchange of letters between Cushing and Cannon, 11 and 13 Jun 1938.

5. Cannon to Stanley Cobb, 11 Apr 1939, in reply to Cobb's letter of Apr 10. See also Jacob Fine to Cannon, 31 May 1938, enclosing a 5-page typescript, "Proposals for a National Organization in Aid of Displaced Foreign Physicians." For Cannon's reaction to the proposals, see his letters to Fine and to Robert B. Osgood, both dated 6 Jun 1938. Edsall had retired from the deanship of H.M.S. in 1935; Mrs. Herman, formerly Susan Hale, had been secretary to Stanley Cobb, with whom her husband, a psychiatrist at the Boston City Hospital, worked until his death in 1935; and Fine was later to become chief of surgery at the Beth Israel Hospital in Boston.

6. "A Program for the Refugee Physicians," *JAMA*, May 13, 1939; reprint issued by the New York Committee for the Resettlement of Foreign Physicians. Dr. Edsall was the honorary chairman of the National Committee. See also Stanley Cobb to Cannon, form letter of 26 Sep 1938, enclosing a set of motions defining the general plan of action of the Boston Committee as well as a membership list; Laurence Farmer, executive secretary of the Emergency Committee, to Cannon, 13 Mar 1939.

7. For the sequence of events concerning the licensing bill in Massachusetts, see Cobb to Cannon, form letter of 27 Oct 1938; Fine to Cannon, form letter of 29 Nov 1938; Fine to Cannon, 8 Feb 1939, with Cannon's reply of Feb 10.

8. Cobb on 4 Mar 1939 asked Cannon to write letters, and Cannon replied on Mar 8. For the field agents, see also Cobb to Cannon, 14 Feb 1939, and Cannon's reply of Feb 16.

9. Bulletin No. 1 of the National Committee for Resettlement of Foreign Physicians, July 1939. Another study center had been organized in New York City by Dr. Irving Graef, associate professor of pathology at the N.Y.U. School of Medicine.

10. Bruno Minz's paper was originally published (in German) in *Arch. Exp. Pathol.* 168 (1932):292–304. He later described his work on the leech-mus-

cle preparation in his monograph, *The Role of Humoral Agents in Nervous Activity* (C.C. Thomas, 1955), pp. 53–57. For an earlier discussion of Feldberg joining Dale's laboratory, see Chapter Sixteen above, pp. 327–329.

11. Minz to Cannon, 6 Apr 1939; Cannon's inquiry to the Boston Committee of 28 Apr 1939 was answered by Dr. Fine on May 5. Minz originally wrote from Paris on 12 Jun 1935, when Cannon was on sabbatical, and Hal Davis (22 Jun 1935) forwarded the letter so Cannon would get it when he arrived in Leningrad; Cannon replied on Oct 5. André Mayer's letter was written on 14 Jan 1938; Cannon wrote to private individuals like David Marine and Alfred E. Cohn on 27 Jan 1938. Correspondence with and about Bruno Minz may be found in fd. 792 of the Cannon Papers.

12. *The Role of Humoral Agents in Nervous Activity* (n.10). See also Berndt E. Hirtz to Cannon, 7 Sep 1940; Fine to Cannon, 3 Mar 1941, and Cannon's reply of Mar 4.

13. See further Cannon and Haimovici, "The sensitization of motoneurones by partial 'denervation'," *Am. J. Physiol.* 126(1939):731–740. Daniel O'Brien of the Rockefeller Foundation notified Cannon of Haimovici's fellowship on 28 Jul 1938 and enclosed a summary of his qualifications. Correspondence with and about H. Haimovici may be found in fds. 725–726 of the Cannon Papers.

14. Haimovici to Cannon, 20 Feb and 20 Apr 1941. Cannon replied on 24 Mar 1941 that he already had written to Alan Gregg of the Rockefeller Foundation and to Jacob Fine of the Boston Committee.

15. Haimovici to Cannon, 5 Jul 1944. Cannon reported the conversation with John Homans in a letter to Fine of 26 Mar 1941; Louis Zetzel wrote to Cannon on May 15. For Cannon's inquiries to the Rockefeller Foundation, see his letter to R.A. Lambert, 16 May 1941; also Lambert to Cannon, 19 May and 18 Jul 1941. Haimovici's *Vascular Surgery: Principles and Techniques* (New York: McGraw-Hill, 1976) had gone through four editions as of 1996. His *Metabolic Complications of Acute Arterial Occlusions and Related Conditions (Myonephropathic-Metabolic Syndrome)* (Mt. Kisco, N.Y.: Futura Publishing Co., 1988) is "dedicated with admiration and reverence to Walter Bradford Cannon."

16. For previous discussion of Otto Loewi, see Chapters Fourteen and Sixteen above, pp. 270–271 and 316–317. Wilhelm Raab had been a Rockefeller fellow in Cannon's department in 1929–30; Jean Reboul had a Chapman fellowship in 1936–37; Julien Maes had been a C.R.B. fellow in 1936–37. Cannon, in a letter to Dr. Fine of 24 Jan 1939, rated David Nachmansohn, Wilhelm Feldberg, and Bruno Minz as three of the younger displaced German physiologists whose work was worthy of support.

17. Cannon to Loewi, 3 Mar 1934 and 11 Jan 1938. For the Dunham Lectureship, see Cannon to Edsall, 29 Dec 1931; also exchange of letters between Cannon and Loewi, 10 and 21 Jun 1933.

18. "An autobiographic sketch," reprinted in Fred Lembeck and Wolfgang Giere, *Otto Loewi: Ein Lebensbild in Dokumenten* (Springer-Verlag, 1969), pp. 168–190; see esp. pp. 185–186.

19. H.H. Dale to Prof. A.E.L. Beznak, 30 Mar 1938, and to Frau Loewi, 11 Apr 1938; in Dale Papers, Royal Society. Hans Loewi's letters to Cannon and to Dale were written on 29 Mar 1938.

20. Messersmith replied on 3 May 1938 to Cannon's letter of Apr 20. See also Cannon to Hans Loewi, to Felix Frankfurter, and to Alfred E. Cohn, all written on 11 Apr 1938.

21. Dale to Cannon, 17 Sep 1938, in reply to Cannon's letter of Sep 8; "An autobiographic sketch" (n.18), p. 186. Dale, in seeking aid for Loewi, had written to A.V. Hill at the Royal Society and to Paul Loewi, a cousin who lived in England.

22. Loewi to Cannon, 24 Nov 1938 and 7 May 1939. See further Bessie Zaban Jones, "To the rescue of the learned: The asylum fellowship plan at Harvard, 1938–1940," *Harvard Library Bulletin* 32 (1984):204–237; esp. pp. 214–219. Mrs. Jones has appended a bibliographical note on pp. 237–238, which lists major books and articles on the intellectual refugees.

23. Ernst Boas to Cannon, 17 Jun 1939, and Cannon's reply of Jun 28; exchange of letters between Boas and Loewi, 2 Aug and 20 Aug 1939. Records in the Nobel Archives indicate that Loewi was awarded a grant of 5,000 Swedish crowns to be paid in 1939 in support of his research because he had not been able to keep his Nobel Prize money. (Agneta Sjövall to A.C. Barger, 16 Feb 1993)

24. Cannon to Boas, 4 Nov 1939. See also Loewi to Cannon, 2–3 Sep 1939; Libman to Cannon, 12 Oct 1939 (and an earlier letter of Aug 31); Boas to Cannon, 25 Oct 1939.

25. Cannon to Loewi, 16 Dec 1939; Loewi to Cannon, 29 Dec 1939. For the offer from N.Y.U., see Homer Smith to Cannon, 16 Nov 1939; Boas to Cannon, 22 Nov 1939; Smith to Cannon, 15 Dec 1939. Homer Smith had been a National Research fellow in Cannon's department for two years from 1923 to 1925.

26. Loewi to Cannon, 6 Jan 1940; Cannon to Loewi, 16 Jan 1940; Loewi to Cannon, 16 Jan 1940. In his second letter, Loewi wrote about a possible cooperative arrangement between N.Y.U. and Cornell, where he might be offered an opportunity to work in Harold Wolff's laboratory.

27. Cannon to Ross Harrison, 23 Jan 1940; Dale to Cannon, 13 Feb 1940; Cannon to Loewi, 21 Feb 1940.

28. Cannon to Dale, 16 Apr 1940; Boas to Cannon, 13 Mar 1940.

29. See further Ernst-August Seyfarth, "Ernst Theodor von Brücke (1880–1941) and Alexander Forbes (1882–1965): Chronicle of transatlantic friendship in difficult times," *Persp. Biol. and Med.* 40(1996):45–54; esp. p. 48. The account here, like Dr. Seyfarth's, has drawn on correspondence in the Alexander Forbes Papers, Countway Library of Medicine.

30. Forbes to Sidgwick, 20 May 1938. See also Brücke's student to Forbes, 16 May 1938; Selig Hecht to Forbes, 17 May 1938; Forbes to Cannon, 17 May 1938.

31. Forbes apprised Cannon of Brücke's situation in a series of lengthy handwritten letters, 28 Jun, 5 Aug, and 21 Aug 1938. See also Cannon to Forbes, 8 Jul (with postscript of 14 Jul) 1938: "A second year would be a gamble, and, in the present state of university affairs in the U. S., a gamble with a not very favorable chance for a man of 59" [actually, Brücke was 57].

32. The letters from Brücke and Brewster Morris to Forbes were both dated 31 Aug 1938.

33. Morris to Forbes, 10 Nov 1938. Forbes reviewed the various steps in the effort to help Brücke emigrate from Austria in a letter to Joseph Erlanger, 10 Oct 1938, before adding: "I feel that in the interests of science and humanity, it is fitting that support for his sustenance should come from his more or less widely distributed circle of friends."

34. Brücke to Forbes, 19 Feb, 24 Feb, and 22 Jun 1939. See also Seyfarth (n.29), p. 50; David S. Wyman, *Paper Walls: America and the Refugee Crisis, 1938–1941*, Univ. of Massachusetts Press, 1968.

35. Forbes to Harlow Shapley, 31 Oct 1939. Brücke authored two articles with Marie Early and Alexander Forbes, "Recovery of responsiveness in motor and sensory fibers during the relative refractory period" and "Fatigue and refractoriness in nerve," which were published in *J. Neurophysiol.* 4(1941):80–91 and 456–472.

36. For Brücke's death, see Forbes's statement of 16 Jun 41 and Arturo Rosenblueth's statement in Cannon's ltr to Brücke's son Franz, 1 Jul 1941. For continuing financial support, see Forbes to Burwell, 11 Mar 1941; letters between Shapley and Forbes, 13–26 Mar 1941; Cannon to Forbes, 24 Apr 1941; Shapley to Cannon, 6 May 1941. In the fall of 1940, Shapley had drafted a statement for establishment of a program of National Research Associates; see further Jones, "To the rescue of the learned" (n.22), pp. 234–235.

37. For Dora Teleky-Brücke, see correspondence in fd. 795 of the Cannon Papers; also Seyfarth (n.29), p. 53.

38. H.H. Dale, "Otto Loewi, 1873–1961," *Biog. Mem. Fellows Roy. Soc.* 8(1962):67–89; see esp. pp. 82–83.

20. From the Body Physiologic . . .

1. Exchange of letters between Ludwig Kast and Cannon, 18 and 21 Sep 1931; also Cannon to the Trustees of the Josiah Macy, Jr. Foundation, 18 Dec 1931. The work of Davis and his collaborators, Cannon wrote, had "extraordinarily important possibilities for tracing the distribution of impulses from the limbs and various sense organs to regions just below the cerebral cortex and to intermediate stations."

Gordon C. Ring joined the department as an instructor in 1931 and remained for most of the decade to help with the teaching load. In 1936–37, voluntary courses were again listed in the course catalog; one on Endocrine Glands was offered by Cannon, Rosenblueth and associates, a second on the Central and Autonomic Nervous Systems was offered by Cannon, Rosenblueth, Davis and associates.

2. For Redfield's resignation, see Cannon to D.L. Edsall, 26 Nov 1930, and Cannon to A.C. Redfield, 11 Feb 1931. For the offer to Bard, see Cannon to Bard, 10 Mar 1931; Bard to Cannon, 12 Mar 1931; Cannon to Bard, 13 Mar 1931. For David Rioch, see exchange of letters between Cannon and George Wislocki, 22 and 26 May 1931.

3. Cannon to S.I. Franz, 9 Dec 1931. Bowditch's remark, quoted by Cannon in *Autonomic Neuro-effector Systems* (Macmillan, 1937), pp. 198–199, was made in "What is nerve-force?" *Proc. Am. Assoc. Adv. Sci.* 35(1886):237–246.

4. C.M. Brooks enclosed his 16-page typescript, "Reminiscences of Dr. Walter B. Cannon at the Department of Physiology at Harvard, 1931–1933," in a letter of 2 Jul 1974 to A.C. Barger; see further Box 189 in the Cannon Papers. It was Brooks who organized the centennial symposium at the Downstate Medical Center, State University of New York, in 1972, later published as *The Life and Contributions of Walter Bradford Cannon.*

5. Cannon to Edsall, 13 Oct 1931 and 23 Mar 1933. Bard's research was reported in an unsigned article on the physiology department in the *Harv. Med. Alum. Bull.,* Nov 1932, pp. 3–6; see esp. p. 4.

6. F.T. Lewis to Cannon, 23 May 1933, in reply to Cannon's letter of May 15. For the counter-offer to Bard, see also Edsall to Cannon, 30 Mar 1933.

7. Cannon to Bard, 8 Apr 1933, in reply to Bard's letter of Apr 2. See also Bard's introductory note to 1933 in a file of his correspondence with Cannon, in Box 199 of the Cannon Papers.

8. Edsall to Cannon, 9 Oct 1934. For Bacq, see his letter to Cannon of 8 Aug 1933; also his *Chemical Transmission of Nerve Impulses:A Historical Sketch* (Pergamon Press, lst English ed., 1975), pp. 38–39. For Rosenblueth, see Louisa Barclay Benton, "Arturo Rosenblueth: Success or Failure?" (undergraduate thesis presented to the Committee on History and Science, B.A. degree, Harvard Univ., 1986), esp. pp. 29–35.

9. Memorandum from Cannon to Davis, 20 Feb 1935; Davis to Cannon, 18 Apr 1935.

10. Cannon to Louis Gross, 14 Dec 1936; see also Cannon to C.S. Burwell, 25 Sep 1935; J.O. Pinkston to Cannon, 5 Dec 1936.

11. For a report on workers in his department at this time, see, e.g., Cannon to Burwell, 13 Feb 1937. See also J.V. Luco to A.C. Barger, 8 Feb 1974, enclosing his reminiscence of Jan 1974; in Box 189 of the Cannon Papers.

12. Ibid. Friedgood's contribution to a symposium on endocrine glands delivered at the Harvard Tercentenary in 1936, "The nervous control of the anterior hypophysis," was later published in *J. Reprod. Fert., Suppl.* 10(1970):3–14; see also Horace Davenport's comment in *Life and Contributions,* (n.4), pp. 230–231, and note on pp. 237–238. Simeone became one of E.D. Churchill's "boys" at the M.G.H. and collaborated in the third edition (1952) of White and Smithwick's *The Autonomic Nervous System.* For Carl J. Wiggers' correspondence with Cannon about his son Harold, see fd. 1705 in the Cannon Papers.

13. For Davis, see Cannon to Burwell, 20 Nov 1935; for Forbes, see Cannon to Burwell, 10 Feb 1936.

14. Wislocki's and Drinker's letters to J.B. Conant were dated 23 Apr 1936; Christian's and Cohn's were 24 Apr 1936; Gasser's letter, like Cannon's, was 1 May 1936. All in Conant Papers, Harvard University Archives.

15. Burwell to Conant, 16 Jan 1936 (in Conant Papers); Cannon to Bradford Cannon, 8 Feb 1936 (in 1936–37 letter-book, Cannon Family Papers).

16. Conant to Cannon, 20 Oct 1937, in reply to Cannon's letter of Oct 19. See also Cannon to Gregg, 24 Nov 1936, in reply to Alan Gregg's letter of Nov 23. For a full report on the use of the first Rockefeller grant, see Cannon to Burwell, 30 Sep 1936; Burwell to Gregg, 1 Oct 1936; and other materials in file 200A of the Rockefeller Foundation records (in Rockefeller Archives Center).

17. Harvey Cushing to Cannon, 2 Aug 1933. The announcement of the Beaumont Lectures appeared in the Jan 1933 *Bull. Wayne St. Med. Soc.;* Cannon's lectures in the Mar-May 1933 issues of the *J. Mich. Med. Soc.* were published in monograph form under imprint, Detroit: Lectureship Foundation Committee, Wayne County Medical Society, 1933. For a report of the 1932 research, see *Harv. Med Alum. Bull* (n.5), p. 4. Gregersen and Cannon's article, "The effect of extirpation of the salivary glands on the water intake of dogs while panting," *Am. J. Physiol.* 102(Nov 1932):336–343, was featured in an editorial in *JAMA,* Mar 11, 1933, pp. 743–744, which reviewed Gregersen's several papers on the subject of thirst.

18. For Cannon's Caldwell Lecture, see "Some reflections on the digestive process," *Am. J. Roentgenol.* 32(1934):575–588; the editorial in the *Brit. Med. J.* appeared on Mar 9, 1935, p. 477. For the Kober Lecture, see "Story of the development of our ideas of chemical mediation of nerve impulses," *Am. J. Med. Sci.* 188(1934):145–159. The *Handbook of General Experimental Psychology,* ed. Carl Murchison, was published by Clark Univ. Press; Cannon's chapter is on pp. 247–263.

19. Cannon to James A. Miller, 29 Sep 1936. For his Hodgen Lecture, see "Significance of the emotional level," *Sci. Monthly* 39(1934)101–110.

20. Elliott Cutler to Cannon, 13 Feb 1934; Cannon to Bradford Cannon, 1 Mar 1934. For Cannon's talk, see "Consideration of possible toxic and nervous factors in the production of traumatic shock," *Ann. Surg.* 100(1934):704–713.

21. J.C. White to Cannon, 13 Jan 1936, in reply to Cannon's letter of Jan 11. See also Cannon's Foreword to the 2nd edition of *The Autonomic Nervous System* (Macmillan, 1941), p. viii.

22. Reviews of the second edition of *The Wisdom of the Body* may be found in the Ideal Scrapbook in the Cannon Papers; the one cited here was written by Chauncey Leake for *Science,* Jul 28, 1939. "Stresses and strains of homeostasis" was published in *Am. J. Med. Sci.* 189(1935):1–14. Cannon's lecture in the Art and Romance of Medicine series was not published, but see his article on homeostasis in the *Cyclopedia of Medicine* (1935), pp. 861–872.

23. The Welch Lectures, delivered at Mt. Sinai Hospital, were published in the hospital's *Journal* 5(1939):587–597 and 598–606. For the article by Brouha, Cannon and Dill, see "The heart rate of the sympathectomized dog in rest and exercise," *J. Physiol.* 87(1936):345–359. Cannon held Bruce Dill in high regard; see, e.g., his letters in Dill's behalf to L.J. Henderson, 8 Feb 1935, to J.B. Conant, 17 Nov 1935, to C.S. Burwell, 30 Mar 1938.

24. Cannon to Bradford Cannon, 30 May 1938; also Mrs. Cannon's letter to Bradford, 24 May 1938; both in 1938 letter-book (in Cannon Family Papers). Materials about the conference and volume on ageing sponsored by the Macy Foundation may be found in fds. 1298–1303 of the Cannon Papers; Cannon's correspondence with Elisha Friedman is in fds. 1888–1890. See further *Problems of Ageing: Biological and Medical Aspects,* ed. E.V. Cowdry, Baltimore: Williams & Wilkins, 1939.

25. Cannon to Gregg, 2 Mar and 29 Apr 1937. See also Cannon to Bradford Cannon, 14 Dec 1935 (in 1935 letter-book), 29 Jun 1936 and 5 Nov

1936 (in 1936–37 letter-book). For more on *Digestion and Health,* see Cannon to Phil Bard, 12 Jun 1936, and reviews in Cannon's Ideal Scrapbook.

26. *Nature,* Feb 12, 1938, p. 266. See also *Lancet,* Oct 30, 1937, p. 1023, and *Brit. Med. J.,* Nov 6, 1937, pp. 912–913.

27. Philip Bard, "Walter B. Cannon's contributions to physiology," in *Walter Bradford Cannon, 1871–1945: A Memorial Exercise* (1945), p. 21; see also Bard, p. viii, in facsimile edition of *The Way of an Investigator* (1968).

28. C.W. Hampel, "The effect of denervation of the sensitivity to adrenine of the smooth muscle in the nictitating membrane of the cat," *Am. J. Physiol.* 111(1935):611–621; Shen and Cannon, "Sensitization of the denervated pupillary sphincter to acetylcholine," *Chinese J. Physiol.* 10(1936):359–371. See further Z.M. Bacq, "Walter B. Cannon's contribution to the theory of chemical mediation of the nerve impulse," in *Life and Contributions* (n.4), esp. section on uptake of catecholamines by neurones, pp. 79ff.

29. Cannon and Rosenblueth, "The sensitization of a sympathetic ganglion by preganglionic denervation," *Am. J. Physiol.* 116 (1936):408–413; see esp. pp. 408–409 and 413. For F.T.P. von Brücke's variant findings, see H.H. Dale to Cannon, 30 Dec 1936; *J. Physiol.* 91(1938):375. Cannon and Rosenblueth's confirmatory study was "The effects of preganglionic denervation on the superior cervical ganglion," *Am. J. Physiol.* 125(1939):276–289.

30. F.A. Simeone, "The effect of the regeneration of the nerve supply on the sensitivity of the denervated nictitating membrane to adrenine," *Am. J. Physiol.* 120(1937):466–474; Simeone, Cannon and Rosenblueth, "The sensitization of the superior cervical ganglion to nerve impulses by partial denervation," *Am. J. Physiol.* 122 (1938):94–100. Cannon's letters to his son were written on 22 Dec 1936 and 6 Feb 1938 (in 1936–37 and 1938 letter-books, Cannon Family Papers).

31. For work with Friedgood on the thyroid, see Cannon's letters to Bradford Cannon of 28 Oct, 26 Nov, and 7 Dec 1935 (in 1935 letter-book), and 22 Dec 1936 (in 1936–37 letter-book). These attempts to replicate the experiments that Cannon did with Binger and Fitz in 1915 were not reported until 1940; see Chapter Twenty-three below, p. 470.

32. Cannon, "Scientific controversy: Its possible harm and value," *Pharos,* May 1938, pp. 5–9; see esp. p. 9. Cannon was clearly irked by Rogoff's more recent statements about the adrenal medulla, especially those expressed at a symposium sponsored by the A.M.A. and published in *Glandular Physiology and Therapy* in 1935.

33. Cannon's address was published in *Science,* Dec 8, 1939, pp. 521–527. See also Cannon to Bard, 1 Aug and 30 Oct 1939.

34. Rosenblueth and Rioch published three articles together in 1933, one on responses of smooth muscle to adrenin and the augmentor action of cocaine, another on electrical excitation of multifibered nerves, the third on temporal and spatial summation in autonomic systems. Forbes's exchange of letters with J.C. Eccles was 10 and 22 Dec 1934; their correspondence may be found in the Forbes Papers, Countway Library of Medicine.

See further Louise H. Marshall, "Instruments, techniques, and social units in American neurophysiology, 1870–1950," in *Physiology in the American Con-*

text, 1850–1940, ed. Gerald L. Geison (Bethesda, Md.: American Physiological Society, 1987), pp. 351–369; esp. the section on axonologists, pp. 358–359. Marshall gives a succinct account of the E.E.G. studies undertaken by Hal Davis and the development of apparatus by Albert Grass and his wife Ellen at Harvard that eventuated in the founding of the Grass Instrument Company.

35. The further exchange between the two men includes a 6-page letter from Forbes to Eccles, 7 May 1935, with copies sent to Dale and Adrian, and Eccles to Forbes, 28 Jul 1935. See also Dale to Cannon, 24 Feb 1937; Eccles' review article, "Synaptic and neuromuscular transmission," in *Ergibn. Physiol.* 38(1936):339–444; *Les Prix Nobel en 1963*, pp. 38–40 and 261–283. Eccles later recalled the epochal chemical-electrical controversy at the Sir Henry Dale Centennial Symposium; see *Notes and Records of the Royal Society* 30(1976):219–230. For an even later review of Eccles' role in the controversy, see Z.M. Bacq, "Chemical transmission of nerve impulses," in *Discoveries in Pharmacology, Vol. I, Psycho- and Neuro-pharmacology* (Elsevier, 1983), pp. 72–75.

36. Cannon to Bradford Cannon, 12 May 1939; in 1939 letter-book (Cannon Family Papers). Cannon's earlier letter to J.T. Lewis was dated 23 Feb 1939. The Hughlings Jackson Lecture was published in *Am. J. Med. Sci.* 198(1939):737–750. An editorial, "Is there a law of denervation?" in *Lancet* (Jan 27, 1940, pp. 179–180), concluded that "the experiments in support of the 'law of denervation' are really independent of any theory of transmission and must be examined on their own merits. They cannot yet be regarded as conclusive, particularly those dealing with the central nervous system, but Cannon has made out an excellent case and put forth a stimulating idea."

37. *New York Times*, Oct 28, 1939; in Cannon's Ideal Scrapbook. See also Cannon and Lissák, "Evidence for adrenaline in adrenergic neurons," *Am. J. Physiol.* 125(1939):765–777; esp. p. 777. For Cannon and Rosenblueth, see *Am. J. Physiol.* 125(1939):276–289; for Cannon and Haimovici, see *Am. J. Physiol.* 126(1939):731–740.

38. Cannon to T.R. Elliott, 14 Oct 1930; Elliott to Cannon, 4 Dec 1930; Cannon to Elliott, 17 Dec 1930.

39. Cannon to Elliott, 30 Jun 1937. The earlier letters are an exchange between Cannon and Elliott, 17 and 29 Dec 1930; Elliott to Cannon, 14 Jun 1932; Elliott to Cannon, 14 Mar 1933, in reply to Cannon's letter of Feb 3. Elliott retired from his chair of medicine at University College Hospital in 1939. Despite his increasing infirmities, he lived into his eighties, his death occuring in 1961.

21. . . . to the Body Politic

1. For Cannon on higher education for women, see his letter to Isabel Hyams, 17 Nov 1930; on Ida Hyde, see Cannon to Ruth Wilson Tryon, 12 May 1939. For the Division of Medical Sciences, see *Walter B. Cannon* (1987), pp. 117–118. See further Division of Medical Sciences file, Harvard Medical Archives; also Eric G. Ball, "The fiftieth anniversary of the founding of the Division of Medical Sciences," *Harv. Med. Alum. Bull.*, May 1958.

2. Cannon to David L. Edsall, 6 Dec 1932 (copy to Ada Comstock), in

reply to Edsall's letter of Dec 3. Cannon sent another letter on Dec 6, enclosing a list of women who had received degrees from the Division of Medical Sciences.

3. Minutes of Administrative Board meeting, Dec 15, 1932 (in Harvard Medical Archives); see also A.L. Lowell to D.L. Edsall, 29 Dec 1932, and Edsall's reply of 3 Jan 1933 (in Dean's Files, Harvard Medical Archives). For other materials about these issues, see fd. 115 (relations with Radcliffe) and fd. 657 (Division of Medical Sciences), in the 1930–33 series, Lowell Papers, Harvard University Archives.

4. Mrs. Cannon's letters of 26 Dec 1932, 15 Feb and 1 Mar 1933 are in a volume of 1932–35 letters to Wilma and John Fairbank in China; in the Cannon Family Papers.

5. See Cannon to George H. Chase, 4 Mar 1933; Chase to Cannon, 6 Mar 1933; Cannon to Chase, 8 Mar 1933. See also exchange of letters between Edsall and Chase, 7 and 9 Mar 1933; in Dean's Files. Cannon had read a statement in protest of the attempt to deprive the Division of Medical Sciences of the right to award the Ph.D. degree at a meeting of the Administrative Board; see minutes of meeting held on Feb 16, 1933, in Harvard Medical Archives.

6. Mrs. Cannon to Wilma and John Fairbank, 16 Jun 1933 (n.4). Earlier, on May 28, she had written: "Did I tell you how pleased we all were over the choice of Prof. Conant for president? . . . He is very friendly to Radcliffe and may be the president who sees Radcliffe finally recognized as an integral part of Harvard instead of a foundling on the doorstep. It must come sometime, and what better time than now?" For President Eliot's last medical faculty meeting, see *Walter B. Cannon* (1987), p. 226; "Eliot's farewell address, May 1, 1909," *Harv. Med. Alum. Bull.*, Spring 1982, pp. 14–16.

7. Cannon to Bradford Cannon, 10 Oct 1933; in 1932–33 letter-book (Cannon Family Papers). For the 150th anniversary, see also "Sesquicentennial celebration" in *Harv. Med. Alum. Bull.*, Dec 1933, pp. 22–29. Conant is quoted in James Hershberg, *James B. Conant: Harvard to Hiroshima and the Making of the Nuclear Age* (N.Y.: Knopf, 1993), p. 64.

8. Edsall's letters to Cannon were written on 19 Feb 1932 and 15 Mar 1933.

9. Cannon to Harvey Cushing, 23 Jun 1933; see also exchange of letters between H.A. Christian and Cannon, 7 and 8 May 1933, with notes of Cannon's speech enclosed. For Cannon on Cushing, see his "Harvey (Williams) Cushing, 1869–1939," *Obituary Notices of the Royal Society*, 1940:277–290.

10. Cannon to F.W. Schlutz, 26 Oct 1934; see also Cannon's letter to Schlutz of Oct 19. Folin's portrait appeared in the Jan 1935 *Harv. Med. Alum. Bull.*; see also pp. 19 and 22. For a personal remembrance of Folin, see Cannon to Gilberto Villela, 24 Mar 1935.

11. Cannon to C.S. Burwell, 28 Feb 1935. At the same time, Cecil Drinker was appointed dean at the School of Public Health, where he had served as assistant dean. For Edsall's retirement, see *Science*, Nov 22, 1935, p. 469; also *Harv. Med. Alum. Bull.*, Oct 1935, with articles by Hans Zinsser, Reginald Fitz, and Cecil Drinker. Cannon presided at the afternoon session of a celebration honoring Edsall; see C.K. Drinker to Cannon, 24 Sep 1935.

12. For Harvard's Tercentenary, see correspondence in fds. 115, 116, 133, and 151 of the Cannon Papers; also his Ideal Scrapbook for an annotated program. Hershberg (n.7) gives a lively account of the gala occasion on pp. 97–99. Among the many publications that came out of the Tercentenary is a volume on *Factors Determining Human Behavior,* containing the contributions of speakers to the symposium chaired by Cannon (Harvard University Press, 1937).

13. "Tercentenary celebration," *Harv. Med. Alum. Bull.,* Jun 1936, pp. 52–61; see esp. pp. 59–60. Cannon's houseguests during the celebration were Joseph Barcroft and B.A. Houssay. For a personal description of the ceremonies, see Mrs. Cannon's family letter of 21 Sep 1936 in the 1936–37 letter-book; in Cannon Family Papers.

14. See Hans Zinsser to Cannon, 28 Jun 1937; Arlie Bock to Cannon, 6 Jul 1937; also exchange of letters between J.B. Conant and Cannon, 18 and 20 Jan 1937. See further folder on honorary degrees, 1936–37, in Conant Papers, Harvard University Archives; also press clippings about the commencement in Cannon's Ideal Scrapbook.

15. For the successor to Richardson, see Edsall to Cannon, 19 Mar 1931; for Elliott Cutler, see, e.g., Cannon to Zinsser, 20 Oct 1931, and to Cutler, 29 Dec 1931. For the Committee on Biochemistry, see Cannon to Edsall, 1 Mar 1935, and Conant to Edsall, 25 Mar 1935 (in Conant Papers); Aub to Cannon, 24 Apr 1935; Cannon to Aub, 23 May 1935; Aub to Cannon, 18 Jun 1935, and report of committee meeting held on Apr 19, 1935 (in J.C. Aub Papers).

16. H.H. Dale to Cannon, 21 Apr 1936, in reply to Cannon's letter of Apr 9. See also Burwell to Conant, 11 Apr 1936; in Conant Papers. Cannon had written Dale, 14 Nov 1935, that the committee was interested in Krayer and asked for information about him.

17. Dale to Cannon, 14 Oct 1933; see further Avram Goldstein, "Otto Krayer, 1899–1982," *National Academy of Sciences Biographical Memoirs* 57(1987), esp. pp. 151–156 and 191.

18. At a meeting of the medical faculty held on Feb 7, 1936, Cannon presented the case of Krayer, which was received favorably due to the strong recommendations of Dale and Verney. For the invitation to Krayer, see Burwell to Cannon, 7 Apr 1936; also Burwell to Conant, 8 May 1936, enclosing Krayer's letter of Apr 20 (in Conant Papers).

19. Exchange of letters between Cannon and Soma Weiss, 27 and 28 Oct 1936; also Burwell to Conant, 15 Oct 1936 (in Conant Papers).

20. Cannon to Dale, 23 Jan 1937. Cannon sent Conant the committee's Proposal with his letter of 21 Jan 1937, which contained a lengthy quote from a statement by Herbert Gasser predicting the future direction of pharmacology; see also Baird Hastings to Conant, 21 Jan 1937 (all in Conant Papers). For the formal offer, see Burwell to Krayer, 26 Jan 1937; also Cannon to Krayer, 25 Jan 1937.

21. Krayer's letter to Cannon was written on 5 Apr 1937, in reply to Cannon's letter of Mar 9, which described the committee's deliberations.

22. For the post at P.U.M.C., see Krayer to Burwell, 14 Jun 1938 (sent by Burwell to Cannon on Jun 16), and to Conant on Jun 27, with explanations; also Goldstein (n.17), pp. 192–193. J.O. Pinkston wrote Cannon on 15 Sep 1938.

23. See further Conant to Burwell, 17 Feb 1939 (referring to letters received from Cutler, Wolbach, Tyzzer and Cobb); Burwell to Conant, 20 Feb 1939, enclosing offer to Krayer of Feb 17 and Krayer's acceptance of Feb 18; Burwell to Conant, 20 Mar 1939; all in Conant Papers. Cannon wrote Pinkston on 26 Feb 1940. Krayer was finally made a full professor in 1951. From that time until his retirement in 1966, he was one of the most influential professors at the Medical School. See further Faculty of Medicine Memorial Minute, May 28, 1986, in *Harvard Gazette,* Jul 11, 1986.

24. For the Baly Medal, see *Lancet,* Nov 21, 1931, p. 1149; for a list of Cannon's honors and awards, see Chronology in the Inventory of the Cannon Papers in the Countway Library.

25. Dale to Cannon, 23 May 1939. The notices about Cannon's election to the Royal Society may be found in his Ideal Scrapbook. See also 1939 letter-book for Mrs. Cannon's letter of May 24; in Cannon Family Papers.

26. For Mrs. Cannon's letter about the Rockefeller Institute's Board of Scientific Directors, see her 1936–37 letter-book; in Cannon Family Papers. See further H.S. Gasser to Cannon, 9 Mar 1936; Cannon to Gasser, 30 Mar 1936.

27. Ibid., see Mrs. Cannon's letter of 29 Apr 1937. For the Endocrinology Committee, see Vernon Munroe to Cannon, 9 Apr 1936; Archie S. Woods to Cannon, 5 May 1937. (Munroe, a member of Cannon's Harvard class of 1896, was an old acquaintance.) Other members of the special committee of the N.R.C. were Earl Engel, Kurt Richter, Oscar Riddle; also Roy Hoskins and Milton Lee, who took care of the secretarial duties.

28. Cannon had earlier declined an invitation to join the national committee of the American Foundation; see his letter to Esther Lape, 14 May 1936, in reply to her letter of May 8. At first he refused appointment to the Medical Advisory Committee, but later reconsidered; see his letters to Truman G. Schnabel, 13 Dec 1937, and to Esther Lape, 17 Dec 1937. See also *New York Herald Tribune,* Apr 26, 1936, for article, "Doctors asked to map future for medicine," describing the survey.

29. John Peters to Cannon, 9 Nov 1937, in regard to a mailing of the Principles and Proposals sent out to a list of subscribers on Nov 7. The editorial in the *New York Herald Tribune* was headed "Medicine and government"; the *New York Times* had an editorial, "Subsidized medicine," in its issue of Oct 25, 1937. There was such an overwhelming demand for further information that Dr. Peters prepared a printed statement, "The Story of the Principles and Proposals for the Improvement of Medical Care," to answer questions and the objections so often expressed.

30. Fishbein's editorial was published in *JAMA,* Oct 16, 1937, pp. 1280–1281; Kilgore's circular letters to "My dear Colleagues" were sent out on 12 Nov 1937. See also exchange of letters between Cannon and Osgood, 22 and 23 Nov 1937; Cannon to Kilgore, 24 Nov 1937. For more on the Principles and Proposals, see further *New Engl. J. Med.,* Nov 11, 1937, pp. 793–794, 798–800, and Nov 18, 1937, pp. 843–844.

31. Cannon to Esther Lape, 7 Feb 1939; see also Cannon to W.W. Norton, 27 Feb 1939. On Cannon as a "systems physiologist," see further Jerome

Lettvin's provocative review, "Musings on *Walter B. Cannon: The Life and Times of a Young Scientist*," in *Persp. in Biol. and Med.* 33(1990):304–310.

32. Mrs. Cannon to her family, 31 Dec 1938, in 1938 letter-book; Cannon to Bradford Cannon, 2 Jan 1939, in 1939 letter-book; both in Cannon Family Papers. Inserted in the 1939 letter-book are a *Science Newsletter* for Jan 7, 1939, and a *New Engl. J. Med.* editorial of Jan 26.

33. Burwell to Cannon, 6 Jan 1939; see also Science Service press release, Dec 30, 1938; E.R. Long in *Science*, Feb 3, 1939, pp. 92–93. Peter J. Kuznick devotes an entire chapter to the A.A.A.S. and the Science and Society movement in his book, *Beyond the Laboratory: Scientists as Political Activiists in 1930s America* (Univ. of Chicago Press, 1987); for Cannon's election to the presidency of the Association, see esp. pp. 101–102.

34. F.R. Moulton's letter to Cannon was dated 7 Jan 1938 [for 1939]; for problems of alcohol, see Cannon to Archie S. Woods, 21 Mar 1939; also Cannon to L.J. Frank, 11 Mar 1941.

35. Cannon to Moulton, 2 Jan 1940. The *Boston Globe* interview and the *New York Times*, Feb 12, 1939, may be found in Cannon's Ideal Scrapbook. See also Cannon's statement for the Boston Lincoln's Birthday Committee sent to K.V. Thimann, 8 Feb 1939. For further discussion of the Lincoln's Birthday Committee, see Kuznick (n.33), pp. 188–194.

36. Elisha Friedman to Cannon, 7 Oct 1940. For more on Friedman, see Chapter Twenty above, pp. 411–412.

37. Cannon to Moulton, 14 Dec 1940; Cannon wired the title of his address to Sam Woodley on Nov 14. About his meeting with Dennison, see Cannon to Friedman, 18 Nov 1940; Cannon to Dennison, 29 Nov 1940.

38. See exchange of letters between John Fairbank and Cannon, 10 and 11 Dec 1940. For other critiques of Cannon's address, see Henry Dennison to Cannon, 12 Dec 1940, and Arthur Holcomb to Cannon, 21 Dec 1940. See also 1940 letter-book for Mrs. Cannon's letter of Dec 31; in Cannon Family Papers.

39. Friedman to Cannon, 23 Jan 1941. Clippings from the *New York Times*, Dec 27, 1940, and *Philadelphia Inquirer*, Dec 28, 1940, are in Cannon's Ideal Scrapbook. "The body physiologic and the body politic" appeared in *Science*, Jan 3, 1941. It was later published as a separate monograph, *The Body as a Guide to Politics*, London: Watts Thinkers' Forum, 1942, and reprinted in *Science and Man*, ed. Ruth Nanda Anshen, N.Y.: Harper, Brace, 1942.

22. A World at War

1. 1939 letter-book, Mrs. Cannon's family letter of Apr 23; see also her letters of Aug 25 and Sep 18, and Cannon's letter to Bradford and Ellen Cannon of Sep 2 (in Cannon Family Papers). Cannon wrote to K. Lissák on 22 Jan 1940.

2. Ibid., Cannon to Bradford and Ellen Cannon, Dec 2. The baby, born in July 1940, was named Walter Bradford Cannon, II. In addition to those in the letter-books, there are four folders (2300–2303) of Cannon's letters to Bradford in St. Louis in the Cannon Papers.

3. 1932–33 letter-book, see esp. Mrs. Cannon's family letter, 26 Nov 1932, and her letter to John Fairbank's mother, Lorena, 8 Dec 1932 (in Cannon Family Papers). See also Cannon to J.L. Yates, 30 Nov 1932 and 27 Dec 1933.

4. See 1938 letter-book for photographs and clippings from the society pages of local newspapers about the wedding at the DeNormandie's home in Lincoln, Mass. (in Cannon Family Papers). The romance took pride of place in Mrs. Cannon's letters this year. According to his sisters, Cannon felt very close and comfortable with Ellen, who bore a resemblance to his mother; see further "Memories of Walter Bradford Cannon, as recalled by Ida, Bernice and Jane, Aug 1959," p. 32 (copy in Cannon Papers).

5. Ibid., Cannon to Bradford Cannon, 7 Jan 1938. Among those in Boston whom Cannon consulted about his son's career were Joe Aub and Elliott Cutler; see, e.g., Cannon to Cutler, 18 Nov 1938; Cutler to Cannon, 29 Nov 1938; Cannon to Cutler 2 Dec 1938. For a reprise of Bradford Cannon's career, see his memoir, "In Retrospect: The flowering of plastic surgery," *JAMA*, Feb 9, 1990, pp. 862–864. In 1986, he and Joseph E. Murray were honored in Boston as leading figures in the fields of plastic and transplant surgery and a videotape of the pair was made for the Leaders in American Medicine series.

6. For the move, see 1939 letter-book, Mrs. Cannon's family letter of May 9. In her letter of May 30, written after the funeral of their old friend Louise Crothers, Cornelia remarked, "Ida is quite heartbroken. I think the combination of her death, Dr. Cabot's death, and her taking up the new apartment alone has combined to make life look pretty difficult to her, but she is a strong person and will rise above it." See also 1940 letter-book, where Marian's wedding is described in a family letter of Aug 12, and there are photographs and press clippings. (In Cannon Family Papers).

7. Exchange of letters between H.H. Dale and Cannon, 13 Feb and 16 Apr 1940; see also letters between Cannon and Dale, 17 May and 6 Jun 1940. For an examination of the unsatisfactory legacy of America's participation in World War I, see M. and S. Harries, *The Last Days of Innocence: America at War, 1917–1918,* N.Y., Random House, 1997.

8. Cannon ended his letter to Magnus Gregersen, 24 May 1940, "I have stated quite frankly my point of view. If it disagrees with yours I am sorry that we shall have to agree to disagree." Phil Bard's comment is in the introductory note to 1940 in his collection of letters with Cannon (in Box 199 of the Cannon Papers). See also Bard's letters of 10 and 27 May 1940, in which he stated his opposition to the idea of peace at any price.

9. A copy of the Peace Resolution, as well as clippings from the *New York Times,* unidentified and undated, and from the *Evening Standard* of London, May 28, 1940, may be found in Cannon's Ideal Scrapbook. On Jun 6, Cannon resigned his sponsorship of the Boston Chapter of the A.A.S.W. The following day, 7 Jun 1940, Cannon wrote A.J. Carlson, "There is evidence of a group here in Boston which has been aggressively arranging for dominance of the [A.A.S.W.] chapter and which is known to be of the extreme left persuasion. Ructions have resulted from the action of this group and members of the Association have resigned in consequence of these high-handed practices."

10. A.V. Hill to Cannon, 28 May 1940, to which Cannon replied on May 29; Cannon to W.W. Norton, 8 Jun 1940. For an account of the A.A.S.W. peace

resolution and its ramifications, see Peter J. Kuznick, *Beyond the Laboratory: Scientists as Political Activists in 1930s America* (Univ. of Chicago Press, 1987), pp. 243–248.

11. Cannon's two letters to Dale were written on 25 Sep and 15 Nov 1940.

12. Cannon to Dale, 27 Jul 1940; in H.H. Dale Papers, Royal Society. On Jul 9, Cannon had informed Dale that there was a U.S. Committee on Refugee Children, which he notified about the Zander children. He also wrote to R.B. Perry, 27 Sep 1940, asking if the Harvard Branch of American Defense had any means of placing the children.

13. Cannon to Gretel Zander, 18 Nov 1941, in reply to her letter of Oct 21. See further correspondence in fd. 794 of the Cannon papers. Two of the Zander children are now residents of Cambridge, Massachusetts: Benjamin is conductor of the Boston Philharmonic Orchestra and his sister Angelica is the wife of Neil Rudenstein, president of Harvard.

14. See further Cannon to L.J. Henderson, 12 and 29 May 1941; Cannon to Hallowell Davis, 22 Oct 1941. See also André Mayer to Cannon, 30 Jun 1944 and [1945] letter in fd. 716 of Cannon Papers. Mrs. Cannon's letter of Mar 29 is in the 1941 letter-book, Cannon Family Papers. Jean Mayer later became a professor of nutrition at Harvard, then president of Tufts University.

15. Exchange of letters between Cannon and Hill, 18 Nov and 25 Dec 1941. See further Cannon's correspondence with Hill in fd. 829 of the Cannon Papers.

16. For the formation of the Shock Committee, see Cannon to L.H. Weed, 22 May 1940; Weed to John Scudder, 25 May 1940; Cannon to C.S. Burwell, 15 Nov 1940. Cannon's letter to Robert Yerkes was written on 10 Apr 1940.

17. Weed to Cannon, 18 Mar 1943. Besides E.J. Cohn, others at Harvard who were concerned with problems bearing on shock and blood substitutes included Cecil Drinker, Joseph C. Aub, Henry K. Beecher and, until his death in 1942, Soma Weiss. For the N.R.C. Shock Committee, see correspondence in fds. 1090 and 1091 in the Cannon Papers. In *The Way of an Investigator,* Cannon reiterated a precept that had come out of the work on shock in World War I: "When there has been loss of blood, the ideal replacement is, of course, blood itself." This time, instead of recommending gum acacia to improve the circulation, he pointed to plasma and albumin as being the most important constituents of blood substitutes (pp. 144–145).

18. Cannon to Robert Lasch, Nieman Foundation, 6 May 1942. Cannon recommended for further reading on the emotional effects of warfare *The Neuroses in War,* ed. by E. Miller (Macmillan, 1940) and *Psychological Effects of War on the Citizen and the Soldier,* by R.D. Gillespie (Norton, 1942); and on shock, along with his own *Traumatic Shock* (D. Appleton, 1923), he cited Virgil Moon's *Shock and Related Capillary Phenomena* (Oxford, 1938) and John Scudder's *Shock (Blood Studies as a Guide to Therapy)* (Lippincott, 1940). See also Cannon's letters to Bard, 22 Aug and 3 Sep 1941.

19. Cannon to Bard, 2 Jan 1942; see also Bard's introductory note for 1942 in his collection of correspondence with Cannon (n.8).

20. "The misery of the Spanish refugees," undated, unsigned mimeo, 4 pp.; "I saw it happen, by Eric Muggeridge," undated mimeo, 2 pp; Muggeridge to Cannon, 16 Mar 1939. See also excerpt of letter from Edgar Mowra to

Elizabeth Gilman of 27 Jan 1939, in which Mowra reacted angrily to an article by H.L. Mencken in the *Baltimore Sun* that quoted a report by Irwin Laughlin, former ambassador to Spain. For more on the Spanish refugees in France, see Howard E. Kershner, *Quaker Service in Modern War: Spain and France, 1939–1940*, New York, Prentice-Hall, 1950.

21. Camille Soula to Cannon, 18 Feb 1939. Years later Cannon wrote to Soula from Mexico (12 Mar 1945); referring to the early months of 1939, he told him: "That was a very trying and critical time, and it was evident that you met the emergency in such a way as to earn the very deep gratitude of those whom you aided. I have heard that you were for a time deprived of your place in the University, and also I have heard that you are now back in the University's service. I hope the latter rumor is indeed correct."

22. Bishop McConnell to Cannon, 24 Mar 1939; see also Cannon to A.S. Coolidge, 28 Mar 1939. For the A.F.S.C., see John Reich to Cannon, 3 Apr 1939; Cannon to Reich, 4 Apr 1939; Reich to Cannon, 6 Apr 1939. See also "Program of the Spanish Refugee Relief Campaign," Sep 1939, 4-page mimeo signed by Bishop McConnell and Rev. Reissig; "National Conference on American Aid to Spanish Republican Refugees," Washington, D.C., Oct 7–8, 1939, 1-page mimeo. For Barsky's new committee, see his letters to Cannon of 31 May and 30 Jun 1939; also Barsky to Cannon, 17 Sep 1939, enclosing draft of revised appeal letter signed by Cannon.

23. For Cannon's meeting with Juan Negrin, see his letter to Bradford and Ellen Cannon, 9 May 1939; also Mrs. Cannon's letter of May 10; both in 1939 letter-book. Among the Spanish refugees whom Cannon attempted to help, largely through the Boston Emergency Committee, were Rafael de los Rios (the nephew of the former Spanish ambassador to the U.S.), Rafael Mendez (who came to work with Otto Krayer at Harvard), Juan Negrin, Jr., and Rosendo Carrasco-Formiguera. Cannon's most frustrating experience by far occurred through his effort to aid J.M. Bellido, who had fled to Toulouse with Carrasco. Despite years of trying, Cannon was forced to admit that problems of exit and transport could not be solved to get Bellido out of France.

24. Herman Reissig to Cannon, 18 Jan 1940, enclosing press release about SERE of Dec 19, 1939. After the Germans occupied France in the spring of 1940, the British government granted Negrin asylum in England, on condition that he abstain from political activity.

25. See Memorandum on the Spanish Refugee Situation, based on reports and decisions of the Pan-American Conference on Spanish refugees held in Mexico City, Feb 15–18, 1940, 6-page mimeo signed by Samuel Guy Inman and Herman Reissig for the S.R.R.C., as well as by John Rich (formerly Reich) of the A.F.S.C. See also Resolutions adopted by the Executive Board of the S.R.R.C., Mar 19, 1940.

26. Reissig to Executive Board, 26 Feb 1940, enclosing his letter to Douglas Jacobs of Feb 24. "I greatly regret the necessity of this action," Reissig wrote Jacobs. "It has, however, been made unavoidable by your conduct at the Mexico City Conference." See also "Inside History of the Spanish Refugee Relief Campaign," draft of statement, undated and unsigned, but evidently written by Reissig in Mar 1940, which has sections on the Mexico City Conference and the Menard Order; in Columbia University Archives.

27. Dr. Barsky's letter to the *New York Post* of Mar 29 (which ended, "We have always fought and will continue to fight for fair treatment of Spanish refugees") was enclosed in his letter to Cannon of 15 Apr 1940. See also Reissig's mimeo letters of 12 Mar and 16 Apr 1940 and "Inside History of the Spanish Refugee Relief Campaign."

28. Exchange of letters between Reissig and Cannon, 22 and 29 Mar 1940. A copy of Cannon's mimeographed letter of resignation, 16 Apr 1940, may be found in his Ideal Scrapbook.

29. Cannon to Edward Barsky, 18 Apr 1940; Barsky replied on May 1. See also Jay Allen to Cannon, 18 Apr 1940.

30. Cannon to Harlow Shapley, 20 Apr 1940; other sponsors of the National Emergency Conference included Alice Hamilton, Haven Emerson, and Wesley Mitchell (a former president of the A.A.A.S). The 4-page mimeo, "Stalinist Outposts in the United States," was Report No. 2 of the Committee for Cultural Freedom; it was supplemented by a 1-page history of Communist-dominated Spanish relief organizations. See also Sidney Hook to Cannon, 24 Apr 1939, enclosing a statement on the formation of the C.C.F.

31. Shapley to Cannon, 30 Apr 1940. For more on Shapley's efforts in behalf of European refugee scholars, see Bessie Zaban Jones, "To the rescue of the learned: The asylum fellowship plan at Harvard, 1938–1940," *Harvard Library Bulletin* 32 (1984):204–237.

32. See Sidney Hook's letters to Cannon of 20 May and 15 Jun 1940. On 24 May 1940, Reissig wrote Cannon at length about Hook and his charges against the Spanish aid organizations. See further Kuznick (n.10), pp. 215–221; Hook's autobiography, *Out of Step: An Unquiet Life in the 20th Century*, New York, Harper & Row, 1987; Christopher Phelps, *Young Sidney Hook: Marxist and Pragmatist*, Cornell Univ. Press, 1997.

33. For testimony about the Medical Bureau's efforts in Spain, see Cannon's correspondence with Reissig, Allen and Puché in fd. 691 and with Eloesser, Ettelson, and Posner in fd. 692 of the Cannon Papers. Posner, an oral surgeon in New York City, gave up his practice to become head of the second unit sent to Spain by the Medical Bureau; Mitchell was his secretary. Pitts, a general surgeon from Texas, and Ettelson, a brain surgeon in Chicago, had been ousted for protesting against Communist political malpractice in the unit. Eloessor was a thoracic surgeon at Stanford. Posner claimed that Barsky was a Communist and that he had been tried in Spain for incompetence and found guilty. All of Posner's charges related to what he had seen in Spain before he left early in 1937.

34. Cannon's vigorous defense was in a letter to Hook of 19 Jun 1940. See also his follow-up letter of Jun 20; Hook to Cannon, 22 Jun 1940; Cannon to Hook, 9 Jul 1940.

35. Hook's offer of an emended version was written on 14 Jul 1940, after Cannon had complied with his request to send on the letters he had received from the doctors serving in Spain. The last letters were Hook to Cannon, 21 Aug 1940; Cannon to Hook, 26 Aug 1940; Reissig to Cannon (about Henson's statements), 27 Aug 1940; Cannon to Hook, 28 Aug 1940.

36. For the E.R.C., see Reissig to Cannon, 3 Sep 1940; Cannon to Dorothy Hickie, 26 Nov 1940. See also Memorandum from Samuel G. Inman re Visit

of Miss Kirchwey, Dr. Cannon, Mr. Del Vayo and Mr. Inman to Washington, on Tuesday, Nov 12, 1940. Their interviews at the State Department included Breckenridge Long and Herbert C. Pell; they also saw Walter T. Prendergast in Nelson Rockefeller's office as well as the Mexican Ambassador. Cannon wrote Reissig on Nov 13, giving his version of the visit, the purpose of which was to find out what could be done to get the Spanish Loyalists remaining in France to Mexico and Latin America.

37. Reissig to Cannon, 18 Nov 1941, with Cannon replying the same day. The exchange of letters between Dorothy Salinger and Cannon were dated 7 and 8 Mar 1941.

38. Cannon to Carrasco, undated handwritten letter from St. Petersburg, Florida, where he and Mrs. Cannon were spending the holidays and the month of January with the Pierces. After arriving in Mexico in Jun 1939, Carrasco, in addition to obtaining a post with La Cara de España en Mexico, was in charge of biological control for an industrial pharmaceutical laboratory; he also gave a series of physiopathological lectures at the universities in Puebla and Guadalajara. By 1941, all of his family had joined him and in the spring he accepted a post at the University of Puebla. Subsequently, Cannon helped him to obtain scientific literature and sent him twenty kymographs for his students.

23. The End of a Career

1. Cannon to Hallowell Davis, 21 Feb 1940. Cannon enclosed the memorandum of his conversation with the dean on Dec 14 in his letter to C.S. Burwell of 19 Jan 1940.

2. Houston Merritt to Cannon, 19 Oct 1940, in reply to Cannon's letter of Oct 17. For the earlier letters, see Cannon to Elizabeth Bright, 21 Sep 1940; Cannon to C.S. Burwell, 22 Sep 1940; Cannon to Gladys H. McCafferty (director of personnel), 24 Sep 1940.

3. Davis to Cannon, 12 Nov 1940; Cannon to Alexander Forbes, 24 Apr 1941. For the war activities, see also Davis to Burwell, 6 Nov 1940, and Cannon to Burwell, 15 Nov 1940.

4. Oscar Ratnoff to A.C. Barger, 17 Jan 1974; in Box 189 of the Cannon Papers. In his letter to Harlow Shapley of 22 Oct 1941, Cannon regretted that Hudson Hoagland had not thus far received the recognition he deserved.

5. Cannon to Burwell, 7 May 1941; Burwell to Cannon, 6 Jun 1941; Cannon to Mrs. McLeod, 30 Jun 1941.

6. See further Rosenblueth and Cannon, "Some features of the early stages of neuromuscular transmission," *Am. J. Physiol.* 130(1940):205–218; Cannon and Rosenblueth, "Some conditions affecting the late stages of neuromuscular transmission," follows on pp. 219–229.

7. For Friedgood and Cannon's paper, see *Endocrinology* 26(1940):142–152; see also Chapter Twenty above, pp. 416–417. In his autobiographical memoir, *The Way of an Investigator,* Cannon ruefully recalled, "I have spent unconscionable time in trying to obtain control of the secretion from the thyroid gland. I fear that something less admirable than steadfastness was manifested in that endeavor." (p. 38)

8. Cannon, "The adrenal medulla," *Bull. N.Y. Acad. Med.* 16(1940)3–13; "The role of chance in discovery," *Sci. Monthly* 50 (1940):204–209. The commencement address at Washington University was unpublished. As recently as 1935, Rogoff had declared that the adrenal medulla does not play an important role in the body and a specific function for adrenaline had still not been proved. "Obviously," Cannon countered, "the adrenal medulla is not essential to existence. There is a difference, however, between being useful and being essential." Decades later, an editorialist in *Lancet* (Mar 9, 1985, p. 561) would allude to adrenaline as "the Sydney Carton of hormones—a bystander except at moments of catastrophe."

9. For the Hitchcock professorship, see Cannon to Burwell, 6 Dec 1940. Cannon obtained leave during the month of January as well, in order to escape the New England winter and work on his lectures at George Pierce's new home in Florida. For clippings and memorabilia about the visits to Berkeley and Stanford, as well as the Friedenwald Medal bestowed on 5 May 1941, see further Cannon's Ideal Scrapbook.

10. See 1941 letter-book for Mrs. Cannon's family letter of Jul 21; also other letters written in July by Cornelia, Walter, and Bradford Cannon (in Cannon Family Papers).

11. Ibid., 28 Sep 1941. For the two papers, see *Am. J. Physiol.* 135(1941):690–741 and 137(1942):681–694.

12. Ibid., 21 Oct 1941. For the Grasses, see George H. Acheson to Saul Benison, 25 Nov 1987, enclosing printed letter from Albert and Ellen Grass "To our friends," celebrating the twentieth anniversary of the Grass Instrument Company, 1935–1955; also pamphlet, "The Electro-encephalographic Heritage until 1960," by Albert M. Grass (Grass Instrument Co., 1984).

13. Cannon to Burwell, 30 Jan 1942, with Burwell's response of the same date; Cannon to J.B. Conant, 2 Feb 1942; Cannon to Burwell, 17 Feb 1942; Conant to Cannon, 17 Feb 1942. The President and Fellows of Harvard College officially accepted Cannon's resignation on 2 Mar 1942.

14. For Mrs. Cannon's comment, see 1942 letter-book, letter of Feb 19; in Cannon Family Papers. See also Cannon to Abe Ravin, 5 May 1942. In addition, Cannon took more care than ever, given the dislocations of wartime, to maintain some sort of continuity for his colleagues at the Medical School. He mourned the passing of some, like Hans Zinsser and Soma Weiss, and welcomed others, like George Thorn, who replaced Dr. Weiss as chief of medicine at the Brigham Hospital, and René Dubos, the new Fabyan Professor of Comparative Pathology.

15. Davis to Burwell, 28 Mar 1942; Cannon to Burwell, 23 Apr 1942. Cannon had earlier written Cecil Drinker (22 Feb 1942) in defense of Rosenblueth, whose alleged contentiousness could be attributed to the fact that "with him an argument is as much a form of sport as a game of tennis would be to another who is more interested in neuromuscular exercise than exercise of the cerebral cortex." See further Louisa Barclay Benton, "Arturo Rosenblueth: Success or Failure?" (undergraduate thesis presented to the Committee on History and Science, B.A. degree, Harvard University, 1986), pp. 62–64.

16. Benton, pp. 51–52, quotes Bard's letter to Drinker of 15 Dec 1939.

17. Ibid., esp. pp. 48–50 and 54–57. Norbert Weiner's *Cybernetics* was published by M.I.T. Press in 1948, and a paper edition was issued in 1961. See also Alex Forbes to Albert P. Usher, corresponding secretary of the A.P.S., 8 Mar 1939; Horace W. Davenport, "Signs of anxiety, rage or distress," *The Physiologist*, Oct/Nov 1981, p. 3.

18. See Cannon to Burwell, 1 May 1942; Burwell to Cannon, 2 May 1942; Cannon to Burwell, 16 Jun 1942. Cannon wrote Elliott Cutler on 19 Jun 1942; see also E.S. Castle to Cannon, 29 Jun 1942, offering a place at the Biological Labs in Cambridge, and Cannon's reply of Jun 30.

19. Alan Gregg to Cannon, 2 Apr 1942, in reply to Cannon's letter of Mar 11.

20. For the 1934–35 correspondence on voodoo death, see fds. 898–901 in the Cannon Papers; for publication of Cannon's article, see also fds. 902–904. For Cannon's report of Norman Freeman's findings, see "Consideration of possible toxic and nervous factors in the production of traumatic shock," *Ann. Surg.* 100(1934):704–713. On 4 Nov 1940, in response to a request for an article on the emotions from Eugene Lyons, the editor of the *American Mercury*, Cannon begged off: "I fear that I have overworked the subject of the influence of emotional excitement on the body—I published a summary of our observations about twenty-five years ago and have republished them since then until I am rather fed up with the subject."

21. See further W.B. Cannon, "'Voodoo' death," *American Anthropologist* 44(1942):169–181. It was reprinted in *Psychosomatic Medicine* 19 (1957):180–190; Richter's article, "On the phenomenon of sudden death in animals and man," (which showed a primary involvement of the parasympathetic, rather than the sympathetic, system) is on pp. 191–198, and the minutes of the meeting held in honor of Cannon are on pp. 179–181. See also clipping from the *Boston Globe*, Jul 2, 1943, and unidentified clipping re voodoo death in Cannon's Ideal Scrapbook. For a more recent treatment of the subject, see Martin A. Samuels, "'Voodoo' death revisited: The modern lessons of neurocardiology," *The Neurologist* 3(1997): 293–304.

22. Houssay wrote to Cannon on 21 Mar 1942. There were other antecedents as well. In 1939, after Secretary of State Cordell Hull informed Cannon that the Division of Cultural Relations had decided to concentrate attention on the development of more effective relations with other American republics, he was invited to attend a conference in Washington, D.C. on inter-American relations. The following year, he became a member of a Harvard committee to further cultural relations with Latin American countries; see J.B. Conant to Cannon, 27 Nov 1940, and Cannon's reply of Nov 29.

23. Cannon replied to Houssay on 2 Apr 1942. See also L.W. Hackett to D.H. Stevens, 21 Apr 1942; exchange of letters between H.A. Moe and Cannon, 29 and 30 Apr 1942.

24. For the N.A.S./N.R.C foreign relations assignments, see Cannon to W. Mansfield Clark, 17 Apr 1942, and to Ross G. Harrison, 21 May 1942. For Cannon's obituary notice of L.J. Henderson, see *Biographical Memoirs of the National Academy of Sciences* 23(1943):31–58; also materials in Box 187 of the Cannon Papers. On 21 May 1942, Cannon wrote William DeB. MacNider: "The world is contracting at present . . . and most of my job will be directed toward

improving relations with Latin American countries. Fortunately, I have a goodly number of students who are now back in Mexico and South America who are occupying positions of influence. I hope that with their help I may be useful in developing closer relations with South American men of science."

25. See Cannon to Moe, 21 May and 5 Jun 1942; also 1942 letter-book for Mrs. Cannon's letter of May 16.

26. Wilma wrote on 28 Jun 1942, and her mother replied on Jul 1; both in 1942 letter-book. Moe's earlier letter to Cannon was 19 Jun 1942.

27. Cannon's letter was dated 18 Aug 1942. See also Ross McFarland to Charles A. Thomson, Chief of the Division of Cultural Relations, 17 Jul 1942; H.A. Moe to Cannon, 23 Jul 1942; Mrs. Cannon's letter of 3 Aug 1942; McFarland to Cannon, 9 Aug 1942; Thomson to Cannon, 11 Aug 1942.

28. Conant to Cannon, 29 Sep 1942, in reply to Cannon's letter of Aug 31 (quoted in Mrs. Cannon's unpublished memoir, "Servant of Science," p. 216A). When Jerome Greene read Cannon's letter, he wrote him (30 Sep 1942), "The Class of 1896 has long regarded you as its brightest ornament, and your classmates have felt the utmost pride in what you have been and what you have done."

29. Cannon to Bard, 13 Jul 1938. Cannon wrote Yerkes about his Bell's palsy on 24 Sep 1942; see also Mrs. Cannon's family letter of Sep 18 in the 1942 letter-book.

30. Cannon to Peyton Rous, 12 May 1939. Dr. Aub added in his case note of 1 Oct 1940 that everyone remarked about Cannon's great vigor and good humor and general well-being. See further "The case of Dr. W.B. Cannon," esp. folder of 1940–46 materials; in Box 8 of the J.C. Aub Papers.

31. Cannon to Bard, 26 and 28 Sep 1942. According to Bard's introductory note for 1942 in his collection of correspondence with Cannon (in Box 199 of the Cannon Papers), a survey committee had been set up by A.N. Richards of the C.M.R. to travel around the country and report what was being done in work on shock. He remembered it as an exhausting job "and there continued to be many meetings which W.B.C. could not attend and he counted on the reports of Magnus Gregersen and me to keep him informed."

32. 1942 letter-book, Mrs. Cannon's family letter of Oct 3. See also Ross Harrison to Cannon, 5 Oct 1942; H. H. Dale to A. N. Richards, 6 Nov 1942 (in Dale Papers, Royal Society).

33. Ibid., 20 and 31 Oct 1942.

24. Medical Aid to China; American-Soviet Friendship

1. Cannon to P.S. Tang, 30 Nov 1937; see also H.C. Chang to Cannon, 14 Jul 1937; Cannon to Robert K.S. Lim, 15 Apr 1938. "Peiping" and "Peking" were both used at this time for the name of the Chinese city now known as Beijing. For Cannon's visit to China in 1935, see Chapter Seventeen above, pp. 337–341.

2. Lim to Cannon, 28 Jun 1938, in reply to Cannon's letters of Apr 15 and May 10. Cannon thought the aid to China provided by the American Red Cross was shameful and wrote President Roosevelt on Mar 24 to complain

about the situation; but in his second letter, Cannon told Lim that prospects for American aid had become more favorable. For the early Chinese aid organizations, see fd. 521 in the Cannon Papers; for Cannon's willingness to serve on the medical committee of the New England Committee, see also William G. Lennox to Cannon, 2 Nov 1937.

3. Marian Exter to Cannon and Cannon to Robert Littell, both 28 Apr 1939. Previously, on 3 Jan 1938, Cannon had written William E. Dodd, Jr., of the China Aid Council that he was personally so deeply concerned with the Spanish situation that he could not take an active part in aiding China. He had, however, publicly promised that the Medical Bureau would offer all the aid it could for instructing the Chinese as to methods used for helping the Spanish Loyalists.

4. Exchange of letters between G. Canby Robinson and Cannon, 8 and 9 Mar 1939. For Cannon's service on the China Medical Board, see correspondence and other materials in Box 41 of the Cannon Papers. See further Mary E. Ferguson, *China Medical Board and Peking Union Medical College: A Chronicle of Fruitful Collaboration, 1914–1951*, N.Y., China Medical Board, 1970.

5. For communications from John Layton Stewart, see fds. 525 and 536; for those from Henry Houghton, see fds. 524, 540, and 541 in the Cannon Papers. The dean of the P.U.M.C., C.E. Lim, met with Lobinstine, Alfred Cohn, Alan Gregg, Canby Robinson and Cannon; see 1-page Memorandum on the Conference with Dr. C.E. Lim. See further Mary Brown Bullock, *An American Transplant: The Rockefeller Foundation and the Peking Union Medical College* (Berkeley: Univ. of California Press, 1980), esp. Chapter 8, "The Challenge of War," pp. 190–208.

6. See, e.g., copy of letter from P.U.M.C. Board of Trustees to John D. Rockefeller, Jr., 12 Dec 1940, in response to Rockefeller's letter of Oct 2, which was sent for discussion by the Trustees in China (his letter did not sit well with them). For events after Pearl Harbor, see E.C. Lobenstine to Cannon, 2 Mar 1942; also "Report on the Peiping Union Medical College," 15 May 1942.

7. Dr. Claude Forkner, a distinguished alumnus of the Harvard Medical School, had served as Cannon's physician during his visit to China in 1935. In the early 1940s, he was affiliated with Cornell Medical College and the New York Hospital. For his appointment as representative in China, see Minutes of meeting of the China Medical Board, Dec 9, 1942; Lobenstine to Cannon, 29 Dec 1942; Forkner to Cannon, 11 Jan 1943.

8. Cannon to James G. Blaine, 7 Apr 1941, in reply to Blaine's offer of the chairmanship of the Medical Division of Apr 2. For the formation of the United China Relief (U.C.R.), see Theodore Roosevelt, Jr., to Cannon, 19 Feb 1941; also mimeo letter signed by Cannon, 24 Apr 1941.

9. See 11-page typescript, "Medical Needs of China's Armies," from T.H. White, Chungking, China, to David Hulburd, Feb 20, 1941, received Mar 12, 1941. White's letter, passed along to Cannon by Helen K. Stevens of ABMAC, was acknowledged in a postscript to a letter Cannon wrote Stevens on 29 Apr 1941. See further White's book, *Thunder Out of China* (N.Y.: William Sloan, 1946), written with his Time-Life colleague Annalee Jacoby, which John Fair-

bank has called "a classic account of wartime China" (*Chinabound: A Fifty Year Memoir,* Harper, 1983, pp. 319–320).

10. Cannon's appeal letter was dated 24 Apr 1941; Carlson's response was quoted in a letter from Helen Stevens to Cannon, 9 May 1941. See also D.D. van Slyke to Cannon, 4 Jun 1941; Stevens to Cannon, 21 Oct 1941 (one of several letters about plans for the future of U.C.R.). United China Relief had been established as a temporary organization for the sole purpose of raising funds. In effect, it represented a struggle for the control of the charity dollar that involved such power-brokers as John D. Rockefeller, Jr. and Henry Luce. Van Slyke's plan was to put U.C.R. on a permanent basis, modeled after the Community Chest—i.e., U.C.R. collecting the money, then giving agencies appropriations to take care of their programs in China.

11. See, e.g., Cannon to Stevens, 16 Jun 1942; Stevens to Cannon, 1 Jul 1942; van Slyke to Cannon, 7 Jul 1942. The earlier exchange of letters between van Slyke and Cannon was 19 and 21 Nov 1941.

12. Telegram from Minna Weiner to John Kingsbury, 1 Jul 1941, in reply to Kingsbury's telegram to Cannon of Jun 30. The rally, hastily put together, was to take place the very next day, on July 2. On Jun 22 the Germans had staged a surprise invasion of the U.S.S.R., which resulted in Stalin doing a turnabout and joining the Allied forces.

13. Cannon to Henry E. Sigerist, 18 Jul 1941, in reply to Sigerist's letter of Jul 16. Cannon did join a Boston group to aid the Soviet Union in the autumn of 1941, however, because his friend Ralph Barton Perry's sponsorship was conditional on his approval; see further Dirk Struick to Cannon, 12 Sep 1941. Dr. Sigerist had visited Russia on several occasions during the 1930s while preparing his book on *Socialized Medicine in the Soviet Union* (London: Victor Gollancz, 1937, with foreword by Sidney Webb; N.Y.: W.W. Norton, 1937). Sigerist's volume was one of the few English-language books that dealt in detail with medical education, medical care and public health in the Soviet Union before World War II.

14. Cannon's message was enclosed in a letter of 8 Oct 1941 to D. Chuvakhin at the Soviet Embassy; a clipping from the Nov 1941 *Soviet Russia Today* with the message is in Cannon's Ideal Scrapbook. See also N.I. Propper to Cannon, 26 Aug 1941, and Cannon's reply of Oct 21. The letter, which Cannon received after considerable delay, was brought to the U.S. by the editor of *P.M.* Propper had been at the Rockefeller Institute in 1936 and visited Cannon in his laboratory and at home.

15. V. Kallistratov to Cannon, 12 May 1942, in reply to Cannon's letter of May 2.

16. Hugh Cabot to Cannon, 6 Oct 1942; see also his letters of Jul 8 and 30. Cannon's attack of Bell's palsy also prevented him from attending the Congress of American-Soviet Friendship held in New York, Nov 7–8, 1942; see further Cannon to Corliss Lamont, 16 Oct 1942.

17. For a history of the training program, see "Report of the Emergency Medical Service Training Schools," Sep 17, 1942, prepared by Dr. Lim for ABMAC. For criticism of Chiang Kai-shek and his Nationalist government,

see, e.g., "1943—CKS Begins to Lose the Mandate," Chapter 20, in Fairbank's *Chinabound* (n.9), pp. 241–264. For Cannon's appointments, see Frank Co Tui to Cannon, 10 Jun 1943; "Tentative alignment of sub-committee membership," which lists Cannon on a subcommittee for Army medical service and on another for medical education.

18. On 29 Apr 1942, Cannon wrote E.C. Lobenstine of the China Medical Board about his role in the election of Lim as a foreign associate of the National Academy of Sciences. See also Minutes of meeting of China Medical Board, Feb 9, 1943, in which the political situation involving Lim was discussed. For Lim's 1944 visit, see D.D. van Slyke to members of the executive committee, ABMAC, 7 Mar 1944; Helen Stevens to Cannon, 5 and 19 Jul 1944.

Besides Lim's activities for the Chinese Red Cross, the training schools, and the Medical Relief Commission, he also built the largest medical center in wartime China. For a colorful description of Lim's war service as well as his career as a physiologist, see further "Robert Kho-Seng Lim, 1897–1969" by Horace W. Davenport, *Biographical Memoirs of the National Academy of Sciences* 51 (1980):281–306. See also "In Memoriam: Robert Kho-Seng Lim" by A.C. Ivy, *Gastroenterology* 58(Apr 1970):580–581. In 1949, on the invitation of Dr. Ivy, in whose lab he had held a fellowship in the 1920s, Lim returned to the U.S. for research posts in various locations until his retirement.

19. Cannon to Mildred Price, 23 Jun 1943; for more on Bethune's letters, see Cannon to Mentana G. Sayers at the China Aid Council, Aug 3 and 7. See also "History of the Blockade," 3-page undated mimeo. Mme. Sun (née Ching-ling Soong) was one of three daughters of Charlie Soong. Her sister Ai-ling had married H.H. Kung and her sister Mai-ling married Chiang Kai-shek. See further Sterling Seagrave, *The Soong Dynasty*, N.Y., Harper & Row, 1985.

20. Cannon to Mrs. E.C. Carter, 26 Nov 1943, in reply to her letter of Nov 19. Mrs. Carter and her husband were old China hands; E.C. Carter was well known as a founder and director of the Institute of Pacific Relations (I.P.R.) and of the United China Relief. See also exchange of letters between Price and Cannon, 23 and 26 Sep 1943; exchange between Cannon and James L. McConaughy, 4 and 6 Oct 1943. McConaughy wrote again on Oct 9, enclosing copy of a letter of Oct 8 from C.L. Hsia, director of the Chinese News Service, containing a detailed report of the situation in China.

21. See exchange of letters between Cannon and Fairbank, 23 and 25 Mar 1944; also copy of letter from Price to McConaughy, 12 Apr 1944, about reports of T.H. White. For the letter to the Chinese Ambassador, see materials in fds. 576 and 577 of the Cannon Papers, which include several drafts of the letter and correspondence with Miss Price and Mrs. Carter.

22. For the Chinese Ambassador's reply, see Wei Tao-Ming to Cannon, 7 Jun 1944. See also John Fairbank to Cannon, 5 Jul 1944, in reply to Cannon's letter of Jun 26; Cannon to Mrs. Carter, 13 Jul 1944. Mildred Price wrote Cannon on Sep 30: "I suppose that we can go on the assumption now that the blockade is still on but goods are being sent through pretty regularly. I therefore do not feel that you should worry any more about giving the letter to the Ambassador to the papers."

23. Sigerist to Cannon, 24 Mar 1943. See also materials in Box 39 of the Cannon Papers, esp. announcements of the formation of the American-Soviet Society and *The American Review of Soviet Medicine* in fd. 497. See further *Making Medical History: The Life and Times of Henry E. Sigerist*, ed. Elizabeth Fee and Theodore M. Brown, Johns Hopkins Press, 1997, esp. Walter J. Lear, "Hot War Creation, Cold War Casualty: The American-Soviet Medical Society, 1943–1949," pp. 259–287.

24. Sigerist to Cannon, 1 Apr 1943, referring to Cannon's letter of Mar 26. After receiving Cannon's letter of 2 Apr 1943, Sigerist wrote again on Apr 7 to tell him, "it is a great relief and a great joy to know that you will serve as president of the American-Soviet Society." For Eloesser, see further Harris B. Shumacker, Jr., *Leo Eloesser, M.D.: Eulogy for a Free Spirit*, New York: Philosophical Library, 1982.

25. Sigerist to Cannon, 13 Oct 1943. For the gala dinner in New York, see Abraham Stone, secretary of the Society, to Cannon, 7 and 25 Jun 1943.

26. For the reception honoring the three Americans taken into the Academy of Sciences of the U.S.S.R., see clipping from *Washington Post*, Aug 13, 1943, and undated article with photograph from the *Information Bulletin*, Embassy of the U.S.S.R., in Cannon's Ideal Scrapbook. Cannon learned of his election to the Russian Academy in May 1942. For the Science Committee and Congress of the National Council, see fds. 498 and 499 in the Cannon Papers. The papers presented at the Congress were published in a volume, *Science in Soviet Russia* (Lancaster, Pa., The Jaques Cattell Press, 1944), with a Preface by Cannon on pp. v-vii.

27. Cannon to Sigerist, 20 Mar 1944. Sigerist reported on the progress of the new organization and its journal in a letter to Elmer McClain, 20 Jun 1944. For expressions of interest in the Society and the *Review*, see, e,g., Chauncy D. Leake to Cannon, 11 Jun 1943; E.B. Krumbhaar to Cannon, 7 Oct 1943; Ross T. McIntire to Robert L. Leslie, 5 Nov 1943.

28. Sigerist to Cannon, 5 Apr 1944, enclosing copy of his memorandum to the executive committee of the Society, 5 Apr 1944; both of these detail the difficulties he faced as editor. See also Roger S. Greene to Cannon, 1 and 5 Apr 1944. Greene, who was then with the State Department in Washington, D.C., passed along some unfavorable comments about the *Review*, which he asked Cannon to look into.

29. For a copy of the Open Letter, enclosed in a letter from Edwin S. Smith to Cannon, 3 Oct 1944, as well as a press clipping about Ambassador Bullitt's statement from the *Daily Worker* of Oct 25, see fd. 505 of the Cannon Papers. This was not the first time Cannon had crossed swords with Luce, a strong supporter of Chiang Kai-shek, who favored the blockade of supplies to the Communist forces in North China that Cannon so staunchly opposed.

30. Henry Luce to Edwin S. Smith, 16 Oct 1944, enclosing transcript of a dispatch from Rome in the *New York Times*, which seemed to confirm much of what Bullitt reported. A draft of Cannon's letter to Luce is dated 27 Oct 1944.

Alfred Kohlberg, on 25 Oct 1944, wrote Cannon, "Imagine my surprise today in having shown to me the 'Daily Worker' with your picture! I read the article and think that somebody has certainly put over a fast one on you." He went on to suggest that Cannon as a signer of the open letter was in embarassing company and hoped that they both would "live long enough to find out from the course of events in Europe whether Bullitt had hit the nail on the head or was completely cock-eyed." Kohlberg was a New York importer of Chinese textiles and a director of ABMAC, who from 1943 began attacking the I.P.R. and the U.C.R. as "soft on communism"; see his correspondence with Cannon in fd. 574 of the Cannon Papers. See also Joseph Keeley, *The China Lobby Man: The Story of Alfred Kohlberg,* New Rochelle, N.Y., Arlington House, 1969.

31. *The Way of an Investigator,* p. 165.

25. On the Wings of Time

1. 1943 letter book, Mrs. Cannon's family letter of Jun 27; see also her letters of Mar 10, May 31 and Jun 19 (in Cannon Family Papers).

2. Ibid., 4 Jul 1943; *Franklin Journal-Transcript,* Jul 8, 1943. See further *The Boston Cocoanut Grove Disaster—A Survey of Treatment,* prepared for the medical profession by the E. L. Patch Co., Boston, Mass., [1943?].

3. John H. Stokes to Cannon, 27 Sep 1943; in Box 8, J.C. Aub Papers. Cannon wrote Robert Yerkes on 15 Sep 1943 (in Yerkes Papers, Yale University); see also 1943 letter-book for Mrs. Cannon's family letter of Sep 28.

4. Cannon to Joe Aub, 6 Oct 1943 (in Aub Papers). Aub replied on Oct 11 that he thought it wise of Cannon to consult Stokes. He approved of Cannon cutting down on work and travel, though he remembered that discontinuing activities in the past had made him restless and dissatisfied.

5. For Cannon's reliance on Bard and Gregersen, see, e.g., Cannon to Philip Bard, 15 Jul and 6 Oct 1943; for Aub's work on shock, see fd. 3, 1940–1946, in Box 8, Aub Papers. For E.J. Cohn, see Douglas Starr, "Again and again in World War II, blood made the difference," *Smithsonian,* Mar 1995, pp. 124–138. Dr. Douglas Surgenor is in process of preparing a full-length biography of Edwin J. Cohn.

6. See further Aub to C.S. Burwell, 26 Aug 1942, and other documents in file, "Physiology: Ad hoc committee to appoint a successor to Dr. W.B. Cannon"; in Box 7, AA 271, Harvard Medical Archives. See also Burwell to J.B. Conant, 10 Feb 1941 (in Conant Papers, Harvard University Archives).

7. C.K. Drinker to Burwell, 15 Nov 1942, and Burwell to Conant, 17 Nov 1942; both in Ad hoc committee file. Another early supporter of Eugene Landis was Walter Bauer; see his letter to Burwell of 30 Jun 1942.

8. Burwell to Conant, 11 Dec 1942, in reply to Conant's letter of Dec 8; in Ad hoc committee file. See further E.L. Landis folder in 1942–43 series, Conant Papers. J.H. Means had little patience with E.J. Cohn's attitude: "He would like a Claude Bernard in his early thirties," Means wrote Conant, "but doesn't know where to find him." Cecil Drinker, referring to the situation in the physiology department, told Conant on Dec 9, "For a number of years, Dr. Cannon has had little direct contact with what is going on and this is expressed through the fact that when I first came here in 1917, Dr. Cannon always had several medical students working with him; whereas, in the past years, this valuable increment to medical thinking has been entirely lacking."

9. For Landis at this time, see materials in Ad hoc committee file (n.6); also "Settee into chair," article in *Harv. Alum. Bull.*, Jan 30, 1943, which gives the progression of professors of physiology at the Medical School.

10. See further Louisa Barclay Benton, "Arturo Rosenblueth: Success or Failure?" undergraduate thesis presented to the Committee on History and Science, B.A. degree, Harvard University, 1986, esp. pp. 65–68; Rosenblueth's letter to Moe is quoted on p. 65. See also Rosenblueth to Cannon, 4 Oct 1943, and other correspondence with and re Rosenblueth in fd. 1626 of the Cannon Papers.

11. Ibid., pp. 68–71; Rosenblueth's letter to Warren McCullough is quoted on p. 68.

12. See further report of the chairman of the Division of Foreign Relations in fd. 1099 of the Cannon Papers. Richard M. Field, a Princeton geologist, was eager to promote a new Division of National Welfare within the N.R.C. and consulted Cannon about his ideas. It became clear that each division had its own agenda as far as international activities were concerned, and had only to keep the chairman of the foreign relations division informed of their projects. This was especially true for the Division of Geology and Geophysics that Field represented—a discipline where exploration and field work had virtually been stopped dead by the war.

13. See Wm. deB. MacNider to Cannon, 11 Nov 1943; also materials re first meeting of the N.R.C. Committee held on Nov 21, in fd. 1092 of the Cannon Papers. MacNider's earlier letter to Cannon was written on 10 Sep 1943.

14. MacNider to Cannon, 25 Jan 1940; also his letter of Apr 6, enclosing Statement from R.E. Coker of the N.R.C., "Biological Processes of Aging." See also Cannon to MacNider, 19 Oct 1939. All in Box 82 of the Cannon Papers.

15. See Cannon to MacNider, 19 Nov 1940, in response to Carlson's letter of Nov 16; Cannon to MacNider, 8 Mar 1941; MacNider to Cannon, 11 Mar 1941. See also Cannon to V. Korenchevsky, 25 Mar 1941.

16. MacNider to members of the Committee, 12 Apr 1944. For funding from the Macy Foundation, see, e.g., MacNider to Cannon, 2 Jan 1942.

17. Christiane Deane to Herbert M. Evans, 4 Nov 1943; copy of her letter in fd. 860 of the Cannon Papers.

18. See Evans to Cannon, 16 Nov 1943 and 31 May 1944; Circular letter to members of FASEB, 10 Aug 1944. Houssay wrote Cannon on 21 Apr 1945 that the crisis was over and he was returning to his Institute of Physiology.

19. R.M. Yerkes to Cannon, 25 Jan 1944, in reply to Cannon's letter of Jan 2. For more on Cannon's visiting professorship at N.Y.U., see Donal Sheehan to Cannon, 12 and 22 Oct and 30 Nov 1943; also other materials in fd. 1359 of the Cannon Papers. Cannon's lecture notes are in Box 179. See also *The Way of an Investigator*, p. 222.

20. Otto Loewi to Cannon, 13 and 26 Oct 1943. For the tribute to Cannon as president of the American-Soviet Medical Society, see program and materials in fd. 503 of the Cannon Papers; also 1944 letter-book for Mrs. Cannon's family letter of Mar 9 (in Cannon Family Papers). Joe DiPalma, the Philadelphia physiologist who was then a medical student at N.Y.U., retained a vivid

memory of his impression of Cannon at the time. What he had seen was a poor little man with fiery red skin, incessantly scratching at himself, and he marveled that this was the great W.B. Cannon, a world-famous medical scientist who was held in such esteem.

21. See further 1944 letter-book for Mrs. Cannon's family letters of Mar 25 and 30; also Cannon to Yerkes, 5 Apr 1944.

22. For the published report, see W.B.Cannon and R.M.Field, *International Relations in Science: A Review of Their Aims and Methods in the Past and the Future,* N.Y., Chronica Botanica, 1945. See also Cannon to Harlow Shapley, 9 Feb 1944; L.C. Dunn to Cannon, 23 Mar and 10 Apr 1944.

23. Cannon to Thomas B. Appleget, 1 Jul 1944. See also Cannon to Frank B. Jewett, president of the National Academy, 1 Sep 1944, about future relations with the A.A.A.S.: "More and more, I believe, we must rely directly or indirectly on public support of scientific effort, either from governmental subventions, or from industry, or from relatively small gifts from many individuals. If I am right, the process involves wide diffusion of knowledge of scientific studies, its achievements and its needs."

24. Cannon to H.S. Gasser, 2 May 1944; Gasser to Cannon, 10 May 1944; Cannon to Gasser, 20 May 1944; all in Rockefeller Archives Center. See also 1944 letter-book for Mrs. Cannon's family letter of May 17.

25. Mrs. Cannon, "Servant of Science," pp. 217; unpublished manuscript in Cannon Papers. On 30 Jan 1944, "Tom" Perry had written Cannon: "I hope you are going to feel like doing your book on 'confessions of a researcher' this summer. Coming from you it would be a great book, both for the layman and for young scholars to whom you have set so inspiring an example."

26. See further 1944 letter-book for Mrs. Cannon's family letters of Aug 12 and 20. On Aug 20, she wrote: "Father's book on the Sensitization of Nerves has reached the tenth of its thirteen chapters—a most dull book from the reader's point of view, but a book that fills me with admiration for scientists, who can spend their days at such boring activities to achieve their desired end. A scientist's work is endless attention to minutiae done over and over. The patience of Job and hope that the end justifies the tedious means must buoy them up. Otherwise, they could not endure it." See also J.V. Luco to A.C. Barger, 8 Feb 1974, enclosing his Jan 1974 reminiscence; in Box 189 of the Cannon Papers.

27. Concerning the Macy grant for Cannon's work at the Biological Laboratories, see, e.g., E.S. Castle to Frank Fremont-Smith, 21 Oct 1944, and to W.H. Claflin, Jr., 30 Nov 1944. See also C.S. Burwell to Cannon, 11 Oct 1944 and Francis Newton to Cannon, 2 Oct 1944. The 1944 "walk notebook" is in Box 174 of the Cannon Papers.

28. 1944 letter-book, Mrs. Cannon's family letters of Dec 20 and 29. See also 1945 letter-book for 21 Dec 1944 memorandum (copy) from Wilma Fairbank to Mr. Brickell about the visit to Mexico of Dr. and Mrs. Cannon; in Cannon Family Papers. According to Louisa Benton (n.10), Rosenblueth had returned to Mexico at the beginning of 1944. There he acquired a student worthy of his collaboration, Juan Garcia Ramos, with whom he established a deep and satisfying friendship. (pp. 69–71.) Benton skims over Rosenblueth's

work with Cannon in Mexico City (pp. 74–75) but devotes a number of pages to another visitor, Norbert Weiner, who spent ten weeks with Rosenblueth in the spring of 1945.

29. 1945 letter-book, see esp. Mrs. Cannon's letters of Feb 3, 8, and 25.

30. Cannon to Edward Barsky, 3 Mar 1945, in reply to Barsky's letter of Feb 20; see also Cannon to Camille Soula, 2 Mar 1945.

31. 1945 letter-book, Mrs. Cannon's family ltr of Feb 25. Cannon wrote his daughter Marian on Mar 3 about his engagements to speak to medical students and to the Mexican Society of Natural History, which had made him an honorary fellow, and also to the Spanish doctors, around 200 of them having reached Mexico. In addition to experimenting almost daily, he had been to a variety of other functions including a reception where he met many Americans—among them, Nelson Rockefeller. For support of the hospital in Mexico, see also Joseph F. Ford to Cannon, 23 Sep, 3 and 9 Nov 1943; Gail Kelvin to Cannon, 15 Oct 1943; other letters in fd. 709 of the Cannon Papers.

32. Ibid., see esp. Mrs. Cannon's letters of Mar 7, 18, and 25. See also Cannon to Henry E. Sigerist, 17 Mar 1945, in Cannon Papers.

33. Ibid., see Mrs. Cannon's letters of Apr 1 and 9.

34. Ibid., Mrs. Cannon's letter of Apr 25. See also Ida Cannon to W.W. Norton, 25 and 28 Apr 1945, in Cannon Papers.

35. Ibid., see Mrs. Cannon's letters of May 21 and 24; also Marian's note of Jun 21.

36. Ibid., see Mrs. Cannon's letter of Jun 30; also Wilma's letter of Jul 7. For *The Way of an Investigator,* see Yerkes to Cannon, 25 Jun 1945; also letters and reviews in Box 171 of the Cannon Papers.

37. Cannon's letter to Aub, 28 Jul 1945, and Mrs. Cannon's earlier letter of Jul 19, are with materials about Cannon's case in Box 8 of the J.C. Aub Papers. See also 1945 letter-book for Mrs. Cannon's family letter of Aug 16.

38. Cannon to Richard Hoffman, 12 and 19 Sep 1945. In his letter of Sep 12, Cannon explained, "For a time the itching was controlled by means of the X-rays, but that agency was used cautiously because already I had overexposed myself to the X-rays when as a medical student (1896–97) I began a study of the digestive functions by mixing with the food a heavy insoluable salt. Later, in cooperation with Dr. Joseph Aub, who has taken care of my general condition, I began heavy doses of 'Navitol' (A and D) and that helped for 4 or 5 years."

39. The letters of Bradford Cannon (dated only Wednesday), Ida Cannon (4 Oct 1945), and Cornelia Cannon (5 Oct 1945), as well as Dr. Aub's case notes, are in Box 8 of the Aub Papers. Also there is a copy of the Peter Bent Brigham Hospital Pathological Service autopsy report on Dr. W.B. Cannon, Oct 1, 1945; see esp. summary on p. 11. See further J.C. Aub, S.B. Wolbach, B.J. Kennedy and O.T. Bailey, "Mycosis fungoides followed for fourteen years: The case of Dr. W.B. Cannon," *Arch. Pathol.* 60 (Nov 1955):535–547, esp. p. 11 of reprint; also transcript of interview of J.C. Aub by Saul Benison, July 19, 1957, pp. 361–362 (in Oral History Research Office, Columbia University).

40. See letters of condolence in Boxes 181 and 182 of the Cannon Papers; also Bradford Cannon to his mother, 26 Oct 1945. For Marian's poignant comment, see *Snatched from Oblivion*, p. 239.

41. See folder of materials re memorial minute and meeting in Box 8 of the Aub Papers; includes Aub to Rosenblueth, 16 Oct 1945, and Rosenblueth's reply of Oct 20. Benton (n.10) notes that news of Cannon's death came as a severe blow to Rosenblueth. In a eulogy, "La Muerte del Dr. Walter Bradford Cannon," he lamented the loss "of a great man and of a great friend." (pp. 78–79) Mrs. Cannon's comment about the H.M.S. memorial meeting is in her family letter of 9 Nov 1945; also in "Servant of Science" (n.25), p. 130.

Index